New Insights into Aerospace Engineering

Volume I

New Insights into Aerospace Engineering Volume I

Edited by **Ted Dunham**

CLANRYE INTERNATIONAL
New Jersey

Published by Clanrye International,
55 Van Reypen Street,
Jersey City, NJ 07306, USA
www.clanryeinternational.com

New Insights into Aerospace Engineering: Volume I
Edited by Ted Dunham

International Standard Book Number: 978-1-63240-380-3 (Hardback)

Printed in the United States of America.

Contents

Preface

If we turn the pages of history all the world over, the first aviation pioneers can be traced to the late 19th to early 20th centuries. One of the most prominent names in the history of aeronautics is of Sir George Cayley, who contributed significantly to this field from the last decade of the 18th to mid-19th century. The first definition of aerospace engineering appeared in the month of February in the year 1958.

There are two main branches of this field of engineering, commonly known as aeronautical engineering and astronautical engineering. There is a vast difference between these two sub branches, as aeronautical engineering deals with aircrafts that are operational in Earth's atmosphere, and astronautical engineering studies about the space craft which are basically operational outside the Earth's atmosphere. Aerospace engineering also studies about the aerodynamic characteristics of a spacecraft.

Given the demanding conditions that flight vehicles are often subjected to, both in the earth's atmosphere and outside it, aerospace engineering has seen some rapid growth in the past few decades. Evolving designs, improved safety features, technical and technological improvements, and comfort, all form part of this branch of engineering. The development and manufacturing of a modern flight vehicle is an extremely complex process. Aerospace engineers design, test, and supervise the manufacture of aircraft, spacecraft, and missiles. Aerospace engineers develop new technologies for use in aviation, defense systems, and space. This branch has crucial uses in both civil and military contexts. This book takes a look at both.

I would like to thank all the contributing authors for their time and efforts. I would also like to thank my family and friends for their constant support.

Editor

1

Recent Experimental Efforts on High-Pressure Supercritical Injection for Liquid Rockets and Their Implications

Bruce Chehroudi

Advanced Technology Consultants, Laguna Niguel, CA 92677, USA

Correspondence should be addressed to Bruce Chehroudi, chehroudib@aol.com

Academic Editor: David Greatrix

Pressure and temperature of the liquid rocket thrust chambers into which propellants are injected have been in an ascending trajectory to gain higher specific impulse. It is quite possible then that the thermodynamic condition into which liquid propellants are injected reaches or surpasses the critical point of one or more of the injected fluids. For example, in cryogenic hydrogen/oxygen liquid rocket engines, such as Space Shuttle Main Engine (SSME) or Vulcain (Ariane 5), the injected liquid oxygen finds itself in a supercritical condition. Very little detailed information was available on the behavior of liquid jets under such a harsh environment nearly two decades ago. The author had the opportunity to be intimately involved in the evolutionary understanding of injection processes at the Air Force Research Laboratory (AFRL), spanning sub- to supercritical conditions during this period. The information included here attempts to present a coherent summary of experimental achievements pertinent to liquid rockets, focusing only on the injection of nonreacting cryogenic liquids into a high-pressure environment surpassing the critical point of at least one of the propellants. Moreover, some implications of the results acquired under such an environment are offered in the context of the liquid rocket combustion instability problem.

1. Introduction

In designs of chemical rocket engines, liquid fuel and oxidizer are often injected as round jets into a hot and elevated-pressure environment of the thrust chamber. The coaxial and impinging jets injectors are two of the well-established designs in liquid rocket engines (LREs). On the other hand, higher specific impulse is a major motivation for operating rocket thrust chambers at progressively higher pressures. Conditions therefore exist in which the injected liquid finds itself near or even above the thermodynamic critical point. Examples are Space Shuttle Main Engine and Vulcain (Ariane 5) with liquid H_2/liquid O_2.

In such cases, major changes occur in some important and key properties of a substance as it approaches the thermodynamic critical point. For example, under thermodynamic equilibrium, the distinction between the liquid and gas phases disappears at and above the critical point and hence it is referred to as a "fluid." Also, large changes in density occur near the critical point. The constant-pressure specific heat becomes very large and surface tension vanishes at and beyond the critical point. As the ambient pressure into which a liquid jet is injected increases, the importance of the solubility of ambient gases into the injected liquid phase increases and one should consider multicomponent phase equilibrium information. For mixtures, determination of the critical conditions, called the "critical mixing temperature or pressure", is a complex process; see Bruno and Ely [1] and Lazar and Faeth [2]. For example, when a pure liquid hydrocarbon fuel drop is introduced into a nitrogen gas, a thin layer on its surface is a mixture of dissolved nitrogen and the fuel which spreads spatially in time; see Umemura [3].

Understanding the behavior of jets under supercritical conditions therefore is critical to design and modeling of the liquid rockets, in particular, cryogenic liquid rockets. For this reason, systematic research programs, both experimental and computations, have been initiated in the past 20 years to understand behaviors of jets under transcritical and supercritical conditions both with and without externally forced (acoustic) excitations. The reason the external excitation is considered stems from the combined experimental/theoretical work suggesting that interaction between acoustic resonance modes of the chamber and the jets could play an important role in combustion instability. The basic premise here is that when an important dynamic feature, such as the injected jet's dark-core or breakup zone, of an injector design becomes sufficiently sensitive to thermofluid parameters of its environment, it is highly likely that this could strengthen the feedback link thought to be critical in the amplification process and hence push the system into an unstable operating regime.

The purpose of this paper is to present an overview of important experimental achievements, characterizing and understanding nonreacting steady liquid jets injected into supercritical conditions, and offer some implications of these results and potential linkages to production engines. It is not the intention of this work to provide a comprehensive review of the subject, rather to present important findings reported in recent decades.

The coverage of the injectors' experimental data in this paper is divided into three parts: single liquid jets, coaxial jets, and impinging jets. However, the treatment for the impinging jets is comparatively shorter due to limited data available extending to supercritical conditions. Two sections are considered for each part. One focuses on jets injected into an environment devoid of any externally imposed acoustic excitations and the other considers the impact of such excitations on jet characteristics. Most cases reviewed here pertain to cryogenic liquid jets with the environment, into which the jets are injected, existing at thermodynamic supercritical temperatures. This is similar to those experienced in cryogenic liquid rocket engines.

2. Single Jet without External Excitation

In this section, relevant experimental work conducted on a single steady round jet injected into an environment lacking any externally imposed acoustic disturbances is considered. The purpose here is to examine the behavior of such jets under high pressures, specifically supercritical conditions. In the selection of candidate cases, the test matrix is chosen to cover supercritical conditions. However, it is preferred that the matrix spans a broader range encompassing both sub- and supercritical conditions because it generally provides a more comprehensive picture of differences and similarities between the two conditions.

Historically, research on supercritical injection of relevance to liquid rockets started with a published work of Newman and Brzustowski [4]. They used a steady CO_2 jet injected into a chamber of pure N_2 and also into mixtures of $CO_2 + N_2$ at both sub- and supercritical pressures and temperatures. Obviously, if the chamber is at a supercritical pressure, the injected jet pressure is higher and must also be at a supercritical pressure. They showed that when the chamber pressure approached just above the critical pressure of the CO_2, injection of CO_2 into mixtures of CO_2 + N_2 (varying initial CO_2 concentration to change mixture density) widened the visual appearance of the jet. This was explained to be due to changes in chamber-to-injectant density ratio. At a higher chamber supercritical pressure, injection of CO_2 into a pure nitrogen gas, but varying temperature (from sub- to supercritical), caused shortening of both the jet visible length and width with chamber temperature.

Newman and Brzustowski [4] also investigated and explained effects of increased chamber temperature on jet appearance. They found that such effects were due to progressive reduction in ambient gas density, hence lowering surface tension to zero at critical temperature, and to increase in liquid CO_2 evaporation. In other experiments, CO_2 was injected into a mixture of $CO_2 + N_2$ with fixed but large initial CO_2 mass fraction in order to reduce jet evaporation. The chamber temperature was fixed at a supercritical value, but its pressure was varied from sub- to supercritical pressures. They hypothesized and conjectured that at supercritical chamber temperatures and pressures the jet may be considered as a variable-density single-phase turbulent submerged gas jet. Finally, assuming self-preserving flow, negligible gravity, zero latent heat of vaporization, ideal gas behavior, and thermal equilibrium between gas and drops, they develop a model for predicting the profile of the outer extent of a supercritical steady jet and its centerline mean axial velocity. Comparison of this model with experiment was very poor near the injector exit area where most important and complex processes take place. Hence, the proposed hypothesis was not backed by this effort and the matter remained unresolved.

After the aforementioned initial study, two organizations (DLR in Germany and AFRL in USA) dominated the field by pursuing systematic research programs to understand jet breakup and dynamics under high chamber pressures at, and specifically exceeding, the critical condition of the injectant. The majority of the experimental works presented here is thus coming from these two organizations plus other satellite universities they collaborated and/or supported. Each organization has constructed a unique facility from which most of the results for nonreacting jets were acquired and presented here. For more details on their facility designs readers are referred to their publications, many of which are listed in the references.

Researchers at DLR began working with the most simplest and fundamental of all cases, that is, the injection of a single nonreacting round jet into a quiescent environment. At AFRL, although initially droplet studies were planned and conducted, a transition to jets at high Reynolds numbers of practical interest was initiated by the author of this paper and his team members; for example, see Chehroudi et al. [5]. The injection of jets were studied at various chamber pressures ranging from subcritical to supercritical (mostly at supercritical temperatures), which included chamber pressures representative of those experienced in typical cryogenic liquid rocket engine (LRE). For safety reasons, and for the H_2/LOX liquid rockets, it was preferred to simulate liquid oxygen with liquid nitrogen. Early studies have shown that at a nonreacting condition the injection behaviors of these two fluids were similar. To avoid complications introduced by mixture effects, however, many of these studies involved injection of cryogenic liquid nitrogen (LN_2) into room temperature gaseous nitrogen (GN_2). Critical pressure and temperature of nitrogen are 3.39 MPa and 126.2 K, respectively. Injections into other ambient gases were also investigated.

The objective of this section of the paper is to present key findings and discuss representative results. In some cases, key conclusions confirmed independently by both organizations (DLR and AFRL) are highlighted when appropriate and if they add to the reader's understanding. Before delving into details of the results, it worth indicating that Mayer et al. [6] were the first to take active steps towards a large-scale facility investigating both single and coaxial cryogenic jets under cold and fired conditions. They used LN_2 jets at 105 K injected into a GN_2 environment at 300 K, but at varying ambient pressures ranging from sub- to supercritical conditions. They reported drastic changes in the jet structure near and above the critical pressure. The jet behaved similar to the classical atomization of liquid fuels, with ligaments and drops, below the critical pressure. Mayer et al. [6] attributed this behavior to a continual decline of surface tension until it vanished at and beyond the critical point.

2.1. Visualization of the Jet Interface. Figure 1 presents results published in a work led by Chehroudi which shows representative images of cryogenic LN_2 jets injected into gaseous nitrogen at 300 K (supercritical in temperature); see Chehroudi et al. [5]. The initial temperature of the jets was measured in a separate experiment and under identical flow conditions with a very small thermocouple. Depending on the flow condition, the measured initial injection temperature varied from 99 K to 110 K, that is, injected at a subcritical value. Pressures in Figure 1 are reported as reduced pressures (P_r), defined to be the chamber pressure divided by the critical pressure of the injected nitrogen. In frames 1 to 4 (of Figure 1), where the chamber pressure is subcritical, the jets have a classical liquid spray appearance. Figure 2 shows magnified images of the three injection cases under sub-, near-, and super-critical chamber pressures. This software magnification is performed to more clearly show the shear layer structure near the injector exit area. As shown in Figure 1, and consistent with the classical liquid jet breakup regimes described by Reitz and Bracco [7], surface instabilities grow downstream from the injector, and very fine ligaments and drops are ejected from the jet (see also the left image in Figure 2). This behavior corresponds to the second wind-induced liquid jet break-up regime described by Reitz and Bracco [7].

Major structural and interfacial changes occur at about $P_r = 1.03$ as shown in frame 5 of Figure 1. Above this chamber pressure, drops are no longer detected, and as characterized by Chehroudi et al. [5], regular "finger-like" entities are observed at the interface. Rather than breaking up into droplets, the interface appears to dissolve at different distances from the dense and dark core. These structures are illustrated at $P_r = 1.22$ in the middle frame of Figure 2. Such a change in morphology of the mixing layer is evidently due to combined effects of the reduction in the surface tension, as the critical pressure is exceeded, and disappearance of the enthalpy of vaporization because of this transition to supercritical pressures.

As the chamber pressure is further increased, the length and thickness of the dense (and dark) core decrease, and the jet begins to appear similar to a turbulent gaseous jet injected into a gaseous environment. This is illustrated in frames 7 and higher in Figure 1. Any further droplet production, and consequently any additional classical liquid atomization, is completely suppressed. These observations were confirmed by Mayer et al. [6, 8], Chehroudi et al. [5], and Roy and Segal [9]. As mentioned earlier, similar results were also found when injecting liquid oxygen instead of the liquid nitrogen.

It is important to indicate that because of the very large density variations between the jet core and the chamber, Chehroudi et al. [5] investigated whether the evolution of the jet within the region of their measurement was affected by the buoyancy forces. Therefore, they calculated the Froude number values under each test condition. As an example, Chehroudi et al. [5, 10] showed that the Froude number ranged from 42,000 to 110,000. To make sense of these values, they looked at the Chen and Rodi [11] results. Chen and Rodi [11] suggested that the flow is momentum dominated when a defined length scale x_b is less than 0.53, while Papanicolaou and List [12] suggested $x_b < 1$. The length scale is given by $x_b = \mathrm{Fr}^{-1/2}(\rho/\rho_\infty)^{-1/4}(x/d)$, where x is the axial distance, d is the initial jet diameter, and ρ and ρ_∞ are the jet and ambient densities, respectively. The Froude number is defined as $\mathrm{Fr} \equiv \rho U^2/gd|\rho_\infty - \rho|$ where U is the velocity difference and g is the gravitational acceleration. Considering a more conservative estimate by Chen and Rodi [11], the jet used by Chehroudi et al. [5, 10] is momentum dominated for distances less than 30 to 40 mm from the injector exit plane. Pictures presented in Figures 1 and 2 cover up to about 5.5 mm (axial distance/diameter ratio of 21.6) from the injector, and hence buoyancy effects can be ignored in favor of inertial forces.

2.2. Length Scale Investigation. Injection of a single LN_2 jet into gaseous N_2 (at 298 K) has been investigated by Branam and Mayer [13] at ambient pressures of 4 MPa and 6 MPa, corresponding to reduced pressures of 1.17 and 1.76, respectively. The initial injection temperature of the nitrogen was near the critical point. They provided a measure of the length scales by analyzing shadowgraph images and called it as the "visible length scale." More details on their image analysis can be found in Branam and Mayer [13].

Figure 3 shows a typical result of the geometrically averaged length scale (average of radial and axial length scales) measured at the x/D of 10 as a function of the radial position. Results from the k-ε computational method are also shown. This suggests that the measured visible length scale is comparable in magnitude to the Taylor length scale determined by the computational method.

The ratio of the axial to the radial length scales indicates whether the visible structures are spherical or more ellipsoidal in shape. Both length scales are shown in Figure 4 for an injection temperature of 123 K, that is, injection of N_2 at a high density. In the near-injector region, the axial length scales are much larger than the radial ones. Further downstream, however, the visible structures become more

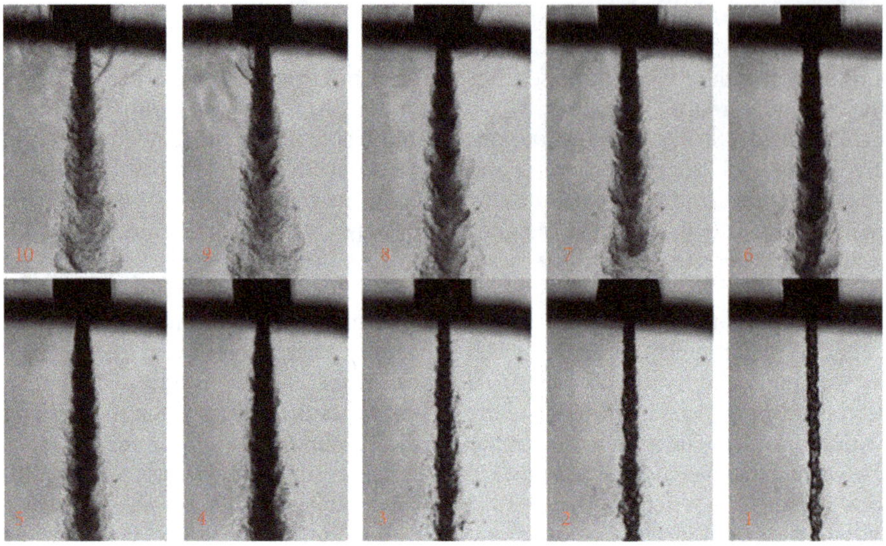

FIGURE 1: Back-illuminated images of a single nitrogen jet injected into nitrogen at a fixed supercritical temperature of 300 K but varying sub- to supercritical pressures (For N_2: $P_{critical} = 3.39$ MPa; $T_c = 126.2$ K). From lower right to upper left: $P_{ch}/P_{critical}$ (frame no.) = 0.23 (1), 0.43 (2), 0.62 (3), 0.83 (4), 1.03 (5), 1.22 (6), 1.62 (7), 2.44 (8), 2.74 (9). Reynolds' number (Re) was from 25,000 to 75,000. Injection' velocity: 10–15 m/s. Froude' number: 40,000 to 110,000. Injectant temperature: 99 to 120 K. Chehroudi et al. [5].

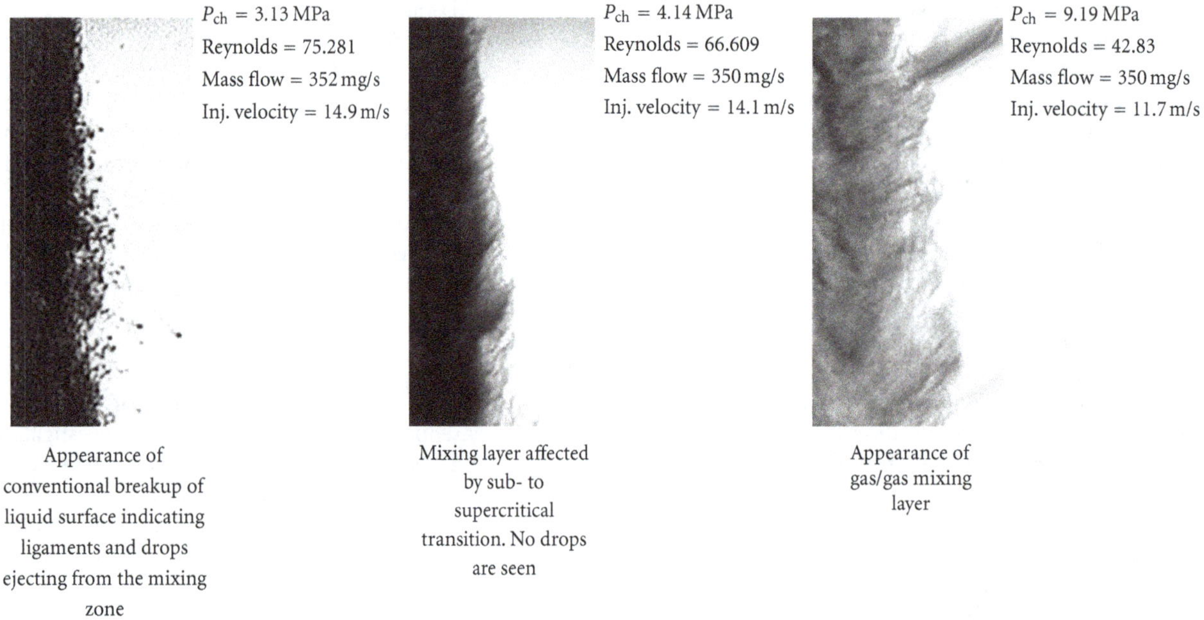

FIGURE 2: Software magnified images of the jets in Figure 1 at their outer boundaries showing transition to the gas-jet-like appearance starting at just below the critical pressure of the injectant. Images are at fixed supercritical chamber temperature of 300 K. Chehroudi et al. [5].

circular in shape. At a higher injection temperature (132 K), the asymmetry between the radial and axial length scales is not as pronounced as that seen under the lower temperature condition (123 K).

2.3. Jet Spreading Angle or Growth Rate. Measurements and estimations of the growth rate of a jet have been a subject of intense research for years because it provides a primary measure of mixing and development of the jet itself. Chehroudi's group was the first to extract quantitative measurements of this physical parameter using the images taken from a cryogenic N_2 jet injected into GN_2 under both subcritical and supercritical pressures; see Chehroudi et al. [5]. These measurements led to important conclusions regarding the character of the growth rate and the behavior of the jet near the injector and under such conditions, specifically at supercritical chamber pressures.

The spreading angle or growth rate was measured from a field of view within 5.5 mm of the injector exit plane (distance-to-diameter ratio of up to 21.6) and was inertially

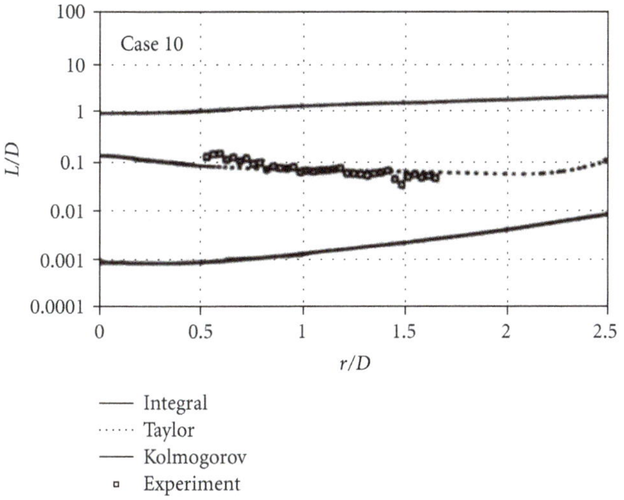

Figure 3: Comparison between calculated and experimental length scales for a single jet of LN_2 injected into GN_2 at $x/d = 10$, chamber pressure of 6 MPa, 1.9 m/s, injected temperature of $T = 132$ K. Branam and Mayer [13].

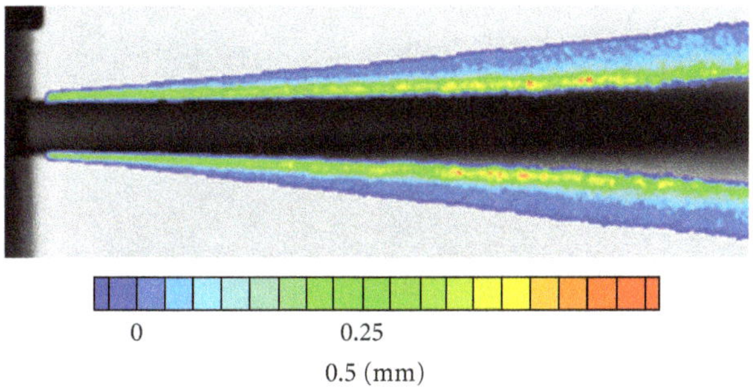

(a) Radial length scale

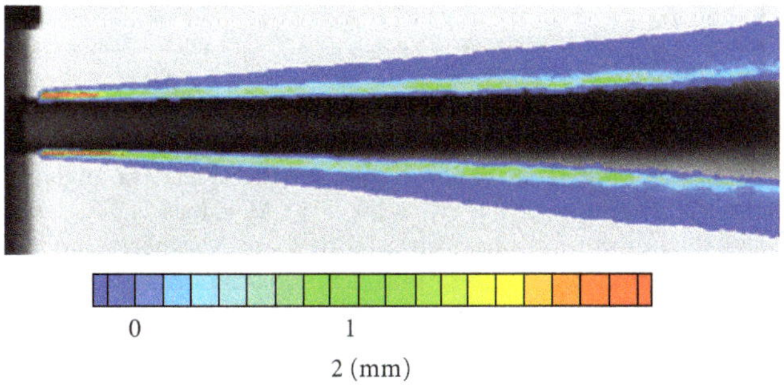

(b) Axial length scale

Figure 4: Experimental length scales for LN_2 into GN_2, chamber pressure of 4 MPa, LN_2 injected temperature of 123 K. Branam and Mayer [13].

dominated as discussed earlier. Chehroudi et al. [5] indicated that their data were also taken from the corresponding and appropriate initial region of the jet to ensure existence of a classical mixing layer. The initial jet spreading angle, or its growth rate, was then measured for all acquired images, and results along with those of others are presented in Figure 5. Of importance in this figure is the justification for the selection of the data sets and the nature of their

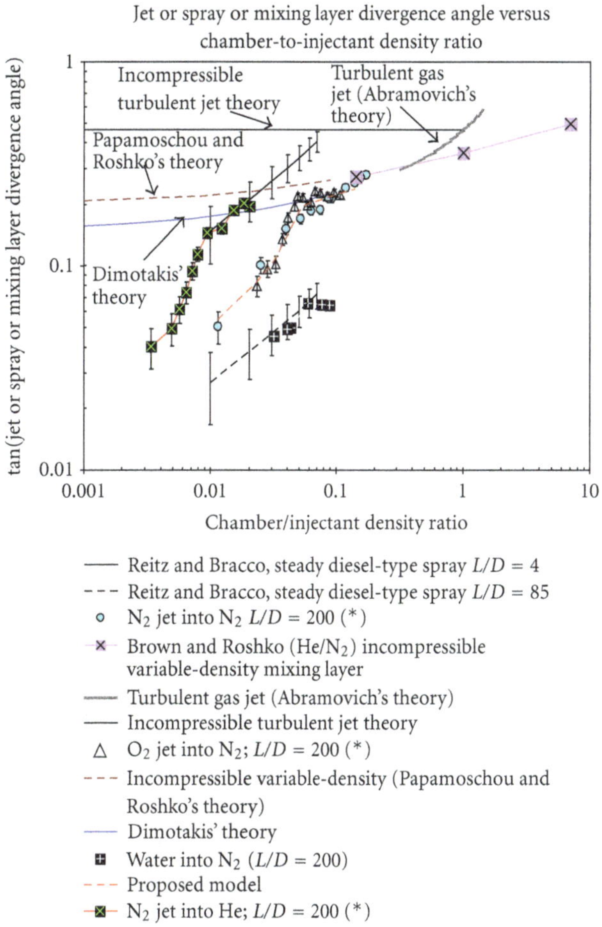

Figure 5: Spreading or growth rate of single jets as a tangent of the visual spreading angle versus the chamber-to-injectant density ratio. Data taken by Chehroudi are indicated by an asterisk (*) in the legend. Chehroudi et al. [5].

measurements by other researchers. They are elaborated at sufficient detail in earlier papers. In order to gain a deeper appreciation of these selections, the reader is referred to Chehroudi et al. [10]. Therefore, they are only mentioned here in brief.

Because the jets investigated by Chehroudi et al. [5] exhibited both liquid-spray-like and gas-jet-like (two-phase and single-phase, resp.) appearances, depending on pressure (see Figure 1), appropriate comparisons with both liquid sprays (injected into a gas) and gas jets (injected into a gas) were justified and hence these results are presented in Figure 5. The simplest is the prediction of the linear growth or constant spreading angle for the turbulent incompressible submerged jet using the mixing length concept. Following Abramovich [14], a semiempirical equation was used which attempts to incorporate the effects of density variations by an introduction of a characteristic velocity (see Chehroudi et al. [10] for more details).

Brown and Roshko [15] measured spreading angles for a subsonic, two-dimensional, incompressible, turbulent mixing layer in which helium and nitrogen were used. Brown [16] (for a temporally growing mixing layer) and Papamoschou and Roshko [17] proposed a theoretical equation for incompressible variable-density gaseous mixing layers. Finally, Dimotakis [18] used the observation that, in general, the entrainment into the mixing layer from each stream was not the same and, in a system moving with a convection velocity, offered a geometrical argument to derive an equation for two-dimensional incompressible variable-density mixing layers. Chehroudi et al. [5] included predictions from these models as shown in Figure 5. Results by Richards and Pitts [19] for variable-density jets are also included.

Because both liquid-spray-like and gas-jet-like visual behaviors were observed, the growth rate for the liquid sprays produced from single-hole nozzles, typical of the ones used in diesel engines, was also incorporated in this figure. Figure 5 covers a density ratio of four orders of magnitude and is regarded as a unique and new plot in its own right. To some extent, and for comparable cases, disagreements between some results in this figure can be attributed to differences in the definition of the mixing layer thicknesses and the adopted measurement methods. For detailed discussion of this figure, see Chehroudi et al. [5, 10].

The important point which was stressed by Chehroudi et al. [10] is that for a range of density ratios in which images exhibit gas-jet-like appearance, the experimental data agrees well with the proposed theoretical equation by Dimotakis [18] and closely follows the trend of the Brown/Papamoschou and Roshko equation as shown in Figure 5. This can be taken as an important quantitative evidence that at supercritical pressures, the injected jets visually behave like a gas. Chehroudi's work appears to be the first time such a rigorous and quantitative evidence had been developed. The fractal dimension results discussed later provide additional evidence in support of this behavior.

Chehroudi et al. [20] also used the Raman scattering studies to measure density distributions. Initially, the growth rate measurements by them, using results acquired during the Raman scattering work, did not provide the same jet thickness values as those determined by the shadowgraphy approach. Apparently, as discussed by Brown and Roshko [15], different thickness definitions exist, and one can explore their relationship. Similar attempts in the context of supercritical jets showed that within the distances investigated, twice the full-width half-maximum (FWHM) of the Raman intensity radial profiles was equivalent to the thickness values measured through shadowgraphy. Realization of this relationship was very critical to consolidate the results from two different methodologies (Raman versus shadowgraphy). Figure 6 shows the growth rate measured using the Raman data in contrast to those determined through shadowgraph images.

These results were subsequently confirmed and extended by Oschwald and Micci [21] through a similar measurement technique. For example, they showed that when twice the FWHM of the Raman radial intensity profiles is used for

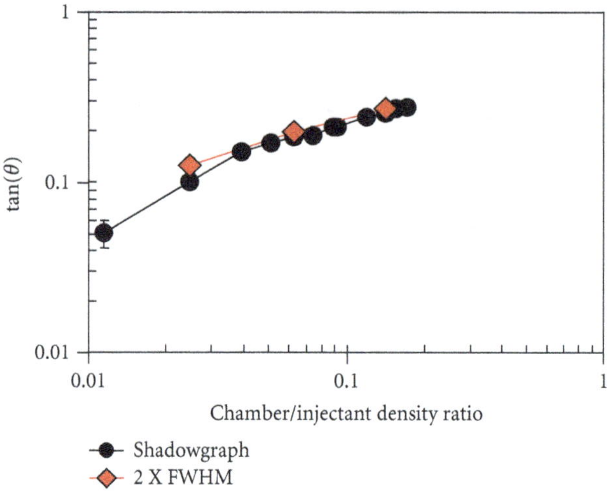

FIGURE 6: Comparison of the tangent of the spreading angle for a single jet of LN_2 injected into GN_2 measured using shadowgraph and Raman's techniques at twice the FWHM values. Chehroudi et al. [20].

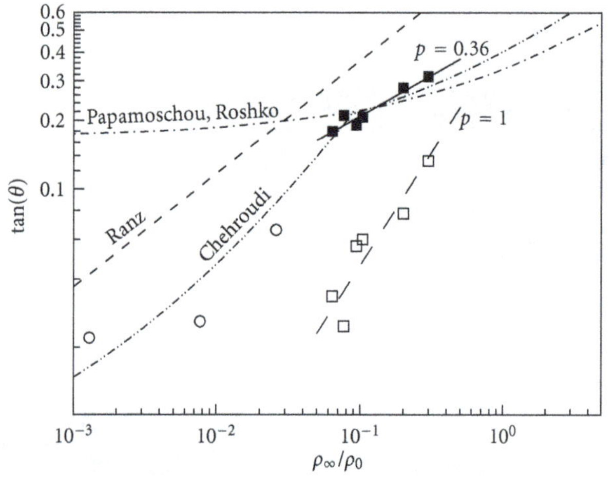

FIGURE 7: Comparison of the tangent of the spreading angle by Raman's techniques using twice the FWHM values (DLR data). Solid squares are for data from x/D of 15 to 32 whereas hollow squares are from x/D of 0.5 to 14. Chehroudi's model is also shown as dash-dot-dot curve. Single jet LN_2 into GN_2 with injector L/D = 11.6. Data from Oschwald and Micci [21]. Open circles are data points from Reitz and Bracco [7].

x/D values within 15 and 32, a good agreement exists between their results and those by Chehroudi's group (see the case designated as "$p = 0.36$" in Figure 7, where "p" is the exponent of the curve fitted through the data). However, the twice FWHM criterion did not fully agree with shadowgraphs for $x/D < 15$ in the study conducted by Oschwald and Micci [21] (see the case identified as "$p = 1.0$" in Figure 7).

Note that Oschwald and Micci [21] measured a larger range of axial distances and found that the criterion was not universal at extended ranges. There are several reasons why different trends might be observed at different distances.

It has to do with the fact that shadowgraphy and Raman scattering approaches measure different physical properties. The Raman signal is taken to be proportional to density whereas shadowgraphy is sensitive to the gradients of the density distribution. Thus a unique relationship between the results measured by the two methods may only be valid for a limited region and/or perhaps specific configurations. Recall that the data discussed above were obtained from injectors with different *L/D* ratios. Another potential cause, and hence discrepancies, is errors in attempting to perform Raman's measurements close to the injector inlet where density variations and thus index of refraction variations can be very large.

2.4. Fractal Dimension of the Interface. Fractals are intimately connected to the concept of self-similarity; see Mandelbrot [22]. The fractal dimension of any curve is between 1 and 2. The more wrinkled and space-filling a curve is, the larger the values of its fractal dimension. Natural curves, such as the outline of a cauliflower, are self-similar only to that within a narrow range of scales. The objective of the analysis here was to measure the fractal dimension of the interface of jets injected into the chamber in order to see if any pattern was uncovered.

The fractal dimension of jets at various pressures ranging from subcritical to supercritical was calculated and compared to results by other researchers. Reference results were taken from Sreenivasan and Meneveau [23] who measured the fractal dimensions of a variety of turbulent gaseous jets, mixing layers, and boundary layers. These results indicated a fractal dimension between 1.33 and 1.38. In addition, the fractal dimensions of a turbulent water jet (Dimotakis et al. [18]) and of a liquid jet in the second wind-induced atomization regime (Taylor and Hoyt [24]) were computed from high-resolution scanned images.

The fractal dimensions from the above reference cases are shown as horizontal lines in Figure 8. Overlaid on top of these lines are discrete points indicating the fractal dimension of LN_2 jets injected into GN_2 at various chamber pressures. One sees that at supercritical chamber pressures, the fractal dimension approaches a value similar to gaseous turbulent jets and mixing layers. As the chamber pressure is decreased, the fractal dimension also decreases. Below a reduced pressure of 0.8, the fractal dimension rapidly declines to a value approximately equal to that of liquid sprays in the second wind-induced liquid jet break-up regime.

A detailed discussion of the above results is also found in Chehroudi et al. [25, 26]. The key conclusion reached by Chehroudi's group is that the results from fractal analysis complement and extend the imaging data they acquired for the initial jet growth rate. At supercritical pressures, jets have a fractal dimension similar to turbulent gas jets, and at subcritical pressures, cryogenic jets have a fractal dimension similar to liquid sprays. The transition occurs at about the same chamber pressure as that when the transition in visual appearance and growth rate data discussed in Figure 5 takes place. Such distinctly different behaviors for jets under sub- and supercritical conditions were first demonstrated by Chehroudi's group in a quantitative manner using fractal analysis.

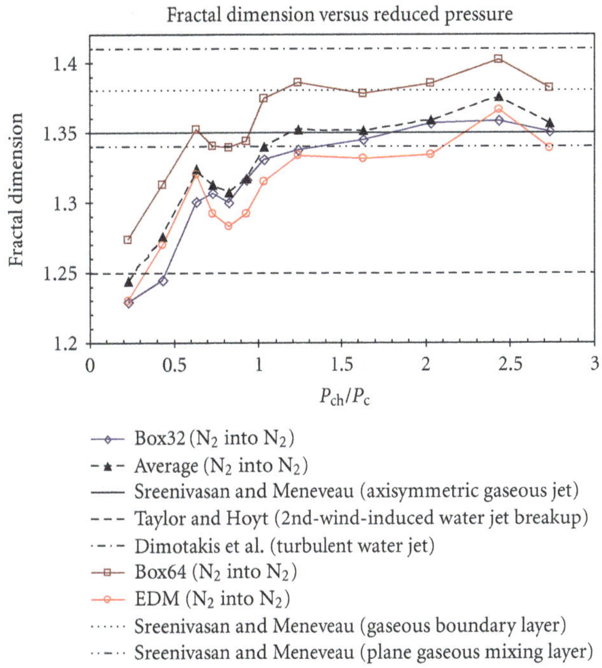

FIGURE 8: Fractal dimensions of the boundaries of various single jets as a function of reduced chamber pressure (chamber pressure divided by the critical pressure of the jet material). Discrete points are data from Chehroudi et al. [25, 26]. Box 32, Box 64, and EDM are different methods of calculating the fractal dimension, giving an impression of the extent of variability; for details, see Chehroudi et al. [25, 26].

2.5. Measurement of the Dark-Core Length. Before presenting their results under supercritical conditions, Chehroudi et al. [25, 26] discussed some measurements from (gaseous and liquid) jets at subcritical conditions to set the stage and provide a contrast to their data. It is therefore useful to summarize what they recalled in their work.

According to Abramovich [14], the length of the "potential core" in isothermal uniform-density axisymmetric and two-dimensional jets is estimated to be about 6 to 10 injector diameters; whereas for nonisothermal cold jets injected into hot environments, it can reach up to about 25 injector diameters depending on jet temperature.

Also, according to Chehroudi et al. [27] the "intact core" length of the liquid sprays similar to the ones used in diesel engines is given by the equation $Cd_j(\rho_l/\rho_g)^{1/2}$ where ρ_l and ρ_g are liquid injectant and chamber gas densities, respectively, d_j is an effective jet exit diameter, and C is a constant between 3.3 and 11. This reflects an intact core length between 33 and 110 injector diameters for the chamber-to-injectant density ratio of 0.01 and between 16.5 and 55 diameters for the chamber-to-injectant density ratio of 0.04. These results are presented in Figures 9(a) and 9(b) for better comparison with what is measured for LN_2 injection into both sub- and supercritical GN_2 environment.

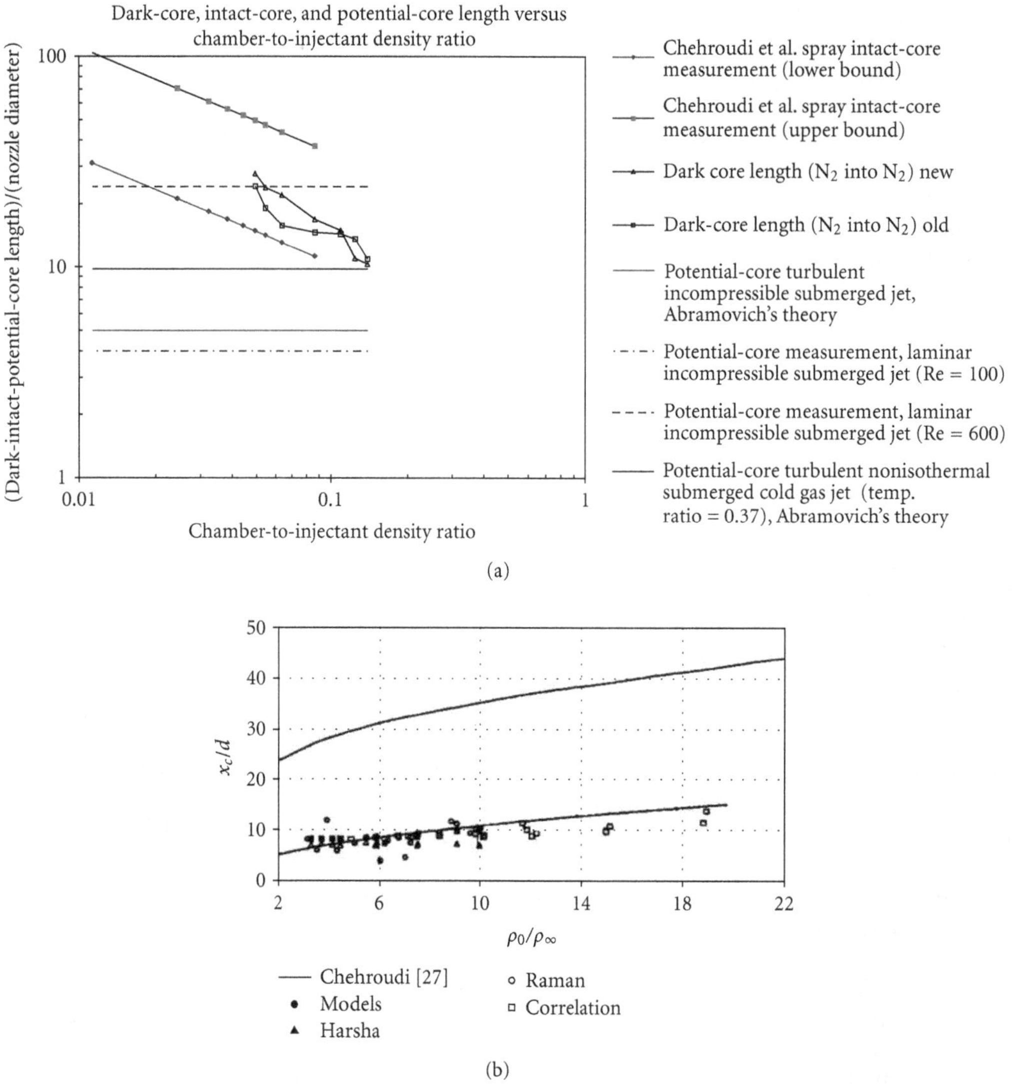

FIGURE 9: Ratio of either the dark-core, intact-core, or potential-core length, depending on the case, divided by the density ratio for single jets. (a): Chehroudi et al. [10], determined by analysis of shadowgraphs, for an injector $L/D = 200$. (b): Branam and Mayer [13], determined via Raman data, injector $L/D = 40$. Models: determined by computer simulation. Correlation: a correlation using a set of Raman's data by Branam and Mayer [13].

Considering that the classical two-stream mixing layer should start from the injector exit and extends to approximately the end of the potential core (or intact core) of the jet, Chehroudi et al. [5, 10] assumed that the jet "dark-core" region seen in their images played a similar role as the intact core or potential core. Figure 9(a) shows the dark-, intact-, or potential-core lengths normalized by the injector hole diameter plotted versus density ratio. By referring to Figure 9(a), Chehroudi et al. [5, 10] indicated that the growth rate data taken by them, which is presented earlier in Figure 1, was indeed from the corresponding and appropriate initial region to ensure existence of a classical mixing layer. They emphasized that it is only then when a valid comparison can be made (as they did) between their results and the two-stream mixing layers available in the literature. Finally, they found that the core length fluctuation levels at supercritical condition were several times lower than those observed at subcritical chamber pressures.

Complementary results by the Branam and Mayer [13] are also shown in Figure 9(b). These results were determined through a log-log plot of the centerline intensity measured by the Raman scattering to distinguish different flow regions. Note that the horizontal axis for Figures 9(a) and 9(b) are inverse of each other. The Branam and Mayer [13] data are at or below the lower bound of the Chehroudi et al. [27] model (i.e., solid curves in Figure 9(b)) which was proposed based on the liquid spray data in various atomization regimes. For chamber-to-injectant density ratios of less than 0.1 in Figure 9(a) (or injectant-to-chamber ratio of greater than 10 in Figure 9(b)), it appears that the Chehroudi et al. [25, 26]

experimental data shown in Figure 9(a) is larger by about a factor of 1.5 to 2 compared to the "correlation" given by Branam and Mayer [13] in Figure 9(b).

Considering that the raw data used by the two groups are from two different injectors and measurement methodologies, the agreement is considered adequate. However, further investigations are warranted.

2.6. Density and Temperature Fields. The main purpose of the Raman scattering measurements was to provide quantitative information and to enable mapping of the jet density field. Temperatures, for example, were calculated assuming application of a suitable equation of state. Radial density profiles were reported by Oschwald and Schik [28] in a normalized fashion. In this section, the centerline density and/or temperature profiles as functions of the axial distance from the injector exit plane and their self-similarity assessment are discussed.

The test conditions were chosen in order to assess the influence of the thermodynamic state of the injected cryogenic N_2 on the jet disintegration process. For example, above the critical pressure, the specific heat is finite but exhibits a maximum at a particular temperature. At this same point, the thermal diffusivity exhibits a minimum value. Three test cases were therefore investigated as shown in Figure 10(a). In test case A, the initial injection temperature is both above the critical temperature and above the temperature where the specific heat assumes a maximum value, whereas for test cases B and C the initial injection temperatures are both below the critical value and the temperature where the specific heat is at a maximum value.

Figures 10(b) and 10(c) show normalized centerline axial profiles of the density and temperature acquired by Oschwald and Schik [28] at a chamber pressure of 4 MPa (near the critical pressure of nitrogen). Note that the density decay behavior becomes slower as the initial injection temperature is decreased. The temperature profile, however, stays flat for up to a normalized distance (x/D) of about 25 to 30. They indicated that the development of the centerline temperature reflects the thermophysical properties of the nitrogen, being specific to the region where the specific heat reaches a maximum. For initial injection temperatures below the temperature where the specific heat reaches a maximum value, as the jet heats up, the fluid has to pass through a state with a maximum specific heat. The fluid temperature can then reach a value where a large amount of heat can be stored without any noticeable increase in temperature. It appears that the maximum specific heat line in a supercritical fluid results in a behavior similar to a liquid at its boiling point. That is, heat transfer to the nitrogen does not increase its temperature but simply expands the fluid (i.e., increases its specific volume). It is also for this reason that the dashed curves in Figure 10(a) are referred to as "pseudo boiling lines." Note that the density of the fluid varies strongly with temperature in this zone. At 6 MPa ($P_r = 1.76$, data not shown), the maximum of the specific heat is much less pronounced and the effects of the pseudo-boiling line is not as distinct as those seen under chamber pressure of 4 MPa. As shown by Oschwald and Schik [28], however, far downstream, it was observed that the temperature of the disintegrating and mixing supercritical fluid jet approached a value representative of a fully mixed jet but at a slower pace than that for the jet density.

The self-similarity of the density field has also been investigated by Chehroudi et al. [20] and the results are presented in Figure 11. According to Wygnanski and Fiedler [29], a fully self-preserved velocity field of a turbulent air jet should be observed at an x/D of greater than 40 when the Reynolds number is near 100,000. So et al. [30] reported self-preservation for x/D values larger than about 20 in a binary gas jet at Reynolds' number (Re) of about 4300. Although it appears that some inconsistencies exist for this criterion, one can see that for the near-critical and supercritical pressures, the density radial profiles approach the similarity model curve shown in Figure 11. The disagreement increases at subcritical pressures where the model is least applicable. It is worth indicating that results from a modeling and computational simulation by Zong et al. [31] also agrees well with Chehroudi's Raman scattering measurements.

Chehroudi et al. [20] using their Raman scattering data determined the FWHM of the radial density profiles at each axial distance from the injector, and the results, along with data by other investigators, are shown in Figure 12. More information on their experimental conditions is given in Table 1. Note that except for Chehroudi et al. [5, 25] and Oschwald et al. [32], all others performed injection of gaseous fluids into an ambient gas at subcritical pressures (based on the injectant critical pressure). Also, the FWHM was determined using the mass fraction profiles in both So et al. [30] and Richards and Pitts [19]. However, reported FWHM values by So et al. [30] using both density and mass fraction profiles were comparable. Chehroudi et al. [20] data in Figure 12 shows an increasingly larger spreading rate as chamber pressure is raised. The data at the supercritical condition ($P_r = 2.03$) approaches that of Richards and Pitts [19] acquired at a density ratio of 1.56 even though this ratio is substantially smaller than Chehroudi et al. [20] data. Results at larger distances were not available for their jet to enable a more comprehensive comparison between the cases.

Richards and Pitts [19] concluded that if care is exercised to ensure that the flow is free of buoyancy and coflow effects, the spreading rate in variable-density jets was independent of the initial density ratio, velocity profile, and turbulence level and conformed with the constant-density results of others. In addition, they proposed a slope in the range between 0.212 and 0.220 for the linear jet growth rate equation; see Figure 12. However, a linear least-square fit to Chehroudi et al. [20] data at $P_r = 2.03$ gives a slope of 0.102, almost half of that by Richards and Pitts [19]. One possible explanation for this difference is that Chehroudi's data covers a range much closer to the injector exit plane than that by Richards and Pitts [19], leading to a lowered growth rate value. A tendency towards higher growth rates can be seen if only the farthest two data points are considered in Chehroudi's data. However, a solid conclusion cannot be drawn based on these two points. It is also possible that at some high enough injectant-to-chamber density ratio, the spread rate universality indicated by Richards and Pitts breaks down and

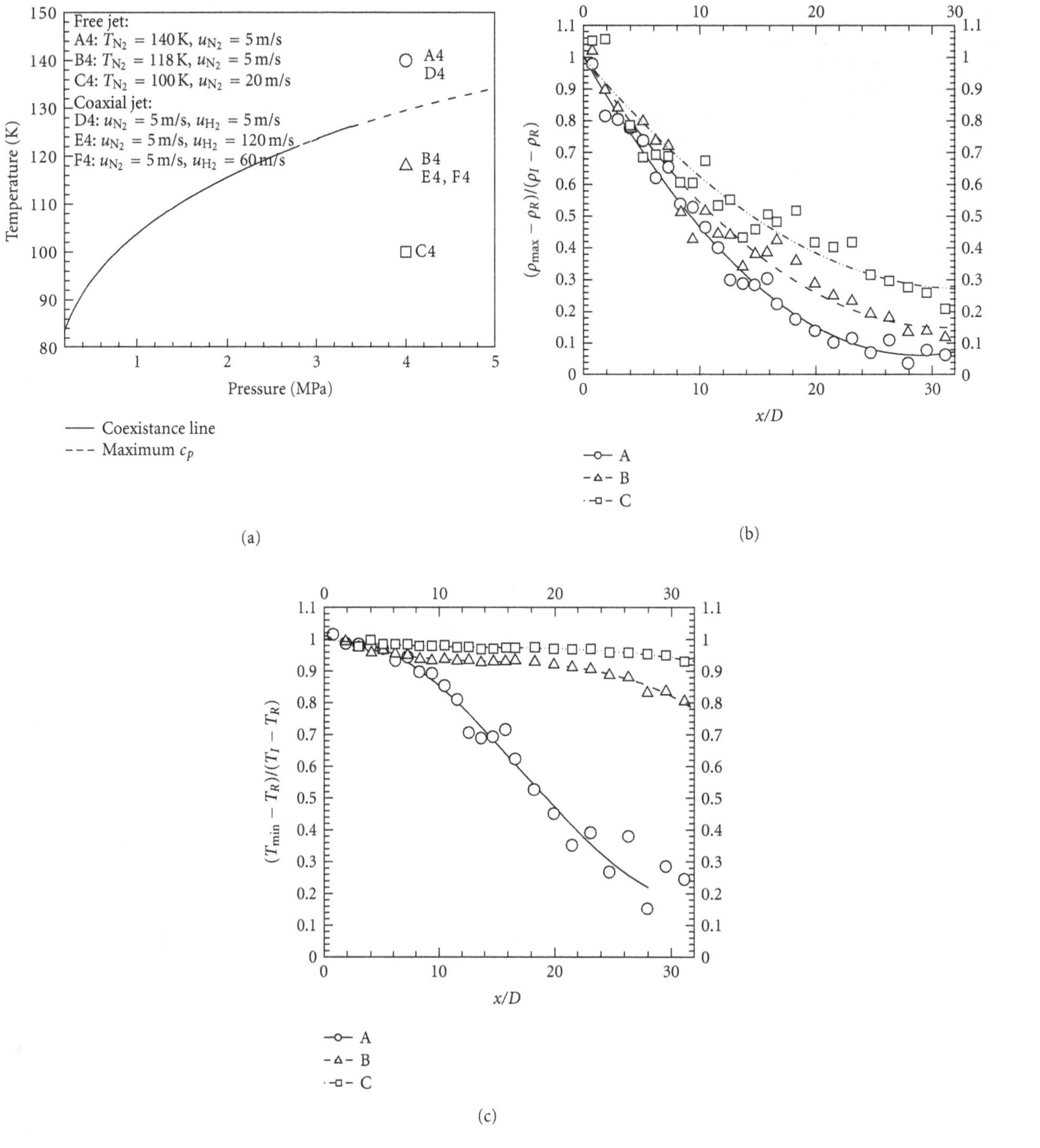

Figure 10: Normalized centerline density (b) and temperature (c) axial profiles of LN_2 injected into GN_2 at three different injection temperatures and a chamber pressure of 4 MPa (i.e., near critical pressure). The plot (a) shows the thermodynamic conditions under which test cases A, B, and C are conducted. The dashed line is the pseudoboiling line. Note, A4, B4, and C4 symbols represent A, B, and C cases, respectively. Oschwald and Schik [28].

one observes a somewhat retarded growth rate for variable-density turbulent jets. Some evidence in support of this position was given by Chehroudi et al. [20].

2.7. Phenomenological Model of the Jet Growth Rate. Using the experimental data collected on the growth of a cryogenic jet, a phenomenological model for the growth rate was proposed by Chehroudi et al. [5, 10] for the first time in the literature. Complete details on the development of this equation are to be found in these references. However, the physical reasoning motivating the proposed model equation is outlined below.

It was noticed by Chehroudi that previous expressions for the growth rate of liquid sprays and of turbulent jets have

TABLE 1: Some information extracted from works by other investigators reported here. T_{inj} and T_{ch} are injection and chamber temperatures. P_{ch} is chamber pressure. L/D is the injector hole-to-diameter ratio, x/D is the normalized distance from injector exit within which measurements were made. Chehroudi et al. [20].

	(Fluid inj./cham.)	T_{inj} K	P_{ch} MPa	T_{ch} K	Reduced pressure P_r	Inj./chamb. density ratio	Diameter D mm	L/D	x/D	Raynolds' number Re	Profile used to measure FWHM
Oschwald et al.	N_2/N_2	118	4	298	1.17	3.34	1.9	11.5	8.42	$1.2E+05$	Density
Oschwald et al.	N_2/N_2	140	4	298	1.17	12.5	1.9	11.5	1.05	$1.3E+05$	Density
Chehroudi et al.	N_2/N_2	95	6.9	295	2.03	7.1	0.505	100	4.8 to 24.4	$3.5E+04$	Density
Chehroudi et al.	N_2/N_2	110	1.5	295	0.43	40.6	0.505	100	4.8 to 24.5	$1.2E+04$	Density
So et al.	(He + Air)/Air	275	0.1	275	0.08	0.64	9.5		5.1	$5.0E+03$	Concentration and density
So et al.	(He + Air)/Air	275	0.1	275	0.08	0.64	9.5		6.4	$5.0E+03$	Concentration and density
Richards and Pitts	He into Air	275	0.1	275	0.44	0.138	6.35	~50	20–80	$4.0E+03$	Mass fraction
Richards and Pitts	C_3H_8 into Air	275	0.1	275	0.02	1.56	6.35	~50	40–120	$2.5E+04$	Mass fraction

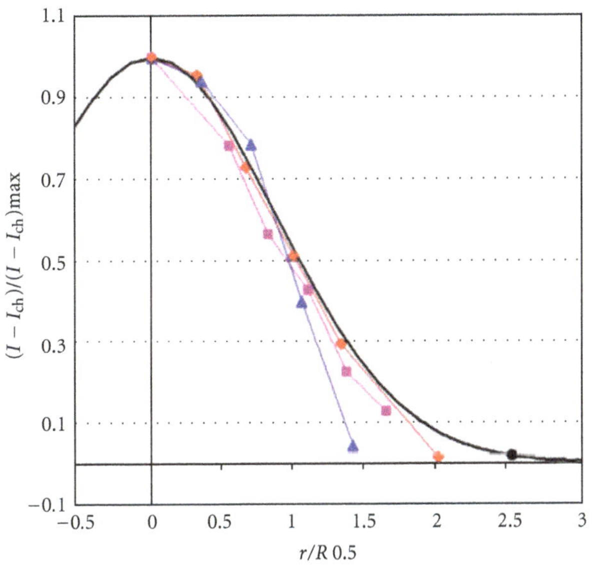

FIGURE 11: Plots of normalized intensity versus normalized radius for a single LN_2 jet injected into GN_2, at $x/D = 12.2$, at subcritical, near-critical, and supercritical pressures (Chehroudi et al. [20]). The solid curve is the self-similar model that represents data from a gaseous jet injected into a gaseous environment. I and I_{ch} symbols are measured intensities in the jet and in the chamber far away from the jet, respectively. Chehroudi et al. [20].

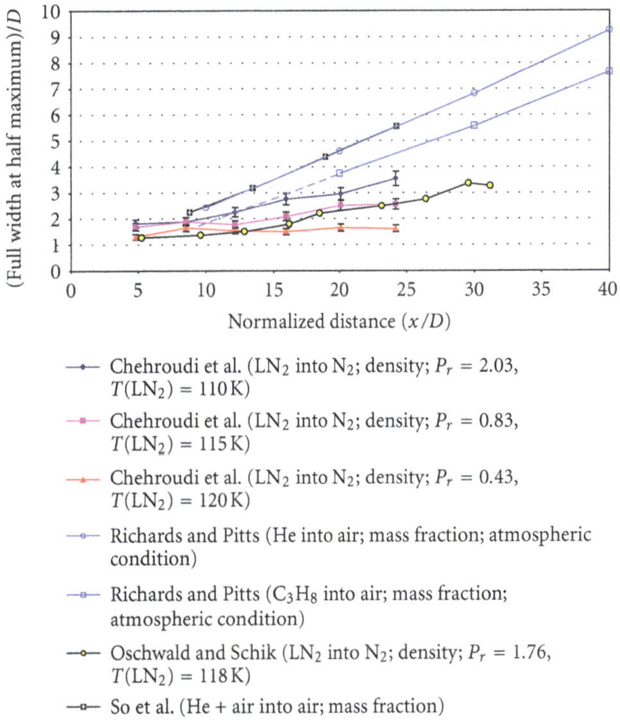

FIGURE 12: Normalized FWHM of the density-surplus radial profiles of single jets plotted as a function of the normalized axial distance from the injector exit plane. From Chehroudi et al. [20].

a remarkably similar form. For example, Reitz and Bracco [7] proposed that the growth rate of isothermal steady liquid sprays could be expressed as

$$\theta \approx 0.27\left[0+\left(\frac{\rho_g}{\rho_l}\right)^{0.5}\right]. \quad (1)$$

The first term in the bracket is the number zero. This zero term was purposely kept to enable a comparison with other equations Chehroudi et al. [5, 10] discussed. They also referred to an equation proposed by Papamoschou and Roshko [17] for incompressible, but variable-density, turbulent gaseous jets:

$$\theta \approx 0.212\left[1+\left(\frac{\rho_g}{\rho_l}\right)^{0.5}\right]. \quad (2)$$

The similarity in the form of these equations is quite astonishing considering vast differences between the cases. Chehroudi then suggested a linkage between the two cases. This is how Chehroudi et al. [5, 10] explained their case.

Imagine a jet that is being injected into a subcritical pressure environment similar to the ones shown in frame 4 of Figure 1 (or the left image in Figure 2). Clearly there are drops, ligaments, and interface "bulges" testifying existence of a surface tension. Chehroudi's group also reported evidence of a thermodynamic phase change under this condition. Hence, one appropriate characteristic time of the problem (at subcritical, $P_r < 1$) is the "bulge" formation/separation time (τ_b) on the interface of the turbulent liquid jet. This time scale, according to Chehroudi et al. [5, 10], characterizes the formation and separation event of bulges from the liquid jet which subsequently produce isolated ligaments and drops. Then they referred to a suggestion made by Tseng et al. [33] that this time scale was equal to $(\rho_l L^3/\sigma)^{1/2}$ for the primary breakup in turbulent liquid jets, where ρ_l, L, and σ are liquid density, characteristic dimensions of turbulent eddies, and surface tension, respectively. The second relevant characteristic time (for subcritical jets) that Chehroudi considered was the gasification time (τ_g). Here, an estimate of this time was calculated through the so-called D-squared law of spherical drop evaporation which is equal to D^2/K, where D and K are drop diameter and vaporization constant, respectively. In addition, Chehroudi also proposed a hypothesis as follows: if the aforementioned characteristic times (calculated for appropriate length scales) were nearly equal in magnitude, then the interface bulges are not able to separate as unattached entities from the jet interface to form ligaments and drops. This is because they are gasified as fast as they desire to be detached. This condition was then defined as the onset of the gas-jet-like behavior. Therefore, the transition between liquid-spray-like and gas-jet-like behaviors would be governed by finding the point at which these characteristic times are approximately equal. The scenario just explained is also supported by the "comb-like" structures seen in the middle image of Figure 2.

Using the above physical description, a model equation was then proposed for the N_2/N_2 system as

$$\theta = 0.27\left[\frac{\tau_b}{\left(\tau_b + \tau_g\right)} + \left(\frac{\rho_g}{\rho_l}\right)^{0.5}\right]. \quad (3)$$

In the limit when $\tau_g \gg \tau_b$ and $\tau_g \to \infty$, this equation collapses to the isothermal liquid spray case. This equation agrees well with the Chehroudi et al. [5, 10] experimental data at subcritical pressures for $\tau_b/(\tau_b + \tau_g) < 0.5$. A constant value of 0.5 was used for this term to predict the spreading rate at higher chamber pressures, including supercritical pressures.

For injection of N_2 into N_2, the characteristic time ratio, $\tau_b/(\tau_b + \tau_g)$, was computed from the experimental measurements of bulge and droplet sizes and calculations of the relevant properties. For N_2 injection into other gases, however, reliable information about the mixture properties at the interface, particularly the surface tension, prevented such a calculation from being performed; see Chehroudi et al. [5, 10]. To model these cases though, they further hypothesized that the characteristic time ratio term is a dominant function of the density ratio; that is, $\tau_b/(\tau_b + \tau_g) = F(\rho_g/\rho_l)$. Chehroudi then referred to the work by Brown and Roshko [15] in which they indicated that such a hypothesis is reasonable. This is because at low Mach numbers, there is no distinction between mixing layers where the two streams have different molecular weights, temperatures, or compressibility effects. Measurements and calculations of $\tau_b/(\tau_b + \tau_g)$ provided a shape of the function "F" for the N_2/N_2 system which is given as a plot in Chehroudi et al. [5, 10]. A curve fit of that plot generates the following equation:

$$F\left(\frac{\rho_g}{\rho_l}\right) = 5.325\left(\frac{\rho_g}{\rho_l}\right) + 0.0288 \quad \text{when } \frac{\rho_g}{\rho_l} < 0.0885$$
$$= 0.5 \quad \text{when } \frac{\rho_g}{\rho_l} \geq 0.0885. \quad (4)$$

It was found that the same function, F, calculated from measurements of the N_2/N_2 system could be made to work for other cases, provided that a case-dependent transformation was made to the density ratio at which the function F is evaluated. The final form of the Chehroudi proposed equation is

$$\theta = 0.27\left[F\left(x\left(\frac{\rho_g}{\rho_l}\right)\right) + \left(\frac{\rho_g}{\rho_l}\right)^{0.5}\right], \quad (5)$$

where $x = 1.0$ for N_2 into N_2, $x = 0.2$ for N_2 into He, and $x = 1.2$ for N_2 into Ar.

The quality of the agreement with experimental data is demonstrated in Figure 13. Hence there are no major changes in the form of the proposed model equation, even for an extreme arrangement such as injection of N_2 into He gas. It is also important to indicate that spreading angle

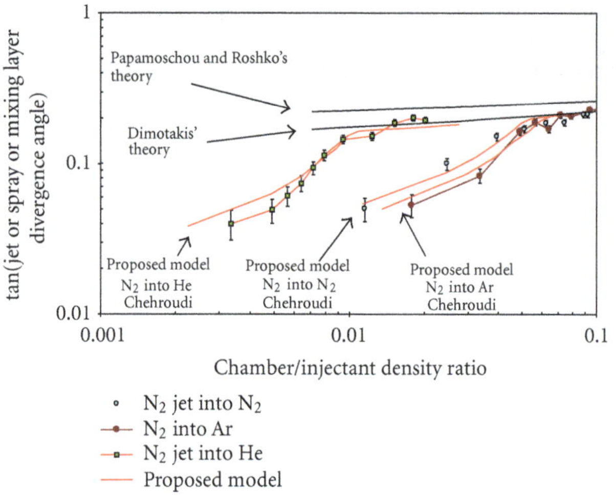

Figure 13: Comparison of the Chehroudi's proposed growth rate model with experimental data for single jets. Chehroudi et al. [5, 10].

under supercritical condition using modeling and computer simulation work by Zong and Yang [34] also agrees with Chehroudi's equation.

3. Single Jet with External Excitation

3.1. Effects of External Acoustic Field. Substantial evidences accumulated in the past attributed combustion instability to a complex interaction of the external acoustic field with the fuel injection processes, leading to incidences of instability in liquid rocket engines. For example, Oefelein and Yang [35] indicated that the near-injector processes in the thrust chamber were generally more sensitive to velocity fluctuations parallel to the injector face than normal to it. For this and other reasons, controlled experimental, analytical, and computational studies have been conducted in the past focusing on the effects of acoustic waves on both gaseous and liquid jets from a variety of injector designs. However, as indicated earlier, with a few exceptions, the scope of this paper is limited to experimental work which covers injection into supercritical conditions.

Chehroudi and Talley [36] used a piezo-siren design capable of generating sound waves with a sound power level (SPL) of up to 180 dB. This was used at three chamber pressures of 1.46, 2.48, and 4.86 MPa (reduced pressures of 0.43, 0.73, and 1.42, resp.). The experimental setup consisted of an acoustic driver and a high-pressure chamber, forming a cavity which resonated at several frequencies. The strongest ones were reported at 2700 and 4800 Hz. This basic setup was also used as the backbone for many of their future studies on coaxial injectors albeit with some improvements in data acquisition and facility plumbing. Chehroudi and Talley [36] injected LN_2 into GN_2 at room temperature under sub- and supercritical chamber pressures. Three different flow rates were considered, and the nature of the aforementioned interaction was documented via a high-speed imaging system using a CCD camera.

Figure 14 shows some sample results from this study where images of the jet are shown at two different perpendicular directions. These pair images at two different angles were taken on different test runs, but the operating conditions were as close as possible to each other. Each composite jet image consists of a mosaic of several images taken from the same test run but at different times and jet axial locations.

Chehroudi and Tally [36] reported that the acoustic field constricted the jet in the wave propagation direction, as seen in Figure 14, and consequently stretched the jet in the direction perpendicular to the propagation. They found that the impact of the acoustic waves on the jet structure was indeed strong at subcritical and near-critical pressures, but weakest at supercritical pressures. This is quite evident in Figure 14. They also observed that the externally imposed acoustic field interacted strongly at low injectant flow rates. The weakest effect was observed at supercritical pressures. This suggests that the mechanisms governing the coupling between acoustic waves and jets may be significantly different for jets injected under supercritical and subcritical conditions. This observation is important and its implications are discussed later in this writing.

To search for a plausible reason for the observed minimal impact at supercritical chamber pressures, Chehroudi and Talley [36] used the information mentioned earlier that at supercritical pressures, the unperturbed-jet spreading rate is the same as that by an incompressible variable-density gaseous jet; see Chehroudi et al. [5, 10]. For this reason, one expects the existence of vortices similar to what has been described, for example, by Rockwell [37] in which transverse disturbances were generated by an externally oscillating plate at a fixed amplitude but different frequencies (1000 to 10000 Hz). A synopsis of Rockwell's conclusions is useful in the context of the discussion here and hence is presented in the next two paragraphs.

A relatively large planar nozzle was designed by Rockwell [37] to inject water and investigate the details of vortex formation and their interactions near the exit area. The effects of the frequency of oscillations on the natural vortex coalescence were classified based on a Strouhal number ($St = fd/U$, where f, d, and U are the externally imposed excitation frequency, jet diameter, and mean jet velocity) versus Reynolds number ($Re = \rho U d/\mu$, where ρ and μ are the injectant density and viscosity) plot, see Figure 15. For the range of Re investigated (1,860 to 10,800), natural breakdown of the jet (i.e., with no external disturbances) was observed to be within a few nozzle widths. The dimensionless frequency (St_N) of natural vortices was found to change with Re number as $St_N = 0.012\,(Re)^{0.5}$. Four regimes were distinguished based on the observed effects of the external disturbance on the nature of the vortex interaction. The reference regime was that which corresponds to the natural breakdown of the jet. The four regimes were named, in order of decreasing St number, as "upper zone," "preservation," "matched excitation," and "forced fusion" regimes. A plot of these regimes is shown in Figure 15 and compared with the values estimated for Chehroudi's study using LN_2 injection into a GN_2 chamber.

In the "upper zone" regime of Figure 15 where the excitation frequencies were higher than about 3 to 4 times the natural breakdown frequency ($St > 3\,St_N$), no effects were observed. In the "preservation" regime, the core flow of the jet tended to be preserved followed by the induction of smaller vortices, the time-averaged velocity profile was narrowed, and the longitudinal turbulence was decreased (relative to undisturbed case). When the excitation frequency was matched with the natural break-down frequency ("matched" regime), the effect was to accelerate the process of vortex formation and growth relative to the undisturbed case. Also, transverse distortion of the jet core was seen due to the vortex growth and coalescence. For the symmetrical jet work of Becker and Massaro [38], symmetrical ring vortices were produced and highest time-averaged widening of the jet has occurred. This was in contrast to the planar jet case in which vortices on the two sides of the jet were 180 degrees out of phase with each other. This regime merges to the "preservation" regime at higher Re number (~10,000). In the "forced fusion" regime (frequencies ~1/3 of natural jet breakdown), the natural break-down vortices were forced to fuse early as a result of the formation of large-diameter applied disturbance vortices. Vaslov and Ginevskiy

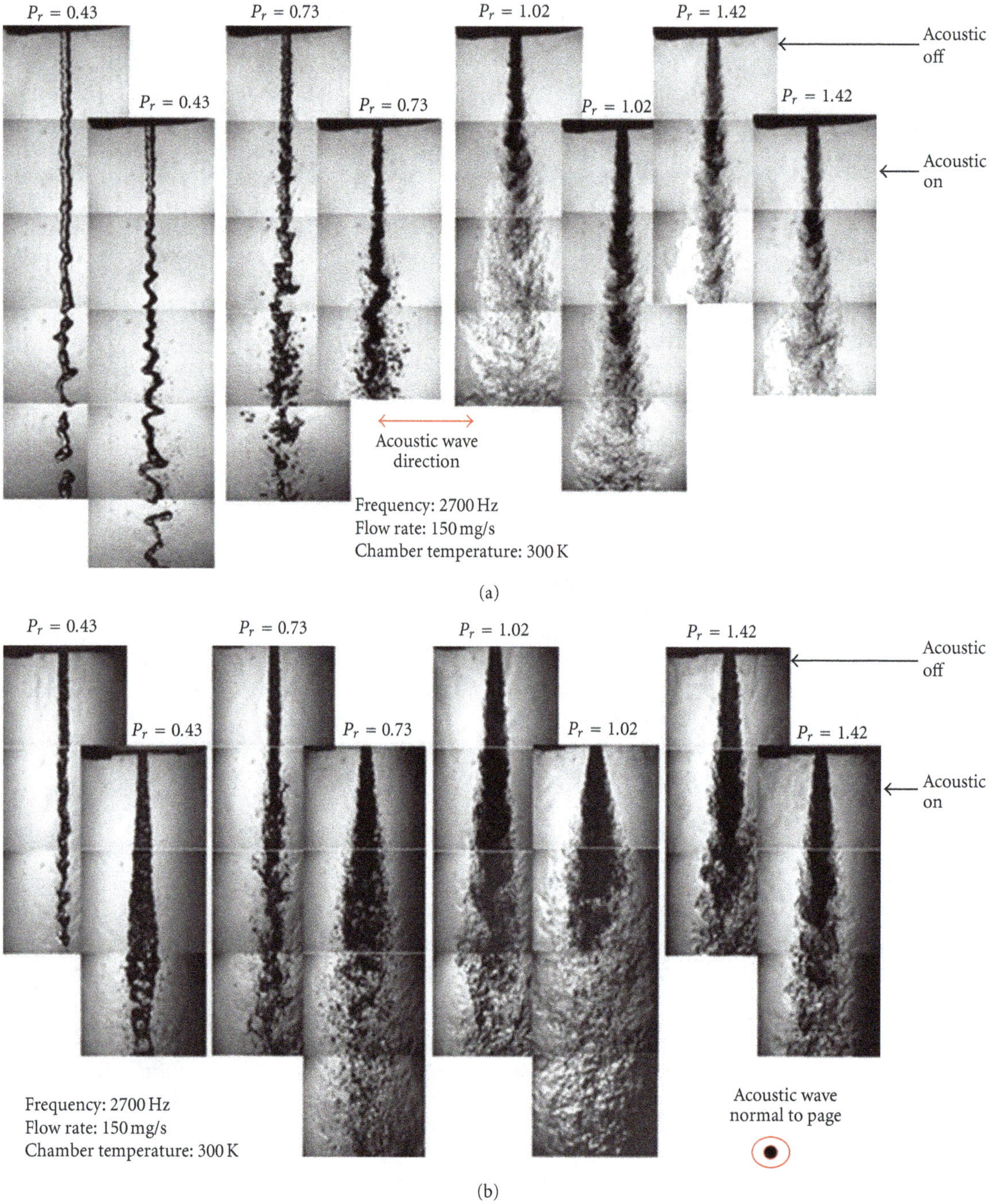

FIGURE 14: Interaction of acoustic waves with a single round liquid nitrogen jet injected into gaseous nitrogen at a range of chamber pressures from sub- to supercritical conditions. (a) Top images show front view; only one acoustic driver located at the left of the jet. The set of images shifted downward are when the acoustic field is turned on as indicated. (b) Bottom images show side views of the same jet (the same acoustic driver is located behind this page). P_r is the reduced chamber pressure (i.e., chamber pressure divided by the critical temperature of the injected nitrogen). Chehroudi and Talley [36].

[39] found that in this regime the time-averaged velocity profile of the jet was broadened. Finally, in the "lower zone" regime (frequencies < 1/10 of natural jet breakdown), the vortex growth was unaffected in their formation region. In the limit, however, as St approaches zero, the jet experienced a quasi-steady deflection process. Rockwell [37] stated that a fallacious averaged jet widening is detected if a time-averaged measurement technique is used downstream the injector, when in reality the jet is being deflected by the applied transverse disturbance.

To see how the results of Chehroudi and Talley [36] relate with Rockwell [37], their supercritical data is superimposed on the plot by Rockwell indicating different regimes discussed in the previous paragraph. Figure 15 shows a

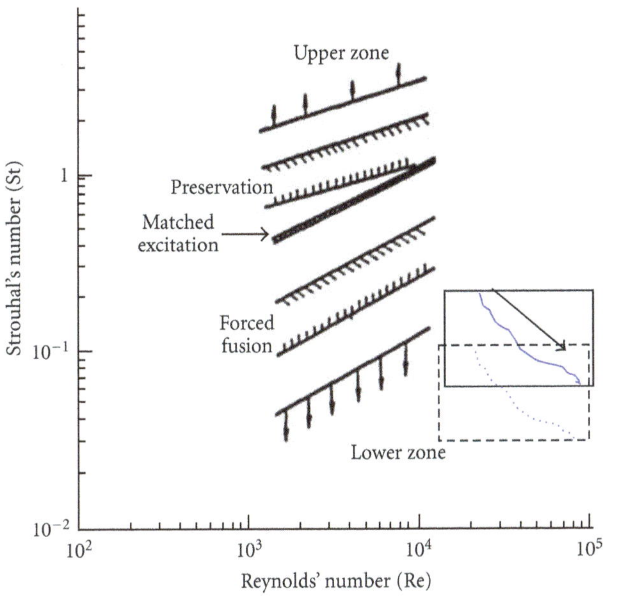

FIGURE 15: Regimes discussed by Rockwell [37] shown on an excitation Strouhal number versus the jet Reynolds number plot. The small solid rectangle to the right indicates the approximate region in which the Chehroudi LN_2 jet data lies when excitation is at 4800 Hz. The curve within this rectangle shows the approximate path as flow rate is increased from the lowest to the highest values, all at supercritical chamber pressures. Similar region for the 2700 Hz excitation is shown as a dashed rectangle. Chehroudi and Talley [36].

plot of the Strouhal number as a function of the Reynolds number. The curve marked as "matched excitation" is the same equation proposed by Becker and Massaro [38] for the natural breakdown of the jet. It can be seen that results from Chehroudi's study are beyond the regions studied by Rockwell [37]. With the caveats of extrapolating Rockwell's results, Figure 15 suggests that the frequencies used by Chehroudi's group are not expected to bring about noticeably large effects of the acoustic waves on supercritical jet because they lie in the region identified by Rockwell [37] as the "lower zone." No effects of the transverse disturbance on the jet were observed in the studies conducted by Rockwell [37] in this regime as described earlier. Note that the lowest flow rate case studied by Chehroudi's group in Figure 15 tends to enter into the "forced fusion" regime where stronger interaction is expected.

4. Coaxial Jets without External Excitation

Coaxial jets injectors have been used in liquid rocket engines since the 1940s and are found in the present-day hardware such as the Space Shuttle Main Engine (SSME); see Hulka and Hutt [40]. Rocket engines such as the SSME use liquid oxygen (LOX) and hydrogen (H_2) as propellants with the LOX flowing through an inner tube (or a center post) of a coaxial injector design and H_2 through an outer annular region. The inner jet goes through the break-up process after it exits the injector and then mixes in the shear layer between the inner and outer streams. Subsequently, combustion and heat release occur to generate the desired engine thrust.

Often the combustion processes are controlled by, or at least intimately related to, the jet breakup and mixing of the oxidizer and fuel streams. Certain aspects of these processes can safely and inexpensively be studied using chemically inert fuel and oxidizer simulants. Moreover, attempts are made for as many parameters as possible, such as velocity ratio, mass flow rate ratio, density ratio, Reynolds' number (Re), Weber's number (We), and Ohnesorge's number (Oh), to represent similar values obtained in real engines. This scaling exercise, to the extent possible, allows reasonable inferences and evaluation of different injector designs before ever being tested under hot fire conditions.

Investigations targeted at enhanced and fundamental understanding of the injection processes and combustion stability, although desirable to be performed under operating engine conditions, are practically problematic and costly for many reasons, including the need for an optically accessible full-scale liquid rocket engine, special material choices, and optical diagnostic limitations. For example, a single-element noncombusting test rig having as much of the key features of a real cryogenic rocket engine as possible, such as cryogenic temperatures, transcritical temperature ranges, supercritical pressures, realistic velocity ratios, and activation of an acoustic field inside its test chamber, is as close as one can get to the actual engine, yet avoiding indicated complications associated with the combustion in a full-scale setup. Therefore, test rigs have been designed in many studies to have such desired characteristics. Combination of the cryogenic liquid N_2 and gaseous N_2 (to eliminate mixture critical phenomena) and N_2/helium (He) system (to better simulate the LOX/H_2 injection issues) were considered in most of these studies.

4.1. Coaxial LN_2/He Injection. Telaar et al. [41] experimentally investigated the influence of the ambient pressure on the atomization phenomena from subcritical to supercritical pressures in a coaxial LN_2/He injection. They used a coaxial injector design with an inner diameter of $d_{LN_2} = 1.9$ mm for the cryogenic nitrogen flow and an annular slit width of 0.2 mm with an outer diameter of 2.8 mm for helium. Two examples of the flow visualization by shadowgraphy near the injector region are shown in Figure 16. The changes in the break-up mechanism at a reduced surface tension are quite evident. At the low chamber pressure of 1 MPa in Figure 16(A), a liquid spray is formed, whereas at the supercritical pressure of 6 MPa in Figure 16(B), a gas-like turbulent mixing of the dense and light fluids is seen. At 6 MPa, the initial temperature of the center LN_2 jet was 97 K, whereas the critical mixing temperature of He/N_2 mixtures at this pressure is 125.7 K. Thus, in the mixing layer between the LN_2 and He gas, transcritical zones may exist. The visual boundary of the LN_2 jet as shown in this figure is assumed to be the layer at which the temperature reaches the critical mixing temperature. At this interface then, the influence of the surface tension forces as compared to the shear forces appears to be negligible.

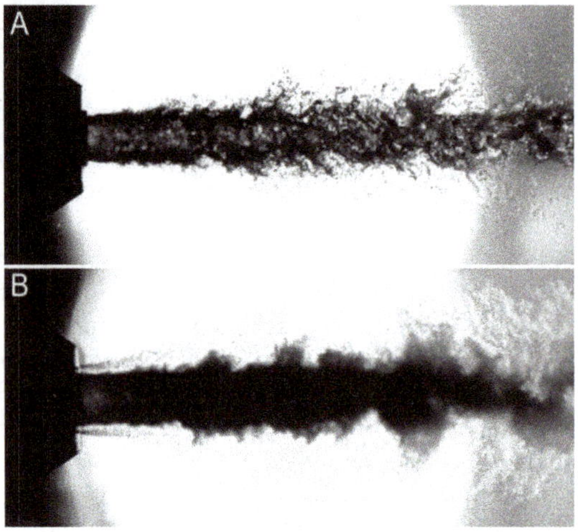

Figure 16: Binary coaxial liquid N_2/GHe system at (A) chamber pressure of 1.0 MPa, (B) chamber pressure of 6.0 MPa. Inner-tube hole diameter, d_{LN_2} = 1.9 mm; liquid nitrogen injection velocity, v_{LN_2} = 5 m/s; helium gas injection velocity, v_{He} = 100 m/s; liquid nitrogen injection temperature, T_{LN_2} = 97 K; gaseous helium injection temperature, T_{He} = 280 K (from Telaar et al. [41]).

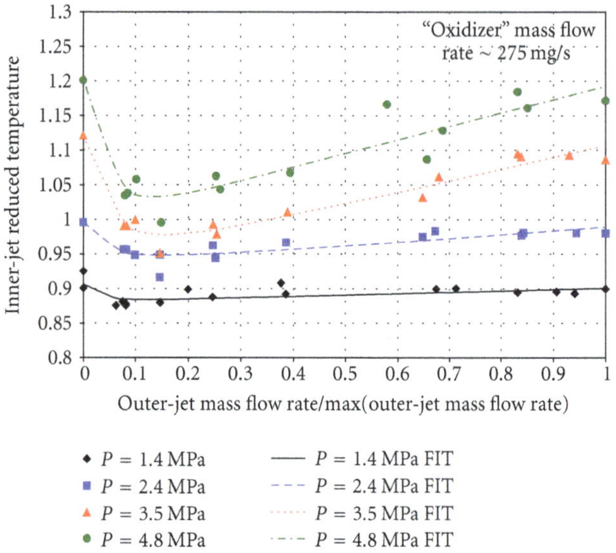

Figure 17: The inner-jet reduced temperature for a coaxial injector plotted versus the annular outer-jet mass flow rate, normalized by its maximum mass flow rate, at four different chamber pressures. Inner-jet mass flow rate is fixed at about 275 mg/s. The lines, called "FIT" on the inset, are curve fits to the raw data. The maximum outer-jet mass flow rates for 1.4, 2.4, 3.5, and 4.8 MPa are 2,995, 2,985, 2,974, and 2,918 mg/s, respectively, Davis and Chehroudi [42].

4.2. Visualization of a Liquid Nitrogen/Gaseous Nitrogen Coaxial Injector. Davis and Chehroudi [42] used a coaxial injector designed to inject a liquid nitrogen, with a coflow of gaseous nitrogen in its annular region, as a part of a program to better understand the nature of the interaction between acoustic waves and liquid fuel jets in cryogenic rocket engines. In their test setup, injection was into a chamber filled with gaseous nitrogen at a supercritical temperature and different pressures spanning from sub- to super critical conditions. However, prior to presenting Davis and Chehroudi's results on acoustic forcing investigations, they provided a wealth of new information on dynamics of the jets (with no acoustic excitations), specifically focusing on the dark-core length of the inner jet in their coaxial injector. This is discussed next.

Davis and Chehroudi [42] argued that in their coaxial injector design, or in any generic one, the outer-jet flow rate played two key roles. First, it assisted the inner-jet breakup at the subcritical condition and enhanced mixing for the supercritical case. Second, because of the temperature differences between the inner jet and the outer jet, there was a heat exchange between the two, both inside and outside the injector. Evidences for the heat transfer behavior were given by examination of the results from jet temperature measurements. Figure 17 shows normalized temperature measurements within the inner jet and at the injector exit plane plotted versus normalized outer-jet mass flow rate at four different chamber pressures covering sub-, near-, and super-critical chamber pressures. Figure 18 shows images indicating typical jet characteristics taken at these same three-chamber reduced pressures. Evidences on the impact of the heat transfer are shown in Figure 17. For instance, at the lowest (subcritical) chamber pressure of 1.4 MPa, Davis and Chehroudi [42] found that the inner-jet exit-plane temperature was fairly insensitive to changes in the annular outer-jet flow rate; see Figure 17. However, at all other elevated pressures, particularly at the supercritical one, an increase in inner-jet exit temperature was measured.

Under zero or no outer-jet flow rate condition in Figure 17, the annular passage of the injector was filled with the warm chamber nitrogen, hence strongly affected the heat transfer not only inside but also outside the injector. Clear evidence for the high heat transfer rate inside the injector was shown from the exit temperature measurements under this no-flow condition; see the y-axis in Figure 17. Here, the observed elevated temperature was reported to be the effects of the warm chamber gases present inside the annular space of the injector. Hence, referring to Figure 18, even though no atomizing (outer-jet) mass flow was at work in frames 11 and 21, one sees a shorter dark-core axial length than those seen in other images presented in frames 13 and 23. A small flow of the colder-than-chamber gaseous nitrogen in the annular passage, though may not assist the jet breakup, lowers the inner-jet exit temperature. Davis and Chehroudi [42] reported that the two processes (i.e., gas-assist liquid atomization and heat transmission) worked in such a way that the net effect was an apparent longer dark core in most cases. For example, compare the images in column one and two (from the left) in the last two rows as shown in Figure 18.

Clearly, the jet behaved differently at different outer-jet mass flow rates. Considering that the outer-jet mass flow rate acted as an atomizing gas, thus accelerating the break-up mechanism, then as the mass flow rate of the outer jet increased slightly (~487 mg/s, frame 3, Figure 18), the inner jet began to break up into relatively large droplets and

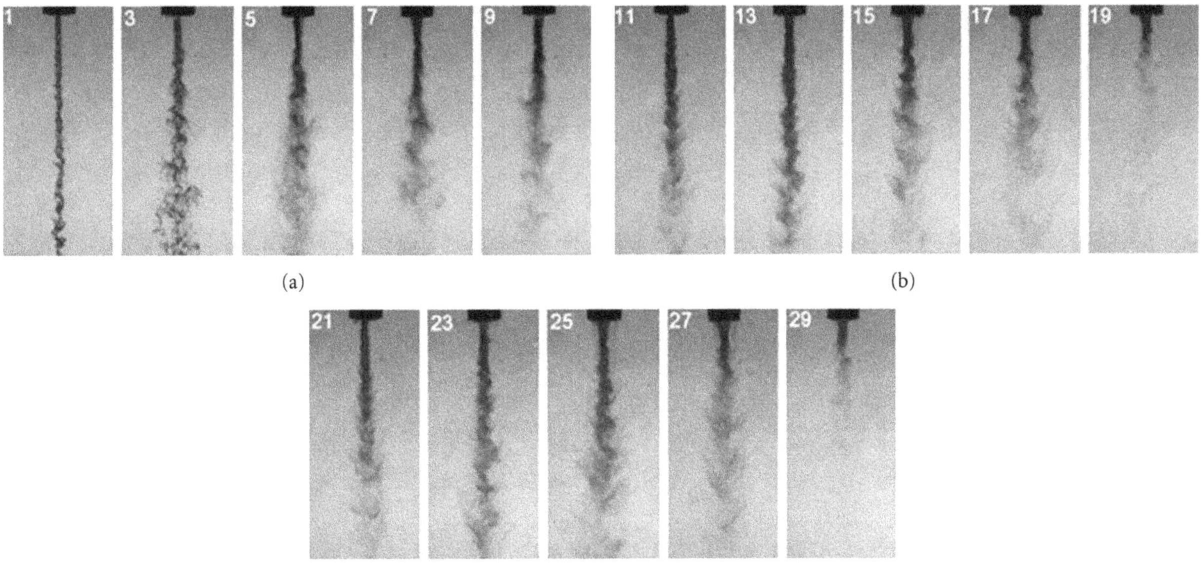

(a) (b) (c)

FIGURE 18: Images of a coaxial jet at approximately the same inner-jet mass flow rates ($\sim$275 mg/s). Columns are at about the same outer-annular-jet mass flow rates. For each row, the annular mass flow rate starts at a zero value to 2800 mg/s and increases from the left column to the right. The chamber pressure levels for images 1–9 are subcritical ($\sim$1.41 MPa), for 11–19 are near-critical ($\sim$3.46 MPa), and for 21–29 are supercritical ($\sim$4.77 MPa). Inner-and outer-tube flow average temperatures at injector exit are 170 K and 112 K, respectively. Davis and Chehroudi [42].

ligaments resulting from its interaction with the atomizing outer-jet flow. When the outer-jet flow rate was increased even further, finer droplets and ligaments were reported forming from the inner jet fluid, and more of its mass from the intact core was converted into droplets and ligaments. This consequently produced a shorter and less dark core for the inner jet. Moreover, as the outer flow rate increased, the heat transfer from the outer jet to the inner one intensified outside the injector, thus lowering the average density of the inner core in the injector near-field area. Under the subcritical condition, Davis and Chehroudi [42] reported that this combined outer-jet-assisted atomization of the inner jet and heat transfer processes created a dark core region more susceptible to external excitations.

4.3. Density Measurements in a Coaxial LN_2/GH_2 Jet. Although use of the Raman scattering is straightforward for a liquid N_2 injection into a gaseous N_2, in a mixture of N_2 and H_2 the Raman signals of the N_2 and H_2 are generated at two different wavelengths. Thus, in principle, utilizing an appropriate filtering technique, the signals from both species can be analyzed independently and the partial densities of each species can be determined.

This diagnostic method has been used to investigate the atomization and mixing of coaxial LN_2/H_2 injection at pressures which were supercritical with respect to the critical pressure of pure nitrogen in the inner jet; see Oschwald et al. [32]. During all the tests, the fluids were injected into a flow channel filled with N_2 at near atmospheric temperature and a pressure of 4 MPa ($P_r = P/P_{crit} = 1.17$). Results from a coaxial LN_2/H_2 injection are shown in Figure 19(b) and are contrasted with a single-jet case presented in Figure 19(a). The decrease of the measured densities at large positive r/D values was reported to be due to refraction of the laser beam experienced as a result of density gradients existing at the H_2/LN_2 interface. Similar radial profiles have also been obtained for a range of axial distances from the injector. Based on these data, two-dimensional species distributions were reconstructed for both H_2 and N_2. For the D4 test case described in Figure 19, such a distribution pertaining to the H_2 is shown in Figure 20.

The maximum of the radial nitrogen density distribution at each axial distance was plotted as a function of the normalized distance (x/D) from the injector; see Oschwald et al. [32]. This was then used to ascertain the evolution of the injection process for the LN_2 jet downstream of the injector, as well as to assess the mixing effectiveness between the H_2 and N_2 species. Compared to the test case without the H_2 coflow, the existence of the coflow decreased the measured nitrogen density much faster from its initial injection value near the injector exit area to a plateau which was a value for the far-field background level. The x/D value at which this plateau occurred was then interpreted as representing a complete mixing of the jet with its background gas.

It was shown in Oschwald et al. [32] that when the inner-jet temperature at the injector exit was above the pseudoboiling temperature, it had a shorter decay distance (x/D) than a jet with an injection temperature initially below it. In both cases, the injected nitrogen was at a supercritical pressure. They also reported that the nitrogen inner jet above the pseudoboiling temperature exhibited a gas-like density; whereas below this temperature it possessed a liquid-like density value. Interestingly, their data suggested that

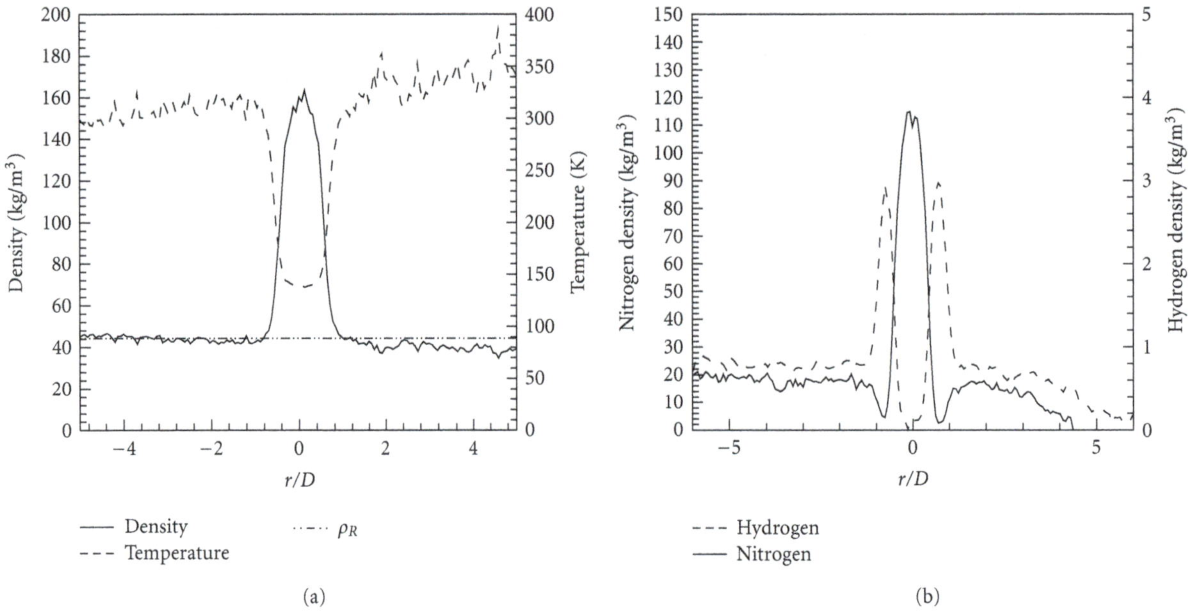

	Test case	V_{N_2} [m/s]	T_{N_2} [K]	V_{H_2} [m/s]	T_{H_2} [K]
single N_2 jet	A4	5	140	—	—
	B4	5	118	—	—
	D4	5	140	60	270
coaxial N_2/H_2 jet	E4	5	118	120	270
	F4	5	118	60	270

Figure 19: (a) Radial N_2 density profile for test case A4 (single jet), 2 mm (x/D = 1.05) downstream of the coaxial injector exit. (b) Radial N_2 and H_2 density profiles for coaxial LN_2/ H_2 injection for test case D4, 2 mm (x/D = 1.05) downstream the injector exit. Laser beam direction is from left to right. Oschwald et al. [32].

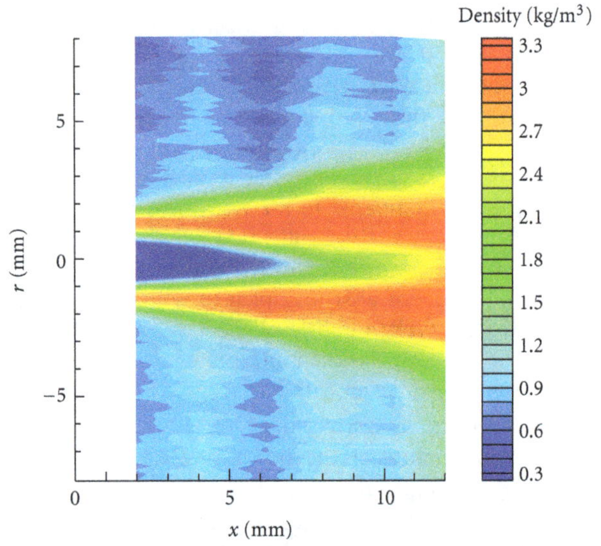

Figure 20: Hydrogen density for a coaxial LN_2/H_2 injection for the test case D4 in Figure 19 (T_{N_2} = 140 K, T_{H_2} = 270 K). Oschwald et al. [32].

the effect of higher H_2 velocity was found not to be as pronounced as this effect of the initial N_2 density.

4.4. Spreading Angle and Core Length Measurements. Two important, and less difficult to measure, geometrical characteristics of such coaxial injectors are the spreading angle of the (combined inner and) outer jet and the length of a dark inner-core of the injected jet. They are readily determined through the high-speed shadowgraphy approach. Both the length of the dark core and the spreading angle provide good measures for mixing characteristic of the jets in coaxial injectors.

Mixing of the propellant streams, if not controlling, is at least intimately related to the combustion process. To a first-order approximation, one measure of the mixing process is the length of the inner jet's liquid and/or dark core, which has been extensively studied by many researchers in the past. For details see Davis and Chehroudi [43]. Experimental core-length correlations, semiempirical theories, and other data for shear coaxial injector studies involving core length were competently summarized by Davis and Chehroudi [42]. Many of the correlations and semiempirical theories reviewed by them indicated dependencies of the core length with parameters such as outer-jet-to-inner-jet velocity ratio (V_r), outer-jet-to-inner-jet momentum flux ratio (M), density ratio (ρ_o/ρ_i), Reynolds' number (Re), and Weber's number (We). One difficulty Davis and Chehroudi reported with applying these relationships at supercritical pressures is that the predicted core length was very small or zero in magnitude. This is because one of the parameters in the equation was (We) and surface tension diminished greatly or vanishes all together. To overcome the difficulty in core length prediction associated with large (We) number, one group of researchers, proposed that M can be used to describe the scaling of the core length for shear-coaxial injectors and need not include (We) number if it was sufficiently high; see Davis and Chehroudi [42, 43] for details.

But before moving forward, it should be mentioned that confusion does exist in the literature especially when vague definitions are used to characterize the core length. The terms potential-core, potential-cone, intact length, intact-liquid length, and break-up length have all been used along with various measurement techniques. To be clear and to remove any possible ambiguities from the data, the dark-core length was defined as the connected dark fluid region between the injector exit area and the first break in the core as defined by an adaptive thresholding procedure; see Davis [44] or Davis and Chehroudi [42] for more details.

A plot of the dark-core length values for a coaxial injector measured by Davis and Chehroudi [43, 44] versus momentum flux ratio is shown in Figure 21. They used liquid nitrogen (in inner jet) with gaseous nitrogen (in outer annular jet) injected into a chamber filled with gaseous nitrogen at various pressures spanning sub- to supercritical values. The chamber temperature was always at a supercritical value. A clear distinction between the subcritical dark-core length (diamond symbols) and this length at the near-critical and supercritical chamber pressures is seen in Figure 21. Subcritical data suggests a much longer length than that at supercritical pressures for a given momentum flux ratio. Davis and Chehroudi indicated that both the near-critical and supercritical pressure conditions produced an appearance of a single-phase coaxial jet even though liquid nitrogen was injected through the inner tube of the coaxial jet. Recall that earlier in the single-jet part of this paper, Chehroudi and his coworkers [5, 10, 26] also found that two-phase-appearing jets were solely observed at subcritical chamber pressures. However, at supercritical chamber pressures, the spreading rate and fractal dimension values were the same as those seen for gaseous jets injected into a gaseous environment (i.e., single phase); that is, these supercritical single jets can be considered as variable-density, single-phase, gaseous jets. This conclusion presented earlier for the single-jet injection is something Davis and Chehroudi [43, 44] also showed to be mostly valid when it comes to the dark-core length observations in coaxial jet injectors. Evidences in support of such a behavior for coaxial injectors are presented in the next few paragraphs.

The dashed line in Figure 21 is a least-square curve fit to the subcritical data, and the dotted line is a similar fit to the near-critical and supercritical data. As indicated by the equations in this figure from Davis and Chehroudi [42, 43], the single-phase-appearing (i.e., at near-critical and supercritical pressures) data has a $M^{-0.5}$ dependence on momentum flux ratio. However, they indicated that the two-phase-appearing subcritical data had a weaker dependence, $M^{-0.2}$, than the single-phase dark-core length at near- and super-critical chamber pressures. Other quantitative differences between the subcritical and supercritical cases have been reported before. They also reported that the $M^{-0.5}$ form of the dependency on M parameter under near- and supercritical pressures is not only valid for gas-gas shear-coaxial jets, but is also for any single-phase (gas-gas or liquid-liquid) shear-coaxial jet.

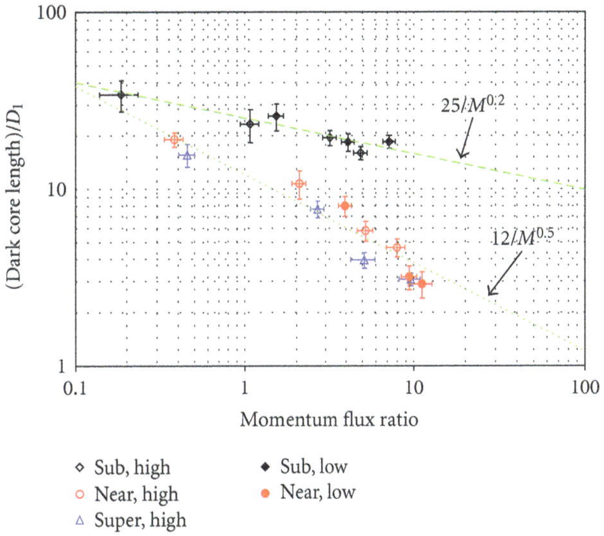

FIGURE 21: Plots of the dark-core length versus momentum flux ratio for a coaxial injector. Liquid nitrogen (inner jet) and gaseous nitrogen (outer annular jet) injected into chamber filled with gaseous nitrogen. The diamond, circle, and up-triangle symbols represent sub-, near-, and supercritical chamber pressure, respectively. The hollow symbols are at a "high" outer-jet temperature (~190 K) and solid symbols are at a "low" outer-jet temperature (~140 K). The dashed line is $25/M^{0.2}$ and the dotted line is $12/M^{0.5}$. Davis and Chehroudi [43, 44].

Figure 22 was constructed by Davis and Chehroudi [42, 43] aiming at a comprehensive comparison of the dark-core length they measured for the LN_2/GN_2 with all other comparable single-phase and two-phase data available in the literature for the potential-core length, intact-core length, and break-up length. This unique figure represents all relevant data in the literature concerning core length spanning 5 orders of magnitude in momentum flux ratio. They also noted that as M approaches zero, one reaches a limit which defines a single round jet configuration because the outer-jet velocity becomes zero. From the information shown in Figure 22, it seems that for $M < 1$, the data points converge and approach the core length range expected for single round jets reported by Chehroudi et al. [27] and Oschwald et al. [45]. The single-phase data by others (i.e., injectants and chamber content all are in one phase, either all gaseous or all liquids) along with the Davis and Chehroudi's

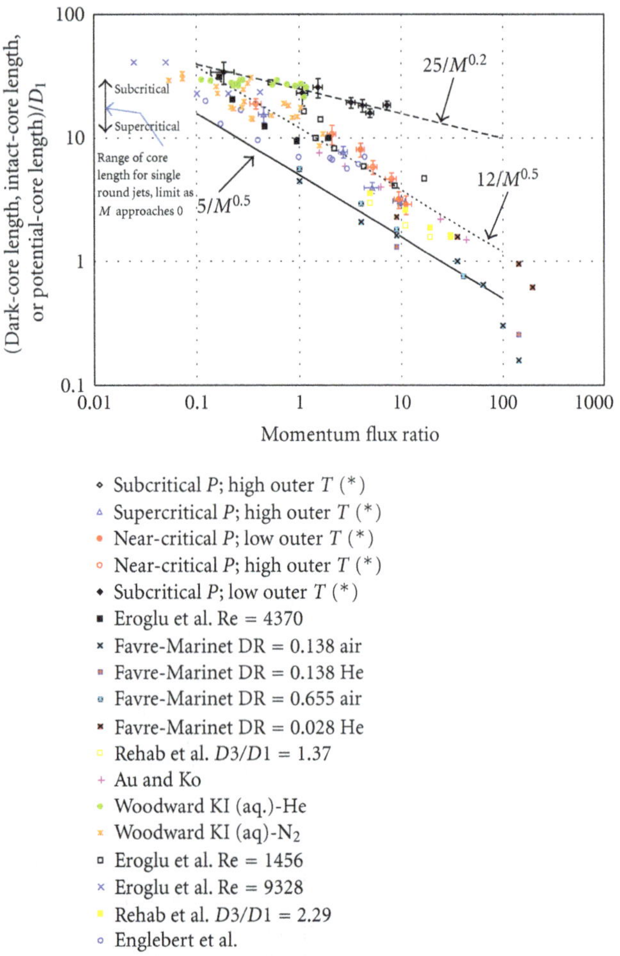

FIGURE 22: This figure shows comparison of the present coaxial-jet dark-core length measurements with all other relevant core length data available in the literature versus momentum flux ratio. Amongst the data reported by others, Eroglu et al. [47], Englebert et al. [48], and Woodward [50] are two-phase flows and the rest are single phase. The range of core length for cryogenic single jet (LN_2/GN_2) is also shown at the left margin. Davis and Chehroudi [42, 43].

LN_2/GN_2 supercritical results are all seen in Figure 22 to follow a dependence close to $L/D_1 = AM^{-0.5}$, where the constant A is between 5 and 12. This information combined with the clear separation of all the two-phase data, including the subcritical core length data seen in Figure 22, essentially suggests a two-phase behavior at subcritical and a single-phase conformance at supercritical chamber pressures.

Even though best efforts were made to use the most relevant data by other investigators, some shortcomings were reported. For example, at $M > 100$ in Figure 22, the experiments of Favre-Marinet and Camano Shettini [46] exhibited a recirculation bubble at the end of the core, and thus the core length decreases as depicted in this figure. The injectors used to produce the two-phase coaxial jets of Eroglu et al. [47] and Englebert et al. [48] had apparently much larger outer-jet gap widths than what is typical of rocket injectors. Additionally, the apparatus of Eroglu et al. [47], reported in Faragó and Chigier [49], did not produce fully turbulent inner jet until $Re > 10^4$. Davis and Chehroudi [42, 43] explained that the lack of a fully turbulent inner jet and the significant differences between their injector and shear-coaxial ones used in rockets could be the reason why the core length measured by Eroglu et al. [47] is shorter than those observed in their work. Englebert et al. [48] reported that the core length scaled with $M^{-0.3}$. The two-phase core length by Woodward [50] for the water potassium iodide solution with helium, however, obeyed very nearly the trends for the subcritical data points (i.e., $25M^{-0.2}$). Considering that the momentum flux ratios near or higher than 10 are of importance for liquid rocket engines (LRE), Davis and Chehroudi [42, 43] indicated that the data for the subcritical (two-phase) case was the only reported information in the neighborhood of the $M = 10$.

Outer-to-inner jet velocity ratio, V_r, has been a design parameter for shear-coaxial injectors, particularly, as a criterion to ensure stable operation of liquid rocket engines (LRE). For LOX/H_2 engines, the design rule of thumb has been to keep the velocity ratio greater than about 10 in order to maintain stable engine operation, as is discussed by Hulka and Hutt [40]. Although this stability criterion has been suggested by the experimental data, no physical explanation has been provided. Related to this is a method to rate a LRE for combustion instability, known as the *temperature ramping*. The temperature ramping of a fired H_2/O_2 LRE is accomplished by lowering the temperature of the H_2 (in the outer jet of the coaxial injector) while maintaining its mass flow rate at a constant value. The lower the H_2 temperature when the onset of the combustion instability is detected, the broader the stability margin of that particular LRE. The temperature ramping is related to and affects the velocity ratio. This is because one of the outcomes of lowering H_2 temperature at constant mass flow rate is the increased density and consequently, lowered injection velocity values. Davis and Chehroudi [42, 43] therefore investigated the impact of the velocity ratio on their nonreacting LN_2/GN_2 coaxial jets, hoping that results may shed some light on the effects of velocity ratio observed in fired production engines. This is explained next.

To determine the effects of the outer-jet temperature (which is GN_2 in Davis and Chehroudi's work) on the coaxial jet, Davis and Chehroudi [42, 43] studied two nominal temperatures of ~190 K and ~140 K, called "high" and "low," respectively. Recognizing that the dark-core length is a dynamic quantity, its basic characterization required measurements of both the average and root mean square (RMS) of length fluctuations. The averaged dark-core lengths were then measured and are shown in Figures 23(a) and 24(a) as a function of velocity ratio for sub-, near-, and super-critical chamber pressures. The RMS of the variations of the dark-core length is also shown in Figures 23(b) and 24(b). Note that Figures 23 and 24 present results for conditions when the external acoustic field is both turned on and off for "high" and "low" outer-jet temperatures. However, it is important to emphasize that in this section only the results when the acoustic field is turned off are discussed, deferring the impact of the externally imposed acoustic field to the next part of this paper.

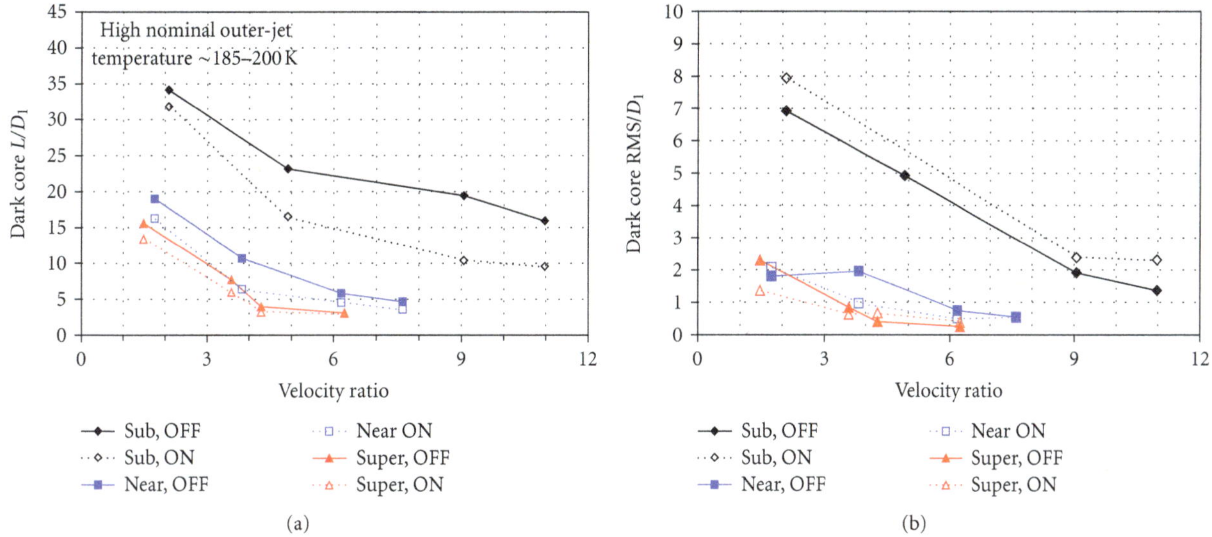

FIGURE 23: Plot of the averaged dark-core length (a) and the root mean square (RMS) of the length variations (b) normalized by the inner diameter of a LN_2/GN_2 coaxial jets injector. The solid symbols and lines represent the data when the acoustic driver is off, and the hollow symbols and dotted lines show the data when the acoustic driver was operated at ~3 kHz. The diamond, square, and up-triangle symbols are sub-, near-, and supercritical chamber pressures, respectively. All cases are for the high nominal outer-jet temperature of ~190 K. In the inset, the words sub, near, and super refer to subcritical, near critical, and supercritical chamber pressures, respectively, and the words OFF and ON refer to the acoustic driver being off and on at ~3 kHz, respectively. Davis and Chehroudi [42, 43].

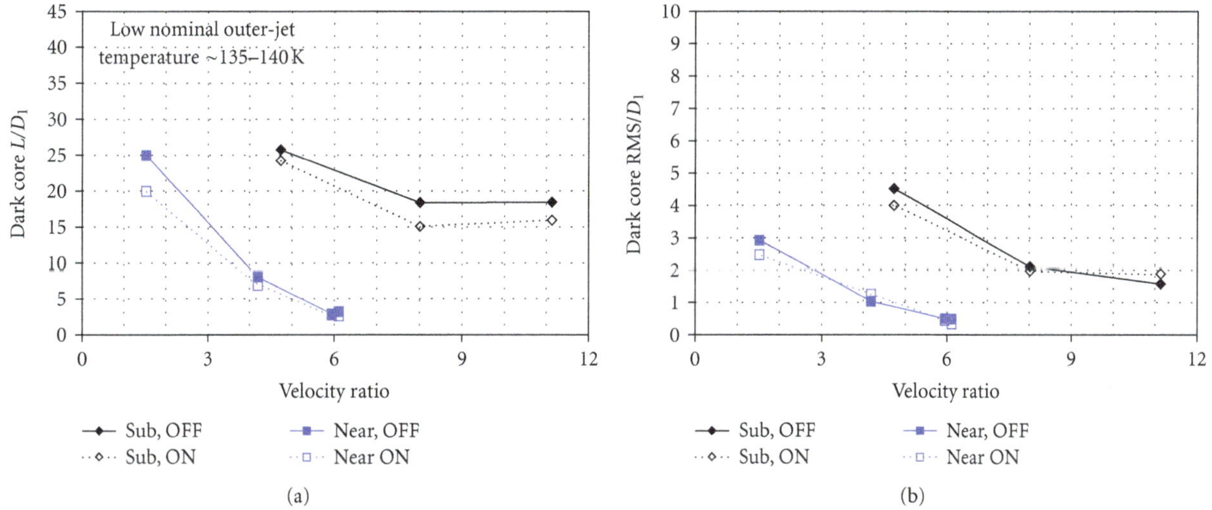

FIGURE 24: Plot of the averaged dark-core length (a) and the root mean square (RMS) of the length variations (b) normalized by the inner diameter of a LN_2/GN_2 coaxial jets injector. The solid symbols and lines represent data when the acoustic driver is off, and the hollow symbols and dotted lines show results when the acoustic driver is operated at ~3 kHz. The diamond and square symbols are for sub- and near-critical chamber pressures, respectively. All cases are for the low nominal outer-jet temperature of ~140 K. In the inset, the words sub and near refer to subcritical and near-critical chamber pressures, respectively, and the words OFF and ON refer to the acoustic driver being off and on at ~3 kHz, respectively. Davis and Chehroudi [42, 43].

Evident in Figures 23(a) and 24(a) is the fact that the length of the dark core decreases as the chamber pressure is increased from sub- to super-critical conditions. The dark core provides an indication of high-density regions of the flow. At a given fixed chamber pressure, as V_r is increased, the length of the dark core decreases and appears approaching a constant value. The RMS values of the dark-core length fluctuations, shown in Figures 23(b) and 24(b), exhibit somewhat similar trends to those seen for the averaged values, that is, substantially lowered at near- and super-critical chamber pressures.

It is known that for a liquid-fueled rocket, atomization and break-up processes, interactions between the propellant jets, droplet formation, and vaporization are all affected by the pressure and, particularly, velocity fluctuations. Also, for any chemically reacting system, it is expected that, to certain but different degrees, the rate at which energy is released be sensitive to the rate of changes in temperature, density, pressure, and, of course, particularly mixture ratio. It is then intuitive to relate, in some form, the RMS values of the dark-core length fluctuations to mixture ratio variations. On the other hand, a low RMS value can be interpreted as the coaxial jet's "*inherent steadiness*" (or insensitivity to external stimuli) and vice versa. Examination of Figures 23(b) and 24(b) clearly shows that this property is drastically reduced as the velocity ratio is increased. Although these results are for one injector (i.e., no interinjector interaction), Davis and Chehroudi [42, 43] proposed a linkage between these nonreacting results and production-engine combustion instabilities observed at low velocity ratios. They stated that "*it is then quite possible that the observed improvement in combustion stability at higher values of velocity ratio is a result of the inability of the jet to generate large mass flow rate fluctuations under these conditions, thus weakening a key feedback mechanism for the self-excitation process.*"

As indicated earlier, in temperature ramping exercises for stability rating of LOX/H_2 engines, the mass flow rate of the outer-jet hydrogen is maintained at a constant value. Davis and Chehroudi [42, 43] also attempted to provide a linkage between the temperature ramping and their experimental data. They said that as the temperature of the H_2 is decreased during a ramping episode, the H_2 becomes more dense, which decreases the injector velocity ratio at a constant mass flow rate. The RMS plots shown in Figures 23(b) and 24(b) suggest that such a decline in this ratio amplifies the jet's *inherent unsteadiness* (i.e., higher RMS values), providing a possible explanation for the engine's eventual arrival into an unstable zone as a temperature ramping test proceeds.

On the spreading angle or growth rate of the coaxial jets, Figure 25 is a compilation of experimental measurements reported earlier by Chehroudi et al. [5, 10] on single jets along with those newly acquired from a coaxial injector and other data. The tangent of the total jet divergence angle is plotted against the chamber-to-injectant (for single jets) or chamber-to-inner-jet (for coaxial jets) density ratio. The data taken by Chehroudi for single jets in the same facility is marked by (*) in the legend. Note that there are also some theoretical expressions presented in this figure for comparison purposes. The coaxial injector data follows the direction of the arrow as chamber pressure is changed from sub- to near- and to super-critical condition. Interestingly, the data for coaxial injection at subcritical condition conforms with the classical incompressible variable-density shear layer (Dimotakis' theory) which is also what is observed for the single jet, but at supercritical chamber pressures. This suggests that under such a condition the outer jet in Chehroudi's coaxial injector grows similar to a variable-density gaseous jet. Coaxial data, although stays nearly at a constant value, departs from the rest in this figure as chamber pressure is raised towards the supercritical condition.

5. Coaxial Jets with External Excitation

Cold flow studies investigating the interaction of acoustic waves with single-jet injectors have recently been extended

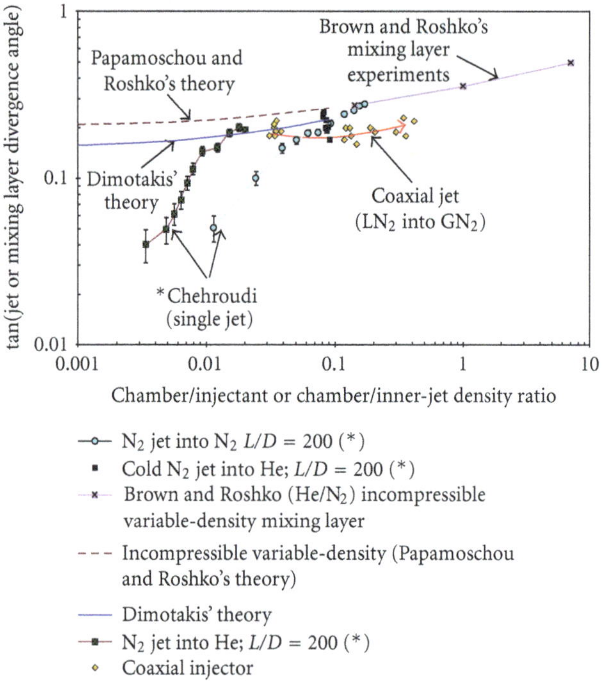

FIGURE 25: Spreading rate of the shear layer versus the chamber/injectant or chamber/inner-jet density ratio for single and coaxial jets compared with different predictions for planar shear layers. For the coaxial data, the chamber pressure increases from sub- to near- and to supercritical conditions in the direction of the arrow.

to include coaxial injectors by Chehroudi et al. [51], Davis and Chehroudi [42] and Leyva et al. [52]. Their objective was to characterize the coaxial jets response and understand how nonreacting injector results are related to historical liquid rocket engine combustion instability data. As it will be shown in the next few paragraphs, the aforementioned observation that the single-jet interaction with the externally imposed acoustic waves was stronger at subcritical pressures than at supercritical pressures appears to remain generally true for the coaxial-jet injector as well, although some influences of the mass flow ratio and initial temperature were observed. Efforts by Chehroudi's group continue in order to better understand how various parameters control the coupling mechanisms.

Figure 26, taken by Davis and Chehroudi [42], shows a sample of ten consecutive images at sub- (rows one and two from top), near- (rows three and four), and supercritical (rows five and six) chamber pressures from a LN_2/GN_2 coaxial injector. Rows one, three, and five are when the acoustic driver was off and the remaining ones are when it was activated at ~3 kHz. The evolution of the jet in time is from left to right in this figure and the time interval between frames is about 55.6 μs. Prominent in all images of the jet is the existence of a dark central region. After close examination of many images at different magnifications, Davis and Chehroudi [42] reported that the dark core under the unexcited subcritical pressures (Figure 25, row one) can be approximated as a cylinder-like structure with unstable

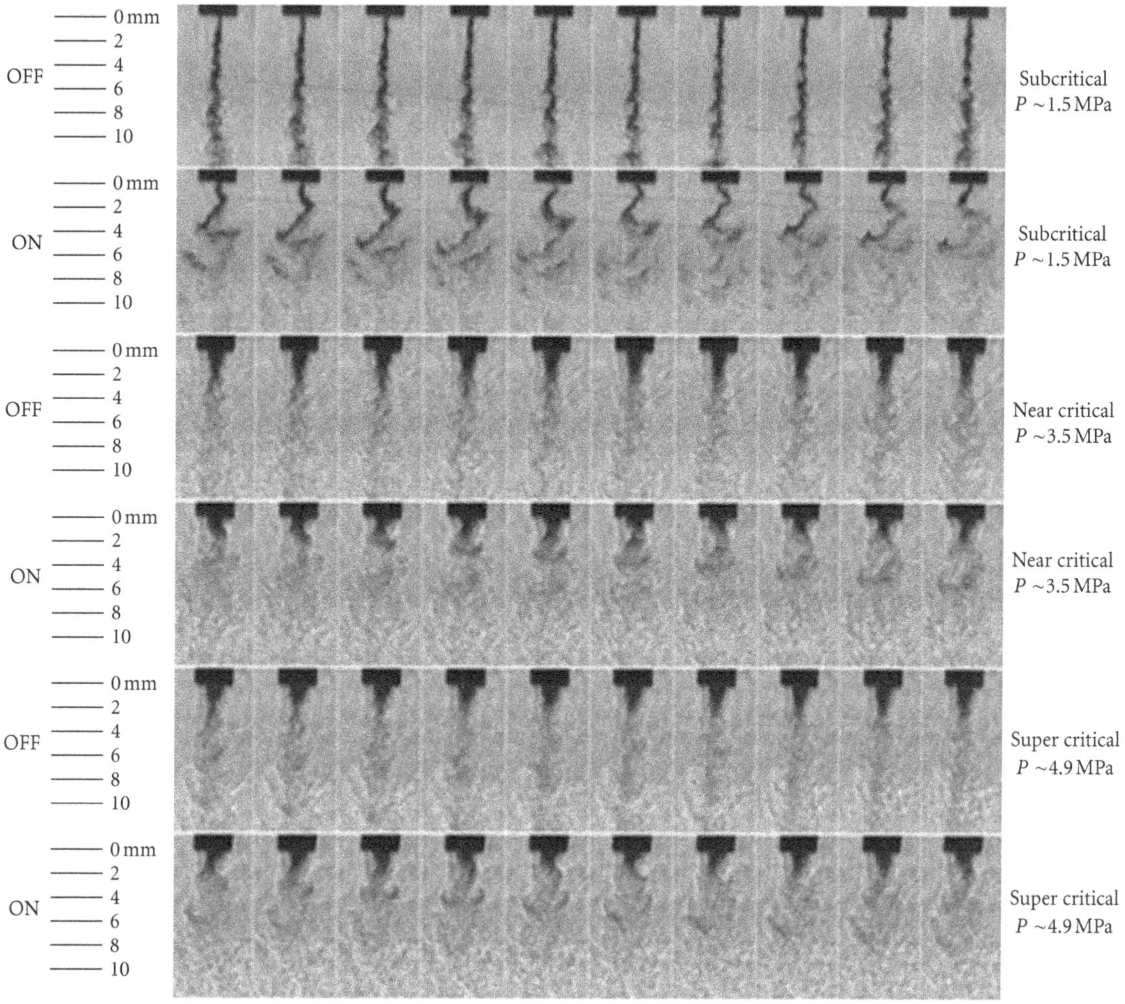

FIGURE 26: Consecutive frames from high-speed shadowgraph movies or a coaxial injector (LN_2/GN_2 into N_2-filled chamber) with the acoustic driver turned off (in rows 1, 3, and 5) and on (in rows 2, 4, and 6) at ~3 kHz. Time increases from left to right with an interval of 55.6 ms between frames. The first two rows are at a subcritical chamber pressure (~1.5 MPa), the third and fourth rows are at a near-critical chamber pressure (~3.5 MPa), and the fifth and sixth rows are at a supercritical chamber pressure (~4.9 MPa). The acoustic driver is turned off for the first, third, and fifth rows and on for the second, fourth, and sixth at ~3 kHz. The light gray lines in the first and second rows connect fluid structure as they evolve in time. Davis and Chehroudi [42].

surface waves of low amplitudes. However, when they increased the chamber pressure to near- and supercritical pressures (rows three and five), the dark-core length became shorter (as indicated earlier) and the structure changed to a more conical shape near the injector exit area. The conical structure of the dark core has been reported before for single-phase coaxial jets by Lasheras and Hopfinger [53] but not at supercritical conditions.

Excitation of the jet with an acoustic driver yielded significantly different behavior of the dark core compared to that of the unexcited one. The strongest effect was observed under subcritical pressures and made clear in Figure 26. Note that the direction of the acoustically induced velocity oscillations is horizontal (towards left and right) in images presented in Figure 26. It is seen that the core of the jet forms large-scale sinusoidal structures as a result of this imposed velocity field oscillations. From the movies they acquired, Davis and Chehroudi [42, 43] explained an observed dominant pattern which is also shown in Figure 26. They said that as a piece of fluid left the injector tip, the momentum from the acoustically induced motion caused a transverse displacement pushing the core of the jet into the higher-speed annular jet area. Then the dense fluid from the core experienced acceleration in the axial direction, which was caused by the high-speed annular jet motion. Upon reversal of the acoustic field, the dense fluid, which was initially from the core but now in the high-speed annular flow, appeared to maintain its transverse component of the momentum imparted upon leaving the injector and hence the dense fluid particle did not reverse its direction. The dense fluid parcel then slowed (both in the axial and transverse directions) as it arrived at the shear layer between the outer jet and the chamber fluid farther downstream, where a "cusp-shaped" structure was formed from the dense

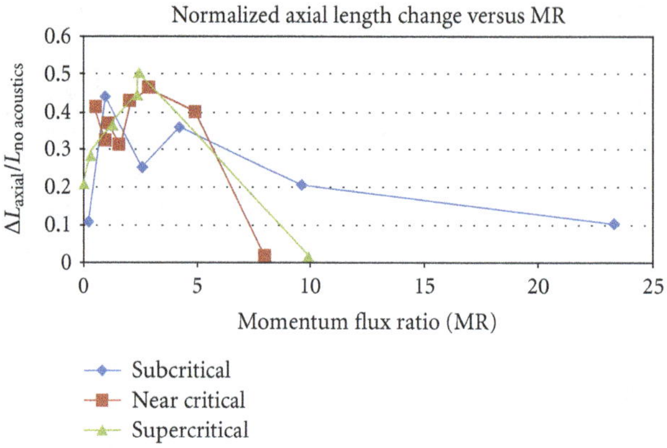

FIGURE 27: Maximum length change between dark-core length without acoustics and dark-core length with acoustics (ΔL_{axial}) divided by the dark-core length without acoustics ($L_{no\ acoustics}$) for each momentum flux ratio case. Results for a coaxial injector. Rodriguez et al. [54].

fluid originating in the core of the jet. Subsequent mixing and heat transfer from the outer-jet to the inner-jet core fluid ultimately cause the fluid parcel to be indistinguishable from the outer-jet fluid.

Figures 23(a) and 24(a) also show the impact of the outer-jet-to-inner-jet velocity ratio (V_r) on the mean dark-core length at three different chamber pressures but under externally imposed acoustic excitation. On an average, when the dark core feels the imposed external acoustic field, its length is shorter than or equal to that when the acoustic driver is turned off. At the near-critical and supercritical chamber pressures, as the V_r increases, the difference between the lengths of the dark core, measured with and without the acoustic field, diminishes. One distinct feature of plots in Figures 23(b) and 24(b) is that, at any given velocity ratio, the RMS of the core fluctuations under subcritical condition is much higher than near- and, particularly, super-critical conditions. And equally important, is that this statement is true with or without the externally imposed acoustic excitation field. Using Davis and Chehroudi's [42, 43] interpretation of the RMS, it appears that the jets under supercritical conditions are more *inherently insensitive* to their environment even when an acoustic field is imposed. Another important feature here is the fact that the magnitude of this RMS of fluctuations drastically declines for all cases, more so for the subcritical case, when velocity ratio is increased. Implications of these unique and important observations for the combustion instability in liquid rocket engines are provided later in this writing.

Rodriguez et al. [54] initiated a set of experiments to analyze the impact of the transverse acoustic wave phase variations on the magnitude of the inner-jet dark-core length. The chosen coaxial jet was exposed to different acoustic conditions by varying the phase between two acoustic sources which faced each other while the jet was located at the midpoint between the two sources, all being inside a high-pressure chamber. The coaxial jet was exposed to a subcritical and a near-critical pressure environment. The measurements were performed on backlit images of the coaxial jet obtained with a high-speed camera. The outer-jet-to-inner-jet momentum flux ratio was varied from 1 to 20 for subcritical conditions and from 0.6 to 5 for near-critical conditions. The resonant frequency of the system was approximately 3 kHz and the maximum pressure variation with respect to total pressure was 3%. Figure 27 shows summary of their results. This figure presents the maximum dark-core length reduction between the case without the external acoustics and that with the acoustics for all phase angles, each normalized with the core length without the acoustics and then plotted as a function of M (momentum flux ratio). The analysis of this data showed a reduction of this length with acoustic excitation within the momentum flux ratio (M) range between 1 and 5 for all chamber pressures, except at the subcritical which extended up to M value of 10. The effect of the acoustic waves at lower or higher M values did not have significant influences on the behavior of the dark-core length.

Rodriguez [55] also examined effects of the injector geometry, specifically the impact of the inner-tube wall thickness on the jet behavior and its interaction with the acoustic field. A new injector design was considered with a thinner inner-tube wall thickness value which consequently produced an inner jet with larger diameter than the original one used in earlier studies. The highest reduction in the dark-core length for the tests with the new coaxial injector geometry took place at moderate M values which supports similar evidence obtained with the original injector possessing a larger inner-tube wall thickness. The results for the new injector also showed that for a given outer-to-inner jet momentum flux ratio, the normalized values of the dark core length between subcritical and nearcritical cases agree reasonably well.

6. Impinging Jets

To the best of this author's knowledge, there is only one impinging jet data set reported under both sub- and supercritical conditions; see Chehroudi [56]. However, the study conducted by Anderson et al. [57] is an example at

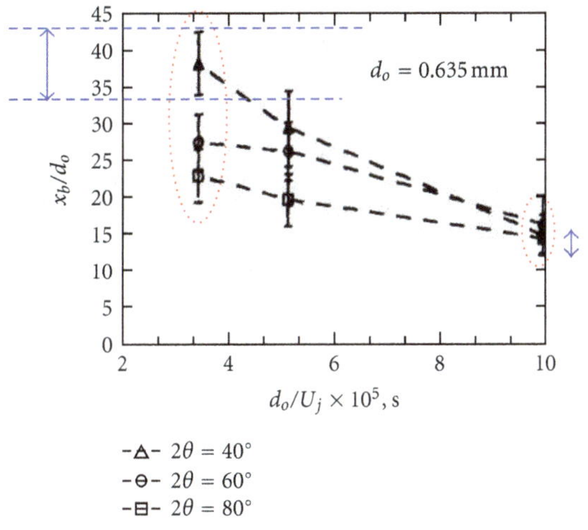

FIGURE 28: This figure shows *sheet break-up length* for an impinging injector as a function of instability parameter at three different impingement included angles. Much higher sensitivity of the *sheet break-up length* is seen with included angle (2θ) at low dn/V ($= d_o/U_j$, in the original article) values. Anderson et al. [57].

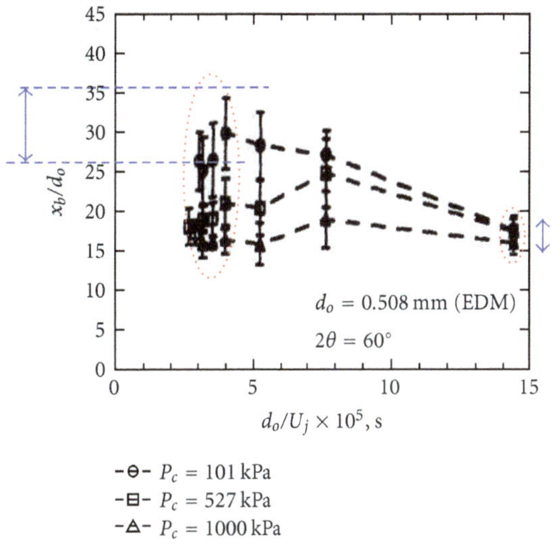

FIGURE 29: This figure shows *sheet break-up length* for an impinging injector as a function of instability parameter at three different chamber pressures. Much higher sensitivity of the *sheet break-up length* is seen with chamber pressure at low dn/V ($= d_o/U_j$, in the original article) values. Anderson et al. [57].

high-pressure subcritical chamber conditions which is of relevance to the discussion in this section.

Figures 28 and 29 taken from Anderson et al.'s [57] studies show measured data for the *sheet break-up* length in a doublet impinging jet injector design plotted against injector-hole-diameter (dn)-to-injection-velocity (V) ratio, that is, dn/V. The dn/V ratio is also called the *stability parameter*. This is because of the dependency of the rocket engine stability region (using this type of injector design) with the dn/V ratio. Large differences between the sheet break-up lengths for different pressures and different impinging-jet included angles (2θ) at low values of the stability parameter are seen in Figure 29. For example, Figure 29 strongly suggests higher sensitivity of the injector when dn/V is reduced through an increase in V (injection velocity) values. This is simply deduced by the enlarged size of the scatter bounds at any given pressure and sensitivity to pressure changes at low dn/V values. Although strictly speaking one should have its frequency response (amplitude and phase) measured, Chehroudi [56] interpreted these results as an indication of injector hypersensitivity under such conditions. At a given pressure or included angle, the *data scatter band* shown in Figures 28 and 29 is also largest at low dn/V values, again and consistently suggesting a more erratic/chaotic dynamic behavior.

The changes in the dark-core (break-up) length were also reported in Figure 30 by Chehroudi [56], showing a progressive increase in chamber pressure up to a supercritical condition for liquid nitrogen injection into gaseous nitrogen environment with no externally imposed acoustic field. The long preimpingement length seen along the jet is expected due to L/dn of about 100 which was intentionally designed to obtain a fully developed condition at the hole exit plane and also to accentuate the effects of chamber pressure on the nature of the impingement. Obviously, shorter dark core is achieved for lower (injector hole) L/dn values used in LRE. It is observed that not only the dark-core length of each individual jet is reduced as supercritical pressures are approached (as before and as expected), but the jet also thickens during this process. Clearly each jet is going through a very similar process discussed earlier in the context of single jet discussion. However, there are important implications as a result of such anatomical changes in an impinging injector which will briefly be discussed in the final part of this paper.

Aside from these results, the author is not aware of any published impinging jet data of relevance under supercritical chamber conditions. The situation is even worse as no systematic investigation or results are available on the impact of external excitation (imposed acoustic field) on impinging-jet injectors.

7. Implications for Combustion Instability

From the review of recent studies presented here, it is clear that our understanding of the jet behavior under transcritical and supercritical conditions has substantially improved in the past 20 years primarily due to pioneering experimental studies reviewed here. Computer simulations have also contributed heavily in the understanding and design process but are not within the scope of this short review. While experimental results have been very valuable for simulations and provided physics-based improvements in computational efforts, the implications of these results towards formulating "*a big-picture*" and efforts in constructing such an image through synergies amongst these pieces of information were needed and are just beginning. One relevant sketch of this *big picture* this author knows is that attempted by Chehroudi [56]. He, in his work, has made an effort to collect pieces of experimental information and offered a compelling

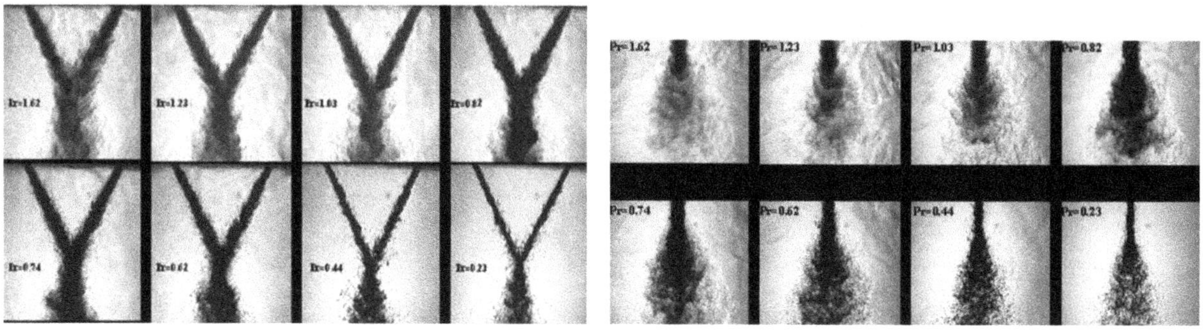

(a) (b)

FIGURE 30: Instant images of sub-, near-, and supercritical impinging jets for LN_2 into GN_2 (room temperature) injection by Chehroudi. The last two rows show the same injector in the first two rows but viewed at a 90-degree angle. $P_{ch} = 0.8, 1.5, 2.1, 2.5, 2.8, 3.5, 4.2, 5.5$ MPa; from lower right to upper left, $P_{ch} = 100, 200, 300, 350, 400, 500, 600, 800$ psig. For nitrogen: $P_{critical} = 3.39$ MPa; $T_{critical} = 126.2$ K. Re = 25,000 to 70,000; holes' length-to-diameter (L/dn) of ~100; no cavitation; injection velocity: 10–15 m/s. Chehroudi [56].

argument to link nonreacting cold-flow studies (in test rigs) to fired subscale engine information and to what occurs in production engine in the context of combustion instability. Of course, as the history of scientific discovery testifies, such efforts are mostly evolutionary in nature and, while they approach full understanding in time, are incomplete in every step taken. Nevertheless, they are important in order to continuously improve the "sketch" and hopefully soon converge to a colorful big picture. Therefore, in this section, the work by Chehroudi [56] published in such a context is summarized to present a sample of an important attempt to tackle the problem through placement of pieces of jigsaw puzzle. But before doing so, a concise description of the combustion instability in liquid rocket engine is offered to provide a context and sufficient background for Chehroudi's work.

Acoustic combustion instability has been one of the most complex phenomena in liquid rocket engines and therefore difficult to fully understand, control, and predict particularly in the design of high-power rockets. The difficulty arises from the emergence of oscillatory combustion with rapidly increasing and large pressure amplitudes. This leads to local burnout of the combustion chamber walls and injector plates which is caused through extreme heat-transfer rates by high-frequency pressure and gas velocity fluctuations; see Harrje and Reardon [58] and Yang and Anderson [59]. It is thought that resonance acoustic modes of the thrust chamber, amongst them the transverse modes being the most troublesome, are excited through the energy provided by the combustion. The amplification process is thought to include a feedback of information from the acoustic field to the injector or near-injector phenomena which in turn tends to reinforce the combustion-to-acoustic-field energy transfer processes.

This energy transfer reasoning alone is the widely cited general principle by Lord Rayleigh [60]. In essence, he made a phasing argument and stated that the interaction between the combustion heat release and the acoustic field is the strongest if heat is added in a region of space and at the time when the acoustic amplitude is the highest. Although this view has been useful to understand a part of the big picture, evidences gathered by past investigations attributed combustion instability to a complex interaction of the external acoustic field with the fuel injection or near-injector processes as a feedback mechanism, thereby leading to incidences of instability in rocket engines. See, for example, Heidemann and Groeneweg [61], Anderson et al. [57], and Hulka and Hutt [40]. For this and other reasons, controlled studies have been conducted probing into the effects of acoustic waves on gaseous and liquid jets from a variety of injector hole designs. A series of investigations concentrated on disturbances induced from within the injection system. They considered the effects of acoustic fields on many phenomena such as flow structure, vortex pairing, and shear layer growth rate in the initial region of the jet (e.g., see a short review article by Kiwata et al. [62]). More relevant to the work reported here are a few reports and articles on gaseous and (in particular) liquid jets under the influence of external (transverse and longitudinal) acoustic fields. The experimental work under high-pressure supercritical condition has already been reviewed here. More information on this and others can be found in Chehroudi and Tally [39] and Davis and Chehroudi [42, 43].

In Davis and Chehroudi's [63, 64] experimental work on a cryogenic nonfired coaxial injector at sub- and supercritical pressures, they have offered a plausible explanation of why in temperature ramping stability rating exercises an engine becomes unstable. In such tests, which are usually conducted in LOX/H_2 cryogenic liquid rocket engines (LRE), they proposed that a progressive reduction of the propellant (H_2) temperature decreases the outer-to-inner jet velocity ratio for shear coaxial injectors and therefore pushes the engine into an unstable operating zone. As indicated earlier, this velocity ratio was found to be a key parameter defining the stability of the engine; see Hulka and Hutt [40]. But what this velocity ratio does to the jet itself and how the connection to combustion instability comes about are explained in the next paragraph. Moreover, in Chehroudi's work, in which an externally imposed acoustic field was used to simulate certain key aspects of their interaction in real engines, it was shown that at subcritical conditions the root mean square (RMS) fluctuation values of the dark-core length were much higher than those at near-critical and supercritical conditions by a factor of 4 to 6 at all velocity ratios; see Figures 23(b) and 24(b). Also, as the outer-to-inner jet velocity ratio declined, the RMS value increased from 1-2 to values of about 7-8 inner-jet hole diameters at subcritical pressures.

Chehroudi [56] interpreted the RMS of the dense dark core as a reflection of mass fluctuations to a first-order approximation; combining it with measurements of a core dominant oscillation frequency consistent with the imposed acoustic field's resonant mode frequency, he then suggested that a connection to rocket combustion instability may be obtained from these data through examination of the RMS of the dark-core length fluctuations. Decreases in the dark-core length fluctuation levels (quantified through the RMS) were then interpreted as the *reduced intrinsic sensitivity* of the jet. Chehroudi [56] then stated the possibility that decreases in the dark-core length fluctuation levels could weaken a key feedback mechanism for the self-excitation process that is believed to drive the combustion instability in cryogenic LRE. This was offered as a possible explanation for the combustion stability improvements experienced in production engines under higher outer-to-inner jet velocity ratios (see also Figures 23 and 24). The effect of temperature ramping was linked to its impact on the outer-to-inner velocity ratio and hence was also explained. More details can be found in Davis and Chehroudi [63, 64], Davis [44], and Leyva et al. [65]. In other words, *the dynamic behavior of the dark-core, specifically its axial length, was considered by* Chehroudi [56] *to be the culprit for coaxial jet injectors.*

Chehroudi [56] has then compared some of his results (from nonreacting test rigs) with those taken in single-element fired rocket engines and demonstrated consistency of results pertaining to the dark core dynamics. For example, he noted that measured mean intact or dark-core length for SSME-like momentum flux ratios by Woodward et al. [66] in a LOX/GH_2-*fired* single-element rocket engine agreed with those of Davis and Chehroudi's [63, 64] nonreacting case. Additionally, Chehroudi [56] referred to the existence of the dark-core length fluctuations reported by Woodward et al. [66]. He also pointed at a recent work by Yang et al. [67] in which they performed tests in a *fired* single-element rocket equipped with a coaxial LOX/CH_4 injector. Their measurements of the dark core length indicated an increasing trend in the level of fluctuations when the outer-to-inner velocity ratio was decreased and the core oscillation spectra showed more high-frequency contents in jet oscillation at lower velocity ratios. Finally, Chehroudi [56], by referring to the work of Smith et al. [68] where subcritical-to-supercritical sweeps were carefully conducted in a *fired* rocket engine, indicated that very low RMS values of the dark-core length at near- and super-critical conditions and high RMS values at subcritical pressures, both measured by Davis and Chehroudi [63, 64] in their *nonreacting* experimental setup, were consistent with the *fired*-engine experimental observations by Smith et al. [68]. Hence, Smith et al. [68] 's reported unstable combustion behavior at subcritical pressures with high core unsteadiness correlates with Davis and Chehroudi's high RMS values at subcritical conditions, which was then interpreted as conditions leading to highly "sensitive" dark-core dynamic response to its surrounding. In short, Chehroudi indicated that these results in such fired engines were consistent with the Davis and Chehroudi [63, 64] conclusions cited above.

Leveraging the observation that some of the key findings in cold nonreacting jets were consistent with subscale fired rocket engine, all using coaxial injectors, Chehroudi [56] then attempted to employ these, particularly the single-jet's, information to offer a compelling argument in addressing the combustion instability in engines using impinging-jet injector designs. Whereas RMS of the core fluctuations was a key factor for coaxial injectors, Chehroudi [56] proposed that both the mean and RMS of the dark core fluctuations in each jet of an impinging-jet injector were key parameters.

On the engine side, Chehroudi [56] used the information that when the so-called stability parameter (dn/V) was reduced in production engines possessing impinging-jet injectors, combustion instability emerged at some point within this change. This stable-to-unstable transition boundary has been discussed in details by Anderson et al. [57] and is referred to as the Hewitt correlation. To offer a possible explanation, Chehroudi [56] had to propose a hypothesis (explained in the next paragraph) which although intuitively acceptable, requires further verification. However, he, by using data from impinging-jet injector designs in production engines and the dark-core lengths measured from single jets reported here (Figure 9), was able to provide some initial support for the hypothesis. Next, the hypothesis itself and some of his supporting information are explained.

For example, Chehroudi [56] showed that a reduction in the dn/V (stability parameter) through changes in either dn or V leads to shortening of the mean dark-core (or break up) length for each jet in an impinging jet injector. He then stated that it was quite possible that as dn/V was reduced in an engine, the mean dark core length reached a critical value ($L_{C,Pth}$) where one intuitively expects *inherently high sensitivity (high RMS)* for an impinging-jet injector "system" to its environmental acoustic field. *Here,* Chehroudi [56] *hypothesized that the Hewitt stable-to-unstable transition point (or line), which is usually observed as dn/V is reduced, was at or near a condition where the distance from the holes exit plane of the impinging injector to the impinging point (i.e., pre-impingement length) reached a critical value* ($L_{C,Pth}$), *creating a situation somewhat similar to what is shown in Figure 31.* The arrangement shown in this figure was considered to be most sensitive to the ambient acoustic field. Chehroudi [56] also demonstrated that a situation shown in Figure 31 could easily take place in the thrust chamber of production engines during operation considering typical geometrical dimensions of impinging jet injectors for these engines. He also provided experimental evidence by Anderson et al. [57] (Figures 28 and 29) from tests of nonreacting impinging jets conducted at elevated chamber pressures suggesting high level of sensitivity for the arrangement shown in Figure 31.

Considering what was discussed for the coaxial jet injector, one implication of the hypothesis Chehroudi proposed is that an impinging jet injector engine should be more

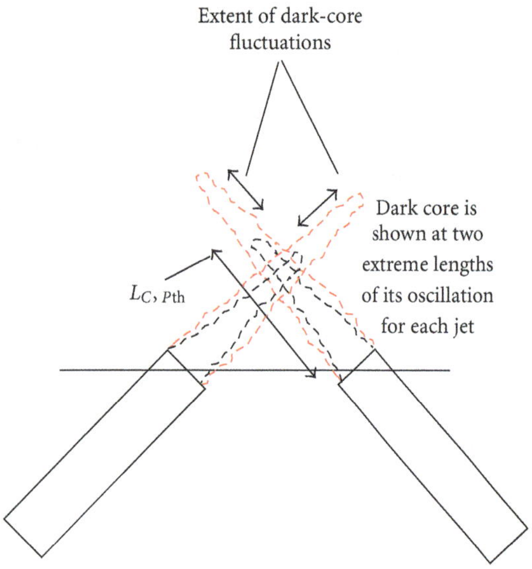

Figure 31: This figure shows the extent of the dark-core (or break-up) length fluctuations for individual jets of an impinging injector design for a situation when the average length ($L_{C,Pth}$) is of the same order as (or less than) the distance from each hole to the impinging point. Chehroudi [56].

stable at sufficiently high pressures, such as supercritical conditions. This is because not only the RMS of the core length fluctuations declines substantially (see Figures 23(b) and 24(b)), but also the length of the core may become adequately shorter than the pre-impingement length depending on the geometrical dimensions of the impinger. The changes in the dark-core (break-up) length can also be inferred by examination of Figure 30, showing a progressive increase in chamber pressure up to a supercritical condition for liquid nitrogen injection into gaseous nitrogen environment with no externally imposed acoustic field. The long pre-impingement length seen along the jet is expected due to L/dn of about 100 which was intentionally designed to obtain a fully developed condition at the hole exit plane and also to accentuate the effects of chamber pressure on the nature of the impingement. Obviously, shorter dark core is achieved for lower (injector hole) L/dn values used in LRE. Not only the dark core length of each individual jet is reduced as supercritical pressures are approached (as before and expected), but the jet also becomes thicker. The impinger is expected to pass through a situation described in Figure 31 as chamber pressure is increased. Hypersensitivity is anticipated at that condition according to Chehroudi's hypothesis. Progressive increase of chamber pressure beyond this point sufficiently thickens each jet and shortens the dark-core length to a situation where the two dark-core lengths are shorter than the preimpingement distance and a gas-like jet is impinging another gas-like jet with enlarged cross-section areas. Based on Chehroudi's hypothesis and given that RMS of the dark core is much lower at supercritical than subcritical conditions, a more robust (targeting and mixing) and less sensitive impinging jet system would be expected at supercritical chamber pressures. However, it is likely that the dynamic behavior of the *potential core* plays a somewhat similar but weaker role under this latter gas-like condition.

Chehroudi's [56] attempt should be considered as the first step, but an important one, towards construction of a "*big picture*" through linkage of the increasing amount of experimental information made available. Essentially he was able to connect pieces of information from nonreacting test rigs, fired subscale engines, and production engines to provide an initial sketch. The sketch proposed by Chehroudi [56] has the advantage, simplicity, and the beauty as well, of identifying the possible weak link in the combustion stability assessment regardless of the design of the injector at least for the two popular cryogenic impinging and coaxial cases described. Much, of course, remains to be investigated, amongst them is the close examination of the historical data on the *dynamic characteristics* of the dark-core (or break-up) length and width for the circular jets forming the impinging injector for the propellant of interest and under the realistic thrust chamber conditions (which are quite rare or nonexistent) to further substantiate that a critical value, $L_{C,Pth}$, is reached when the onset of instability is detected in an engine. Also, the *dynamic* characterization of each jet forming the impinging injector and when the two jets meet, in the presence of an externally imposed acoustic field, is highly desirable to assess sensitivity of the dark-core or break-up length of the jet to relevant design and operating variables.

8. Conclusions

It is evident from this relatively short and targeted review that much has been learned on jet breakup and relevant processes involved in cryogenic liquid propellant injection into high-pressure supercritical condition over the past twenty years. In a quest to understand coaxial jet injectors used in H_2/O_2 cryogenic liquid rocket engines, systematic research efforts have been initiated which consider a logical progression from single jet injected into a quiescent environment to such jets under externally imposed acoustic field and to coaxial injectors with and without acoustic forcing, all under nonreacting conditions.

It is now clear that the impact of progressively lowered surface tension, as environmental pressure approaches the injectant's critical pressure from subcritical and exceeds to supercritical condition, suppresses classical liquid breakup and drop/ligament formation processes and a gas-like jet behavior emerges at or above the critical pressure of the injectant. It is also shown, quantitatively, and for the first time by Chehroudi's group that single liquid jets injected into a supercritical pressure and temperature behave similar to variable-density incompressible jets in visual growth rate. The gas-like nature of these single jets under such conditions has also been demonstrated through fractal analysis of the injected jet boundary. Based on physical arguments and visualizations, an effective model, and the only one known to this author, for the growth rate of such jets was proposed by Chehroudi, a model which mimics the experimental data quite well from sub- to supercritical chamber pressures and for different fluids.

Visual length scales were measured which suggest conformance with the Taylor microscale. The Raman scattering measurements were conducted both for density profiles and growth rates. Based on this information, relationships between the same jet parameters when measured by shadowgraphy and by the Raman scattering were established. Similarity behavior in the radial density profile was demonstrated for the near- and super-critical single jets when the gas-like appearance emerged. Jet growth rates under supercritical pressures, as identified by density profiles, showed tendency to converge with those extrapolated for gaseous jets near the injector exit plane. Temperature estimates from the Raman data suggested existence of a pseudo-boiling behavior when the injection temperature is below the pseudo-boiling line at which the fluid-specific heat becomes very large (mathematical infinity). When the injected liquid possessing temperatures below the pseudo-boiling line warms up under supercritical chamber pressures, it eventually reaches this line, at which time most of the absorbed heat is taken up by the large specific heat value of the fluid, thus exhibiting a pseudo-boiling behavior.

Dark-core length dynamics of cryogenic single jets injected into both sub- and super-critical conditions were measured and compared with those of liquid fuel sprays and single-phase jets. At subcritical conditions, larger fluctuations of this length were reported than those under supercritical chamber pressures. Investigations on the response of the single jets to externally imposed acoustic waves indicated high sensitivity of the jet at subcritical pressures which then substantially diminished at supercritical pressures. Such a lack of response or jet's inherent insensitivity at supercritical pressures was explained through comparison with results extracted from similar studies conducted on single-phase (liquid and gaseous) jets.

Coaxial-jet studies revealed the existence of a dark core originating from the central hole exit area, similar to the single-jet case, but for the coaxial case it was surrounded by a cushion of the outer jet promoting liquid atomization at subcritical and mixing at supercritical pressures. The dark core interface assumed the appearance of a gaseous jet again under supercritical conditions similar to what was observed in a single-jet case. However, the possibility of delay in such a transition exists depending on the solubility of the inner jet and the critical mixing condition.

The mean dark-core length measurements in coaxial injectors with no externally imposed acoustic field exhibited a dual-character behavior depending on whether it experienced subcritical or supercritical chamber pressures. Results for the LN_2 injection to GN_2 were compared with a vast number of pertinent cases from single-phase gaseous and liquid flows to two-phase flows and found that the dark-core length scales with the outer-to-inner jet momentum flux ratio to the power of -0.2, for the subcritical, and to -0.5 for supercritical pressures. Interestingly, as outer-to-inner jet momentum flux ratio approaches zero (i.e., turning into a single jet), the dark-core length approaches values corresponding to the single-jet measurements. The growth rate for the coaxial jets, as measured by the visual spreading angle of the outer (GN_2) jet, agreed with that of supercritical pressures measured for the single jet but then declined below this value as one approached the supercritical pressures.

The impact of the external acoustic forcing on the coaxial jet was seen to be the highest at the subcritical pressures, similar to the single jet and substantially declined when reaching critical and supercritical chamber pressures. Effects of increases in outer-to-inner jet velocity ratio, an important design parameter for stable engine operation and recommended to be greater than 10, were seen to dramatically reduce the mean dark-core length at subcritical pressures while having a minimal impact on this length under supercritical pressures. The dynamic behavior of the core was examined by measurements of the RMS of the core length fluctuations and interpreted as a factor influencing or contributing towards the propellant mass flow oscillations in a rocket thrust chamber. This too behaved very similar to the dark-core length mean value and was distinctively much higher at subcritical pressures than any others. This RMS behavior was then used to offer possible explanations for the empirical observation that stable engine operation required higher outer-to-inner jet velocity ratios (greater than about 10) and also for temperature ramping stability rating tests. Finally, for the injector used, the overall impact of the acoustic field was seen to be important only within a certain region of momentum flux ratios from about 1 to 5.

Implications of a subset of these results were discussed in the context of the rocket combustion instability. Essentially, evidences were shown which suggest that the RMS of the dark-core length and its mean value are important for dynamic behavior of an impinging jet injector. It appears that for the coaxial jets, only RMS values are of relevance in this context.

A hypothesis was proposed that associates the stable-to-unstable transition border in the Hewitt stability plot (for liquid rockets using impinging jets designs) to an event which is defined as an arrangement seen at a moment when the mean core length of one or more of the single jets in an impinging-jet injector design reaches the pre-impingement distance. Some evidences were presented in support of this hypothesis. Essentially, a new alternate and compelling explanation for the Hewitt transition border was offered.

References

[1] T. J. Bruno and J. F. Ely, *Supercritical Fluid Technology: Review in Modern Theory and Applications*, CRC Press, 1991.

[2] R. S. Lazar and G. M. Faeth, "Bipropellant droplet combustion in the vicinity of the critical point," in *Proceedings of the 30th Symposium (International) on Combustion*, p. 801, The Combustion Institute, 1971.

[3] A. Umemura, "Supercritical droplet gasification combustion," in *Proceedings of the IUTAM Symposium on Theories Combustion on Droplets and Sprays*, Taiwan, December 1994.

[4] J. A. Newman and Brzustowski, "Behavior of a liquid jet near the thermodynamic critical region," *AIAA Journal*, vol. 9, no. 8, pp. 1595–1602, 1971.

[5] B. Chehroudi, D. Talley, and E. Coy, "Initial growth rate and visual characteristics of a round Jet into a sub- to supercritical

environment of relevance to rocket, gas turbine, and diesel engines," in *Proceedings of the 37th AIAA Aerospace Science Meeting and Exhibit*, Reno, NV, USA, January 1999, AIAA 99-0206.

[6] W. Mayer, A. Schik, C. Schweitzer, and M. Schaffler, "Injection and mixing processes in high pressure LOX/GH2 rocket combustors," in *Proceedings of the 32nd AIAA/ASME/SAE/ASEE Joint Propulsion Conference & Exhibit*, Lake Buena Vista, Fla, USA, 1996, AIAA Paper no. 96-2620.

[7] R. D. Reitz and F. B. Bracco, "On the dependence of spray angle and other spray parameters on nozzle design and operating condition," in *Proceedings of the SAE International Congress and Exposition*, Detroit, MI, USA, February-March 1979, SAE Paper no. 790494.

[8] W. Mayer, A. Ivancic, A. Schik, and U. Homung, "Propellant atomization in LOX/GH2 rocket combustors," in *Proceedings of the 34th AIAA/ASME/SAE/ASEE Joint Propulsion Conference & Exhibit*, Cleveland, Ohio, USA, July 1998, Paper no. 98-3685.

[9] A. Roy and C. Segal, "Experimental study of subcritical to supercritical jet mixing," in *Proceedings of the 47th AIAA ASM Meeting, Paper AIAA-2009-809*, Orlando, Fla, USA, January 2008.

[10] B. Chehroudi, D. Talley, and E. Coy, "Visual characteristics and initial growth rates of round cryogenic jets at subcritical and supercritical pressures," *Physics of Fluids*, vol. 14, no. 2, pp. 850–861, 2002.

[11] C. J. Chen and W. Rodi, *Vertical Turbulent Buoyant Jets: A Review of Experimental Data*, Pergamon Press, 1980.

[12] P. N. Papanicolaou and E. J. List, "Investigations of round vertical turbulent buoyant jets," *Journal of Fluid Mechanics*, vol. 195, pp. 341–391, 1988.

[13] R. Branam and W. Mayer, "Characterization of cryogenic injection at supercritical pressure," *Journal of Propulsion and Power*, vol. 19, no. 3, pp. 342–355, 2003.

[14] G. N. Abramovich, *The Theory of Turbulent Jets*, MIT Press, Cambridge, UK, 1963.

[15] G. L. Brown and A. Roshko, "On density effects and large structure in turbulent mixing layers," *Journal of Fluid Mechanics*, vol. 64, no. 4, pp. 775–816, 1974.

[16] G. Brown, "The entrainment and large structure in turbulent mixing layers," in *Proceedings of the 5th Australian Conference on Hydraulics and Fluid Mechanics*, pp. 352–359, 1974.

[17] D. Papamoschou and A. Roshko, "The compressible turbulent shear layer: an experimental study," *Journal of Fluid Mechanics*, vol. 197, pp. 453–477, 1988.

[18] P. E. Dimotakis, "Two-dimensional shear-layer entrainment," *AIAA Journal*, vol. 24, no. 11, pp. 1791–1796, 1986.

[19] C. D. Richards and W. M. Pitts, "Global density effects on the self-preservation behaviour of turbulent free jets," *Journal of Fluid Mechanics*, vol. 254, pp. 417–435, 1993.

[20] B. Chehroudi, R. Cohn, D. Talley, and A. Badakhshan, "Raman scattering measurements in the initial region of sub- and supercritical jets," in *Proceedings of the 36th Joint Propulsion Conference*, Huntsville, AL, USA, 2000, AIAA 2000-3392.

[21] M. Oschwald and M. M. Micci, "Spreading angle and centerline variation of density of supercritical nitrogen jets," *Atomization and Sprays*, vol. 12, no. 1–3, pp. 91–106, 2002.

[22] B. B. Mandelbrot, *The Fractal Geometry of Nature*, W. H. Freeman and Company, San Francisco, Calif, USA, 1983.

[23] K. R. Sreenivasan and C. Meneveau, "The fractal facets of turbulence," *Journal of Fluid Mechanics*, vol. 173, pp. 357–386, 1986.

[24] J. J. Taylor and J. W. Hoyt, "Water jet photography—techniques and methods," *Experiments in Fluids*, vol. 1, no. 3, pp. 113–120, 1983.

[25] B. Chehroudi, D. Talley, and E. Coy, "Fractal geometry and growth rate of cryogenic jets near critical point," in *Proceedings of the 35th AIAA/ASME/SAE/ASEE Joint Propulsion Conference*, Los Angeles, Calif, USA, June 1999, AIAA Paper 99-2489.

[26] B. Chehroudi and D. Talley, "The fractal geometry of round turbulent cryogenic nitrogen jets at subcritical and supercritical pressures," *Atomization and Sprays*, vol. 14, no. 1, pp. 81–91, 2004.

[27] B. Chehroudi, S.-H. Chen, F. V. Bracco, and Y. Onuma, "On the intact core of full-cone sprays," SAE Transaction 850126, 1985.

[28] M. Oschwald and A. Schik, "Supercritical nitrogen free jet investigated by spontaneous Raman scattering," *Experiments in Fluids*, vol. 27, no. 6, pp. 497–506, 1999.

[29] I. Wygnanski and H. E. Fiedler, "The two-dimensional Mixing region," *Journal of Fluid Mechanics*, vol. 41, no. 2, pp. 327–361, 1970.

[30] R. M. C. So, J. Y. Zhu, M. V. Ötügen, and B. C. Hwang, "Some measurements in a binary gas jet," *Experiments in Fluids*, vol. 9, no. 5, pp. 273–284, 1990.

[31] N. Zong, H. Meng, S.-Y. Hsieh, and V. Yang, "A numerical study of cryogenic fluid injection and mixing under supercritical conditions," *Physics of Fluids*, vol. 16, no. 12, pp. 4248–4261, 2004.

[32] M. Oschwald, A. Schik, M. Klar, and W. Mayer, "Investigation of coaxial LN_2/GH_2-injection at supercritical pressure by spontaneous raman scattering," in *Proceedings of the 35th AIAA/ASME/SAE/ASEE Joint Propulsion Conference and Exhibit*, Los Angeles, Calif, USA, June 1999.

[33] L.-K. Tseng, G. A. Ruff, P.-K. Wu, and G. M. Faeth, "Continuous- and dispersed-phase structure of pressure-atomized sprays," in *Proceedings of the Progress in Astronautics and Aeronautics: Recent Advances in Spray Combustion*, 1995.

[34] N. Zong and V. Yang, "Cryogenic fluid jets and mixing layers in transcritical and supercritical environments," *Combustion Science and Technology*, vol. 178, no. 1-3, pp. 193–227, 2006.

[35] J. C. Oefelein and V. Yang, "Comprehensive review of liquid-propellant combustion instabilities in F-1 engines," *Journal of Propulsion and Power*, vol. 9, no. 5, pp. 657–677, 1993.

[36] B. Chehroudi and D. Talley, "Interaction of acoustic waves with a cryogenic nitrogen jet at sub- and supercritical pressures," in *Proceedings of the 40th Aerospace Sciences Meeting and Exhibit*, Reno, NV, USA, 2002, AIAA 2002-0342.

[37] D. O. Rockwell, "External excitation of planar jets," *Journal of Applied Mechanics*, vol. 39, no. 4, pp. 883–891, 1972.

[38] H. A. Becker and T. A. Massaro, "Vortex evolution in a round jet," *Journal of Fluid Mechanics*, vol. 31, pp. 435–448, 1968.

[39] Y. V. Vaslov and A. S. Ginevskiy, "Acoustic effects on aerodynamic characteristics of a turbulent jet," Tech. Rep. FTD-MT-24-232-68, Foreign Technology Division, Air Force Systems Command, 1968.

[40] J. Hulka and J. J. Hutt, "Instability phenomena in liquid oxygen/hydrogen propellant rocket engines," in *Liquid Rocket Engine Combustion Instability*, V. Yang and W. E. Anderson,

Eds., AIAA Progress in Astronautics and Aeronautics, pp. 39–71, 1995.

[41] J. Telaar, G. Schneider, and W. Mayer, *Experimental Investigation of Breakup of Turbulent Liquid Jets*, ILASS-Europe 2000, Darmstadt, Germany, 2000.

[42] D. Davis and B. Chehroudi, "The effects of pressure and acoustic field on a cryogenic coaxial jet," in *Proceedings of the 42nd AIAA Aerospace Sciences Meeting and Exhibit*, pp. 10741–10759, January 2004.

[43] D. Davis and B. Chehroudi, "Behaviour of a rocket-Like coaxial injector in an acoustic field, ILAS america," in *Proceedings of the 19th Annual Conference on Liquid Atomization and Spray Systems*, Toronto, Canada, May 2006.

[44] D. W. Davis, *On the behavior of a shear-coaxial Jet, spanning sub- to super-critical pressures, with and without an externally imposed transverse acoustic field [Ph.D. thesis]*, Department of Mechanical and Nuclear Engineering, The Pennsylvania State University, 2006.

[45] M. Oschwald, J. J. Smith, R. Branam et al., "Injection of fluids into supercritical environments," *Combustion Science and Technology*, vol. 178, no. 1–3, pp. 49–100, 2006.

[46] M. Favre-Marinet and E. B. Camano Schettini, "Density field of coaxial jets with large velocity ratio and large density differences," *International Journal of Heat and Mass Transfer*, vol. 44, no. 10, pp. 1913–1924, 2001.

[47] H. Eroglu, N. Chigier, and Z. Farago, "Coaxial atomizer liquid intact lengths," *Physics of Fluids A*, vol. 3, no. 2, pp. 303–308, 1991.

[48] C. Englebert, Y. Hardalupas, and J. H. Whitlaw, "Article usage statistics center," *Proceedings of the Royal Society A*, vol. 451, pp. 189–229, 1995.

[49] Z. Farágó and N. Chigier, "Morphological classification of disintegration of round liquid jets in a coaxial air stream," *Atomization and Sprays*, vol. 2, pp. 137–153, 1992.

[50] R. D. Woodward, *Primary atomization of liquid jets issuing from rocket engine coaxial Injectors [Ph.D. thesis]*, Pennsylvania State University, Department of Mechanical Engineering, University Park, Pa, USA, 1993.

[51] B. Chehroudi, D. Davis, and D. Talley, "Initial results from a cryogenic coaxial injector in an acoustic field," in *Proceedings of the 41st Aerospace Sciences Meeting and Exhibit*, Reno, NV, USA, 2003, AIAA 2003-1339.

[52] I. A. Leyva, J. I. Rodriguez, B. Chehroudi, and D. Talley, "Preliminary results on coaxial jet spread angles and the effects of variable phase transverse acoustic fields," in *Proceedings of the 46th AIAA Aerospace Sciences Meeting and Exhibit*, Reno, Nv, USA, January 2008.

[53] J. C. Lasheras and E. J. Hopfinger, "Liquid jet instability and atomization in a coaxial gas stream," *Annual Review of Fluid Mechanics*, vol. 32, pp. 275–308, 2000.

[54] J. I. Rodriguez, I. A. Leyva, B. Chehroudi, and D. Talley, in *Proceedings of the 21st Annual Conference on Liquid Atomization and Spray Systems (ILASS '08)*, Orlando, Fla, USA, May 2008.

[55] J. I. Rodriguez, *Acoustic excitation of liquid fuel droplets and coaxial jets [Ph.D. thesis]*, University of California at Los Angeles, 2009.

[56] B. Chehroudi, "Physical hypothesis for the combustion instability in cryogenic liquid rocket engines," *Journal of Propulsion and Power*, vol. 26, no. 6, pp. 1153–1160, 2010.

[57] W. E. Anderson, H. M. Ryan, R. J. Santoro, and R. A. Hewitt, "Combustion instability mechanisms in liquid rocket engines using impinging jet injectors," in *Proceedings of the 31st AIAA/ASME/SAE/ASEE Joint Propulsion Conference and Exhibit*, San Diego, Calif, USA, July 1995, Paper AIAA-95-2357.

[58] T. D. Harrje and H. F. Reardon, "Propellant rocket combustion instability," NASA Report, 1972, NASA SP-194.

[59] "Liquid rocket engine combustion instability," in *Proceedings of the AIAA Progress in Astronautics and Aeronautics*, V. Yang and W. E. Anderson, Eds., vol. 169, p. 577, 1995.

[60] L. Rayleigh, "The explanation of certain acoustical phenomena," in *Proceedings of the Royal Institution*, vol. 8, pp. 536–542, London, UK, 1878.

[61] M. F. Heidemann and J. F. Groeneweg, "Analysis of the dynamic response of liquid Jet atomization to acoustic oscillations," NASA Technical Note, 1969, NASA TN D-5339.

[62] T. Kiwata, A. Okajima, and H. Ueno, "Effects of excitation on plane and coaxial jets," in *Proceedings of the 3rd Joint ASME/JSME Fluid Engineering Conference*, pp. 18–22, San Francisco, Calif, USA, 1999.

[63] D. W. Davis and B. Chehroudi, "Measurements in an acoustically-driven coaxial jet under supercritical conditions," *Journal of Propulsion and Power*, vol. 23, no. 2, pp. 364–374, 2007.

[64] D. W. Davis and B. Chehroudi, "Shear-coaxial jets from a rocket-like injector in a transverse acoustic field at high pressures," in *Proceedings of the 44th AIAA Aerospace Sciences Meeting*, pp. 9173–9190, Reno, Nv, USA, January 2006, Paper No. AIAA-2006-0758.

[65] I. A. Leyva, B. Chehroudi, and D. Talley, "Dark core analysis of coaxial injectors at sub-, near-, and supercritical pressures in a transverse acoustic field," in *Proceedings of the 43rd AIAA/ASME/SAE/ASEE Joint Propulsion Conference and Exhibit*, pp. 4342–4359, Cincinnati, Ohio, USA, July 2007.

[66] R. D. Woodward, S. Pal, S. Farhangi, G. E. Jensen, and R. J. Santoro, "LOX/GH2 shear coaxial injector atomization studies: effect of recess and non-concentricity," in *Proceedings of the 45th AIAA Aerospace Sciences Meeting and Exhibit*, pp. 6925–6946, Reno, Nv, USA, January 2007.

[67] B. Yang, C. Francesco, L. Wang, and M. Oschwald, "Experimental investigation of reactive liquid Oxygen/CH4 coaxial sprays," *Journal of Propulsion and Power*, vol. 23, no. 4, pp. 763–771, 2007.

[68] J. J. Smith, M. Bechle, D. Suslov, M. Oschwald, O. J. Haidn, and G. M. Schneider, "High pressure LOx/H2 combustion and flame dynamics preliminary results," in *Proceedings of the 40th AIAA/ASME/SAE/ASEE Joint Propulsion Conference & Exhibit*, Fort Lauderdale, Fla, USA, July 2004, AIAA-2004-3376.

2

A DES Procedure Applied to a Wall-Mounted Hump

Radoslav Bozinoski[1] and Roger L. Davis[2]

[1] *Thermal/Fluid Science & Engineering Department, Sandia National Laboratories, Livermore, CA 945501, USA*
[2] *Mechanical and Aero Engineering Department, University of California at Davis, Davis, CA 95616, USA*

Correspondence should be addressed to Radoslav Bozinoski, rbozinoski@gmail.com

Academic Editor: Linda L. Vahala

This paper describes a detached-eddy simulation (DES) for the flow over a wall-mounted hump. The Reynolds number based on the hump chord is $\text{Re}_c = 9.36 \times 10^5$ with an in-let Mach number of 0.1. Solutions of the three-dimensional Reynolds-averaged Navier-Stokes (RANS) procedure are obtained using the Wilcox $k - \omega$ equations. The DES results are obtained using the model presented by Bush and Mani and are compared with RANS solutions and experimental data from NASA's 2004 Computational Fluid Dynamics Validation on Synthetic Jets and Turbulent Separation Control Workshop. The DES procedure exhibited a three-dimensional flow structure in the wake, with a 13.65% shorter mean separation region compared to RANS and a mean reattachment length that is in good agreement with experimental measurements. DES predictions of the pressure coefficient in the separation region also exhibit good agreement with experiment and are more accurate than RANS predictions.

1. Introduction

Simulation of "steady" and "unsteady" flow of aerodynamic bodies has matured a great deal over the past decade. Aerodynamic performance and flow structures can be predicted with acceptable accuracy except in the complex flow regions of mixing and at off-design conditions near stall. In these regions, flow structures with multiple eddies that mingle and mix are often not predicted well due to inadequate computational grid density and a breakdown of turbulence models.

The focus of the current effort has been to validate the Navier-Stokes code and to further develop the detached-eddy viscosity model using the wall-mounted hump grid. This case proved difficult for most of the participants of the NASA workshop, especially those using two or even three-dimensional RANS techniques [1–3]. Most, if not all, participants using Reynolds-averaged Navier-Stokes predicted too large a separation region, which clearly showed the limitations presented with simply using a RANS procedure. Efforts by Morgan et al. [1] to surpass the limitations of RANS models with high-order spacial and temporal methods applied to the Navier-Stokes and $k - \epsilon$ equations showed little improvement, especially in the separation region of the flow. Similar results were obtained by Balakumar [3] using a fifth-order accurate weighted essentially nonoscillatory (WENO) scheme for space discretization and a third-order, total variation diminishing (TVD) Runge-Kutta scheme for time integration. Israel et al. [4] evaluated a direct numerical simulation (DNS) as well as a flow simulation methodology (FSM) where the RANS closure equations are combined with a contribution function that is used to determine the magnitude of the turbulent stress term. They observed good agreement between FSM, DNS, and the experimental data and suggest that differences between the numerical predictions and experiment can be attributed to discrepancies in the numerical representation of the experiment rather than deficiencies in the numerical methodology. Miller and Seitz [5] compared a SAS model proposed by Menter, Kuntz, and Bender [6] as well as DES and found that both the methods produced results in good agreement with experiment and prove to be more appropriate for industrial applications since they are less computationally intensive than DNS and LES simulation.

2. Governing Equations

The unsteady, Favre-averaged governing flow-field equations for an ideal, compressible gas in the right-handed, Cartesian coordinate system using primary variables are used. The three-dimensional continuity, momentum, and energy equations can be written in conservative form as follows:

$$\frac{\partial \rho}{\partial t} + \frac{\partial(\rho u_j)}{\partial x_j} = 0,$$

$$\frac{\partial \rho u_i}{\partial t} + \frac{\partial(\rho u_j u_i)}{\partial x_j} = -\frac{\partial P}{\partial x_i} + \frac{\partial \hat{\tau}_{ij}}{\partial_j},$$

$$\frac{\partial E}{\partial t} + \frac{\partial(\rho u_j H)}{\partial x_j} = \frac{\partial}{\partial x_j}\left[u_i \hat{\tau}_{ij} + \left(\frac{\mu}{\mathrm{Pr}} + \frac{\mu_T}{\mathrm{Pr}_t}\right)\frac{\partial h}{\partial x_j}\right], \tag{1}$$

$$\hat{\tau}_{ij} = \tau_{ij} + \tau_{ij}^R = (\mu + \mu_T)S_{ij},$$

$$S_{ij} = \frac{1}{2}\left(\frac{\partial u_i}{\partial x_j} + \frac{\partial u_j}{\partial x_i}\right) - \frac{2}{3}\frac{\partial u_k}{\partial x_k}\delta_{ij}.$$

Since we are dealing with compressible flows, we require an equation of state to relate the energy with pressure and enthalpy

$$E = \rho\left[e + \frac{1}{2}V^2\right],$$

$$H = h + \frac{1}{2}V^2 = \frac{\gamma}{\gamma - 1}\frac{P}{\rho} + \frac{1}{2}V^2 = \frac{E + P}{\rho}. \tag{2}$$

Additional governing equations as developed by Wilcox [7, 8] are used for the transport of turbulent kinetic energy and turbulence dissipation rate in regions of the flow where the computational grid or global time-step size cannot resolve the turbulent eddies. The equations for the turbulent mixing energy and specific dissipation rate can be seen below and are Wilcox 2006 [7] model

$$\mu_T = \frac{\rho k}{\omega}, \tag{3}$$

$$\frac{\partial \rho k}{\partial t} + \frac{\partial \rho u_j k}{\partial x_j} = \tau_{ij}^R \frac{\partial u_i}{\partial x_j} - \beta_k \rho \omega k + \frac{\partial}{\partial x_j}\left[(\mu + \sigma^* \mu_T)\frac{\partial k}{\partial x_j}\right], \tag{4}$$

$$\frac{\partial \rho \omega}{\partial t} + \frac{\partial \rho u_j \omega}{\partial x_j} = \left(\frac{\gamma \omega}{k}\right)\tau_{ij}^R \frac{\partial u_i}{\partial x_j} - \beta_\omega \rho \omega^2 + \frac{\partial}{\partial x_j}\left[(\mu + \sigma \mu_T)\frac{\partial \omega}{\partial x_j}\right], \tag{5}$$

where

$$\beta_k = \frac{9}{100}, \qquad \beta_\omega = \frac{3}{40},$$

$$\sigma^* = \frac{1}{2}, \qquad \sigma = \frac{1}{2}. \tag{6}$$

In regions of the flow where the larger-scale eddies can be resolved with the computational grid, techniques borrowed from large-eddy simulation are used to represent the viscous shear and turbulent viscosity. The large-eddy subgrid model described by Smagorinsky [9] is modified according to the detached-eddy considerations described by Strelets [10] and Bush and Mani [11]

$$\mu_T = \rho l_{le}\sqrt{k}, \tag{7}$$

where l_{le} is an eddy length scale proportional to the grid/time-step filter width, Δ:

$$l_{le} = \min\left(\frac{\sqrt{k}}{\omega}, \beta_k C_{\mathrm{des}}\Delta\right). \tag{8}$$

In addition, the dissipation term, $\beta_k \rho k \omega$, of the turbulent kinetic energy transport equation (4) is limited by the eddy length scale, l_{le}, according to

$$\beta_k \rho k \omega \Longrightarrow \beta_k \rho k * \max\left(\omega, \frac{\sqrt{k}}{\beta_k C_{\mathrm{des}}\Delta}\right), \tag{9}$$

where C_{des} is a proportionality coefficient

$$\Delta = \max\left(dx, dy, dz, u * dt, \sqrt{k} * dt\right). \tag{10}$$

Equation (10) determines the smallest eddies that can be resolved. The first three terms determine what the minimum eddy size the grid can support. The two time terms are the convection velocity and subgrid scale turbulence, respectively. They are included to ensure that the time step is small enough to resolve the unsteady effects. Equation (7) can be made to resemble (3) using the following:

$$\omega_B = \max\left(\omega, \frac{\sqrt{k}}{\beta_k C_{\mathrm{des}}\Delta}\right),$$

$$\mu_T = \frac{\rho k}{\omega_B}. \tag{11}$$

3. Numerical Techniques

The conservation of mass, momentum, and energy equations are solved using a Lax-Wendroff control-volume, time-marching scheme as developed by Ni [12], Dannenhoffer [13], and Davis et al. [14, 15]. Numerical solutions of unsteady flows can be performed with either the explicit [12] or a dual time-step procedure [16]. These techniques are second-order accurate in time and space. A multiple-grid convergence acceleration scheme [12] is used for steady, Reynolds-averaged Navier-Stokes solutions and the inner convergence loop of unsteady simulations using the dual time-step scheme. The approach is called MBFLO and has two-dimensional [17], axi-symmetric [18] (with and without swirl), and three-dimensional [19] versions. The three-dimensional procedure and results for a DES using the Bush and Mani algorithm for flow over a wall-mounted hump are described here.

The combined second- and fourth-difference dissipation model of Jameson [20] is used in the current procedure for both the mean flow and turbulence equations. The

TABLE 1: Typical speed-up and efficiencies.

Processes	Speed Up	Efficiency	Pts/Process
1	1.00	100%	1,780,440
2	2.00	100%	892,440
4	4.00	100%	448,440
8	7.79	97%	226,440
16	14.60	91%	115,440
20	18.06	90%	93,240
40	31.39	78%	48,840

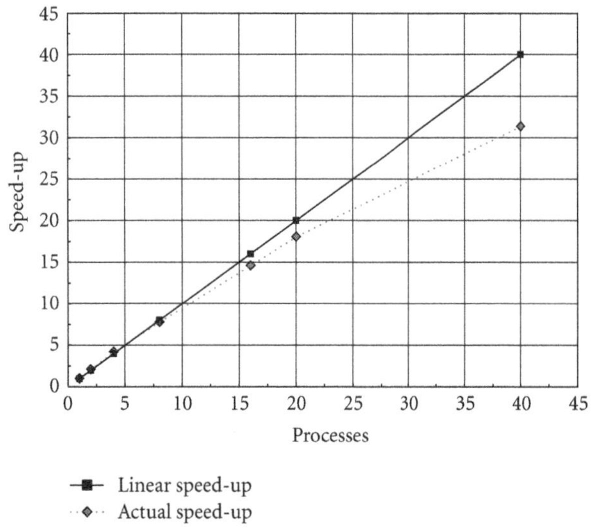

FIGURE 1: Typical Speed-Up factor as a function of number of processes.

fourth-difference dissipation is scaled by the inverse of the absolute value of the mean strain rate squared. This function decays the numerical dissipation in all viscous flow regions, including boundary layers, wakes, large eddies, and secondary flows.

Parallelization is performed using the Message Passing Interface (MPI) library [21]. Figure 1 and Table 1 show the typical speed-up and associated efficiencies as functions of the number of processors. The configuration used to generate this data was similar to that shown below in the results section where the computational grid consisted of 1.80 million grid points. The data was generated on a Linux cluster consisting of 3.6 GHz Intel Xeon processors. Figure 1 shows that a speed-up factor of 18.06 is realized with 20 processes yielding a 90% parallel efficiency. In Table 1, we can see that efficiencies of 90% and higher can be obtained if no less than 100, 000 grid points per process are used.

4. Results

The Navier-Stokes codes has been verified with analytical data for a series of standard test cases such as steady inviscid flow over circular bumps, turbulent flow over airfoils, and laminar and turbulent flow over a flat plate, as well as axi-symmetric flows [17–19]. The focus of the current investigation is to demonstrate and validate the DES model for a separated flow and to determine what advantage may exist, in terms of accuracy, for DES compared to three-dimensional URANS and its applicability in the design process.

The simulation of a turbulent flat plate, constructed in such a way as to develop the inflow boundary conditions for the wall-mounted hump, was initially performed. The length of the flat plate was determined through preliminary CFD tests, so the predicted boundary layer thickness was essentially that of the experiment. The flow conditions were also set to match the wall-mounted hump with Ma $= 0.1$ and $\mathrm{Re}_c = 9.36 \times 10^5$.

For the flat plate mesh, the computational domain extended from $-2.80 \le x/c \le 4.25$ in the streamwise direction and from $0.0 \le y/c \le 0.909$ in the normal direction. There were 121 points used in the streamwise direction with a maximum stretching ratio of $\Delta x = 1.2$. The wall spacing normal to the surface was set to 3.6×10^{-6} and corresponds to $\Delta y^+ = 0.25$. A maximum stretching ratio of 1.2 was achieved with 177 points clustered near the flat plate surface.

The simulation was performed by initializing the inlet with a uniform flow field where the nondimensional velocity components were set to $u = 1, v = 0$, and $w = 0$ and nondimensional density to $\rho = 1$. In the area over the flat plate, the velocity profile was initialized using the log-law profile, and density and energy were recomputed to keep the pressure at the freestream. During the computation, the inflow boundary condition held total pressure and total temperature constant, while at the exit, static pressure was held constant. Along the flat plate surface from the inlet to the leading edge of the flat plate, an inviscid wall boundary condition was imposed with an adiabatic no-slip wall over the entire flat plate surface. The upper wall of the flow domain was also set to an inviscid wall.

To verify the predicted solution, the velocity profile was plotted in terms of the inner layer velocity and length variables u^+ and y^+, respectively, and compared to the empirical viscous sublayer, log-law, and 1/7th power law relations as shown in Figure 2(a). Figure 2(b) shows the predicted velocity and experimentally measured profiles at $x/c = -2.8$, which corresponds to the inlet of the wall-mounted hump.

Once the inlet conditions were generated using the flat plate, the wall-mounted hump test case for turbulent flow

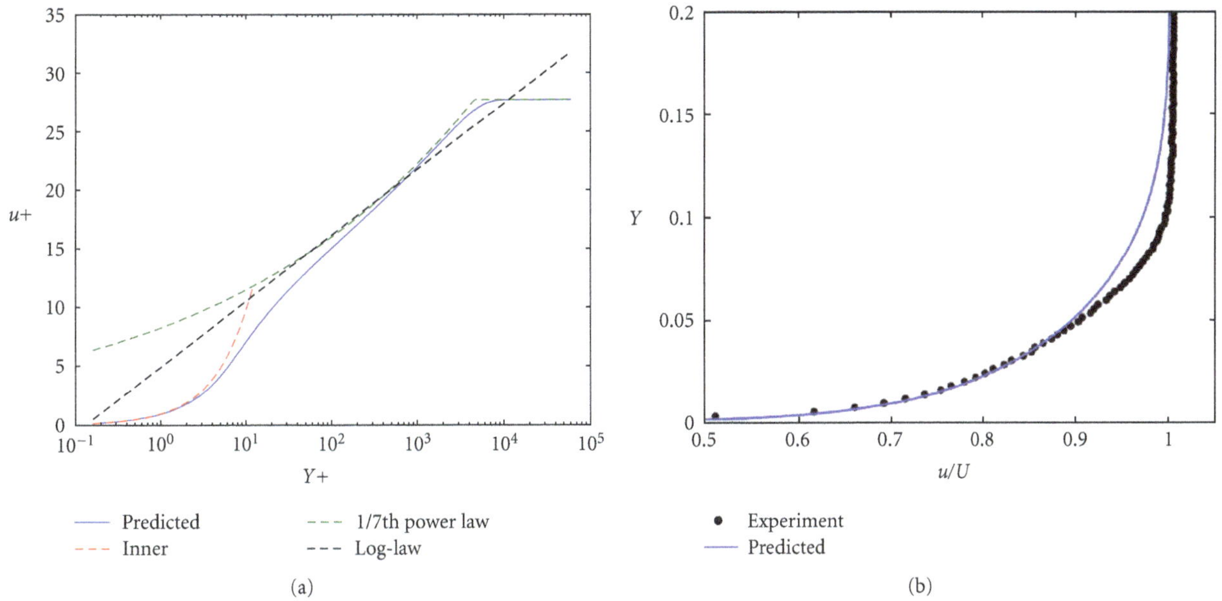

FIGURE 2: (a) Inflow U-velocity profile in inner-wall coordinates. (b) A turbulent RANS U-velocity profile comparison.

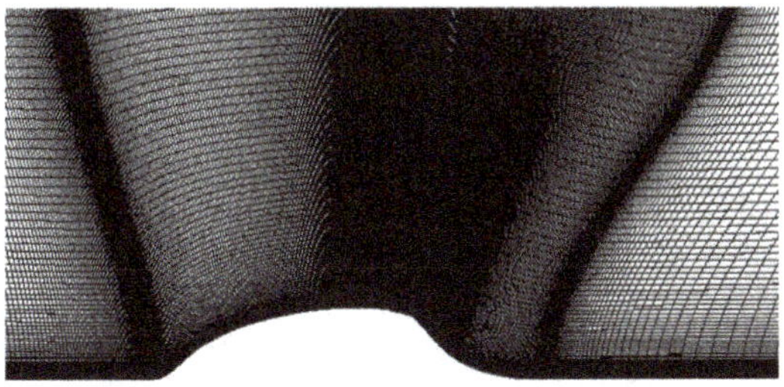

FIGURE 3: Computational grid for wall-mounted Hump (note: every fifth point is shown).

FIGURE 4: Surface grid for 3D unsteady cases (note: 5 spanwise planes are shown).

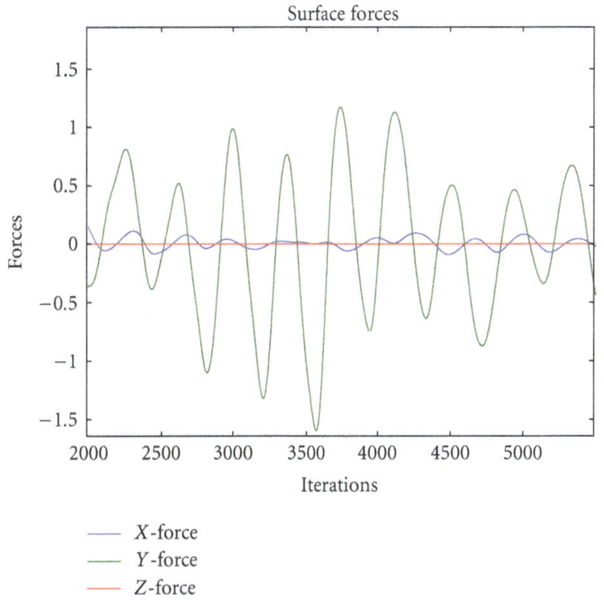

FIGURE 5: Lift and drag signals for three-dimensional simulation.

was simulated and compared to experimental data from the NASA workshop [22] that focused on synthetic jets and turbulent separation control. The hump geometry was constructed to simulate a 20% thick Glauert-Goldschmied airfoil with a chord length of 0.472 m (1.38 ft.), a maximum height of 0.054 m (0.176 ft.), and a span of 0.584 m (1.92 ft.). The experimental data was obtained for $\text{Ma}_\infty = 0.10$ and $\text{Re}_c = 9.36 \times 10^5$ based on the chord. The test case that was investigated was without flow control, where there was no blowing or suction in the slot. The numerical results obtained used a three-dimensional structured grid smoothed over the slot and with the top wall shape adjusted to approximately account for side plate blockage, as recommended by the workshop [22]. The grid shown in Figure 3 extended from $x/c = -9.20$ to 3.5 in the streamwise direction, with the hump located from $x/c = 0.0$ to 1.0 and from $y/c = 0.0$ to 0.9 in the normal direction. The mesh contained 155, 937 grid points with 881 in the streamwise direction and 177 normal to the wall. An initial RANS simulation was performed using three planes in the spanwise direction for a grid that consisted of 467, 811 points. This allowed a nominally two-dimensional flow simulation to be performed and compared to the experiment. For true three-dimensional simulations, the original 881×177 grid described was used and extruded in the spanwise direction 0.15 chord lengths and can be seen

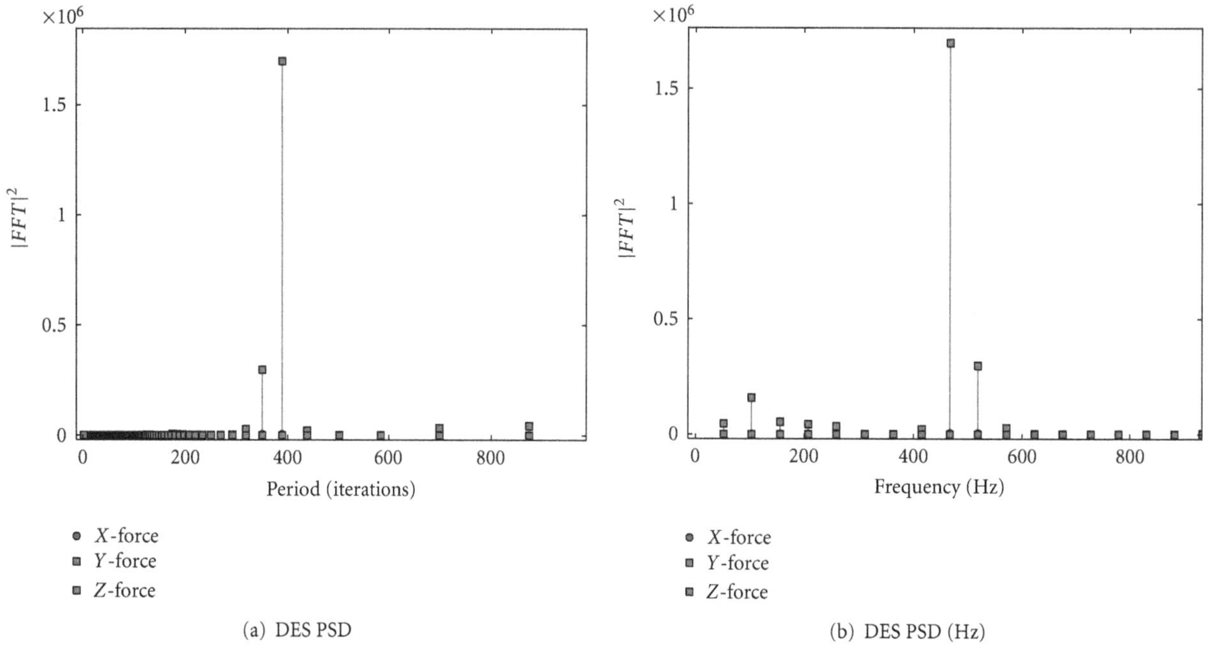

FIGURE 6: Lift and drag signals for three-dimensional simulation.

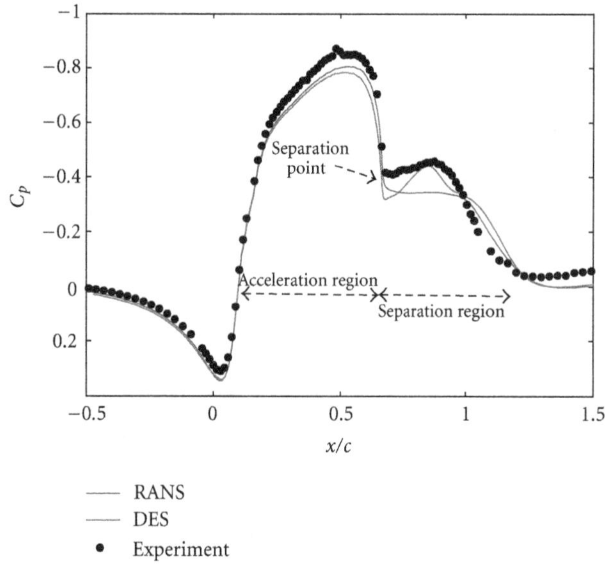

FIGURE 7: Pressure coefficient comparison.

in Figure 4. The spanwise direction was meshed using 49 points with uniform spacing and corresponds to roughly two boundary layer thicknesses based on the inlet.

With the aid of the turbulent flat plate simulation, the inflow velocity and density profiles for the wall-mounted hump case were generated and used to initialize the flow domain over the wall-mounted hump and helped decrease the computational time required to obtain a solution. The top of the flow domain was set with an inviscid wall boundary condition with an adiabatic no-slip wall along the south boundary starting at the inlet and leading over the hump section. Static pressure was once again held constant at the exit of the flow domain. For the nominally two-dimensional case using the steady RANS solver, inviscid walls were imposed for the spanwise boundaries, whereas periodic boundary conditions were used for the three-dimensional unsteady simulation. The turbulent freestream intensity, Tu, and the dissipation length scale, l_{dis}, were set to 1.00% and 0.06, respectively. The DES coefficient used was that suggested by the original authors of a value of 0.6 [11].

The temporal periodicity and unsteady behavior was initially studied for the three-dimensional DES case. The information obtained was then used to determine the number of time-steps necessary to resolve a minimum of 10 periodic cycles for the time-averaged DES at the given global time-step. Figures 5 and 6(a) show the signal history for the instantaneous x-, y-, and, z-surface forces as well as the Power Spectral Density (PSD) as a function of the number of time-steps. The PSD of surface forces was plotted here as a function of the number of time-steps to more readily see periodicity in the flow. In Figure 6(a), we see that the peak spectral signal is repeated approximately every 400 time-steps, which corresponds to a frequency of 420 Hz. The corresponding Strouhal number based on the hump height and freestream speed is approximately 0.65. Once the period was determined, this case was run for approximately 10 periodic cycles and time-averaged.

Figure 7 shows the time-averaged pressure coefficient distribution resulting from the RANS and time-averaged DES simulations. The numerical results predicted using the RANS procedure produced higher pressure levels in the separation region, $0.65 \leq x/c \leq 1.3$, when compared to the experimental data. This is consistent when compared to other RANS simulations [1–3] in the NASA workshop.

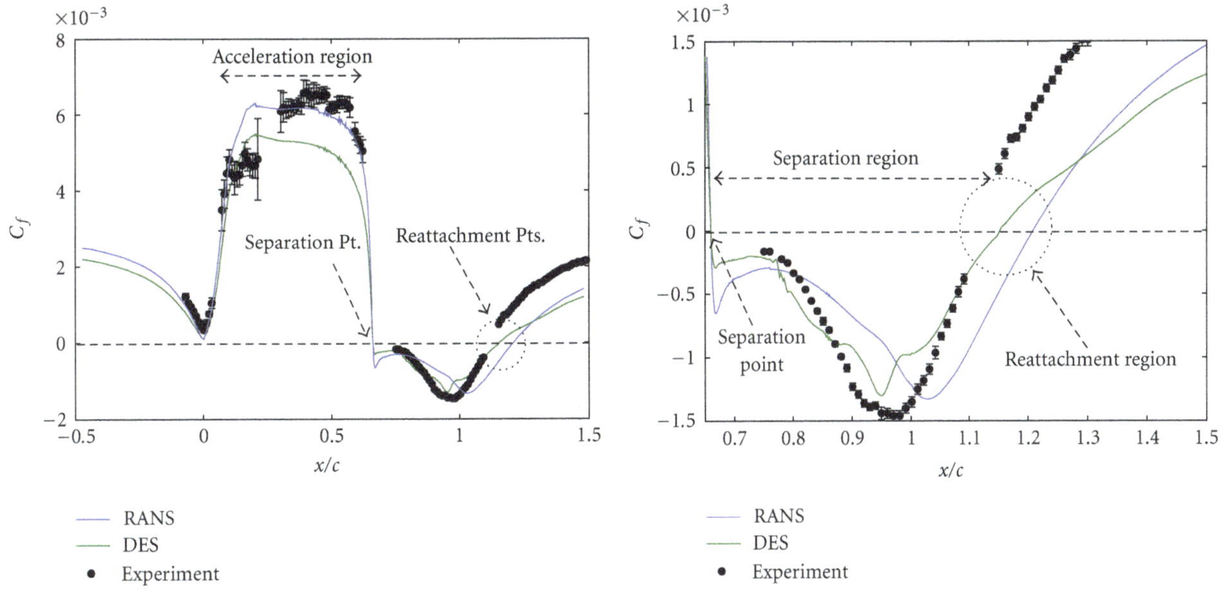

(a) RANS and DES

(b) DES in separation region

Figure 8: Predicted skin friction of time-averaged RANS and DES.

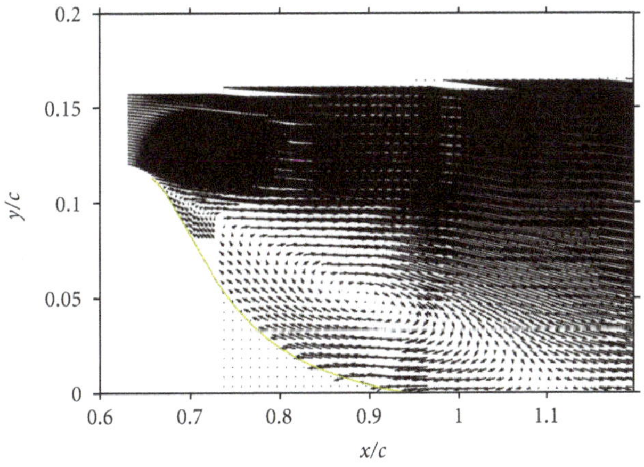

(a) Experimental PIV

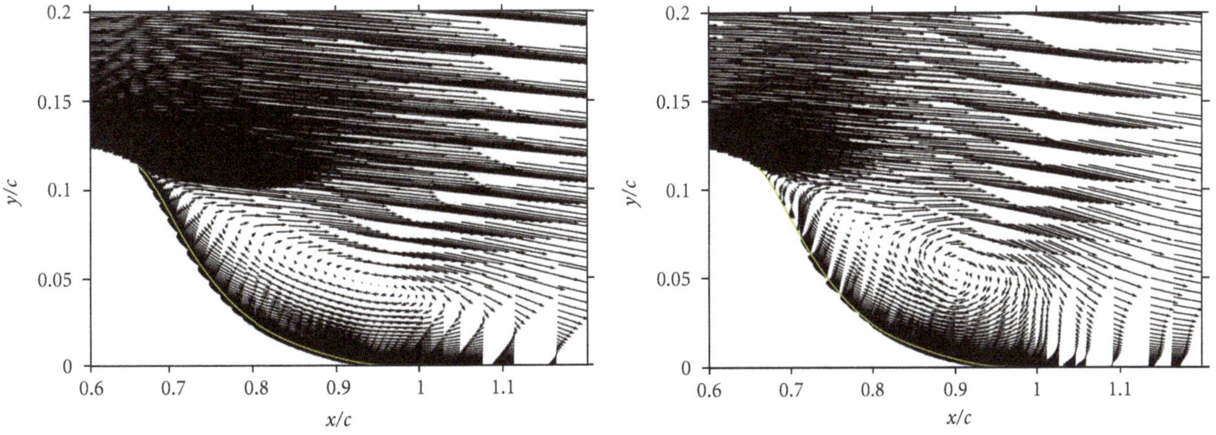

(b) RANS

(c) Time-averaged DES

Figure 9: Comparison of U-velocity contours.

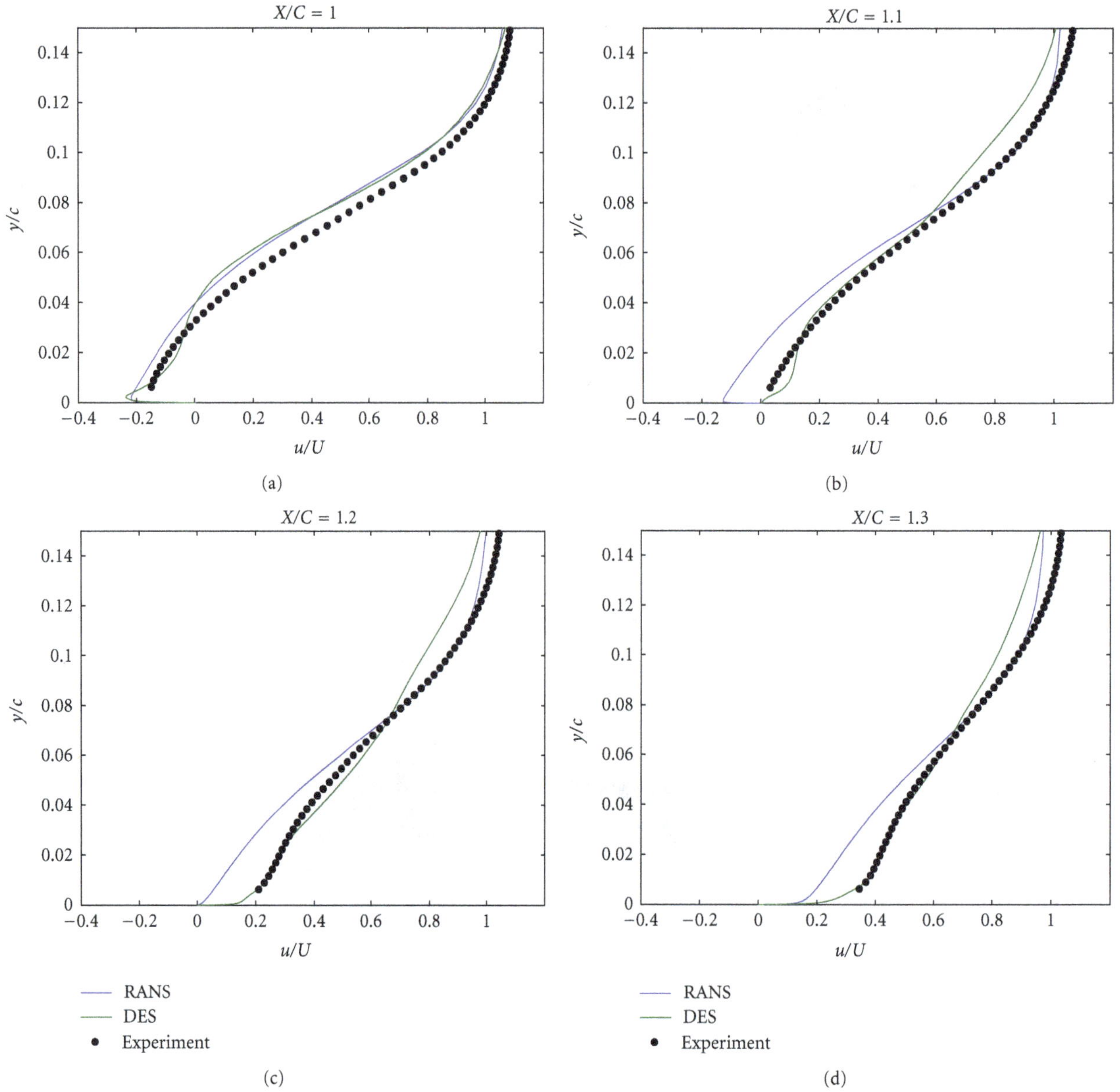

Figure 10: Mean U-velocity profile comparison at x = (a) 1.0, (b) 1.1, (c) 1.2, and (d) 1.3.

Researchers who used shear-stress transport (SST) RANS models showed similar predictions for the pressure leading up to the hump as well as higher pressures in the separation region. The baseline RANS cases run by Morgan, Rizzetta, and Visbal [1] also showed similar pressure predictions for the flow leading up to the hump and in the separation region. RANS simulations run by Krishnan, Squires, and Forsythe [2] also showed similar predictions as well as those made by Balakumar [3] and Šarić et al. [23]. The current DES case showed similar comparison with the experimental data in the acceleration region of the flow, $0.00 \leq x/c \leq 0.65$, with significant improvement in the separation region located between $0.65 \leq x/c \leq 1.3$.

A comparison of the time-averaged skin-friction coefficient, C_f, is shown in Figure 8. Once again, there is good agreement with the steady RANS prediction upstream of the hump to $x/c = 0$ and over the hump to the beginning of the separation region $x/c = 0.65$. The DES case slightly underpredicts the skin friction in the acceleration region of the flow with a max value of approximately 0.005. Similar results were shown by Morgan et al. [24] using an implicit large-eddy simulation. Both simulations accurately predicts the onset of the separation region at approximately $x/c = 0.65$. In the separation region, we see the predicted reattachment point for the RANS case is at $x/c = 1.2$ and the time-averaged DES at $x/c = 1.105$. The DES matches the experimental data in the separation region much better than the RANS procedure. Table 2 shows an overview of the separation and reattachment locations for the RANS and DES procedures compared to the experimental data. Here,

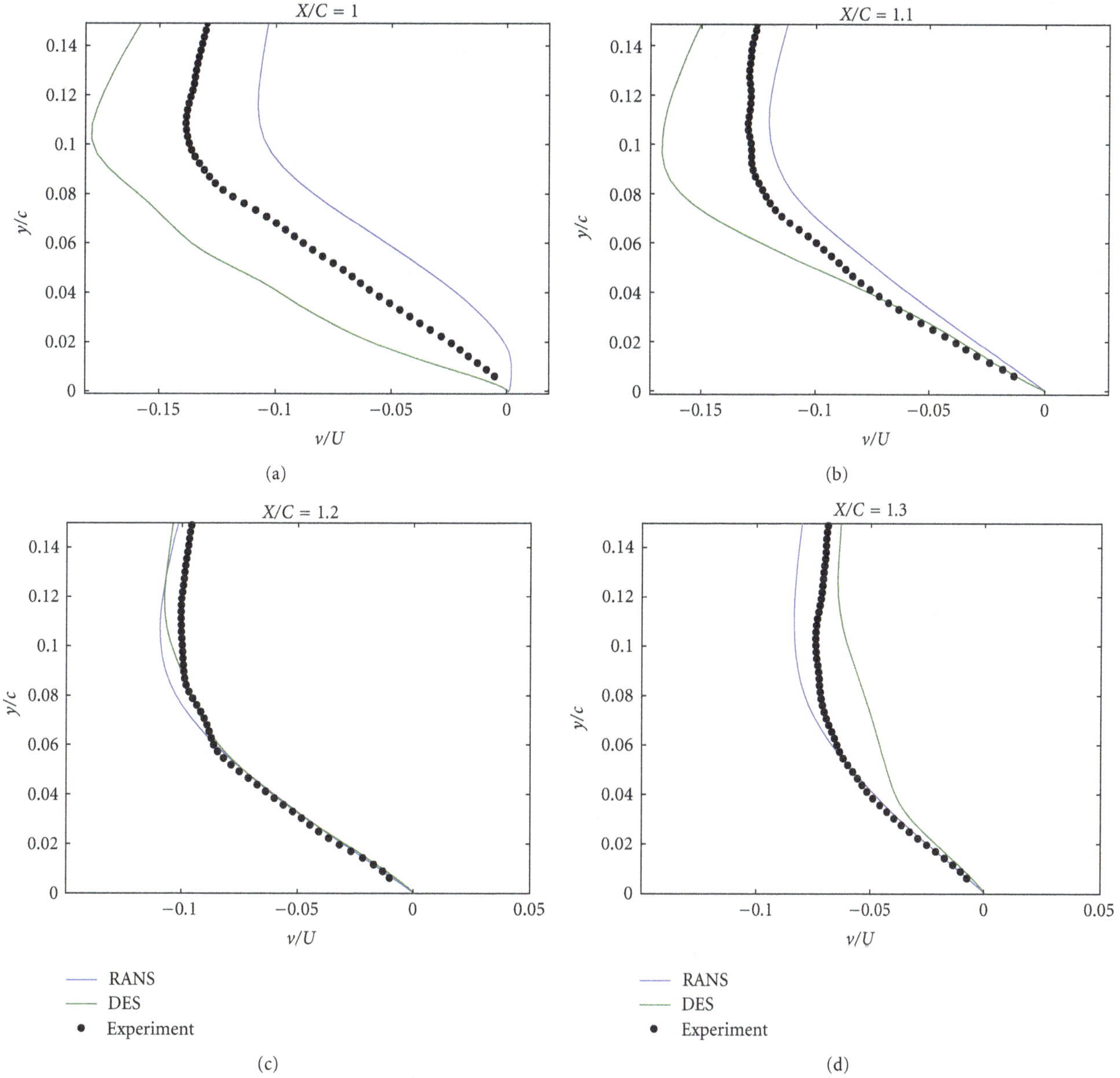

FIGURE 11: Mean V-velocity profile comparison at x = (a) 1.0, (b) 1.1, (c) 1.2, and (d)1.3.

TABLE 2: RANS and DES Comparison.

	Separation (x/c)	Reattachment (x/c)
RANS	0.660	1.200
TA DES	0.656	1.105
Experiment	0.680	1.110 ± 0.003

we can see that both procedures accurately predict the onset of separation with the RANS procedure overpredicting the size of the separation bubble. The DES, however, properly predicts the reattachment point and the separation bubble size.

Figure 9 shows a stream trace velocity comparison. Here we can more easily see the flow reattachment locations and the effect the DES has on the flow. In Figure 9(b), we see that the streamline plot clearly shows that the RANS solution has a longer separation bubble than that observed experimentally in Figure 9(a). The time-averaged DES streamlines, Figure 9(c), show significant improvement for the predicted separation bubble length, with a 13.65% shorter mean separation region compared to RANS and a mean reattachment length that is in good agreement with experimental measurements.

Time-averaged velocity profiles at x/c = 1.0, 1.1, 1.2, and 1.3, corresponding to locations within the separation region and slightly downstream of it, were also compared with experimental data. The peak reverse flow velocity predicted at x/c = 1.0 in Figure 10(a) is close to the experimentally measured velocity and slightly lags in the region away from

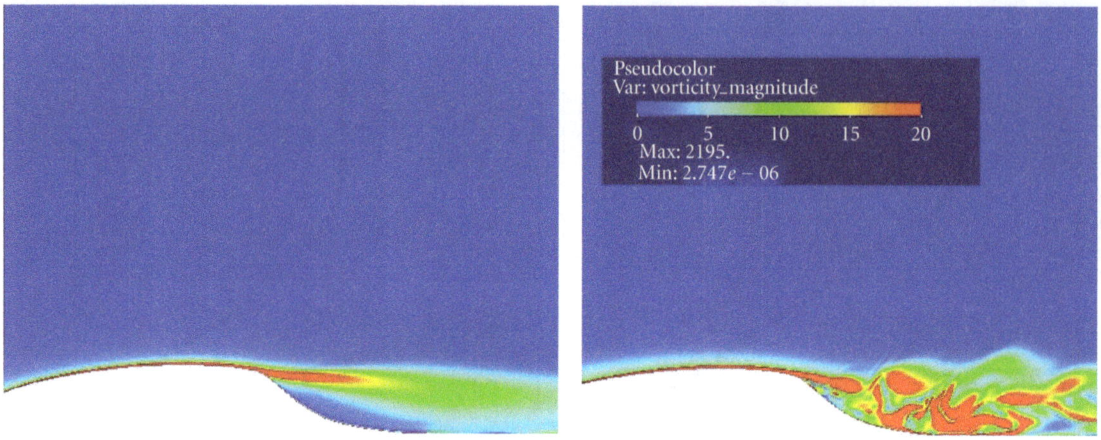

(a) RANS

(b) Instantaneous DES

FIGURE 12: Spanwise averaged vorticity magnitude contours.

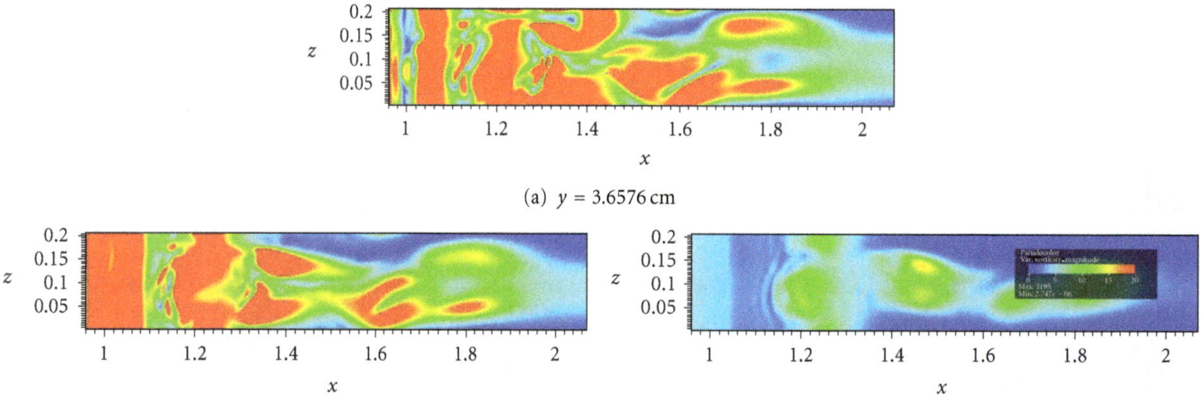

(a) $y = 3.6576$ cm

(b) $y = 4.2672$ cm

(c) $y = 5.4864$ cm

FIGURE 13: Stream-slice vorticity magnitude contours at different distances away from wall.

the wall. Referring to the velocity profiles at $x/c = 1.1$ and 1.2 in Figures 10(b) and 10(c), it can be seen that the experimentally measured flow reattachment point is at $x/c \approx 1.1$, while MBFLO RANS and DES procedures predicted the reattachment point at $x/c \approx 1.2$ and 1.105, respectively. In the near wall region of the flow, the DES procedure more accurately compares with the experimental velocity profiles at all four locations.

The DES and RANS profiles of the v-velocity component at $x/c = 1.0$ and 1.1 in Figure 11 show good qualitative comparison at $x/c = 1.0$ and 1.1 and match the experimentally measured values very well at $x/c = 1.2$ and 1.3. It should be noted that the v-component of the velocity is a magnitude smaller than the u-component and small changes can more readily be seen.

Figure 12 shows contour plots comparing the spanwise spatially averaged vorticity and for the RANS and instantaneous DES. The instantaneous vorticity contours in Figure 12(b) show a large range of resolved eddies consistent with large-eddy simulation treatment.

Instantaneous spanwise vorticity contour slices at $y = 3.6576$, 4.2672, and 5.4864 cm. normal to the wall are shown in Figure 13. Here, we can more easily see the three-dimensionality that has formed in the wake of the wall-mounted hump. Figure 14 presents instantaneous vorticity isosurfaces for the DES prediction. Here we can clearly see the separated shear layer downstream of the hump.

The detached-eddy simulation procedure, described by Bush and Mani [11], has been shown to be consistently more accurate than standard RANS. It predicts well the experimentally measured flow quantities such as pressure coefficient, surface skin friction, reattachment length, and mean velocity profiles. The RANS procedure predicted a delayed reattachment point, which indicates reduced turbulent mixing inside the separation region. Attempts at using higher-order numerical techniques, applied to RANS procedures [1, 3], have shown similar results. It should be noted that the DES procedure used a single constant model coefficient of 0.6; although additional computations using various coefficients would be essential to further examine

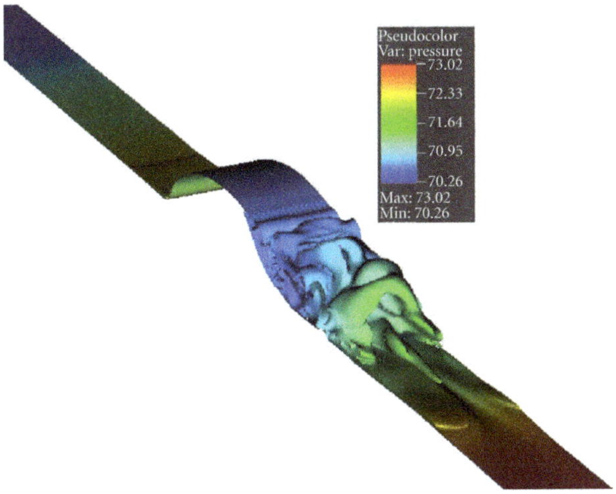

Figure 14: Vorticity isosurfaces (shaded by pressure) for DES prediction.

the DES procedure, it was beyond the scope of the current investigation.

5. Conclusion

A general Reynolds-averaged/detached-eddy simulation procedure was applied to the prediction of the flow over a wall-mounted hump. The initial results using the RANS and DES procedures compared well with experiment as well as other participants of NASA's 2004 Computational Fluid Dynamics Validation Workshop. Like other participants using RANS models, the onset of separation was accurately predicted and the reattachment point was overpredicted. The RANS procedure also overpredicted the mean pressure, skin friction, and velocity profiles in the separation zone. The DES procedure using the Bush and Mani model showed much better results. The three-dimensional structures resolved in the wake of the DES improved the local flow physics in the separation region and the predictions of the mean pressure distribution, skin friction, and streamwise velocity.

Nomenclature

E: Total energy
e : Internal energy
H : Total enthalpy
h : Static enthalpy
k : Turbulent kinetic energy
p : Pressure
Pr : Prandtl number
Pr_t: Turbulent Prandtl number
S_{ij}: Mean strain-rate tensor
u_i: Velocity component
V : Velocity magnitude
$\beta_{k,\omega}$: Constants defined by $k - \omega$ equations
μ : Coefficient of viscosity
μ_T: Turbulent coefficient of viscosity
ω : Turbulent dissipation frequency
ω_B: DES filtered turbulent dissipation frequency
$\hat{\tau}_{ij}$: Total shear stress tensor
τ_{ij}: Laminar shear stress tensor
τ_{ij}^R: Reynolds shear stress tensor
ρ : Density
Tu : Freestream turbulence level as %
l_{le}: Eddy length scale
Δ : DES filter width
σ, σ^*: Constants defined by $k - \omega$ equations
C_{des}: DES coefficients.

Acknowledgments

The authors would like to thank Dr. John Clark and the managers of the turbine branch at the Wright-Patterson Air Force Research Laboratory in Dayton, Ohio, for their support of this effort under contract 09-S590-0009-20-C1. Sandia National Laboratories is a multi-program laboratory managed and operated by Sandia Corporation, a wholly owned subsidiary of Lockheed Martin Corporation, for the U.S. Department of Energy's National Nuclear Security Administration under contract DE-AC04-94AL85000.

References

[1] P. E. Morgan, D. P. Rizzetta, and M. R. Visbal, "High-order numerical simulation of turbulent flow over a wall-mounted hump," *AIAA Journal*, vol. 44, no. 2, pp. 239–251, 2006.

[2] V. Krishnan, K. D. Squires, and J. R. Forsythe, "Prediction of separated flow characteristics over a hump," *AIAA Journal*, vol. 44, no. 2, pp. 252–262, 2006.

[3] P. Balakumar, "Computations of flow over a hump model using higher order method with turbulence modeling," in *Proceedings of the 43rd AIAA Aerospace Sciences Meeting and Exhibit*, Reno, Nev, USA, January 2005, AIAA Paper No. 2005-1270.

[4] D. M. Israel, D. Postl, and H. F. Fasel, "A flow simulation methodology for analysis of coherent structures and flow control," in *Proceedings of the 2nd AIAA Flow Control Conference*, 2004.

[5] S. J. Miller and P. A. Seitz, "The interaction between a fluidic actuator and main flow using SAS turbulence modeling," in *Proceedings of the 3rd AIAA Flow Control Conference*, San Francisco, Calif, USA, June 2006.

[6] F. R. Menter, "A scale-adaptive simulation model for turbulent flow predictions," in *Proceedings of the 41st AIAA Aerospace Sciences Meeting*, Reno, Nev, USA, 2003.

[7] D. C. Wilcox, *Turbulence Modeling for CFD*, DCW Industries, 3rd edition, 2006.

[8] D. C. Wilcox, "Reassessment of the scale-determining equation for advanced turbulence models," *AIAA Journal*, vol. 26, no. 11, pp. 1299–1310, 1988.

[9] J. Smagorinsky, "General circulation experiments with the primitive equations," *Monthly Weather Review*, vol. 91, no. 3, pp. 99–164, 1963.

[10] M. Strelets, "Detached eddy simulation of massively separated flows," in *Proceedings of the 39th Aerospace Sciences Meeting and Exhibit*, Reno, Nev, USA, January 2001, AIAA Paper No. 2001-879.

[11] R. H. Bush and M. Mani, "A two-equation large eddy stress model for high sub-grid shear," in *Proceedings of the 15th AIAA Computational Fluid Dynamics Conference*, Anaheim, Calif, USA, June 2001, AIAA Paper No. 2001-2561.

[12] R. H. Ni, "Multiple-grid scheme for solving the euler equations," *AIAA Journal*, vol. 20, no. 11, pp. 1565–1571, 1982.

[13] J. F. Dannenhoffer, *Grid adaptation for complex two-dimensional transonic flows [Ph.D. thesis]*, Massachusetts Institute of Technology, 1987.

[14] R. L. Davis, R.-H. Ni, and J. E. Carter, "Cascade viscous flow analysis using the Navier-Stokes equations," *Journal of Propulsion and Power*, vol. 3, no. 5, pp. 406–414, 1987.

[15] R. L. Davis, D. E. Hobbs, and H. D. Weingold, "Prediction of compressor cascade performance using a Navier-Stokes technique," *Journal of Turbomachinery*, vol. 110, no. 4, pp. 520–531, 1988.

[16] A. Jameson, "Time dependent calculations using multigrid with applications to unsteady flows past airfoils with wings," in *Proceedings of the 10th Computational Fluid Dynamics Conference*, p. 14, Honolulu, Hawaii, USA, June 1991, AIAA Paper No. 1991-1596.

[17] R. L. Davis and J. F. Dannenhoffer, "A detached-eddy simulation procedure targeted for design," in *Proceedings of the 46th AIAA Aerospace Sciences Meeting and Exhibit*, Reno, Nev, USA, January 2008, AIAA Paper No. 2008-534.

[18] A. J. Andrade, R. L. Davis, and M. A. Havstad, "A RANS/DES numerical procedure for axisymmetric flows with and without strong rotation," in *Proceedings of the 46th AIAA Aerospace Sciences Meeting and Exhibit*, Reno, Nev, USA, January 2008, AIAA Paper No. 2008-702.

[19] R. Bozinoski and R. L. Davis, "General three-dimensional, multi-block, parallel turbulent Navier-Stokes procedure," in *Proceedings of the 46th AIAA Aerospace Sciences Meeting and Exhibit*, Reno, Nev, USA, January 2008, AIAA Paper No. 2008-756.

[20] A. Jameson, W. Schmidt, and E. Turkel, "Numerical solution of the Euler equations by finite volume methods using Runge Kutta time stepping schemes," in *Proceedings of the 14th Fluid and Plasma Dynamics Conference*, Palo Alto, Calif, USA, June 1981, AIAA Paper No. 1981-1259.

[21] W. Gropp, E. Lusk, and A. Skjellum, *MPI: A Message-Passing Interface Standard*, Scientific and Engineering Computation Series, The MIT Press, 1994.

[22] C. L. Rumsey, T. B. Gatski, W. L. Sellers, V. N. Vatsa, and S. A. Viken, "Summary of the 2004 computational fluid dynamics validation workshop on synthetic jets," *AIAA Journal*, vol. 44, no. 2, pp. 194–207, 2006.

[23] S. Šarić, S. Jakirlić, A. Djugum, and C. Tropea, "Computational analysis of locally forced flow over a wall-mounted hump at high-Re number," *International Journal of Heat and Fluid Flow*, vol. 27, no. 4, pp. 707–720, 2006.

[24] P. E. Morgan, D. P. Rizzetta, and M. R. Visbal, "Large-eddy simulation of separation control for flow over a wall-mounted hump," *AIAA Journal*, vol. 45, no. 11, pp. 2643–2660, 2007.

3

Indirect Optimization of Satellite Deployment into a Highly Elliptic Orbit

Francesco Simeoni,[1] Lorenzo Casalino,[1] Alessandro Zavoli,[2] and Guido Colasurdo[2]

[1] *Dipartimento di Ingegneria Meccanica e Aerospaziale, Politecnico di Torino, Corso Duca degli Abruzzi 24, 10129 Torino, Italy*
[2] *Dipartimento di Ingegneria Meccanica e Aerospaziale, Sapienza Università di Roma, Via Eudossiana 18, 00184 Roma, Italy*

Correspondence should be addressed to Lorenzo Casalino, lorenzo.casalino@polito.it

Academic Editor: C. B. Allen

The analysis of the optimal strategies for the deployment of a spacecraft into a highly elliptic orbit is carried out by means of an indirect optimization procedure, which is based on the theory of optimal control. The orbit peculiarities require that several perturbations are taken into account: an 8×8 model of the Earth potential is adopted and gravitational perturbations from Moon and Sun together with solar radiation pressure are considered. A procedure to guarantee convergence and define the optimal switching structure is outlined. Results concerning missions with up to 4.5 revolutions around the Earth are given, and significant features of this kind of deployment are highlighted.

1. Introduction

Space trajectory optimization has gained importance thanks to the development of digital computers and the demanding features of space missions that are currently envisaged. Payload maximization is fundamental to ensure mission feasibility and to reduce costs; sometimes, flight-time minimization is sought to comply with operational requirements. Numerical methods for trajectory optimization can be in general classified into three main groups: indirect methods, direct methods, and evolutionary algorithms. The last group is receiving a great attention, as these algorithms are intrinsically apt to multidisciplinary and multiobjective optimization and in principle are capable of achieving the global optimum in a very large search space. When low-thrust trajectories are considered, evolutionary algorithms typically rely on approximations to maintain the computational time at reasonable values, as a large number of evaluations are required to achieve the solution; for these reasons, they usually provide only an estimate of the optimal solution and a refinement is required to obtain the optimum; direct methods are often used for this purpose. As a consequence, low-thrust trajectories are often dealt with by either direct or indirect methods that typically perform single-objective optimization, attaining a local optimum close to a tentative solution; additional effort is necessary to assure the achievement of the global optimum.

An accurate comparison of direct and indirect methods is found in Betts [1]. A wide number of similarities between the methods is highlighted. Direct methods introduce a parametric representation of the control and/or state variables; the large number of variables, which are required to accurately describe the problem, usually leads to long computational times that can be reduced by taking the matrix sparsity into account. Indirect methods are fast due to the reduced number of variables and may offer higher accuracy and interesting theoretical insight into the problem characteristics. However, three main drawbacks of indirect techniques need to be underlined [1]: analytic expressions for the optimum necessary conditions must be derived, the region of convergence for a root-finding algorithm may be small, and, for problems with path inequalities, it is necessary to guess the sequence of constrained and unconstrained subarcs. It is, however, important to note that also direct methods rely on a tentative solution and may not converge to the optimal solution. However convergence difficulties prevent indirect methods from finding a solution, whereas direct methods find at least a suboptimal solution.

The authors have been using an indirect method for many years and have developed a procedure that mitigates the drawbacks of this approach; in particular the formulation of the optimization problem is made quite simple and attention can be paid to strategies to achieve convergence. In the past two decades the procedure has been tested on different and often difficult problems of spaceflight mechanics and very accurate results have been obtained [2–8]. Recently, the authors had to employ their procedure in a complex application concerning the finite-thrust deployment of a satellite into a highly elliptic orbit (HEO). Dynamical model and problem data were assigned; numerical results were requested for the sake of comparison. HEOs are gaining interest in the scientific community because they represent a less expensive alternative to halo orbits around the Earth-Moon Lagrangian points; in fact, they combine great semimajor axis with large eccentricity, thus providing long stays far from the Earth disturbances at the apogee, where spacecraft velocity is very low. For these reasons they are envisaged as proper operational orbits of spacecraft devoted to deep-space observation. An example is the now-canceled Simbol-X mission [9], which was designed to employ a two-spacecraft formation flying on a HEO to create a new-generation X-ray telescope.

HEOs present low perigee and high apogee; thus, perturbations due to the Earth asphericity and gravitational attraction of Sun and Moon are important and cannot be neglected. In the literature great attention is paid to the behavior of a satellite in such perturbed environment in order to find fuel-saving station-keeping strategies and improve lifetime. Less attention has been paid to the optimization of the deployment maneuver. Typically, these complex problems are faced using direct methods [10, 11]; indirect optimization methods have been widely applied to interplanetary trajectories (due to the simple dynamics), whereas their application to geocentric problems (involving perturbations) is less frequent. In these cases the attention is often focused on continuous-thrust solutions [12–14], and minimum-time transfers are generally sought; minimum-fuel missions with coast arcs would be too long when a high-I_{sp} very-low-thrust engine is employed. A very-low-thrust multirevolution transfer was analyzed [15] using averaging techniques to deal with perturbations and a continuation scheme to achieve the optimal solution.

Spacecrafts often use chemical engines for orbit deployment. In these cases the thrust level is not low enough for a convenient continuous-thrust transfer and coast arcs are necessary to maintain the fuel consumption within acceptable levels. Several authors [16–18] have dealt with this kind of bang-bang problems but perturbations are usually neglected. In a recent article [19], Thevene and Epenoy include J2 perturbations, whose effects are exploited to reduce the propellant consumption in reconfiguring a four-spacecraft formation; Chuang et al. [20] discuss the effects of atmospheric drag and Earth oblateness for a fixed-duration transfer.

Table 1: Initial and final orbit characteristics.

	a, km	e	i, deg	Ω, deg	ω, deg	ν
Initial	98922	0.931985	5.2	90.0	270.0	0.0
Final	106247	0.798788	—	—	—	180.0

In the present paper the multirevolution finite-thrust deployment of a satellite into an assigned HEO is optimized, taking the relevant perturbations into account. The dynamical model considers an 8×8 model of the Earth gravitational potential, solar radiation pressure, and gravitational perturbations from Moon and Sun, whose positions are obtained via ephemerides. The maneuver is essentially a perigee raising and atmospheric drag, which is negligible, can be omitted; however the procedure described here could include aerodynamic forces [4, 21]. The perigee radius exhibits significant variations, even during a ballistic orbit, due to the influence of Sun and (mostly) Moon; these variations are sensitive to the position of the perturbing bodies, thus depending on the departure date. During each apogee passage, the engine and the luni-solar gravitation modify the perigee height; the orbital period is also changed, and this has influence on the position of the perturbing bodies at the following apogees. The effect on the burn structure (i.e., the order of the perigee and apogee burns), which minimizes the fuel consumption, is hard to predict.

The analytic formulation of the indirect optimization problem is quite simple, and the necessary conditions for optimality are derived with little effort. Numerical convergence to the optimal solution is instead difficult to achieve and the dependence of the burn structure on the departure date makes the use of continuation techniques on the departure date unfeasible. The present paper introduces a proper stepped procedure to find out the optimal burn structure for any assigned departure date; numerical examples are presented. A fast convergence to the optimal solution is obtained for a wide range of departure dates; the switching structure of the optimal trajectory highlights how perturbations influence the deployment and how their effects can be exploited to save propellant. Results prove that an indirect method can be effective in analyzing this kind of maneuvers; an accurate problem formulation and a suitable solution procedure help to mitigate the drawbacks of the indirect approach.

2. Dynamic Problem

The paper considers the transfer of a satellite from an elliptic parking orbit with low perigee to the final operational HEO. Reference is made to the now-canceled Simbol-X mission; the characteristics of the initial and final osculating orbits are given in Table 1. The mission starts at the perigee of the initial orbit, where the launcher has released the spacecraft, and ends when the apogee of the final orbit is reached. Initial position, velocity, and mass are known. Only semimajor axis a, eccentricity e, and true anomaly ν are assigned at the final point; no constraints are imposed on the value of the other orbital parameters (deployment to an assigned final orbit was dealt with in [22]). The initial mass is 960 kg, and the final mass is maximized. For operational reasons, thrust cannot be used during the first revolution (orbit acquisition) and in the proximity of the last apogee passage (start of operations).

The spacecraft is modeled as a point with variable mass. Position $\mathbf{r}$, velocity $\mathbf{v}$, and mass m of the spacecraft are the problem state variables, described by differential equations

$$\frac{d\mathbf{r}}{dt} = \mathbf{v},$$
$$\frac{d\mathbf{v}}{dt} = -\frac{\mu\mathbf{r}}{r^3} + \frac{\mathbf{T}}{m} + \mathbf{a}_p, \quad (1)$$
$$\frac{dm}{dt} = -\frac{T}{c}.$$

The trajectory is controlled by the thrust vector $\mathbf{T}$ (the effective exhaust velocity c is assumed constant). The perturbing acceleration

$$\mathbf{a}_p = \mathbf{a}_J + \mathbf{a}_{\text{lsg}} + \mathbf{a}_{\text{srp}} \quad (2)$$

is composed of the perturbations due to the Earth asphericity $\mathbf{a}_J$, luni-solar gravity $\mathbf{a}_{\text{lsg}}$, and solar radiation pressure $\mathbf{a}_{\text{srp}}$.

The Earth Mean Equator and Equinox of Epoch J2000 reference frame (i.e., EME2000) is adopted; $\mathbf{I}$, $\mathbf{J}$, and $\mathbf{K}$ are unit vectors along the axes of EME2000. Precession and nutation are neglected. Position is described by radius r, right ascension ϑ, declination φ as

$$\mathbf{r} = r\cos\vartheta\cos\varphi\,\mathbf{I} + r\sin\vartheta\cos\varphi\,\mathbf{J} + r\sin\varphi\,\mathbf{K}. \quad (3)$$

The topocentric reference frame, identified by unit vectors $\boldsymbol{\imath}$ (radial), $\boldsymbol{\jmath}$ (eastward), and $\mathbf{k}$ (northward), is introduced. One has

$$\begin{Bmatrix} \boldsymbol{\imath} \\ \boldsymbol{\jmath} \\ \mathbf{k} \end{Bmatrix} = \begin{bmatrix} \cos\vartheta\cos\varphi & \sin\vartheta\cos\varphi & \sin\varphi \\ -\sin\vartheta & \cos\vartheta & 0 \\ -\cos\vartheta\sin\varphi & -\sin\vartheta\sin\varphi & \cos\varphi \end{bmatrix} \begin{Bmatrix} \mathbf{I} \\ \mathbf{J} \\ \mathbf{K} \end{Bmatrix}. \quad (4)$$

The position vector in the topocentric frame is $\mathbf{r} = r\boldsymbol{\imath}$, and the velocity vector is expressed as

$$\mathbf{v} = \dot{\mathbf{r}} = u\boldsymbol{\imath} + v\boldsymbol{\jmath} + w\mathbf{k} \quad (5)$$

with u, v, and w being radial, eastward, and northward components, respectively. The scalar state equations are easily derived:

$$\frac{dr}{dt} = u,$$
$$\frac{d\vartheta}{dt} = \frac{v}{(r\cos\varphi)},$$
$$\frac{d\varphi}{dt} = \frac{w}{r},$$
$$\frac{du}{dt} = -\frac{\mu}{r^2} + \frac{(v^2+w^2)}{r} + \frac{T_u}{m} + (a_p)_u,$$
$$\frac{dv}{dt} = \frac{(-uv + vw\tan\varphi)}{r} + \frac{T_v}{m} + (a_p)_v,$$
$$\frac{dw}{dt} = \frac{(-uw - v^2\tan\varphi)}{r} + \frac{T_w}{m} + (a_p)_w,$$
$$\frac{dm}{dt} = -\frac{T}{c}, \quad (6)$$

where subscripts u, v, and w denote the components along $\boldsymbol{\imath}$, $\boldsymbol{\jmath}$, and $\mathbf{k}$, respectively. It is important to note that the state equations are relatively simple with this set of variables in comparison, for instance, to the use of equinoctial elements. This fact facilitates the analytical derivation of the necessary condition for optimality.

2.1. Earth Potential Model. The Earth potential description is based on the Earth Gravitational Model EGM2008, which provides normalized spherical harmonic coefficients for the Earth gravitational potential; the "Tide Free" system is used [23]. The developed code can be quickly modified to consider higher-degree terms or the "Zero Tide" system. The Earth's rotation is assumed to be uniform, neglecting precession and nutation. The EME2000 reference frame is adopted. The gravity model is described in detail in [24].

According to EGM2008, the potential corresponding to the Earth asphericity is expressed as

$$\Phi = -\frac{\mu}{r}\sum_{n=2}^{N}\left(\frac{r_E}{r}\right)^n \sum_{m=0}^{n}(C_{nm}\cos m\lambda + S_{nm}\sin m\lambda)P_{nm}(\sin\varphi), \quad (7)$$

where μ is the Earth gravitational parameter and r_E is the semimajor axis of the Earth ellipsoid. In this paper N is chosen equal to 8. The associated Legendre functions $P_{nm}(\sin\varphi)$ and the spherical harmonic coefficients C_{nm} and S_{nm} are used in the unnormalized form that permits faster computations. Normalized quantities would allow for a greater accuracy, which is not necessary for the present application.

The terrestrial latitude coincides with declination φ, as nutation is neglected. The terrestrial longitude λ is obtained as $\lambda = \vartheta - \vartheta_{\text{Gref}} - \omega_E(t - t_{\text{ref}})$, where ϑ_{Gref} is the Greenwich right ascension at the reference time t_{ref} (51544.5 MJD) and ω_E is evaluated on the basis of the sidereal day, neglecting precession.

The perturbing acceleration due to the Earth asphericity is the gradient of $-\Phi$, and its components in the topocentric frame are thus evaluated as

$$(a_J)_u = -\frac{\partial\Phi}{\partial r},$$
$$(a_J)_v = -\frac{(\partial\Phi/\partial\vartheta)}{(r\cos\varphi)}, \quad (8)$$
$$(a_J)_w = -\frac{(\partial\Phi/\partial\varphi)}{r}.$$

Differentiation with respect to r and ϑ is straightforward; derivatives with respect to φ require the derivatives of the associated Legendre functions, which are obtained recursively, exploiting the properties of the Legendre polynomials. Derivatives are evaluated directly with respect to φ (some authors use the colatitude $\pi/2 - \varphi$, the only difference being a sign change of the derivatives); one has, posing $P_{nm} = 0$ for $m > n$,

$$\frac{\mathrm{d}P_{nm}}{\mathrm{d}\varphi} = \begin{cases} P_{n1} & \text{for } m = 0, \\ \dfrac{\left[P_{n(m+1)} - (n+m)(n-m+1)P_{n(m-1)}\right]}{2} & \text{for } m > 0. \end{cases} \quad (9)$$

Further details can be found in [25–27].

2.2. LuniSolar Perturbation. Moon and Sun positions are evaluated using DE405 JPL ephemeris [28], which directly provide the body position in rectangular coordinates x_b, y_b, and z_b (with either subscript $b = s$ for Sun or $b = l$ for Moon) with respect to the Earth in the International Celestial Reference Frame and therefore in the EME2000 frame (differences between these frames are very small and can be neglected in the present problem). The perturbing acceleration on the spacecraft, which is caused by a body with gravitational parameter μ_b and position vector with respect to the Earth $\mathbf{r}_b = x_b\mathbf{I} + y_b\mathbf{J} + z_b\mathbf{K}$, is given by the difference of the gravitational accelerations that the perturbing body causes on spacecraft and Earth, that is,

$$\mathbf{a}_{bg} = -\left(\frac{\mu_b}{R^3}\right)\mathbf{R} - \left(\frac{\mu_b}{r_b^3}\right)(\mathbf{r}_b), \quad (10)$$

where $\mathbf{R} = \mathbf{r} - \mathbf{r}_b$ is the spacecraft relative position vector with respect to the perturbing body (and $-\mathbf{r}_b$ is the Earth relative position), as shown in Figure 1.

The acceleration is projected onto the topocentric frame (based on the spacecraft position) to easily obtain

$$\begin{aligned} \left(a_{bg}\right)_u &= \left(\frac{\mu_b}{R^3}\right)[(r_b)_u - r] - \left(\frac{\mu_b}{r_b^3}\right)(r_b)_u, \\ \left(a_{bg}\right)_v &= \left(\frac{\mu_b}{R^3}\right)(r_b)_v - \left(\frac{\mu_b}{r_b^3}\right)(r_b)_v, \\ \left(a_{bg}\right)_w &= \left(\frac{\mu_b}{R^3}\right)(r_b)_w - \left(\frac{\mu_b}{r_b^3}\right)(r_b)_w \end{aligned} \quad (11)$$

with $R = \sqrt{[r - (r_b)_u]^2 + (r_b)_v^2 + (r_b)_w^2}$. The position components of the perturbing body in the spacecraft topocentric frame are

$$\begin{aligned} (r_b)_u &= x_b \cos\vartheta\cos\varphi + y_b \sin\vartheta\cos\varphi + z_b \sin\varphi, \\ (r_b)_v &= -x_b \sin\vartheta + y_b \cos\vartheta, \\ (r_b)_w &= -x_b \cos\vartheta\sin\varphi - y_b \sin\vartheta\sin\varphi + z_b \cos\varphi. \end{aligned} \quad (12)$$

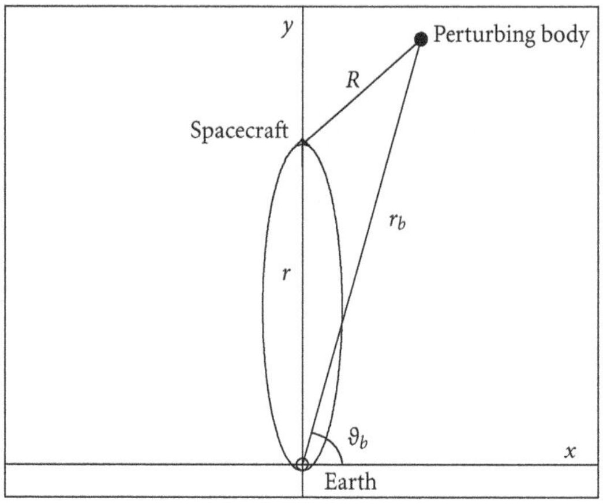

FIGURE 1: Schematic geometry of gravitational perturbations.

The perturbing acceleration is thus written as a function of time and state variables (viz., only r, ϑ, and φ, as gravity forces only depend on position). The lunisolar perturbation is the sum of the gravitational perturbations due to Moon ($b = l$) and Sun ($b = l$).

2.3. Solar Radiation Pressure. The photon pressure at distance R from the Sun is $p = L_S/4\pi R^2 c_{\text{light}}$, where L_S is the total power radiated by the Sun and c_{light} is the speed of light; the photon pressure at $R^* = 1\,\text{AU}$ is $p^* = 4.55682 \cdot 10^{-6}\ \text{N/m}^2$. Assuming reflectivity $\eta = 0.7$ the acceleration on a spherical body of mass m and cross-section S is

$$\mathbf{a}_{\text{srp}} = (1+\eta)p^*\left(\frac{R^*}{R}\right)^2\left(\frac{S}{m}\right)\frac{\mathbf{R}}{R} = \frac{\Gamma\mathbf{R}}{mR^3}, \quad (13)$$

whose components are

$$\begin{aligned} \left(a_{\text{srp}}\right)_u &= \left[\frac{\Gamma}{(mR^3)}\right][(r_s)_u - r], \\ \left(a_{\text{srp}}\right)_v &= \left[\frac{\Gamma}{(mR^3)}\right][(r_s)_v], \\ \left(a_{\text{srp}}\right)_w &= \left[\frac{\Gamma}{(mR^3)}\right][(r_s)_w]. \end{aligned} \quad (14)$$

The effect of solar radiation pressure is therefore an acceleration in the Sun-spacecraft direction, inversely proportional to the squared distance of the two bodies. This acceleration and the solar gravity acceleration show the same dependence on distance and are parallel but with opposite directions; the similarity with the first term on the right-hand side of (11) allows one to treat them simultaneously. One should note that the perturbing acceleration in (13) depends also on the instantaneous mass; this fact introduces an additional term in the time derivative of the mass adjoint variable.

A conical shadow of the Earth is considered to determine the eclipses when $(r_s)_u < 0$ (Sun and spacecraft on opposite sides with respect to the Earth). The relevant quantities are sketched (not to scale) in Figure 2. The Earth determines a shadow cone with semiangle $\gamma_{\text{shadow}} = \sin^{-1}(r_E/r_s)$, where r_E

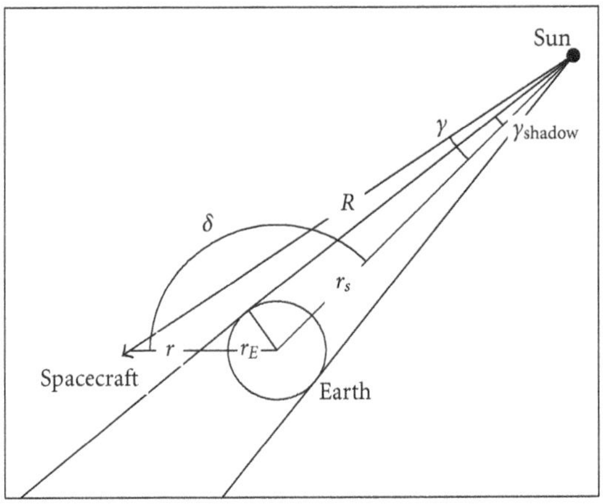

FIGURE 2: Schematic geometry of the Earth shadow.

is the Earth radius. The spacecraft is instead on the surface of a cone (centered at the Sun with axis on the Earth-Sun connecting line) with semi-angle $\gamma = \sin^{-1}(r \sin\delta/R)$, where the angle δ between the Earth centered vectors pointing to the Sun and spacecraft is evaluated as $\delta = \cos^{-1}[(r_s)_u/r_s]$. The spacecraft is in the Earth shadow when $(r_s)_u < 0$ and $\gamma < \gamma_{shadow}$. Results show that, in the present case (m = 960 kg, S = 5.7 m^2), solar radiation pressure has a negligible influence on performance, causing variations of the final mass of a few grams.

3. Indirect Optimization Method and Optimal Controls

Optimal control theory (OCT) [29] is applied to the problem described above to determine the optimal solution. The state variables are collected in vector $\mathbf{x} = (r\ \vartheta\ \varphi\ u\ v\ w\ m)^T$. Adjoint variables $\boldsymbol{\lambda} = (\lambda_r\ \lambda_\vartheta\ \lambda_\varphi\ \lambda_u\ \lambda_v\ \lambda_w\ \lambda_m)^T$ are introduced, and the Hamiltonian is defined as

$$H = \boldsymbol{\lambda}^T \dot{\mathbf{x}} = H' + \boldsymbol{\Lambda}^T \mathbf{T} - \lambda_m \left(\frac{T}{c}\right), \tag{15}$$

where H' collects all the terms that do not contain the control (i.e., $\mathbf{T}$ in the present problem), and

$$\boldsymbol{\Lambda} = \lambda_u \boldsymbol{\imath} + \lambda_v \boldsymbol{\jmath} + \lambda_w \mathbf{k} \tag{16}$$

is the adjoint vector to the velocity, which is named primer vector in the literature [30].

OCT provides the Euler-Lagrange equations for the adjoint variables

$$\frac{d\boldsymbol{\lambda}^T}{dt} = -\frac{\partial H}{\partial \mathbf{x}}. \tag{17}$$

The perturbing terms depend on the state variables and therefore influence the derivatives of the adjoint variables. As far as the geopotential is concerned, differentiation with respect to r and ϑ is again straightforward. The recursive scheme, which is outlined in Section 2.1, is again used to evaluate the derivatives with respect to declination φ. Explicit expressions are derived for the gravitational perturbations of Moon and Sun (which depend on r, ϑ and φ) and solar radiation pressure (which also depends on m). The equations of the derivatives, which are tedious but simple to obtain, are omitted for the sake of conciseness.

Pontryagin's maximum principle (PMP), which states that the optimal control must maximize the Hamiltonian, is used to determine the optimal controls (thrust magnitude and direction). The thrust $\mathbf{T}$ must be parallel to the primer vector and the Hamiltonian, which is rewritten as

$$H = H' + T\left(\frac{\Lambda}{m} - \frac{\lambda_m}{c}\right), \tag{18}$$

is linear with respect to T; a bang-bang control arises and the thrust magnitude must be either maximum or minimum (i.e., zero) depending on the sign of the switching function $S_F = \Lambda/m - \lambda_m/c$ (Λ indicates the primer vector magnitude)

$$T = \begin{cases} T_{max} & \text{for } S_F > 0, \\ 0 & \text{for } S_F < 0. \end{cases} \tag{19}$$

When indirect procedures are used, the thrust level is usually decided during integration on the basis of the sign of the switching function. It is instead convenient to split the trajectory into f arcs that are joined at the switch points, that is, where the control (here the thrust magnitude) is discontinuous; in other problems a similar split is enforced where state variables are discontinuous and/or their values are subject to constraints. The time lengths of these arcs become f additional unknown parameters of the present problem. This scheme avoids integration instabilities, assures high accuracy, and improves the convergence of the numerical procedure. The switching structure (i.e., number and order of thrust and coast arcs) is specified a priori; after the achievement of the numerical solution, the switching function history is checked and the switching structure is modified if PMP is violated (except at departure and arrival where, according to mission constraints, the engine cannot be used even though it would be beneficial).

OCT also provides the boundary conditions for optimality [3], which depend on performance index and boundary conditions on the state variables. Mayer formulation is adopted, and the performance index is written as

$$\phi\left(\mathbf{x}_{(j-1)_+}, \mathbf{x}_{j_-}, t_{(j-1)_+}, t_{j_-}\right) \quad j = 1, \dots, f \tag{20}$$

with subscripts $j-$ and $j+$ indicating the values just before and after point j. Boundary conditions on the state variables are collected in the form

$$\boldsymbol{\psi}\left(\mathbf{x}_{(j-1)_+}, \mathbf{x}_{j_-}, t_{(j-1)_+}, t_{j_-}\right) = 0, \quad j = 1, \dots, f. \tag{21}$$

Optimality requires

$$-\boldsymbol{\lambda}_{j_-}^T + \frac{\partial \phi}{\partial \mathbf{x}_{j_-}} + \boldsymbol{\mu}^T \left[\frac{\partial \boldsymbol{\psi}}{\partial \mathbf{x}_{j_-}}\right] = 0, \quad j = 1, \dots, f, \tag{22}$$

$$\boldsymbol{\lambda}_{j_+}^T + \frac{\partial \phi}{\partial \mathbf{x}_{j_+}} + \boldsymbol{\mu}^T \left[\frac{\partial \boldsymbol{\psi}}{\partial \mathbf{x}_{j_+}}\right] = 0, \quad j = 0, \dots, f-1, \tag{23}$$

$$H_{j_-} + \frac{\partial \phi}{\partial t_{j_-}} + \boldsymbol{\mu}^T \frac{\partial \boldsymbol{\psi}}{\partial t_{j_-}} = 0, \quad j = 1, \dots, f, \tag{24}$$

$$-H_{j_+} + \frac{\partial \phi}{\partial t_{j_+}} + \boldsymbol{\mu}^T \frac{\partial \boldsymbol{\psi}}{\partial t_{j_+}} = 0, \quad j = 0, \dots, f-1. \tag{25}$$

The constant Lagrange multipliers $\boldsymbol{\mu}$ are eliminated from (22)–(25); the resulting boundary conditions for optimality and the boundary conditions on the state variables, given by (21), are collected in a single vector in the form

$$\boldsymbol{\sigma}\left(\mathbf{x}_{(j-1)_+}, \mathbf{x}_{j_-}, \boldsymbol{\lambda}_{(j-1)_+}, \boldsymbol{\lambda}_{j_-}, t_{(j-1)_+}, t_{j_-}\right) = 0, \quad j = 1, \dots, f, \tag{26}$$

which, together with state and adjoint differential equations, defines a multipoint boundary value problem (MPBVP).

In the present problem, at the initial point ($j = 0$), t_0 and all state variables are assigned; at the final point ($j = f$) the apogee radius r_A and orbit semilatus rectum p are given, and

$$\begin{aligned} r_f - r_A &= 0, \\ u_f &= 0\,, \\ v_f^2 + w_f^2 - \frac{\mu p}{r_A^2} &= 0 \end{aligned} \tag{27}$$

are the conditions imposed on the state variables. The performance index to be maximized is the final mass, that is, $\phi = m_f$. The boundary conditions for optimality provide

$$\lambda_{\vartheta f} = 0, \tag{28}$$

$$\lambda_{\varphi f} = 0, \tag{29}$$

$$\lambda_{vf} w_f - \lambda_{wf} v_f = 0, \tag{30}$$

$$\lambda_{mf} = 1. \tag{31}$$

The final time is free and (24) with $j = f$ provides the transversality condition

$$H_f = 0. \tag{32}$$

Application of (24) and (25) at every switch point prescribes the Hamiltonian continuity; state and adjoint variables are continuous and, as a consequence, the switching function must be null at the switch points

$$S_{Fj} = 0 \quad j = 1, \dots, f-1. \tag{33}$$

The numerical problem consists of 14 differential equations represented by (6) and (17). Initial values of the seven adjoint variables and time lengths of f coast and burn arcs are unknown in the present problem; an equal number of boundary conditions, given by (27)–(33), completes the MPBVP. The problem is homogeneous in the adjoint variables, and (31) can be replaced by assigning the initial value $\lambda_{m0} = 1$ in order to reduce the number of unknowns. The unknown parameters are collected in a vector $\mathbf{p}$.

4. Numerical Solution

The solution of the problem outlined in the previous section is a difficult task; many aspects of the relevant numerical procedure deserve attention. Variables are normalized using the Earth equatorial radius, the corresponding circular velocity, and the spacecraft initial mass as reference values. Differential equations are integrated by using a variable-order variable-step integration scheme, based on the Adams-Moulton formulas, in order to carry out a fast and very accurate integration.

A single-shooting technique is adopted to solve the MPBVP, which is transformed into a series of initial value problems (IVPs) leading to convergence by means of Newton's method [31], according to the following shooting algorithm: at the rth iteration (1) using the current tentative values $\mathbf{p}^r$ for the unknown parameters, the related IVP is solved numerically; (2) the errors $\boldsymbol{\sigma}$ on the boundary conditions are evaluated; (3) the new tentative values $\mathbf{p}^{r+1}$ are evaluated according to Newton's rule

$$\mathbf{p}^{r+1} = \mathbf{p}^r - \left(\frac{\partial \boldsymbol{\sigma}}{\partial \mathbf{p}}\right)^{-1} \boldsymbol{\sigma}. \tag{34}$$

This process is repeated until all errors are smaller than an assigned value (here set equal to 10^{-7}).

At any step, the sensitivity matrix $\partial\boldsymbol{\sigma}/\partial\mathbf{p}$ is numerically evaluated by means of a first-order forward finite-difference scheme. Each unknown parameter is in turn perturbed by a small amount δp_i; the new IVP is solved, and the change of the errors on the boundary conditions $\delta\boldsymbol{\sigma}$ is evaluated. The choice of the perturbation step δp_i is important to achieve a proper evaluation of the sensitivity matrix. A rule of thumb suggests δp_i of the same order of the square root of the absolute tolerance used by the integrator ($\delta p_i = 5 \cdot 10^{-6}$ in this application).

The initial guess $\mathbf{p}^1$ is very important for the convergence of the numerical procedure. The time at the switch points (i.e., where the engine is turned on or off) may exhibit significant changes (the orbital period is modified by the maneuvers that are performed). The corresponding right ascension instead can be easily guessed as thrust arcs are in correspondence of the orbit apsides, either perigee or apogee, and ϑ is preferred as the independent variable in place of time t. Moreover, ϑ is replaced by a nondimensional variable ε in order to "fix" the integration intervals; inside each arc

$$\varepsilon = j - 1 + \frac{\vartheta - \vartheta_{j-1}}{\vartheta_j - \vartheta_{j-1}}, \tag{35}$$

ε is zero at the initial point, assumes consecutive integer values at the switch points, and is equal to the total number of arcs f at the final point. The f right-ascension values ϑ_j at the switch points and at the final point are the problem additional unknowns that replace the corresponding unknown times ($\vartheta_0 = 1.5\pi$ is an assigned initial condition).

The features of the present indirect approach, in particular, the peculiar treatment of the switch conditions, widen the convergence radius of the numerical procedure. On the other hand, the MPBVP solution is correct only if the tentative switching structure corresponds to that of the optimal solution. If the tentative solution lacks a burn arc, a suboptimal solution is usually found, with the switching function greater than zero during a coast arc. The request of adding a further thrust arc is clear, but, unfortunately, this sometimes implies the ensuing removal of another propelled arc. If the tentative structure presents one more burn than the optimal solution, (33) cannot be fulfilled at the two extremities of the superfluous arc, and the code would provide an unfeasible solution (e.g., a burn arc with negative length) or the MPBVP solution process would not converge at all. The suggestion that an arc should be removed is clear, but no replacement arc is necessary, and this case is easier to manage.

An experienced user is able to guess the unknown parameters and a suitable burn structure with a trial and error process, which is repeated until PMP is satisfied everywhere along the trajectory. This manual procedure is time-consuming, and an automated procedure is useful if the analysis has to cover a large number of different cases (e.g., a wide launch window). In the present problem the optimal switching structure is largely dependent on Moon and Sun positions and therefore on the departure date. A stepped procedure is devised to overcome the convergence difficulties related to the search for the optimal solution for an assigned departure date. The procedure exploits continuation techniques and is based on the observation that, in general, it is more difficult to introduce a new arc than to remove an existing one. It requires a tentative solution with the maximum number of thrust arcs that the problem could present. A suitable solution is obtained by assuming a simplified dynamical model that considers only the gravitational perturbation related to the second zonal harmonic; the optimal solution for this problem (termed *J2 problem* in the following) is independent of the departure date.

4.1. J2 Problem. The procedure starts using a dynamical model that considers only the gravitational perturbation related to the Earth oblateness, that is, C_{20} is the only nonzero harmonic coefficient in (7). This problem is independent of the departure date.

According to the Keplerian model, the initial conditions of the spacecraft would permit ballistic attainment of the desired final apogee and the deployment would require only the perigee raising. The effect of the Earth oblateness on semimajor axis and eccentricity is null after a complete orbit; the perigee is unchanged, but the actual apogee of the initial orbit is lower. Actual differences between final and initial orbit in terms of perigee and apogee radii are 14650 km and 9172 km, respectively, when J2 is considered. The mission requires to raise both perigee and apogee, but the apogee maneuver is very expensive from a propulsive point of view, in comparison to the very small perigee maneuver. The optimal impulsive strategy [32] prescribes two burns: a perigee impulse, followed by an apogee impulse (PA is used to describe this burn sequence).

In the finite-thrust case it is convenient to split the impulsive maneuvers into multiple burn arcs centered at the apsides, in order to reduce the propulsive losses. The perigee burn, which should precede all the apogee burns, is very short and, if the number of revolutions is limited, the split of the longer apogee burn is preferable in terms of propellant consumption. The optimal burn structure consists of a single perigee burn (to raise the apogee) followed by burns at every apogee passage. When 4.5 revolutions are permitted, only three apogee burns can be performed (engine cannot be used in the proximity of departure and arrival apsides): the optimal burn sequence is therefore PAAA.

After this suitable switching structure is assumed, one can easily guess the right ascension at the engine switches. The physical meaning of the primer vector suggests that the adjoint vector to velocity should be parallel to the thrust direction, which is essentially horizontal and in the orbital plane for the initial burn. Latitude has a minimal influence, and longitude has no influence at all; the corresponding adjoint variables are therefore (roughly) zero. From a practical point of view, the magnitude of the primer vector Λ and the adjoint variable to radius λ_r are the only parameters difficult to estimate; convergence is easily obtained.

4.2. Stepped Procedure. When the complete dynamical model is considered, the optimal trajectories (and the associated control laws) are dependent on time and their burn structure becomes more difficult to assess. Moon and Sun with their attraction are capable of varying significantly the perigee radius; the Moon, in particular, can raise/lower the perigee even by 200 km in a single ballistic revolution, depending on its position. Thus, the optimal length of the apogee burns is dictated by the need not only of containing the propulsive losses but also of modifying the orbital periods; the time of passage at each consecutive apogee is anticipated or delayed, in order to find the perturbing bodies in more favorable or less unfavorable positions.

As a consequence, the optimal solution of the complete problem does not exhibit a uniform split of the three apogee burns. The optimal split depends on the departure date: in many cases one burn (occasionally two burns) may vanish. The switching structure of the mission departing on an assigned date cannot be guessed a priori. A suitable procedure that autonomously fixes the optimal switching structure is here outlined. Starting from the solution of the J2 problem, which is independent of the departure date and has engine firings during all apogee passages, a fraction P_f of the remaining perturbations is introduced, and the solution of the corresponding MPBVP is searched for; the fraction P_f is progressively increased in a discrete number of steps (k_P) until it reaches unity.

At each step, P_f is not imposed as an assigned constant but is considered as an unknown parameter, with an additional constraint that enforces it to the desired value. The iterative procedure progressively moves P_f towards its target value, while the other unknowns are simultaneously adjusted to satisfy all the boundary conditions. In the first k_R iterations a relaxation of the Newton scheme, which uses only a fraction of the correction in (34), is used to avoid that abrupt changes of the unknown parameters prevent convergence.

Convergence is usually reached if the step ΔP_f is chosen small enough. Solution is not achieved (or an unfeasible solution is found) only when the optimal switching structure has one burn less than the previous step. In these cases, the optimal solution, which was obtained with the previous level of P_f, suggests which thrust arc has to be removed. In particular, the maximum values of the switching function during every burn arc are compared, and the burn arc with the smaller value is removed. The right ascensions at the extremes of the arc to be removed are imposed to coincide with the closer apsis (thus imposing null length of the burn arc); these new boundary conditions replace (33) at the same extremes. The last converged solution is still used as initial guess; at the first iteration the only unsatisfied boundary conditions concern perturbation fraction and right ascensions at the extremes of the arc to be removed. An analysis of the achieved solution in the light of PMP is performed to confirm that the new switching structure is optimal.

The suitable values for k_P and k_R are problem dependent; in general, one can expect to reduce issues related to convergence when these two values are increased, but this would also increase the overall computation time. For the problem under investigation a good balance is given by choosing $k_P = 7$ (with $P_f = 0.05, 0.1, 0.2, 0.4, 0.6, 0.8, 1$) and $k_R = 4$ (1/5, 1/4, 1/3, and 1/2, in this order, are the fractions used). The computational time to obtain a full-perturbation 4.5-revolution solution is about 30 seconds on an Intel i7 CPU at 2.67 GHz.

The procedure, which is fully automated, has been tested for 366 departure dates between December 1, 2015, and November 30, 2016, with one-day steps. An analysis of the results has shown that the optimal solution according to PMP is achieved almost in every case. In a single instance the procedure removed a very tiny arc that is present in the optimal solution. This error was due to insufficient values for k_P and k_R; convergence was immediately obtained by adding 4 additional steps and increasing k_R to 9. Moreover, some of the solutions without the first apogee burn (P0AA) are only suboptimal. The corresponding optimal solutions present two ballistic revolutions, followed by a perigee and two apogee burns (0PAA); however, final masses differ only by a few grams. The correct solution can be achieved manually; an autonomous convergence procedure was not developed due to the scarce number and interest of these low-performance opportunities.

The applicability of this procedure is more general and may concern any other parameter whose progressive change reduces the number of burn arcs. For instance, an analogous procedure could be adopted to deal with changes of the thrust magnitude; in particular, when the thrust is low enough, the engine is on at every apsis passage, but burn arcs may vanish when the thrust magnitude is increased: the analysis of the switching function suggests which arc has to be removed. It was verified that an 8 N solution for the deployment here considered can be obtained from the 1 N solution departing on the same date, by progressively increasing the thrust magnitude (e.g., in 1 N steps).

5. Numerical Results

The deployment of the spacecraft into the target HEO was studied in details using the indirect procedure previously described. Perturbations (except J2) were initially neglected, in order to assess the general characteristics of the optimal strategy. Two thrust levels were considered with the aim of better understanding how thrust acceleration and number of revolutions influence the propellant consumption. No burns are permitted in the first revolution and propulsion cannot be employed at the final apogee insertion. The maneuver requires at least 2.5 revolutions around the Earth; a maximum of 4.5 revolutions was assumed.

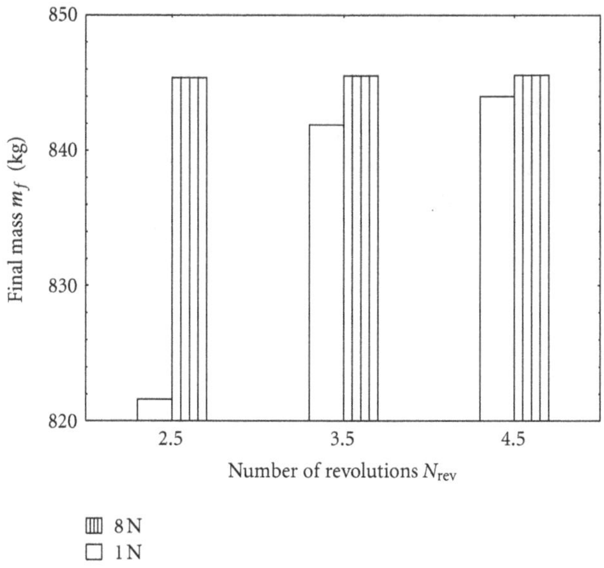

FIGURE 3: Final mass for different numbers of revolutions (J2 only).

5.1. J2 Solution. Figure 3 shows that the final mass increases with the number of revolutions and burns; this effect is more important when the thrust acceleration is low. In fact, the number of burns has no influence for impulsive thrust, as impulses can be applied exactly at apogee and perigee passages, without velocity losses; these instead arise for the finite-thrust maneuvers as thrust is exploited far from the apsides. A larger number of revolutions permit the split of the propulsive effort between shorter arcs centered at the apsides, with lower losses. One should also remember that a larger number of burns are preferable as they allow an easier correction of the errors that may occur during the maneuvers.

Table 2 provides some details of these maneuvers. The different time length of the thrust arcs during the same transfer is related to the mass consumption; as the spacecraft becomes lighter, thrust arcs are shortened in order to have an almost uniform split of the total velocity change ΔV between the burns. Note that the velocity is larger at each later apogee and the spacecraft sweeps a longer arc even though the corresponding time length is shorter. This effect is again more evident for the lowest thrust level. One should also note that the time length of the perigee burn increases with the number of revolutions. This happens because, during the long apogee arc (A1) of a fast transfer, the apogee height is slightly increased, reducing the requirement for the perigee burn, as shown in Figure 4, which presents the evolution

TABLE 2: Time length and angular length of burn arcs (J2 only).

			$T = 1\,\text{N}$			
N_{rev}		P1	A1	P2	A2	A3
2.5	Δt, h	2.66	76.50	3.77	—	—
	$\Delta\vartheta$, deg	250.15	58.73	154.38	—	
3.5	Δt, h	0.45	35.55	—	34.79	—
	$\Delta\vartheta$, deg	121.27	18.40	—	23.15	—
4.5	Δt, h	0.65	23.28	—	22.95	22.63
	$\Delta\vartheta$, deg	152.88	10.90		12.97	15.02
			$T = 8\,\text{N}$			
N_{rev}		P1	A1	P2	A2	A3
2.5	Δt, h	0.09	8.50	—	—	—
	$\Delta\vartheta$, deg	28.50	4.67	—	—	—
3.5	Δt, h	0.09	4.29	—	4.20	—
	$\Delta\vartheta$, deg	28.97	2.04	—	2.61	—
4.5	Δt, h	0.09	2.87	—	2.83	2.79
	$\Delta\vartheta$, deg	29.05	1.30	—	1.55	1.80

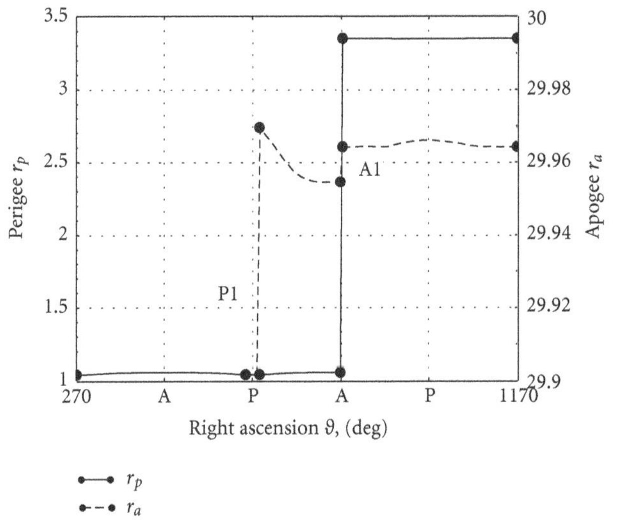

FIGURE 4: Perigee and apogee evolution for the 2.5-revolution 8 N transfer (J2 only).

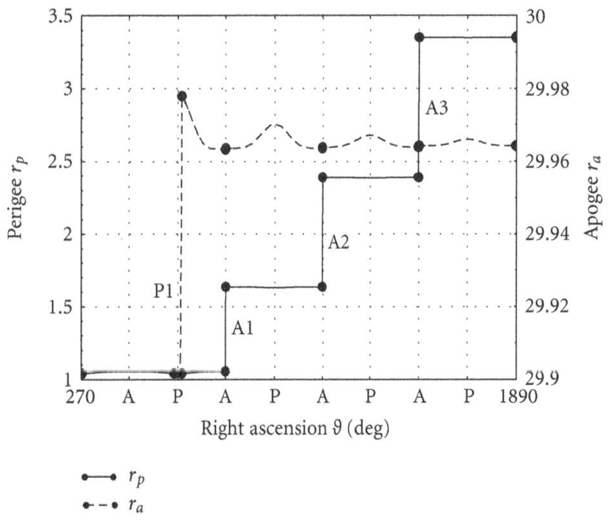

FIGURE 5: Perigee and apogee evolution for the 4.5-revolution 8 N transfer (J2 only).

of perigee and apogee for the 2.5-revolution transfer with $T = 8\,\text{N}$. The apogee change during the apogee burns (A1, A2, and A3) when 4.5 revolutions are performed is instead negligible, as shown in Figure 5. The r_a and r_p variations, which occur during the coast arcs, are due to the J2 effect.

The shortest transfer with $T = 1\,\text{N}$ is the only maneuver with two perigee burns as the time length of the apogee firing would otherwise not be sufficient to obtain the required perigee rise. For this reason, two perigee burns are necessary: the first one (P1) increases the apogee above the required value (see Figure 6) to have a longer available time for an efficient perigee rise during the apogee burn; the second perigee burn (P2) eventually reduces the apogee to the required value.

5.2. Lunisolar Perturbation. Perturbations other than J2, mainly solar and lunar gravity, change the performance of the deployment maneuver and the optimal switching structure. The final mass for the 4.5-revolution transfer with $T = 8\,\text{N}$ is shown in Figure 7 as a function of the departure date in the 1-year launch window starting December 1, 2015 (MJD 57357), which is studied in detail. Figure 8 shows the final mass in the first three months of the launch window and indicates the burn structure of each optimal transfer.

A simplified analysis permits an estimation of the most significant effect of the third body gravitation, which affects the spacecraft perigee height, as a function of the perturbing body position. Since the orbit is highly elliptic, the spacecraft spends most time at the apogee ($\nu = 180\,\text{deg}$, $\vartheta = 90\,\text{deg}$)

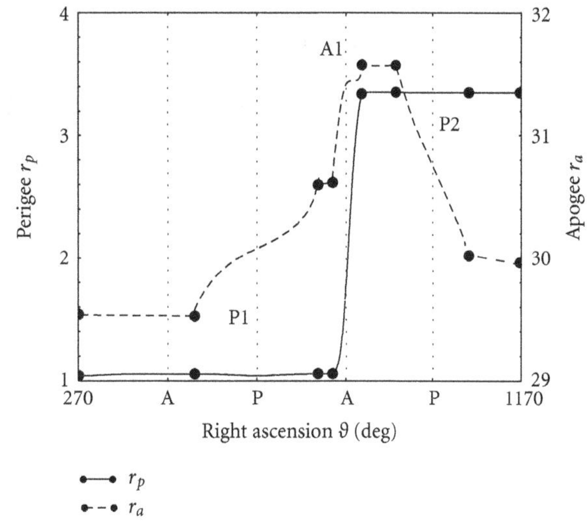

FIGURE 6: Perigee and apogee evolution for 2.5-revolution 1 N transfer (J2 only).

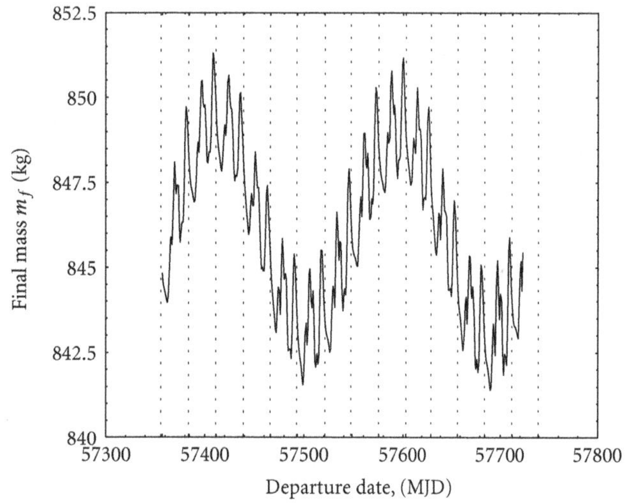

FIGURE 7: Final mass for 4.5-revolution 8 N transfers.

where, in addition, the perturbing acceleration is larger due to the greater Earth-spacecraft distance. Therefore, only this point is considered in the following analysis. An increase of the perigee altitude is the mission main requirement, and the acceleration component parallel to the apogee velocity, that is, the tangential component a_t, is the main cause of the investigated effect. If coplanar orbits are assumed, one easily determines (see Figure 1)

$$a_t = -\frac{\mu_b}{r_b^2}\left[\left(\frac{r_b}{R}\right)^3 - 1\right]\cos\vartheta_b \tag{36}$$

with the spacecraft distance from the perturbing body expressed as

$$R^2 = r_b^2 + r^2 - 2r_b r \sin\vartheta_b. \tag{37}$$

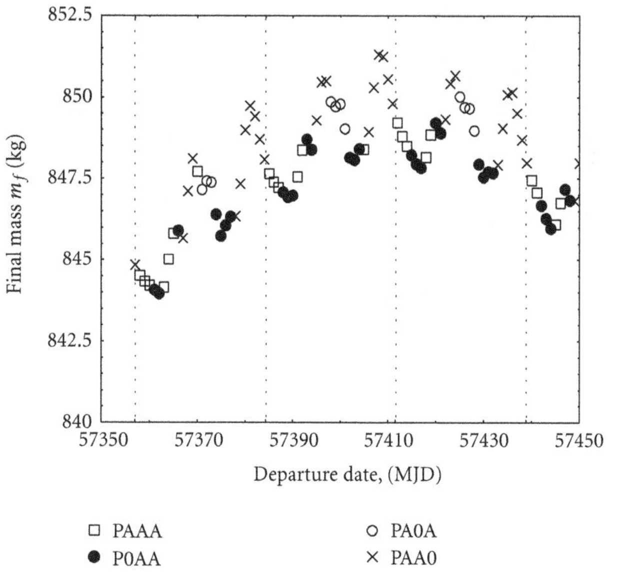

FIGURE 8: Switching structure for 4.5-revolution 8 N transfers.

When the Sun is the perturbing body, $r \ll r_b$ and only first-order terms are retained to obtain $(r_b/R)^3 \approx 1 + 3(r/r_b)\sin\vartheta_b$ and

$$a_t \approx -\frac{\mu_b}{r_b^2}\frac{3}{2}\sin(2\vartheta_b) \tag{38}$$

with maximum positive values at $\vartheta_b = 135$ and 315 deg (the most favorable positions of the Sun) and maximum negative values at $\vartheta_b = 45$ and 225 deg (the most unfavorable positions). If short-period oscillations due to the lunar gravity are ignored (two cycles are found in each sidereal period of the moon), the final mass in Figure 7 closely follows this trend, with two sinusoids in the 1-year launch window. The peak values occur for departure on January 24, 2016, (MJD 57411, right ascensions of the Sun at departure and arrival are 307 and 323 deg) and July 27, 2016 (MJD 57596, 127 and 143 deg).

When the Moon is considered, the spacecraft distance from the Earth becomes comparable to the Earth-Moon distance ($r/r_b \approx 0.5$) and the previous simplification does not hold. The symmetry of the result with respect to the x axis is broken and the effects of the third-body perturbation are enhanced when spacecraft apogees and Moon are on the same side with respect to the Earth, that is, when $\sin\vartheta_b > 0$. The maximum benefit occurs when $\vartheta_b \approx 115$ deg (with a less pronounced beneficial effect at $\vartheta_b \approx 330$ deg), whereas the largest negative effect is at $\vartheta_b \approx 65$ deg (with a less remarkable effect at $\vartheta_b \approx 210$ deg). Figure 9 shows the performance in terms of final mass for the 2.5- and 4.5-revolution transfers, when the Moon is held fixed during the whole maneuver and the solar attraction is neglected. The longest missions exploit significant lunar assists during five apogee passages, two more than in the shortest missions: variations of the final mass with respect to the average value show roughly the same 5/3 ratio.

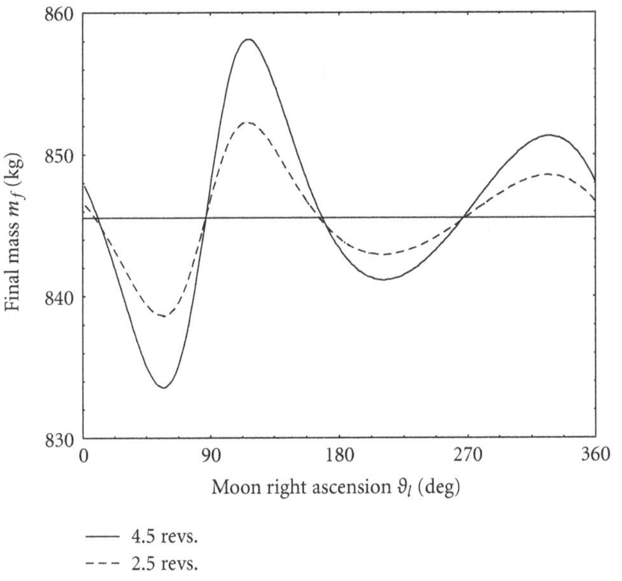

FIGURE 9: Final mass as a function of the Moon fixed position for 2.5- and 4.5-revolution 8 N transfers (Sun neglected).

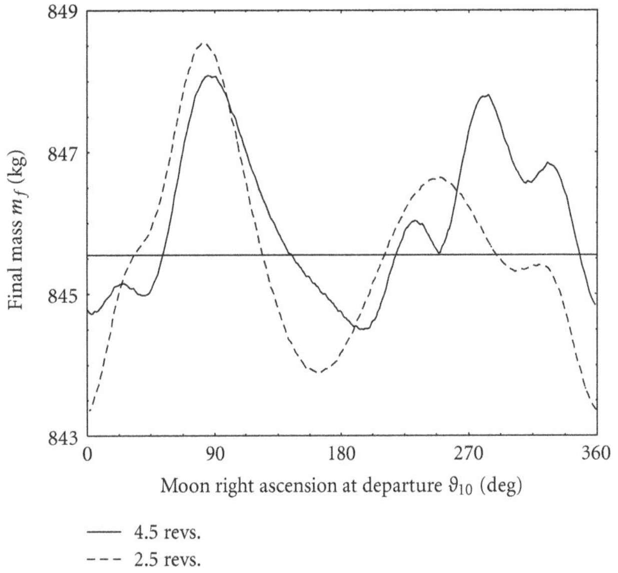

FIGURE 10: Final mass as a function of the Moon initial position for 2.5- and 4.5-revolution 8 N transfers (Sun neglected).

The moon influence becomes more complex when its actual motion is considered. The final mass is shown in Figure 10 as a function of the Moon position at the beginning of the maneuver; the solar gravitational attraction is again neglected. In this case, the final mass presents smaller variations than in the fixed-Moon case (y-axis scale is different from Figure 9). During each spacecraft revolution the Moon moves about 40–55 degrees (changes are due to the Moon eccentricity and to the increase of the spacecraft orbital period caused by apogee burns) and will approximately occupy an unfavorable position two revolutions after a favorable configuration and vice versa.

The Moon is at the most favorable position (ϑ_b = 115 deg) at the first apogee passage of the best 2.5-revolution transfer, offsetting the penalty at the last one (205 deg). On the contrary, the worst performance occurs when the three apogee passages find the Moon at about 20, 70, and 125 degrees, with the most unfavorable configuration at the second passage.

The Moon moves about 210 degrees during the best 4.5-revolution transfer and is in a favorable position at the first (115 deg) and last (about 305 deg) apogee passages (with a single unfavorable position, at about 205 degrees, during the third passage). However, the spacecraft adjusts the burn lengths and varies the orbital period during each revolution to put forward or push back the passages in order to enhance/reduce the effects of favorable/unfavorable geometrical configurations. On average, the final mass is larger in comparison to the shortest transfers, but the maximum achievable mass is lower.

The switching structure of the optimal 4.5-revolution missions (see Figure 8) changes with a clear regularity according to the departure date. The best missions require the removal of the last burn arc (PAA0), whereas the removal of the first arc (P0AA) is beneficial in the worst cases. Optimization prescribes the removal of the last two arcs in four cases, and the switching structure becomes PA00. The same optimal structures repeat at roughly 14-day intervals, corresponding to a half revolution of the Moon around the Earth. For an assigned departure date, the thrust strategy has almost no capability of phasing the initial apogee passage with the Moon position. However, a longer thrust arc at the first apogee increases the total time of flight. On the contrary, when the first apogee burn vanishes, the following orbital periods are shorter and the whole mission is faster. The trip time may differ more than 12 hours (about 6 degree in angular position of the Moon). Figure 11 shows that the mission departing on December 12 delays the last apogee passage to find the Moon in a more favorable position. On the contrary, but with a similar aim, the mission starting on December 5 anticipates the fourth apogee passage. The Moon complex influence on the spacecraft trajectories suggests that further analyses could provide interesting hints for all the missions that exploit lunar resonance.

5.3. Example Case Analysis. The deployment with departure on December 1, 2015, is analyzed in detail to describe the continuation procedure, which progressively increases the perturbation magnitude taken into account. Figure 12 shows an enlargement of the switching function history as a function of the nondimensional variable ε, for different levels of perturbation fraction P_f; this representation is chosen for the sake of clarity, even though it hides the actual arc lengths (provided in Table 3). Thrust would be required at departure and arrival where PMP is not satisfied ($S_f > 0$) but is not allowed for operational reasons. The odd intervals correspond to coast arcs, whereas the second arc is the perigee burn and the other even intervals correspond to three apogee burns.

TABLE 3: Characteristics of the 4.5-revolution transfer with departure on December 1, 2015.

P_f	$t_2 - t_1$ hr	$t_4 - t_3$ hr	$t_6 - t_5$ hr	$t_8 - t_7$ hr	m_f kg
0.0	0.09	2.87	2.83	2.79	845.57
0.2	0.09	3.59	2.75	2.16	845.40
0.4	0.09	4.27	2.79	1.45	845.25
0.6	0.09	4.97	2.96	0.59	845.10
0.8	0.09	5.51	3.02	—	844.96
1.0	0.09	5.84	2.70	—	844.83

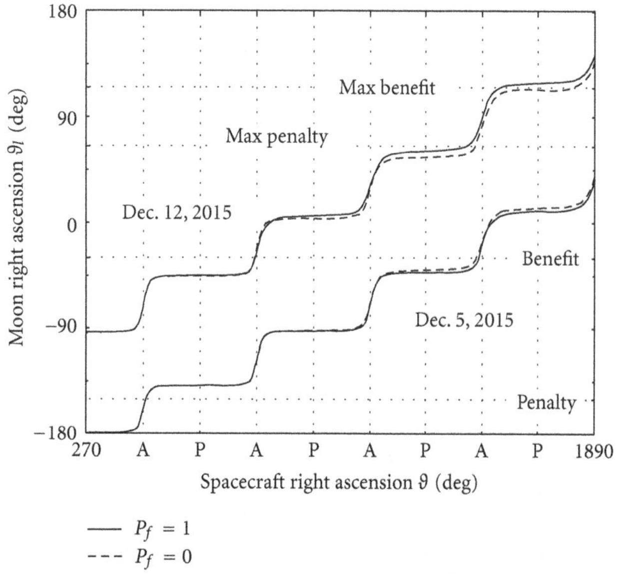

FIGURE 11: Moon angular position during 4.5-revolution transfers.

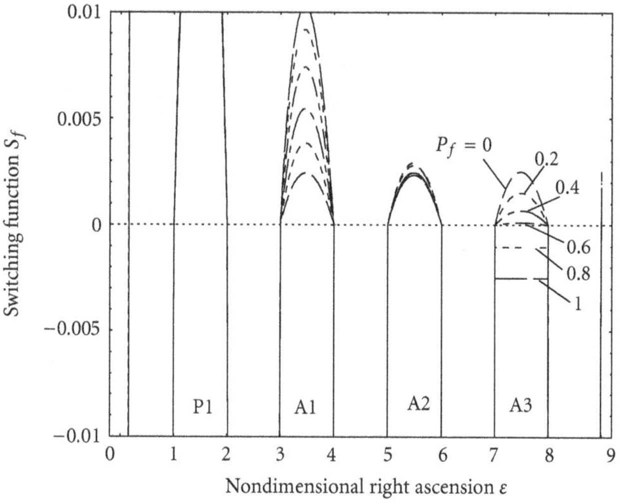

FIGURE 12: Switching function history for 4.5-revolution transfer departing on December 1, 2015.

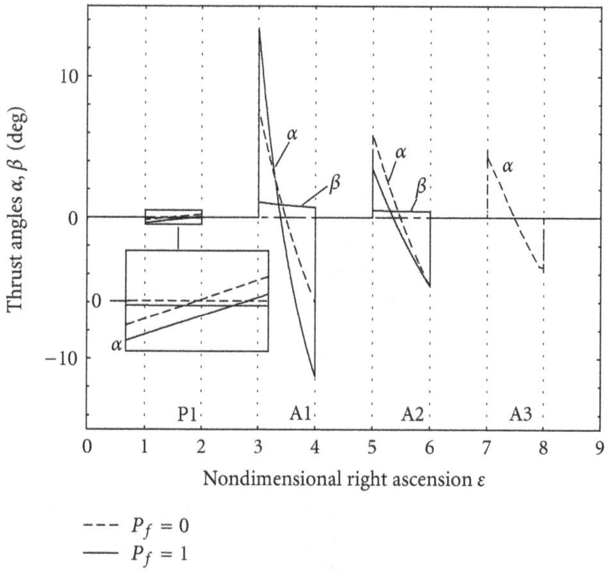

FIGURE 13: Thrust angles of the 4.5-revolution transfer with departure on December 1, 2015.

At the beginning of the stepped procedure, when no perturbation except J2 is considered ($P_f = 0$), the switching function is almost the same during every apogee burn. For this particular departure date, when the lunisolar perturbation is increased, the first apogee burn becomes more convenient, and in this arc the switching function peak grows, while the burn length increases; on the other hand, the last thrust arc is shorter, and there the switching function peak decreases. The switching structure remains the same until $P_f = 0.6$ but provides an unfeasible solution for $P_f = 0.8$ and must be changed. The lowest peak of the switching function for $P_f = 0.6$ occurs in the third apogee burn, which is therefore removed. Convergence assuming the new PAA0 structure is obtained, and PMP confirms the optimality of this solution. Arc time lengths and relevant masses for the December 1 departure are summarized in Table 3.

Figure 13, which refers to the deployment with departure on December 1, 2015 (8 N thrust, 4.5 revolutions), shows the angle β between thrust and orbit plane (positive values tend to increase inclination) and the angle α between the spacecraft velocity and the thrust projection onto the orbit plane (positive values correspond to thrust towards the Earth). When all perturbations are considered, a small out-of-plane thrust component is introduced during each burn in order to slightly change the orbit plane and improve the beneficial effect of the luni-solar gravitational perturbation. In the orbit plane, the thrust is directed inward with respect to the Earth when the spacecraft is moving outward ($0 < \nu < 180$ deg, i.e., after perigee and before apogee) and vice-versa, to reduce the orbit eccentricity in agreement with the variational equations for orbital parameters.

6. Conclusions

The maneuver for the deployment of a satellite into a highly elliptic orbit has been analyzed by means of an indirect optimization method, which handles perturbations from Earth asphericity, lunar and solar gravity, and solar radiation pressure. The optimal thrust strategy depends greatly on the departure date, and is difficult to assess "a priori," especially for multirevolution missions. A continuation scheme based on the gradual introduction of the relevant perturbations is used to converge to the optimal solutions starting from a single initial solution, obtained only considering J2 effect. The switching function is checked during the procedure to maintain the optimal burn structure. The algorithm proves to be fast and reliable and convergence is almost always obtained automatically, without any user's action. Simple adjustments of the procedure parameters may be necessary when, in rare cases, convergence to the optimal solution is not directly obtained.

The availability of a powerful tool permitted a detailed analysis of this kind of deployment. The influence of thrust level and number of revolutions on the transfer performance, that is, mass delivered to the final orbit, has been investigated. The benefit of an increased number of burns and the performance drop that is experienced when a fast transfer is combined with low thrust level have been quantified. As far as the lunisolar perturbation is concerned, the position of the perturbing bodies in correspondence of the spacecraft passages at apogee has a major influence on performance. Due to its slow apparent motion, the Sun position that gives the maximum benefit can be exploited by properly choosing the launch date, without significant changes of the thrust strategy, that is, of the switching structure. On the contrary, the Moon moves rapidly and occupies different positions at the consecutive spacecraft apogee passages; in the search for the optimal maneuver, the time lengths of coast and thrust arcs change (in limit cases, some burn arcs may disappear and the optimal switching structure is modified) in order to find the Moon in the most favorable positions. Results show that the proper selection of the departure date and the use of an optimal strategy, which exploits luni-solar perturbations, can save several kilograms of propellant.

Acknowledgment

This work was supported by the Centre National d' Etudes Spatiales (CNES; Contract no. 93333/00).

References

[1] J. T. Betts, "Survey of numerical methods for trajectory optimization," *Journal of Guidance, Control, and Dynamics*, vol. 21, no. 2, pp. 193–207, 1998.

[2] L. Casalino, G. Colasurdo, and D. Pastrone, "Optimization procedure for preliminary design of opposition-class mars missions," *Journal of Guidance, Control, and Dynamics*, vol. 21, no. 1, pp. 134–140, 1998.

[3] L. Casalino, G. Colasurdo, and D. Pastrone, "Optimal low-thrust escape trajectories using gravity assist," *Journal of Guidance, Control, and Dynamics*, vol. 22, no. 5, pp. 637–642, 1999.

[4] L. Casalino, "Singular arcs during aerocruise," *Journal of Guidance, Control, and Dynamics*, vol. 23, no. 1, pp. 118–123, 2000.

[5] G. Colasurdo and L. Casalino, "Missions to asteroids using solar electric propulsion," *Acta Astronautica*, vol. 50, no. 11, pp. 705–711, 2002.

[6] G. Colasurdo and L. Casalino, "Optimal control law for interplanetary trajectories with nonideal solar sail," *Journal of Spacecraft and Rockets*, vol. 40, no. 2, pp. 260–265, 2003.

[7] L. Casalino and G. Colasurdo, "Optimization of variable-specific-impulse interplanetary trajectories," *Journal of Guidance, Control, and Dynamics*, vol. 27, no. 4, pp. 678–684, 2004.

[8] M. La Mantia and L. Casalino, "Indirect optimization of low-thrust capture trajectories," *Journal of Guidance, Control, and Dynamics*, vol. 29, no. 4, pp. 1011–1014, 2006.

[9] P. Gamet, R. Epenoy, and C. Salcedo, "SIMBOL-X: a formation flying mission on HEO for exploring the universe," in *Proceedings of the 20th ISSFD Conference*, NASA Technical Report Service, Annapolis, Md, USA, September 2007, ID: 20080012679.

[10] I. M. Ross, Q. Gong, and P. Sekhavat, "Low-thrust, high-accuracy trajectory optimization," *Journal of Guidance, Control, and Dynamics*, vol. 30, no. 4, pp. 921–933, 2007.

[11] Y. Gao, "Near-optimal very low-thrust earth-orbit transfers and guidance schemes," *Journal of Guidance, Control, and Dynamics*, vol. 30, no. 2, pp. 529–539, 2007.

[12] X. Yue, Y. Yang, and Z. Geng, "Indirect optimization for finite-thrust time-optimal orbital maneuver," *Journal of Guidance, Control, and Dynamics*, vol. 33, no. 2, pp. 628–634, 2010.

[13] J. A. Kéchichian, "Analytic expansions of luni-solar gravity perturbations along rotating axes for trajectory optimization: part 1: the dynamic system," *Acta Astronautica*, vol. 68, no. 11-12, pp. 1947–1963, 2011.

[14] J. A. Kéchichian, "Analytic expansions of luni-solar gravity perturbations along rotating axes for trajectory optimization-part 2: the multipliers system and simulations," *Acta Astronautica*, vol. 68, no. 11-12, pp. 1914–1930, 2011.

[15] S. Geffroy and R. Epenoy, "Optimal low-thrust transfers with constraints—generalization of averaging techniques," *Acta Astronautica*, vol. 41, no. 3, pp. 133–149, 1997.

[16] D. C. Redding and J. V. Breakwell, "Optimal low-thrust transfers to synchronous orbit," *Journal of Guidance, Control, and Dynamics*, vol. 7, no. 2, pp. 148–155, 1984.

[17] R. Bertrand and R. Epenoy, "New smoothing techniques for solving bang-bang optimal control problems—numerical results and statistical interpretation," *Optimal Control Applications and Methods*, vol. 23, no. 4, pp. 171–197, 2002.

[18] T. Haberkorn, P. Martinon, and J. Gergaud, "Low-thrust minimum-fuel orbital transfer: a homotopic approach," *Journal of Guidance, Control, and Dynamics*, vol. 27, no. 6, pp. 1046–1060, 2004.

[19] J. B. Thevene and R. Epenoy, "Minimum-fuel deployment for spacecraft formations via optimal control," *Journal of Guidance, Control, and Dynamics*, vol. 31, no. 1, pp. 101–112, 2008.

[20] J. C. H. Chuang, T. D. Goodson, and J. Hanson, "Multiple-burn families of optimal low- and medium-thrust orbit transfers," *Journal of Spacecraft and Rockets*, vol. 36, no. 6, pp. 866–874, 1999.

[21] L. Casalino, D. Pastrone, and G. Colasurdo, "Integrated design of hybrid rocket upper stage and launcher trajectory," in *Proceedings of the 45th AIAA/ASME/SAE/ASEE Joint Propulsion Conference and Exhibit*, Denver, Colo, USA, August 2009, Paper AIAA 2009-4843.

[22] A. Zavoli, F. Simeoni, L. Casalino, and G. Colasurdo, "Optimal cooperative deployment of a two-satellite formation into a highly elliptic orbit," in *Proceedings of the 2011 AAS/AIAA Astrodynamics Specialist Conference*, Girdwood, Alaska, USA, August 2011, Paper AAS 11-641.

[23] Anonymous, EGM 2008 Model coefficients-Original Release, National Geospatial-Intelligence Agency, http://www.xmarks.com/site/earth-info.nga.mil/GandG/wgs84/gravitymod/egm2008/egm08%5C_wgs84.html, 2011.
[24] Anonymous, "World geodetic system 1984, its definition and relationships with local geodetic systems," NIMA Technical Report TR8350.2, Department of Defense, Third Edition, National Geospatial-Intelligence Agency, Washington, DC, USA, 2000.
[25] "Legendre functions," in *Handbook of Mathematical Functions with Formulas, Graphs, and Mathematical Tables, 9th Printing*, M. Abramowitz and I. A. Stegun, Eds., chapter 8, pp. 331–339, Dover, New York, NY, USA, 1972.
[26] "Orthogonal polynomials," in *Handbook of Mathematical Functions with Formulas, Graphs, and Mathematical Tables, 9th Printing*, M. Abramowitz and I. A. Stegun, Eds., chapter 22, pp. 771–802, Dover, New York, NY, USA, 1972.
[27] W. Bosch, "On the computation of derivatives of legendre functions," *Physics and Chemistry of the Earth A*, vol. 25, no. 9–11, pp. 655–659, 2000.
[28] http://ssd.jpl.nasa.gov/?planetephexport, 2011.
[29] A. E. Bryson and Y.-C. Ho, *Applied Optimal Control*, Hemisphere Publications, Washington, DC, USA, 1975.
[30] D. F. Lawden, *Optimal Trajectories for Space Navigation*, Butterworths, London, UK, 1963.
[31] G. Colasurdo and D. Pastrone, "Indirect optimization method for impulsive transfer," in *Proceedings of the AIAA/AAS Astrodynamics Conference*, pp. 558–563, Scottsdale, Ariz, USA, August 1994, Paper AIAA 94-3762.
[32] G. A. Hazelrigg, "Globally optimal impulsive transfers via Green's theorem," *Journal of Guidance, Control, and Dynamics*, vol. 7, no. 4, pp. 462–470, 1984.

4

Boron Particle Ignition in Secondary Chamber of Ducted Rocket

J. X. Hu, Z. X. Xia, W. H. Zhang, Z. B. Fang, D. Q. Wang, and L. Y. Huang

College of Aerospace and Materials Engineering, National University of Defense Technology, Changsha 410073, China

Correspondence should be addressed to Z. X. Xia, xiazhixun@nudt.edu.cn

Academic Editor: Valsalayam Sanal Kumar

In the secondary chamber of ducted rocket, there exists a relative speed between boron particles and air stream. Hence, the ignition laws under static conditions cannot be simply applied to represent the actual ignition process of boron particles, and it is required to study the effect of forced convective on the ignition of boron particles. Preheating of boron particles in gas generator makes it possible to utilize the velocity difference between gas and particles in secondary chamber for removal of the liquid oxide layer with the aid of Stoke's forces. An ignition model of boron particles is formulated for the oxide layer removal by considering that it results from a boundary layer stripping mechanism. The shearing action exerted by the high-speed flow causes a boundary layer to be formed in the surface of the liquid oxide layer, and the stripping away of this layer accounts for the accelerated ignition of boron particles. Compared with the King model, as the ignition model of boron particles is formulated for the oxide layer removal by considering that it results from a boundary layer stripping mechanism, the oxide layer thickness thins at all times during the particle ignition and lower the ignition time.

1. Introduction

Systematic advances in missile propulsion systems technology have provided large increases in missile performance capabilities. Replacement of much of solid propellant rocket oxidizer with free stream air, as in the ducted rocket concept, offers over 5 to 1 increase in missile range capability [1]. Boron has been considered for many years as a prime candidate used for increasing the ducted rocket capabilities based on its high potential energy release on both a volumetric and gravimetric basis coupled with a high energy of combustion, high combustion temperature, and low-molecular-weight products [2]. These properties make boron an attractive material for use in ducted rocket propellants [3]. In order for these advantages to be realized, however, the boron particles must ignite and burn completely within a very limited residence time. Since boron particles are generally initially coated with an oxide layer with inhibits combustion and since boron has an extremely high boiling point, which necessitates surface burning subsequent to oxide removal, this can become difficult, particularly under adverse operating conditions [4].

In previous work, two scenarios for boron ignition and combustion have been identified, which differ in the controlling element in the overall process mechanism [5]. Two controversial theories remain to be resolved for describing boron ignition (removal of oxide coating) in terms of the rate limiting steps [4, 6, 7]. Previous experiments of Macek and semple [8] and Li and Williams [7] have demonstrated that when the boron particle temperature is high (e.g., 1800 K), the rate of boron diffusion exceeds the rate of oxygen diffusion across the oxide layer. Recent experimental observations indicating significant agglomeration and liquefaction of boron at 1213 K in both O_2 and Ar environments have also been interpreted as evidence supporting the diffusion of boron instead of oxygen within the oxide coating as the limiting process [3, 9].

Because ignition and combustion of boron particles involve the release of heat, layer evolution assumes the form of a thermo-hydrodynamic process, coupled with the solid boron substrate underneath and with the ambient oxidizing atmosphere, which is assumed to be stagnant [5]. It is important to note that most the current models of boron particles ignition constrain the symmetry to be conserved. That is to say that the oxidizing atmosphere is assumed to be stagnant and uniform if the particle and its initial oxide layer are given to be spherically symmetric. In the secondary combustion chamber of ducted rocket, there always exists a relative speed between the boron particle and air stream [10]. Hence, the ignition laws under static conditions cannot be simply applied to represent the actual ignition process of boron particles, and it is required to study the effect of forced convective on the ignition of the boron particle. Much research has been conducted to study the ignition of the boron particle, both experimentally and theoretically, in pure oxygen and other atmospheres, with relatively little work available on ignition of the boron particle under forced convection conditions [11, 12]. Dirk Meinkohn developed a thin model to allow the oxide flow field to be derived in the creeping-flow approximation and demonstrated that boron particle ignition may then be caused by the Marangoni effect, which is shown to entail the spreading of punctures and ruptures in the oxide layer, leading to layer thinning or even complete layer removal [13, 14]. Preheating of the boron particles in gas generator makes it possible to utilize the velocity difference between gas and particles in secondary chamber for removal of the liquid oxide layer with the aid of Stokes' forces. Povitsky and Goldman demonstrated there is indication of regime of fast ignition enhanced by Stokes' forces by experimentally and numerical simulation [15].

The object of this study is to investigate the oxide layer movement and to characterize its influence on ignition of boron particles in the case of high initial relative particle velocity. An ignition model of boron particles is formulated for the oxide layer removal by considering that it results from a boundary layer stripping mechanism.

2. Formulation

2.1. Boundary Layer Stripping Analysis

2.1.1. Assumptions. The movement of the oxide layer on the front half-surface of the boron particle is shown schematically in Figure 1, using a spherical coordinate system with the particle center as origin. x is the curvilinear coordinate along the interface separating the two fluids and y is the coordinate perpendicular to it. $y = 0$ represents the interface between the gas layer and liquid oxide one. $y = +\infty$ represents the value of gas boundary layer. $y = -\infty$ represents the value of liquid boundary layer. u_g, u_l represent fluid velocity in gas or liquid boundary layer. R is the radius of the boron particle. X is the thickness of liquid oxide layer. U_∞ represents gas freestream velocity. $U(x)$ represents gas fluid velocity around the boron particle.

In order to solve the problem of liquid oxide layer movement, the following assumptions are adopted.

(1) The temperature inside and around the boron particle is uniformity.

(2) The flow is steady and incompressible.

(3) There is no flow separation at the interface between solid particle and liquid oxide layer.

(4) The value of liquid oxide boundary layer is less than the thickness of liquid oxide layer.

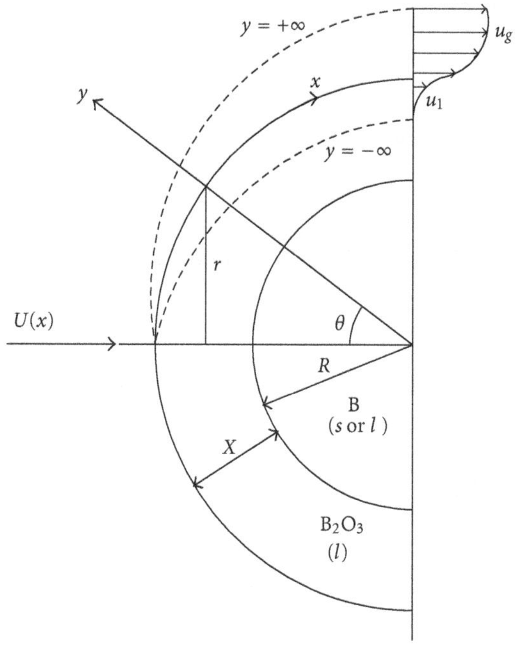

FIGURE 1: Scheme of oxide layer movement relative to surface of flowing boron particles.

(5) The boron particle shape can be approximated by a sphere for the simplified analysis here, the oxide layer thickness is much smaller than particle radius, and hence the gas fluid velocity around the boron particle can be represented in the following expression:

$$U(x) = 1.5\ U_\infty \sin\frac{x}{R}. \tag{1}$$

(6) Equating the shear stress in the gas layer to that in the liquid layer at the interface yields the following equation:

$$\left(\mu_g \frac{\partial u_g}{\partial y}\right)_{y=0} = \left(\mu_l \frac{\partial u_l}{\partial y}\right)_{y=0}. \tag{2}$$

(7) The pressure gradient in the gas layer and the liquid layer is equal:

$$\left(\frac{\partial P}{\partial x}\right)_l = \left(\frac{\partial P}{\partial x}\right)_g. \tag{3}$$

2.1.2. Equations and Solution. An approximate solution to the two-boundary-layer problem can be obtained by assuming arbitrary simple velocity distributions containing a few parameters and the using the momentum differential relations to determine three parameters [16]. If we assume that the flow is steady and incompressible, then it yields the boundary-layer differential equations for the gas and for the liquid.

The basic equations for the gas is given by

$$\frac{\partial(u_g r)}{\partial x} + \frac{\partial(v_g r)}{\partial y} = 0,$$
$$u_g \frac{\partial u_g}{\partial x} + v_g \frac{\partial u_g}{\partial y} = -\frac{1}{\rho_g}\frac{\partial P}{\partial x} + \frac{\mu_g}{\rho_g}\frac{\partial^2 u_g}{\partial y^2}. \tag{4}$$

The boundary conditions for equations above is given by

$$y = 0 : v_g = 0, \quad u_g = u_g(x), \quad \mu_g \frac{\partial u_g}{\partial y} = \mu_l \frac{\partial u_l}{\partial y},$$
$$y = \infty : u_g = U(x), \tag{5}$$

where the pressure gradient in the gas boundary layer is given by

$$-\frac{1}{\rho_g}\frac{\partial P}{\partial x} = U\frac{\partial U}{\partial x}. \tag{6}$$

Then (4) can be given by

$$u_g \frac{\partial u_g}{\partial x} + v_g \frac{\partial u_g}{\partial y} = U\frac{\partial U}{\partial x} + \frac{\mu_g}{\rho_g}\frac{\partial^2 u_g}{\partial y^2}. \tag{7}$$

The basic equations for the liquid are given by

$$\frac{\partial(u_l r)}{\partial x} + \frac{\partial(v_l r)}{\partial y} = 0, \tag{8}$$

$$u_l \frac{\partial u_l}{\partial x} + v_l \frac{\partial u_l}{\partial y} = \frac{\rho_g}{\rho_l}U\frac{\partial U}{\partial x} + \frac{\mu_l}{\rho_l}\frac{\partial^2 u_l}{\partial y^2}. \tag{9}$$

The boundary conditions for equations above are given by

$$y = 0 : v_l = 0, \quad u_l = u_g(x), \quad \mu_l \frac{\partial u_l}{\partial y} = \mu_g \frac{\partial u_g}{\partial y},$$
$$y = -\infty : \mu_l \frac{\partial u_l}{\partial y} = 0. \tag{10}$$

In order for equations solution, the stream function, ψ, is defined by

$$u = \frac{L}{r}\frac{\partial \psi}{\partial y}, \qquad v = -\frac{L}{r}\frac{\partial \psi}{\partial x}, \tag{11}$$

where L denotes the latent length.

Some normalized variable parameters are defined by

$$\xi = \int_0^x \frac{r^2(x)U(x)}{L^2 U_\infty}dx, \qquad \eta = \frac{yU(x)r(x)}{L\sqrt{2\mu_g U_\infty \xi/\rho_g}},$$
$$\psi = G(\eta)\sqrt{\frac{2\mu_g U_\infty \xi}{\rho_g}}. \tag{12}$$

Then the equations for the gas can be simplified as

$$G''' + GG'' + \beta(\xi)\left(1 - G'^2\right) = 0, \tag{13}$$

where $\beta(\xi) = 2\xi U_\infty((L^2/r^2)dU/dx(1/U^2))$.

The boundary conditions for equations above can be simplified as

$$\eta = 0 : G = 0, \quad G' = \overline{u}(x),$$
$$\eta = \infty : G' = 1. \tag{14}$$

Then the equations for the liquid can be simplified as

$$L''' + LL'' + \beta(\xi)\left(\frac{\rho_g}{\rho_l} - L'^2\right) = 0. \tag{15}$$

The boundary conditions for equations above can be simplified as

$$\eta = 0 : L = 0, \quad L' = G', \quad L'' = G''\Gamma,$$
$$\eta = -\infty : L'' = 0, \tag{16}$$

where $\Gamma = \sqrt{\rho_g \mu_g/(\rho_l \mu_l)}$, gas viscosity $\mu_g = 6.2 \times 10^{-5}$ Pa · s, liquid viscosity $\mu_l = 3.2 \times 10^{-4}$ Pa · s.

After what may seem like an interminable manipulation of the governing equations, we have finally set up that particular form of the equations that will be most appropriate as well as convenient for the gas and liquid boundary layer flow. In order to solve (13) and (15), $G'(0)$ and $G''(0)$ should be first got. According to literature [17], $G'(0) = \overline{u}(x) = (\rho_g \mu_g/\rho_l \mu_l)^{1/3}$. Firstly, $G''(0)$ is given or estimated, the classical four-order Runge-Kutta method is used to solve these ordinary differential equations. If the obtained boundary values satisfy the actual boundary conditions at the outer edge; namely, $G'(+\infty) = 1$ and $L''(-\infty) = 0$, the calculation can be stopped, and the solutions have been found. Otherwise, we have to revise the estimated value of $G''(0)$ and repeat the numerical integrating procedure until all the boundary conditions at the outer edge are satisfied.

2.1.3. The Conditions of Aerodynamic Shattering. During aerodynamic shattering, the tangential friction stress is propitious to sweep the liquid oxide layer from solid particle, and the surface tension is unpropitious to sweep the liquid oxide layer.

The tangential friction stress can be expressed by

$$\tau = \left(\mu_g \frac{\partial u_g}{\partial y}\right)_{y=0} = \mu_g \frac{U^2 r G''(0)}{L\sqrt{2v_g U_\infty \xi}}. \tag{17}$$

For a spherical particle, the surface tension can be expressed by

$$F = \frac{C_1 \sigma}{R}. \tag{18}$$

According to literature [15], the constant $C_1 = 2.0$ and the coefficient of surface tension $\sigma = 0.01$.

We and Re can be expressed by

$$\text{We} = \frac{\rho_g u_r^2 d_p}{\sigma}, \qquad \text{Re} = \frac{\rho_g u_r d_p}{\mu_g}, \tag{19}$$

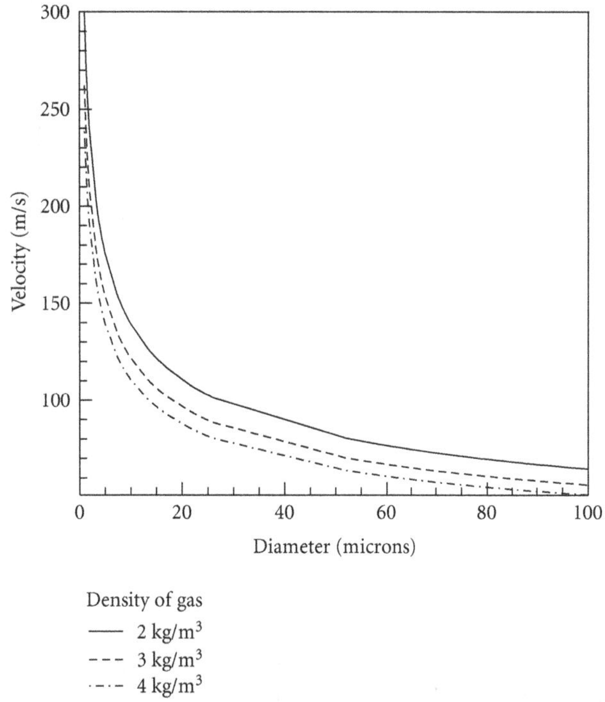

FIGURE 2: Critical velocity versus particle diameter for various gas densities.

where $u_r = U_\infty - W$ represents the relative velocity between gas flow and boron particles. W represents the velocity of boron particles. d_p represents the diameter of boron particles.

While the surface tension is less than the tangential friction stress, aerodynamic shattering may happen. That is to say, part of liquid oxide layer is detached from the particle if the following equation comes into existence:

$$\frac{\mathrm{We}}{\sqrt{\mathrm{Re}}} \geq \frac{8}{9}\frac{C_1}{G''(0)}. \tag{20}$$

The critical velocity $u_{r\min}$ can be reduced by the relation

$$u_{r\min} = \left(\frac{64C_1^2\sigma^2}{81G''(0)\rho_g\mu_g d_p}\right)^{1/3}. \tag{21}$$

Beyond $u_{r\min}$, part of liquid oxide layer is detached from the particle and it is shifted to the rear. The formula above permits determination of the range of critical velocities for boron particle coated by a thin liquid oxide layer. The critical velocity versus particle diameter for various gas densities is shown in Figure 2. It may be seen that it is difficult for aerodynamic shattering to happen when the particle diameter is less than 15 microns.

The mass of fluid in the circumferential liquid oxide layer being swept along by the gas steam at a distance $x = \pi(R + X)/2$ from the stagnation point is

$$\frac{dm}{dt} = \pi d_p\rho_l\int_{-\infty}^{0} u_l\, dy = \pi d_p\rho_l\sqrt{\frac{d_p u_r\mu_l}{\rho_l}}\int_{-\infty}^{0} L'(\eta)d\eta. \tag{22}$$

We can assume that the thickness of liquid oxide layer of boron particle is still uniformity along the uniformity after aerodynamic shattering. The following formula can be reduced

$$\frac{dm}{dt} = \rho_l\pi d_p^2\frac{dX}{dt}. \tag{23}$$

The change in the liquid oxide layer of boron particles due to aerodynamic shattering can be reduced:

$$\frac{dX}{dt} = \sqrt{\frac{u_r\mu_l}{(d_p\rho_l)}}\int_{-\infty}^{0} L'(\eta)\, d\eta. \tag{24}$$

2.2. Ignition Model of Boron Particle in High-Speed Flow. The results obtained above are used for developing an ignition model of boron particles, adopting a spherical form for the particles. For nonspherical particles, the aerodynamic shattering of liquid oxide layer is the same in principle, but the relevant expressions are more complicated.

In the analysis below we resort to King's model, based on the assumption that the exothermic reaction between boron and oxygen takes place at the boron-boron oxide interface as a result of oxygen diffusion across the melting oxide layer [4, 15]. As the oxygen converts the boron into boric oxide, thickens the oxide layer, and releases heat, the oxide simultaneously vaporizes at the outer surface, this second process is endothermic and reduces the heating rate. Heat is added to or removed from the particle by convection and radiation, while inside the particle it is transferred by conduction from the surface to the center. Ignition occurs when the liquid oxide layer at the leading point of the particle is completely removed by the tangential friction stress effects [4, 15].

Step 1.

$$\mathrm{B(s)} + \frac{3}{4}\mathrm{O_2} \longrightarrow \frac{1}{2}\mathrm{B_2O_3(l)} + Q_{RX},$$
$$\mathrm{B(l)} + \frac{3}{4}\mathrm{O_2} \longrightarrow \frac{1}{2}\mathrm{B_2O_3(l)} + Q_{RX2}, \tag{25}$$

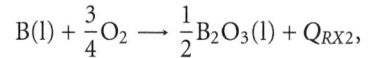

where $Q_{RX} = 6.132 \times 10^5$ J/mole, $Q_{RX2} = 6.342 \times 10^5$ J/mole.

$$w_{\mathrm{B}} = \frac{64.8 \times 10^{-8}(R+X)^2 T_p\, P_{\mathrm{O_2}} e^{-22600/T_p}}{X}\ (\text{mole/s}), \tag{26}$$

where w_{B} is the molar consumption rate of boron.

Step 2.

$$\mathrm{B_2O_3(l)} \longrightarrow \mathrm{B_2O_3(g)} - \Delta H_{\mathrm{vap}}, \tag{27}$$

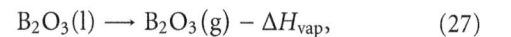

$$w_E = \frac{1.005 \times 10^{10}(R+X)^2\alpha e^{-44000/T_p}}{T_p^{0.5}\left(1 + 4.5 \times 10^7\alpha P(R+X)/\left(T_p\mathrm{Nu}\right)\right)}\ (\text{mole/s}), \tag{28}$$

where w_E is the molar rate of vaporization of boron, $\Delta H_{\mathrm{vap}} = 3.78 \times 10^5$ J/mole.

Step 3.

$$B_2O_3(l) + H_2O \longrightarrow 2HBO_2 - \Delta H_H, \quad (29)$$

$$Ew_H = 9.15 \times 10^{-7} \frac{Nu}{P} (R+X) T_p^{1/2} \exp\left[18.1\left(1 - \frac{2100}{T_p}\right)\right] \times \left\{ -0.15 + \left[0.0225 + P_{H_2O} \times \exp\left(-18.1\left(1 - \frac{2100}{T_p}\right)\right)\right]^{0.5} \right\} (\text{mole/s}), \quad (30)$$

where w_H is the molar rate of removal of boric oxide, $\Delta H_H = 3.15 \times 10^5$ J/mole. The equation for particle radius and oxide layer thickness dynamic can be written as

$$\frac{dR}{dt} = -\frac{w_B M_B}{4\pi R^2 \rho_B}, \quad (31)$$

$$\frac{dX}{dt} = \frac{(w_B/2 - w_E - w_H) M_{B_2O_3}}{4\pi R^2 \rho_{B_2O_3}} - \sqrt{\frac{u_r \mu_l}{(d_p \rho_l)}} \int_{-\infty}^{0} L'(\eta) d\eta. \quad (32)$$

The first term on the right-hand side of formula (33) is due to King's model, and the second refers to our model of liquid oxide layer movement. For u_r less than $u_{r\,\min}$, the second term vanishes. The heat balance of liquid oxide layer is described by the equations (note that three different enthalpy balances must be employed depending on whether the particle temperature is less than, equal to, or greater than the melting point of boron, 2450 K):

$$\frac{dT_p}{dt} = \frac{Q_1}{\left(4\pi R^3 \rho_{B c_{pBs}}/3\right) + 4\pi R^2 X \rho_{B_2O_3} c_{pB_2O_3}} \quad \left(T_p < 2450\,\text{K}, f = 0\right),$$

$$\frac{df}{dt} = \frac{Q_1}{4\pi R^3 \rho_B \Delta H_M / 3} \quad \left(T_p = 2450\,\text{K}, 0 < f < 1\right), \quad (33)$$

$$\frac{dT_p}{dt} = \frac{Q_2}{\left(4\pi R^3 \rho_B c_{pBl}/3\right) + 4\pi R^2 X \rho_{B_2O_3} c_{pB_2O_3}} \quad \left(T_p > 2450\,\text{K}, f = 1\right),$$

where f is fraction of boron in the liquid phase, T_0 is free stream gas temperature, T_p is particle temperature, T_{RAD} is surrounding radiation temperature, Q_1, Q_2 are heat transfer between particle and the surroundings, $Q_1 = w_B Q_{RX} - w_E \Delta H_{vap} - w_H \Delta H_H + 4\pi(R+X)^2 [h(T_\infty - T_p) + \sigma\varepsilon\alpha_R(T_{RAD}^4 - T_p^4)]$, $Q_2 = w_B Q_{RX2} - w_E \Delta H_{vap} - w_H \Delta H_H + 4\pi(R+X)^2 [h(T_\infty - T_p) + \sigma\varepsilon\alpha_R(T_{RAD}^4 - T_p^4)]$. The following values were used for other parameters appearing in equations above: $M_B = 10.82$, $\rho_B = 2.33 \times 10^3\,\text{kg/m}^3$, $M_{B_2O_3} = 69.64$, $\rho_{B_2O_3} = 1.85 \times 10^3\,\text{kg/m}^3$, $\Pr = 0.72$, $\varepsilon = 0.8$, $\alpha_R = 1$, $\Delta H_M = 2.1 \times 10^3$ J/mole, $h = 1.46 \times 10^{-2} \times \text{Nu} \times T_\infty^{0.8}/(R+X)$ J/(m$^2 \cdot$ s $\cdot$ K), $\text{Nu} = 2 + 0.6\,\text{Re}^{1/2}\Pr^{1/3}$, $c_{pBs} = 2130 + 0.294\,T_p$ J/(kg $\cdot$ K), $c_{pBl} = 2.84 \times 10^3$ J/(kg $\cdot$ K), $c_{pB_2O_3} = 1.84 \times 10^3$ J/(kg $\cdot$ K).

3. Results and Discussion

The model of boron ignition developed in this study involves nine independent parameters, value of which must be input to the resulting numerical computer program for prediction of particle ignition time and minimum gas temperature required for particle ignition [18]. There variables are relative velocity between gas flow and boron particles, initial oxide thickness, initial particle temperature, initial particle size, ambient temperature (surroundings gas temperature), effective surroundings radiation temperature, ambient pressure, water vapor mole fraction, and oxygen mole fraction. As part of this study, each of the above independent variables has been systematical varied to determine its effect on whether the particle will ignite and, if so, what the ignition time will be.

3.1. Effect of Ambient Temperature. In Figure 3 plots of oxide layer thickness, particle temperature, particle radius, and fraction boron melted versus time presented for a typical case in which the numerical analysis predicts particle ignition. The particle treated is 10 microns in radius with an initial oxide layer thickness of 0.1 micron and initial temperature of 1800 K. The pressure is 0.5 MPa, the oxygen mole fraction is 0.2, the water vapor mole fraction is zero, and the relative velocity between gas flow and boron particles is 100 m/s. The ambient temperature and effective surroundings radiation temperature are both 2100 K.

As may be shown in Figure 3, once the particle temperature exceeds approximately 2000 K, the oxide evaporation and the change due to aerodynamic shattering rise above the oxide generation rate, resulting in thinning of the oxide layer. The boron particle temperature continue to increase as the oxide layer thins while the boron melting point is reached. At this point the oxide thickness and particle temperature remain constant until the boron melts. After the boron particle has melted, the particle temperature resumes its rise and the oxide layer continues to thin until the particle ignite. Compared with King's model, as the ignition model of boron particles is formulated for the oxide layer removal by considering that it results from a boundary layer stripping mechanism, the oxide layer thickness thins at all times during the particle ignition and lower the ignition time.

In Figure 4, similar results are presented for a case identical except for reduction of the ambient temperature and effective surroundings radiation temperature by 200 K to

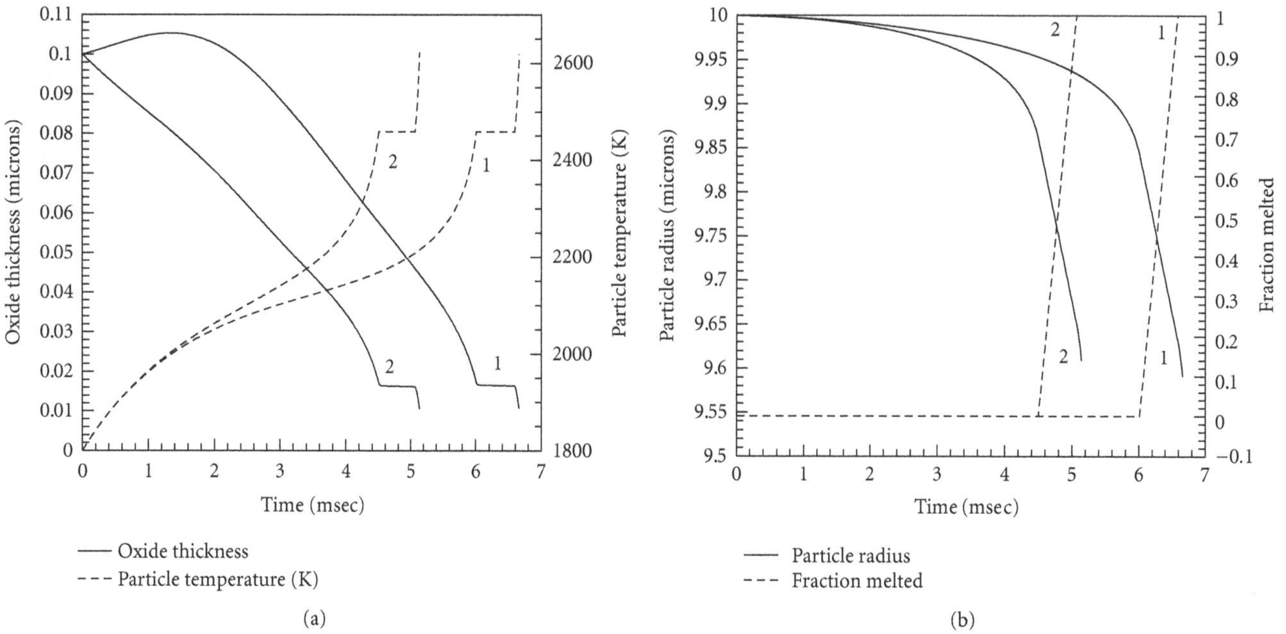

FIGURE 3: Predicted time dependence of important variables for particle which ignites at ambient temperature of 2100 K ((a) King's model (b) our model with film moving).

1900 K. The initial stages of the ignition process are quite similar. However, heat losses to the surroundings when the particle temperature rises above the 1900 K ambient temperature are sufficient to retard the evaporation rate sufficiently that it drops back to the generation rate with the result that the oxide layer ceases thinning. In this case, a stable quasi-steady-state is reached. The boron particle temperature remains a steady value, and the ignition does not occur.

3.2. Effect of Total Pressure and Oxygen Mole Fraction. The effects of total pressure and oxygen mole fraction in the gas-free stream on minimum gas temperature required for particle ignition and ignition time are shown in Figures 5 and 6. These calculations were all performed for particles of 10 microns radius with an initial temperature of 1800 K and initial oxide thicknesses of 0.1 microns. The calculations were carried out for a 100 m/s gas stream in the surroundings with water vapor mole fraction of zero at the conditions that ambient temperature and effective surroundings radiation temperature are both 2100 K. Similar trends were noted with other particle sizes and initial oxide thicknesses.

As shown in Figure 5, decreased total pressure lowers the gas temperature required for ignition at low oxygen mole fractions but raises the required gas temperature at high oxygen mole fractions. These fairly complex dependencies result from interaction of the effect of oxygen mole fraction on the oxide generation rate and the effect of total pressure on the oxide removal rate. Increase of the total pressure intensified oxygen diffusion through the oxide layer but reduced the oxide vaporizing rate. It also leads to increased gas density, with the attendance opposite influence on tangential friction stress [15]. Similar effects on the ignition time at a fixed gas temperature and velocity as functions of total pressure and oxygen mole fraction are shown in Figure 6. At high oxidizer mole fractions, ignition time reduces with reducing total pressure while at low oxidizer mole fractions the pressure dependency is reversed.

3.3. Effect of Particle Radius. The effect of particle radius on ignition time was also studied. In Figure 7, predictions of ignition time versus particle radius for particle which are predicated to ignite are presented for several gas temperatures ranging from 2050 K to 2200 K, and the radiation surroundings temperature is set equal to ambient gas temperature. The particles (radius size range from 5 to 20 microns with an initial oxide layer thickness of 0.1 micron) were assumed to be preheated to 1800 K prior to ejection gas stream. The pressure is 0.5 MPa, the oxygen mole fraction is 0.2, the water vapor mole fraction is zero, and the relative velocity between gas flow and boron particles is 100 m/s. As shown in Figure 7, the ignition time decreases monotonically with decreasing particle radius due to the decrease in particle mass-to-surface area ratio with decreasing radius. For the higher ambient gas temperature, ignition time is shorter.

3.4. Effect of Water Vapor Mole Fraction. The effect of water vapor mole fraction in the gas-free stream on ignition time is shown in Figure 8. These calculations were all performed for particles of 10 microns radius with an initial temperature of 1800 K and initial oxide thicknesses of 0.1 microns. The pressure is 0.5 MPa, the oxygen mole fraction is 0.2, and the relative velocity between gas flow and boron particles is 100 m/s. The ambient temperature and effective surroundings radiation temperature are both 2100 K. As shown in Figure 8, the ignition time decreases monotonically with increasing water vapor mole fraction due to treating water

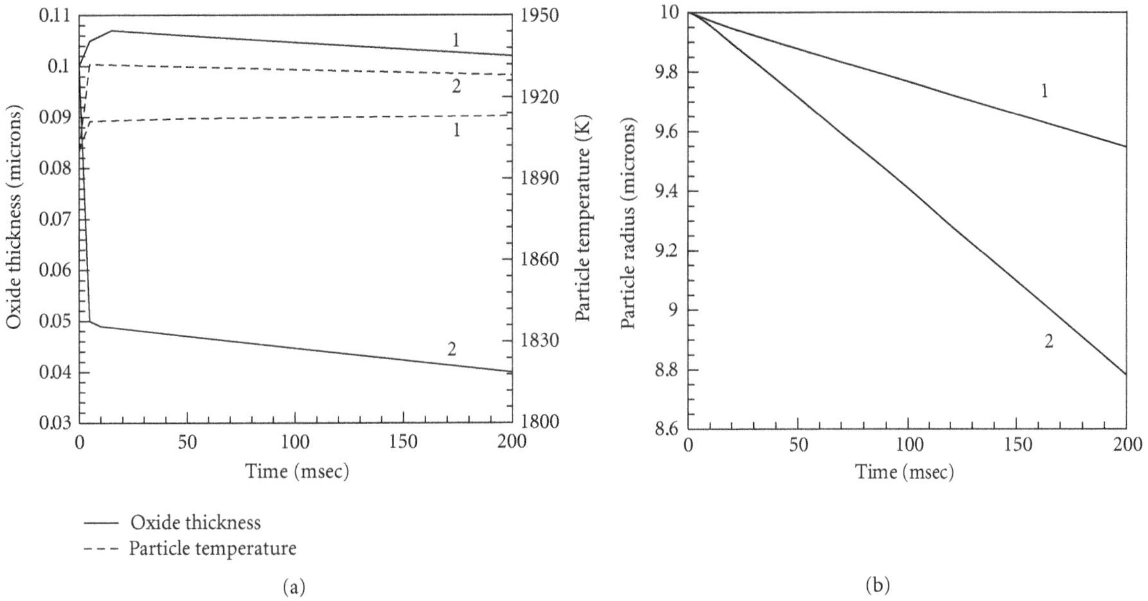

FIGURE 4: Predicted time dependence of important variables for particle which ignites at ambient temperature of 1900 K (a) King's Model (b) our model with film moving).

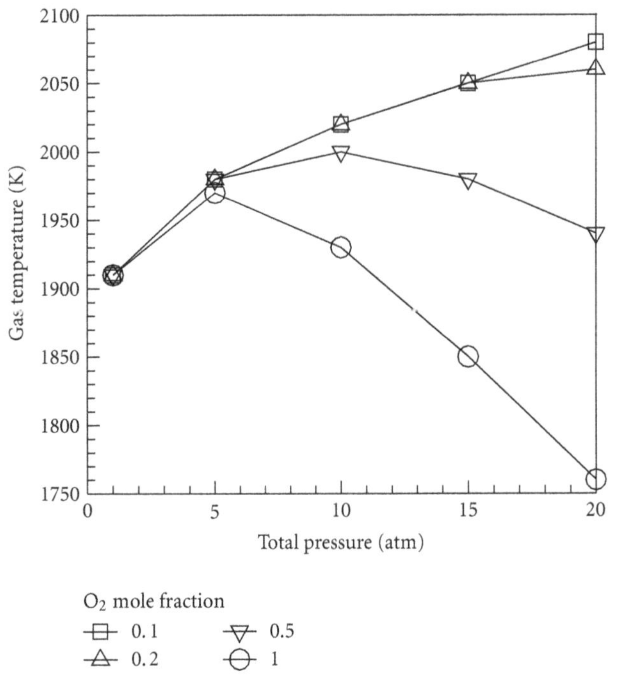

FIGURE 5: Minimum gas temperature required for ignition versus total pressure for various oxygen mole fractions.

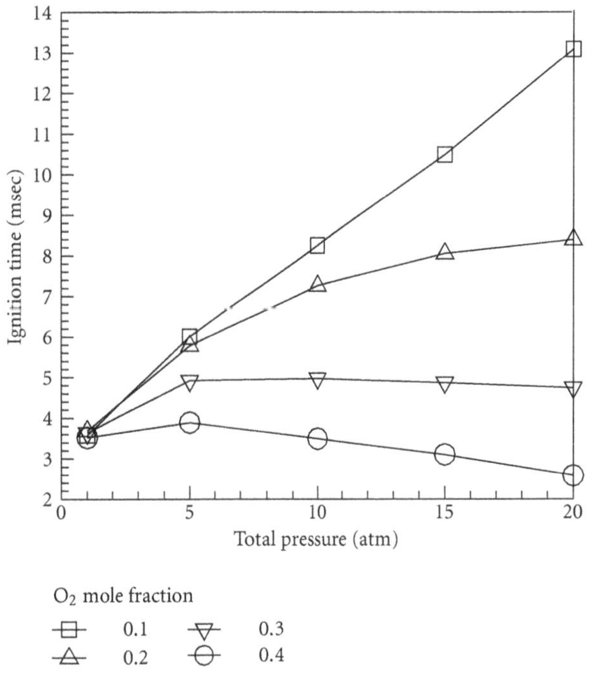

FIGURE 6: Ignition time versus total pressure for various oxygen mole fractions.

vapor as reacting by a diffusion-limited reaction with the boric oxide to aid in removal of the oxide layer.

4. Conclusions

The shearing action exerted by the high-speed flow causes a boundary layer to be formed in the surface of the liquid oxide layer, and the stripping away of this layer accounts for the accelerated ignition of the boron particle. It is shown that migration of the liquid oxide layer on the front half-surface of the particle can be derived. The rate of disintegration is founded by integrating over the thickness of the liquid boundary layer to determine the mass flux leaving the boron particle surface at its equator. An ignition model of

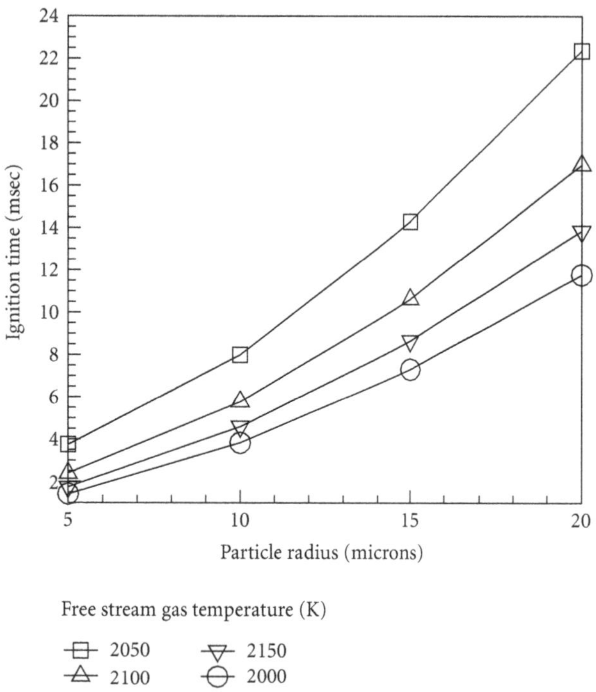

FIGURE 7: Ignition time versus particle radius for various ambient gas temperatures.

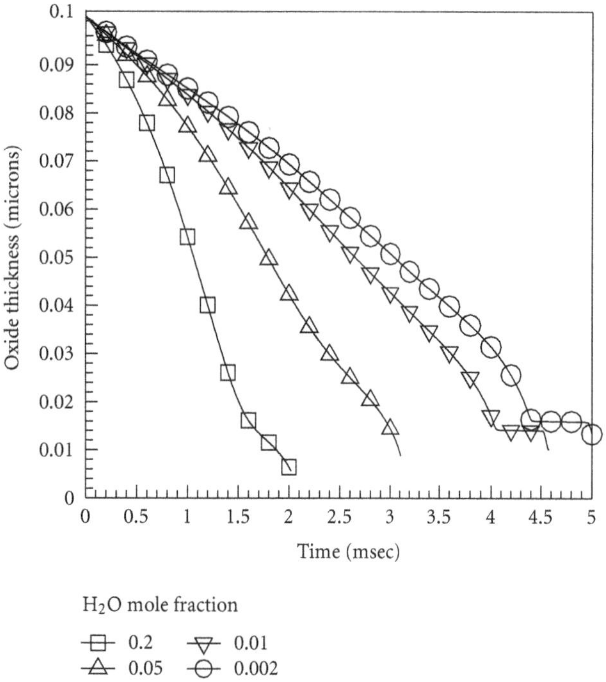

FIGURE 8: Oxide layer thickness versus time for various water vapor mole fractions.

boron particles is formulated for the oxide layer removal by considering that it results from a boundary layer stripping mechanism. Effects of various parameters on boron particle ignition were studied with this model. Compared with King's model, as the ignition model of boron particles is formulated for the oxide layer removal by considering that it results from a boundary layer stripping mechanism, the oxide layer thickness thins at all times during the particle ignition and lower the ignition time. The present results support a physical mechanism through which preheated boron particles in gas generator of ducted rockets can be ignited in the secondary chamber by ejection into a high-speed stream.

Acknowledgments

The support of the China Sponsorship Council (CSC) (Grant no. 51006118) is gratefully acknowledged. J. X. Hu thanks Benveniste Natan of Technion-Israel Institute of Technology for the help and encouragement.

References

[1] R. S. Fry, "A century of ramjet propulsion technology evolution," *Journal of Propulsion and Power*, vol. 20, no. 1, pp. 27–58, 2004.

[2] B. Natan and A. Gany, "Ignition and combustion boron particles in the flow field of a solid fuel ramjet," *Journal of Propulsion and Power*, vol. 7, no. 1, pp. 37–43, 1991.

[3] C. L. Yeh and K. K. Kuo, "Ignition and combustion of boron particles," *Progress in Energy and Combustion Science*, vol. 22, no. 6, pp. 511–541, 1996.

[4] M. K. King, "Ignition and combustion of boron particles and clouds," *Journal of Spacecraft and Rockets*, vol. 19, no. 4, pp. 294–306, 1982.

[5] D. Meinköhn, "Boron particle ignition and the marangoni effect," *Combustion Science and Technology*, vol. 176, no. 9, pp. 1493–1536, 2004.

[6] G. Mohan and F. A. Williams, "Ignition and combustion of boron in O_2/inert atmospheres," *AIAA Journal*, vol. 10, no. 6, pp. 776–783, 1972.

[7] S. C. Li and F. A. Williams, "Ignition and combustion of boron particles," *International Journal of Energetic Materials and Chemical Propulsion*, vol. 2, no. 1–6, pp. 248–271, 1991.

[8] A. Macek and J. M. Semple, "Combustion of boron particles at atmospheric pressure," *Combustion Science and Technology*, vol. 1, no. 3, pp. 181–191, 1969.

[9] W. Zhou, R. A. Yetter, F. L. Dryer, H. Rabitz, R. C. Brown, and C. E. Kolb, "Multi-phase model for ignition and combustion of boron particles," *Combustion and Flame*, vol. 117, no. 1-2, pp. 227–243, 1999.

[10] J. X. Hu, Z. X. Xia, and D. Q. Wang, "Study on burning rate of boron particles in secondary chamber of ducted rocket under forced convection conditions," *Journal of Solid Rocket Technology*, vol. 30, no. 1, pp. 21–25, 2007.

[11] R. O. Foelsche, R. L. Burton, and H. Krier, "Ignition and combustion of boron particles in hydrogen/oxygen explosion products," *AIAA 97-0127*, 1997.

[12] M. A. Gurevich, I. M. Kir'yanov, and E. S. Ozerov, "Combustion of individual boron particles," *Combustion, Explosion, and Shock Waves*, vol. 5, no. 2, pp. 150–153, 1969.

[13] D. Meinköhn, "Liquid oxide surface layers in metal combustion," *Combustion Theory and Modelling*, vol. 8, no. 2, pp. 315–338, 2004.

[14] D. Meinköhn, "The effect of particle size and ambient oxidizer concentration on metal particle ignition," *Combustion Science and Technology*, vol. 181, no. 8, pp. 1007–1037, 2009.

[15] A. Povitsky and Y. Goldman, "Boron particle ignition in high-speed flow," in *Proceedings of the 29th AIAA Joint Propulsion Conference*, vol. 2202, AIAA paper 93, 1993.

[16] A. A. Ranger and J. A. Nicholis, "Aerodynamic shattering of liquid drops," *AIAA Journal*, vol. 7, no. 2, pp. 285–290, 1969.
[17] M. Zhou and F. C. Zhuang, "Aerodynamic shattering of liquid drops using a boundary layer stripping mechanism," *Journal of National University of Defense Technology*, vol. 13, no. 3, pp. 29–33, 1991.
[18] M. K. King, "Boron particle ignition in hot gas stream," *Combustion Science and Technology*, vol. 8, no. 5-6, pp. 254–273, 1974.

5

FE Analysis of Dynamic Response of Aircraft Windshield against Bird Impact

Uzair Ahmed Dar, Weihong Zhang, and Yingjie Xu

Laboratory of Engineering Simulation and Aerospace Computing, Northwestern Polytechnical University, Xi'an, Shaanxi 710072, China

Correspondence should be addressed to Uzair Ahmed Dar; uzair@mail.nwpu.edu.cn

Academic Editor: Hong Nie

Bird impact poses serious threats to military and civilian aircrafts as they lead to fatal structural damage to critical aircraft components. The exposed aircraft components such as windshields, radomes, leading edges, engine structure, and blades are vulnerable to bird strikes. Windshield is the frontal part of cockpit and more susceptible to bird impact. In the present study, finite element (FE) simulations were performed to assess the dynamic response of windshield against high velocity bird impact. Numerical simulations were performed by developing nonlinear FE model in commercially available explicit FE solver AUTODYN. An elastic-plastic material model coupled with maximum principal strain failure criterion was implemented to model the impact response of windshield. Numerical model was validated with published experimental results and further employed to investigate the influence of various parameters on dynamic behavior of windshield. The parameters include the mass, shape, and velocity of bird, angle of impact, and impact location. On the basis of numerical results, the critical bird velocity and failure locations on windshield were also determined. The results show that these parameters have strong influence on impact response of windshield, and bird velocity and impact angle were amongst the most critical factors to be considered in windshield design.

1. Introduction

High velocity bird impact is one of the most significant hazards to both civilian and military aircrafts [1]. When fighter military aircrafts are operated at lower altitude and higher speed, the probability of bird impact increases and proves lethal to the safety of pilot and critical aircraft components. Windshield is the exposed part of aircraft and prone to bird impact. In order to ensure the safety of aircraft, the windshield must be capable of withstanding high velocity impact threats. Design of an optimum impact resistant windshield is a challenge and requires extensive experimental testing. The advanced numerical techniques are being adopted as an effective tool to simulate the bird strike event and provide a substitute to excessive costly experimentations. Moreover, it allows analyzing the most stringent impact conditions that cannot be considered in the experiments and provides detailed insight to the impact process which is difficult to observe during experimental testing.

Several researchers such as Zang et al. [2], Samuelson and Sornas [3], and Boroughs [4] carried out finite element analysis to investigate the impact response of windshield against bird strike. McCarty et al. [5, 6] used Materially and Geometrically Nonlinear Analysis (MAGNA), a nonlinear finite element analysis program for designing of windshield and canopy of military aircrafts. Wang et al. [7–9] simulated the failure of aircraft windshield against bird impact by using a modified nonlinear viscoelastic constitutive model together with Zhu-Wang-Tang (ZWT) damage model for the PMMA-based windshield. The constitutive model and its failure criterion effectively predicted the failure of windshield under a range of impact velocities. The authors also examined different critical factors which affect the impact response of aircraft windshield against bird strike. The results proposed that material model for windshield, boundary conditions, mesh density, surrounding structure of windshield, and bird velocity are the critical factors that must be taken into account for FE analysis of aircraft anti-bird design. Guida et al. [10] numerically examined the influence of geometric parameters of windshield and impact parameters under high speed bird impact by using explicit FE code. The results showed that windshield panel dimensions, thickness, and curvature as

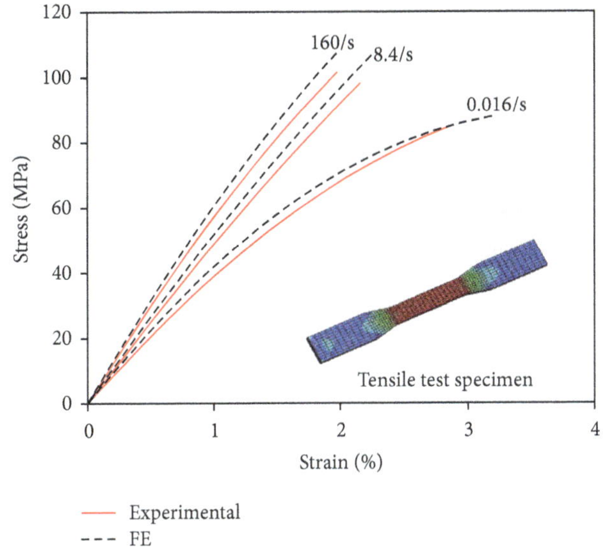

FIGURE 1: FE validation of windshield material model.

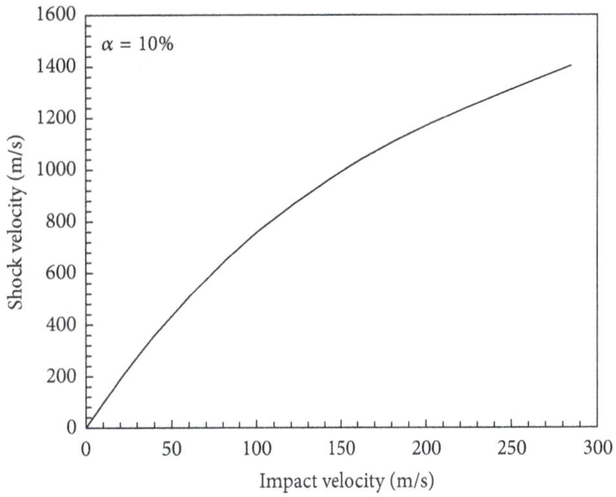

FIGURE 2: Shock velocity as function of impact velocity.

well as bird velocity, size, and impact angle considerably affect the impact response of windshield. Liu et al. [11] applied smooth-particle-hydrodynamics (SPH) based approach to model the bird against impact on windshield structure by using explicit FE solver. The maximum displacements of various points on camber line of windshield were compared with experimental results and were found to be in good agreement. The results also indicated that SPH solver gives better comparison than Lagrangian solver. Zhu et al. [12] numerically studied the bird impact response on windshield by employing user-defined material subroutine in explicit FE solver. Numerical results of failure modes of the windshield, deformation, displacement, and strain curves of the measured points on the windshield agreed well with the experimental results. Yang et al. [13] carried out FE and experimental investigation to study the structure strength of windshield subjected to bird impact. Based on numerical scheme, the critical impact velocities on certain locations on windshield are determined and verified through experimental results.

The present work aims to numerically model the impact response of windshield and examine the effect of various factors that contribute to optimize the design of certain windshield. Several critical factors such as mass and shape of bird, impact velocity, angle of impact, and impact location were studied and their influence on the dynamic response of windshield was assessed. The numerical model was developed and implemented in explicit finite element hydro code ANSYS AUTODYN.

2. Numerical Modeling

2.1. Windshield Material Model. The windshield considered in this study is monolithic, uniform cross-section, PMMA-based aviation organic glass. The dynamic properties of this material have been extensively studied in the literature and several material models are available for modeling the impact response of this material [7, 14–17]. In this study, the elastoplastic material model along with maximum principal strain failure criterion was defined to predict damage and failure of windshield. The material model was implemented by incorporating isotropic hardening using Von Mises yield criterion together with rate-dependent Cowper Symonds plasticity law. Cowper Symonds strength model defines the yield strength of isotropic strain hardening of strain rate dependent material [10]. The yield surface can be defined as

$$Y = \left(A + B\varepsilon_{\text{pl}}^{n}\right)\left[1 + \left(\frac{\dot{\varepsilon}^{\text{pl}}}{D}\right)^{1/q}\right], \qquad (1)$$

where Y is yield stress of the material, A is yield stress at zero plastic strain, B is strain hardening coefficient, n is strain hardening exponent, and D and q are strain rate hardening coefficients. Von Mises is simple and convenient criterion to apply as it defines a smooth and continuous yield surface with good approximation at high stresses. At given principal stresses σ_1, σ_2, and σ_3, the yield criterion is defined as

$$(\sigma_1 - \sigma_2)^2 + (\sigma_2 - \sigma_3)^2 + (\sigma_3 - \sigma_1)^2 = 2Y^2. \qquad (2)$$

The maximum principal strain criterion implies that if the maximum tensile principal strain exceeds the prescribed limits, then material will instantaneously fail. Failure is predicted when either of the principal strains ε_1 or ε_2, resulting from the principal stresses σ_1 or σ_2, equals or exceeds the maximum strain corresponding to the yield strength σ_y of the material in uniaxial tension or compression. For yielding in tension the minimum principle strain ε_1 would equal the yield strain in uniaxial tension. If the strains are expressed in terms of stress, then

$$\begin{gathered}\varepsilon_1 = \frac{\sigma_1}{E} - \frac{\nu}{E}(\sigma_2 + \sigma_3), \\ \sigma_1 - \nu(\sigma_2 + \sigma_3) \leq \sigma_y, \\ \varepsilon_{\text{fail}} = \varepsilon_{\text{total}} - \frac{\sigma_{\text{total}}}{E} \quad \sigma > Y.\end{gathered} \qquad (3)$$

TABLE 1: Material properties of windshield.

Density ($Kg{\cdot}m^{-3}$)	Elastic modulus (GPa)	Tangent modulus (MPa)	Poisson's ratio	Yield strength (MPa)	Ultimate strength (MPa)	Failure strain
1186	3.2	230	0.4	68	78	0.067

The material properties of windshield used in the analysis are given in Table 1.

2.2. Bird Material Model. The actual bird is combination of flesh, blood, and bones, and it is difficult to implement the actual bird constitutive model in numerical program. Numerous researchers used different approaches to closely approximate the material response of bird. Some authors modeled the bird with elastoplastic material law along with certain failure criteria [12, 18–20], while others used equation of state approach for constitutive modeling with pressure volume behavior of water [21–24]. Bird is mostly composed of water. It behaves like a soft body and acts as a fluid on the structure during the impact. Water like hydrodynamic response by using Mie-Gruneisen equation of state (EOS) with negligible strength effects was implemented to model bird in this study. Mie-Gruneisen EOS correlates the material volumetric strength and pressure to density ratio, and it is easy to establish Gruneisen equation based on the shock Hugoniot [25]

$$P_H = \frac{\rho_o C^2 \cdot \mu(\mu+1)}{[1-(s-1)\mu]^2}, \quad (4)$$

where

$$\mu = \frac{\rho}{\rho_o} - 1, \quad (5)$$

where ρ and ρ_o are the initial and instantaneous densities of material, C is the intercept, and s is linear Hugoniot slop of shock velocity (v_s) and particle velocity (v_p) relationship. Gruneisen equation describes a linear relation between shock and particle velocity, where

$$v_s = s \cdot v_p + C. \quad (6)$$

Gruneisen form of equation of state based on shock Hugoniot is

$$P = P_H + \Gamma\rho(E - E_H). \quad (7)$$

It is assumed that $\Gamma\rho = \Gamma_o\rho_o$ = constant and

$$E_H = \frac{1}{2}\frac{P_H}{\rho_o}\left(\frac{\mu}{\mu+1}\right). \quad (8)$$

For $s > 1$, this formulation gives a limiting value of compression because the pressure approaches to infinity as the term $[1-(s-1)\mu = 0]$ in (4) becomes zero and the pressure therefore becomes infinite and gives maximum density of $\rho = \rho_o s(s-1)$. Also long before this regime is approached, the assumption of $\Gamma\rho = \Gamma_o\rho_o$ = constant is not valid. Moreover, the assumption of a linear relationship between the shock velocity v_s and the particle velocity v_p does not hold for too large compression. And at high shock strengths, some nonlinearity in this relationship is apparent. To incorporate the nonlinearity in numerical model, two linear fits to the shock velocity and particle velocity relationship were established: one holding at low shock compressions defined by $v > v_b$ and other at high shock compressions where $v < v_e$. The region between v_b and v_e is covered by a smooth interpolation between the two linear relationships [25]. The equation of state has been further improved to include a quadratic shock and particle velocity relationship

$$v_s = s_1 v_p + s_2 v_p^2 + C, \quad (9)$$

where s_1 and s_2 are coefficients of slop, Γ is Gruneisen parameter, and E is internal energy. In the present numerical scheme, the water Gruneisen EOS parameters $\Gamma = 0.28, C = 1483$, and $s = 1.75$ were used to model bird as soft projectile and other parameters were set to zero [25]. The density of the bird was taken as 900 kg/m^3 as it has been taken by most of the researchers before.

2.3. Validation of Numerical Model

2.3.1. Model for Windshield. In order to validate the numerical model for windshield material, FE simulations of simple uniaxial tensile test were performed. A standard dumbbell-shaped tensile test specimen with dimensions of 200 × 20 × 2.67 mm and gauge length of 50.8 mm was built. The specimen was meshed with Lagrangian solid elements and the material properties were taken as described in Section 2.1. One end of the specimen was fixed while constant velocity load was applied at the other end. The rate of loading was adjusted according to rate of change of strain. The results from FE simulations were compared with available results [17] in terms of stress strain curves at different strain rates as shown in Figure 1. The results show that the model predicts the stress-strain behavior of PMMA at different strain rates with fair accuracy.

2.3.2. EOS for Bird. To validate the Gruneisen EOS implemented in FE solver, the Hugoniot and stagnation pressures were compared with theoretically obtained pressure values and Wilbeck [26] experimental results. Wilbeck experimental results carry significant importance in bird strike analysis, and many researchers used his experimental data for comparison of numerical results. For theoretical calculations, one-dimensional Hugoniot analysis was carried out in which bird impact is characterized in two stages. The first stage is initial shock called Hugoniot pressure which gives the maximum possible value of pressure during impact, and the

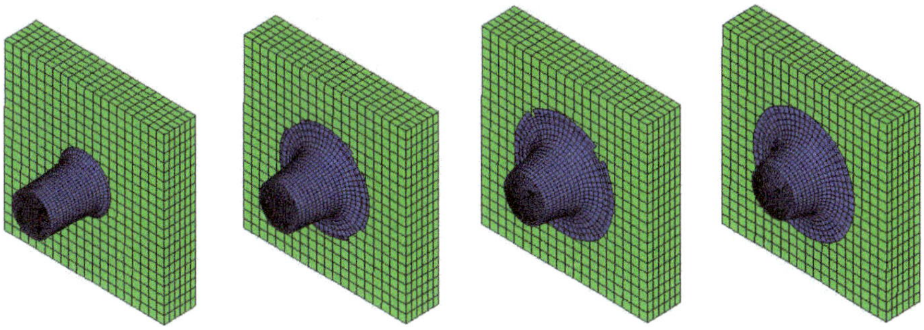

FIGURE 3: Deformation of bird on impacting rigid plate.

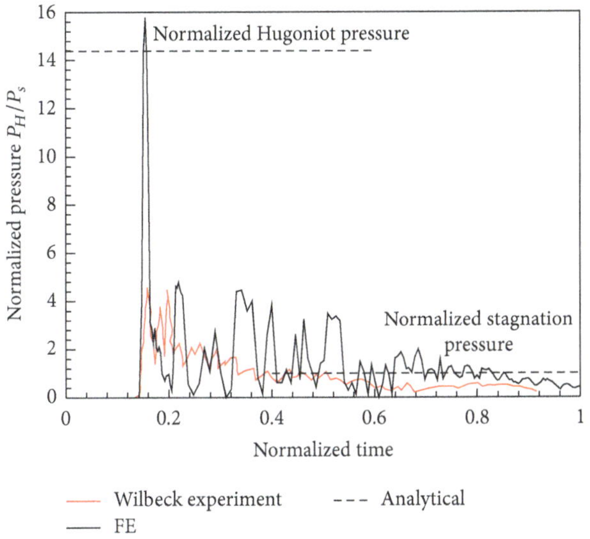

FIGURE 4: Comparison of results for Hugoniot pressure.

other stage is steady state flow called stagnation pressure in which pressure stabilizes with time. The maximum pressure, that is, Hugoniot pressure, can be estimated by implying hydrodynamic impact theory and equations of conservation mass and momentum

$$\rho v_s = \rho_o (v_s - v_p),$$
$$P_1 + \rho v_s^2 = P_2 + \rho_o (v_s - v_p)^2, \quad (10)$$

where ρ, ρ_o, P_1, and P_2 are the initial and instantaneous density and pressure values. v_s is shock propagating velocity and v_p is particle velocity behind shock. The particle velocity v_p is assumed to be equal to the initial impact velocity v_i of bird. The Hugoniot analytical pressure can be defined as

$$P_H = P_2 - P_1 = \rho v_s v_i. \quad (11)$$

The shock velocity v_s is function of initial impact velocity v_i and can be obtained by solving the nonlinear equation [24]

$$\frac{v_s}{v_s - v_i} = (1 - \alpha) \left[\frac{v_s v_i (4k - 1)}{C^2} \right]^{-1/(4k-1)}, \quad (12)$$

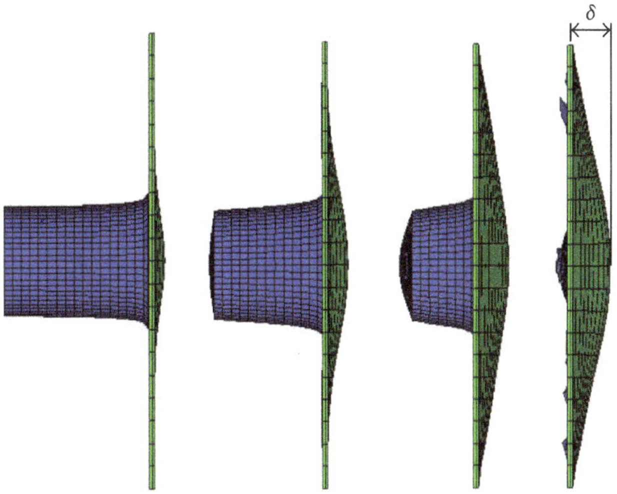

FIGURE 5: Bird impact on flexible aluminum plate.

where C is velocity of sound in the material, α is porosity of material (for 10% porosity $\alpha = 0.1$), and value of k is determined experimentally. The variation of shock velocity with initial impact velocity is shown in Figure 2.

The steady flow pressure can be estimated by using Bernoulli relationship

$$P_s = \frac{1}{2} \rho v_i^2. \quad (13)$$

For FE validation, a square rigid plate of 0.5 m side was modeled in Lagrangian grid fully constrained at sides. The bird is modeled as right cylinder of 0.18 m length and 0.06 m radius with Lagrangian solid elements and material properties described in Section 2.2. In the analysis, bird with initial impact velocity of 116 m/s was impacted against rigid plate, and values for Hugoniot and stagnation pressures at central point of impact were obtained as shown in Figure 3. In order to compare the results with experimental data, the velocity of bird was taken as 116 m/s used by Wilbeck in his experiments. The results were normalized by dividing pressure with stagnation pressure and time with total impact duration and compared with experimental and analytical results.

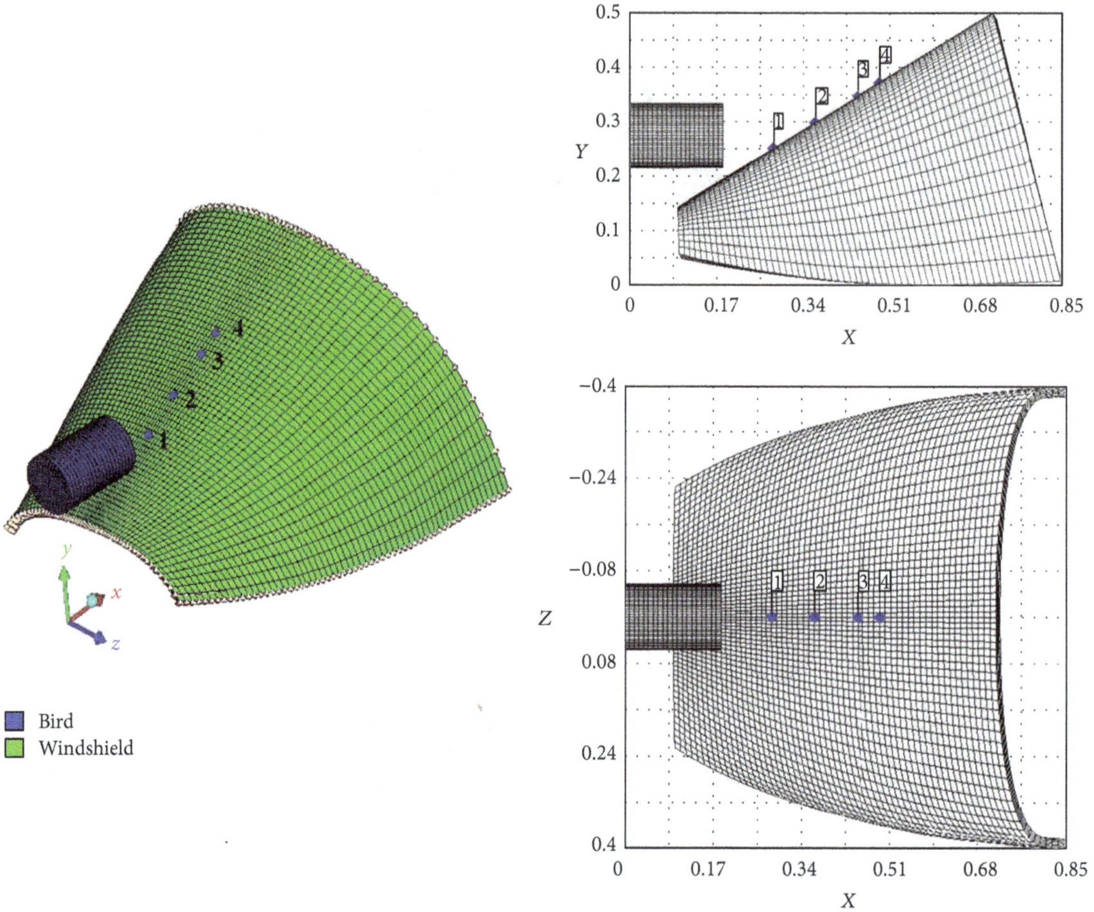

Figure 6: Finite element model of bird impact on windshield: Isometric view (Left) Side and top views (Right).

From analysis, the Hugoniot pressure was predicted to have a maximum value of 102.5 MPa and the stagnation pressure is 6.5 MPa, giving normalized values of 15.7 and 1, respectively. The analytical values of Hugoniot pressure and stagnation pressure were calculated as 86.3 MPa and 6 MPa, which after normalizing gives 14.3 and 1, respectively. The comparison of FE, analytical, and experimental values of normalized pressure is shown in Figure 4. The trend of the plot is consistent with experimental results where a sudden peak pressure value was observed at initial shock and the pressure then stabilized with time. The duration of pressure decay was also in accordance with experimental results. However, the FE simulation predicted peak pressure is higher than the experimental results. The overprediction in pressure peak can be credited to bird orientation during impact, because in real bird impact experiments the initial pressure peak is highly dependent on orientation of the bird as described by Hedayati and Ziaei-Rad [24] in their work. Moreover, the Hugoniot peak pressure occurs in a very short time about 4-5 μs and the capturing of the accurate pressure peak is a challenging experimental task. The modern high frequency state of the art pressure transducers must be employed to record the accurate pressure peak during impact event, and latest experimental data in this regard is required for better comparison of FE and experimental results.

However, for the present work, the bird model was further verified by impacting the bird on a flexible metallic plate and the resultant plastic deformation after impact was determined. Welsh and Centonze [27] experimental results were considered for this purpose in which the bird with initial velocity of 146 m/s was impacted on 6.35 mm thick T6061-T6 aluminum plate and corresponding plastic deformation of the plate was measured. FE simulations were performed in accordance with experimental parameters and the residual plastic deformation δ of the plate after impact was determined as shown in Figure 5. The FE predicted deformation of 43.68 mm was in close agreement with experimentally determined value of 41.275 mm. The validated material models for bird and windshield were then used to build a full scale model for bird-windshield impact problem.

2.3.3. 3D FE Model of Impact Problem. The windshield and bird were modeled with solid elements by using Lagrangian grid as shown in Figure 6. The windshield consists of 14,400 elements with 60 elements along the curvature, 40 elements along sides, and 4 through thickness elements. More refined elements distribution is adopted around the area of impact as most of the deformation takes place at this particular impact region. The 1.8 Kg bird was modeled as right cylinder of 60 mm radius and 180 mm length. The bird was modeled

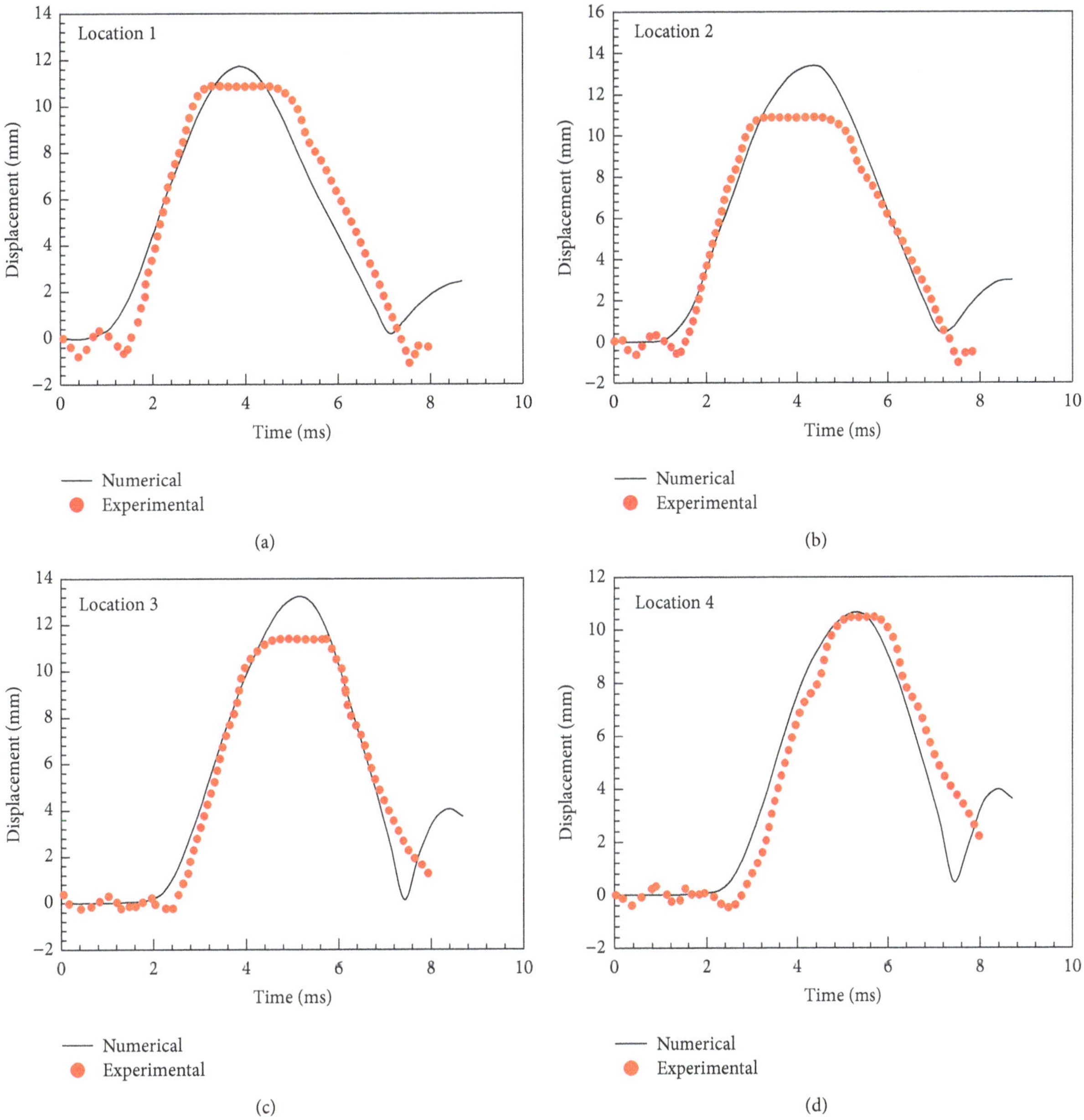

FIGURE 7: Displacement time plots for various locations on windshield.

as soft body with 10 elements across radius and 20 elements along length of cylinder.

The elements in the mesh undergo severe distortion due to high rate of deformation. These distorted elements must be removed from the mesh in order to run program smoothly and to avoid reduced time step problem. The deletion of distorted elements is carried out by using geometric strain erosion model provided in AUTODYN. During the impact process, the bird highly deforms, perishes, and loses most of its mass due to elements removal process which causes inaccurate results of numerical calculations. This problem was overcome by employing the option of retaining inertia of eroded nodes in which the solver ascribes the removed cell mass to their nodal points. The contact between windshield and bird was defined by using Lagrange/Lagrange interaction option of AUTODYN. In this option, an automatic detection zone is defined around each interacting Lagrangian subgrid (independent or with itself). When a node enters into this zone, it is automatically repelled. The edges of the windshield are fully constrained to provide fixed boundary conditions. Four gage points 1, 2, 3, and 4 (Figure 6) were marked on the windshield central line to record the values of displacement, stress, and strain and to compare them with experimental results.

2.3.4. Experimental Confirmation. In the experimental results [11, 19], 1.8 Kg bird with velocity of 64.4 m/sec was impacted on windshield at four different locations (1, 2, 3, and 4). The displacement profiles for all the four locations were recorded during experiments. Numerical simulations were

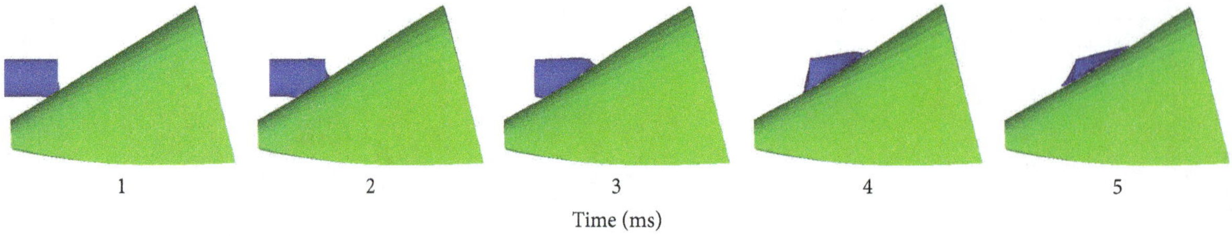

FIGURE 8: Different stages of deformation of windshield at bird impact velocity of 64.4 m/sec.

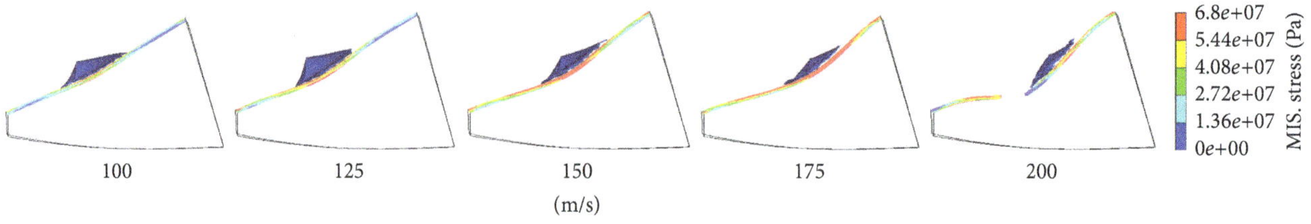

FIGURE 9: Impact response of windshield with increase in impact velocity.

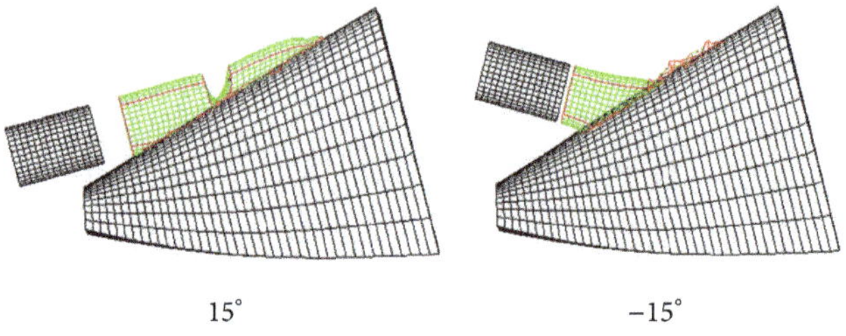

FIGURE 10: Two different angles of impact on windshield.

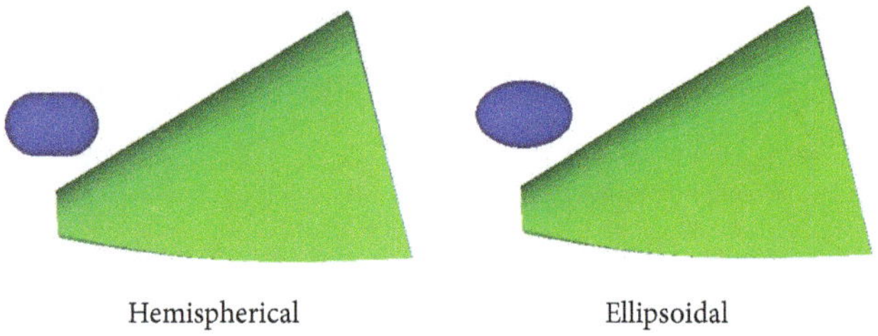

FIGURE 11: Response of windshield for different bird shapes.

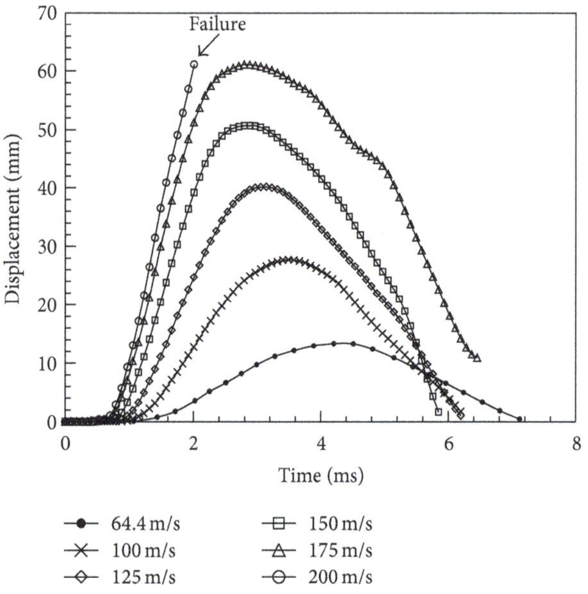

FIGURE 12: Displacement time plot at Location 2.

performed, and results were compared with experimental data and found in good agreement which prove the validity of numerical model as shown in Figure 7. This verified that numerical model was then further employed to investigate the effect of various parameters on windshield behavior.

3. Simulations

Numerical simulations were carried out to predict the dynamic response of windshield for range of impact velocities, impact angle and location of impact, and bird mass and shape. Figure 8 depicts various modes of deformation of windshield at different time intervals when impacted at location 2 with velocity of 64.4 m/sec. The windshield remains in the elastic state and keeps its original shape after impact. No sign of damage or failure was observed at this impact velocity. Also, at this velocity, the bird fails partially and slides along the surface of windshield.

When the velocity of bird is increased, the stresses escalate due to higher impact force and windshield tends to deform plastically. With further increase in impact velocity windshield shield reaches its elastic limit and suffers from permanent deformation leading to its complete failure. The equivalent stresses along cross-section of windshield for different velocities are shown in Figure 9. At higher velocity, the bird deforms severely and gets fragmented during sliding along the surface of windshield. Two different orientations

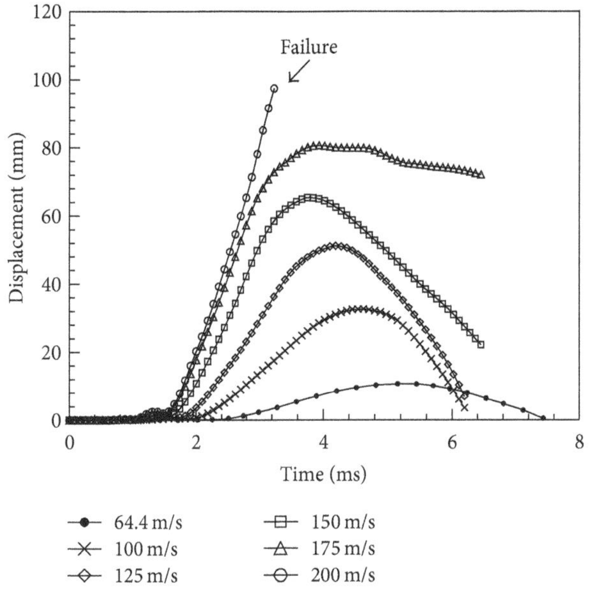

FIGURE 13: Displacement time plot at Location 4.

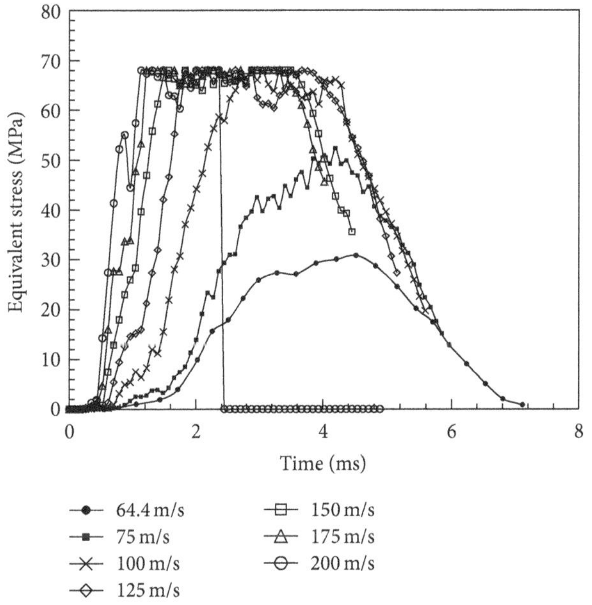

FIGURE 14: Equivalent stress history at Location 2.

of bird, that is, 15° and −15° from its axis, were studied to see the effect of impact angle on response of windshield and shown in Figure 10. The change of impact angle showed significant effect on normal displacement, stress, strain, and impact force. For 15° impact angle, partial bird failure occurs and whole bird body slides along the windshield surface in its way of impact.

At −15° angle, the bird transfers most of its energy to windshield giving higher values displacement and stress at point of impact. Most of the body of bird fails during this event.

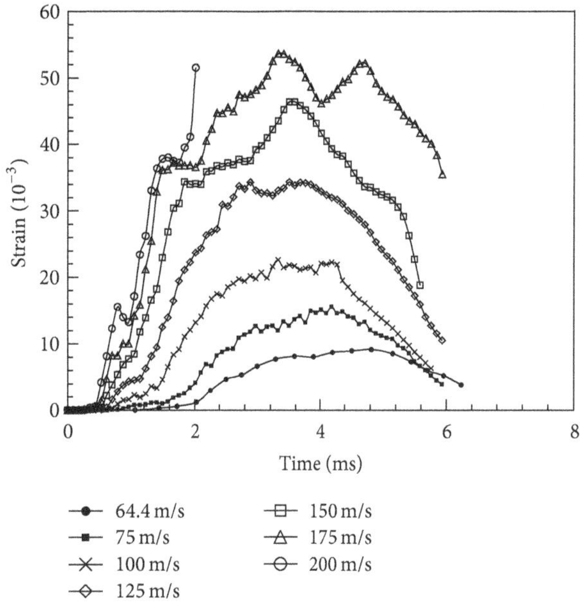

FIGURE 15: Effective strain history at Location 2.

Impact response for two additional bird geometries (hemispherical and ellipsoidal of similar length to diameter ratio) was also simulated and shown in Figure 11. For all bird shapes, the maximum length to diameter ratio and mass of the bird were taken constant. The simulations results show that impact The simulations results show that impact due to cylindrical-shaped bird produces slightly higher deformation on the windshield due to its increased contact area.

4. Results and Discussion

The impact response of windshield for various impact velocities was considered. It was noted that normal displacement at all gage locations increases with the increase of velocity. The time to reach maximum displacement at locations 3 and 4 increases as they lie farther from point of impact (location 2). The displacement time plots for different impact velocities at locations 2 and 4 are shown in Figures 12 and 13. Increase in velocity caused more deformation at the upper half of windshield because more of the bird mass slide and transferred more energy to the upper end. An instantaneous failure at the point of impact occurs when velocity was increased to 200 m/s.

Figures 14 and 15 show the plots for equivalent stress and corresponding strain at impact point. The amplitude of stress increases with the increase of velocity; at 64.4 m/s the value of maximum stress is 30 MPa which rises up to 50 MPa at 75 m/s velocity. At 100 m/s impact velocity, the stresses in the windshield become higher than the yield stress. On further increasing the velocity to 125 m/s, the ultimate stress limit is reached and upper end of windshield fails. At 200 m/s impact velocity, the windshield failed instantly at the point of impact in 1.9 ms.

Therefore, at velocity of 100 m/s and higher, the plastic deformation starts to prevail in windshield and it remains

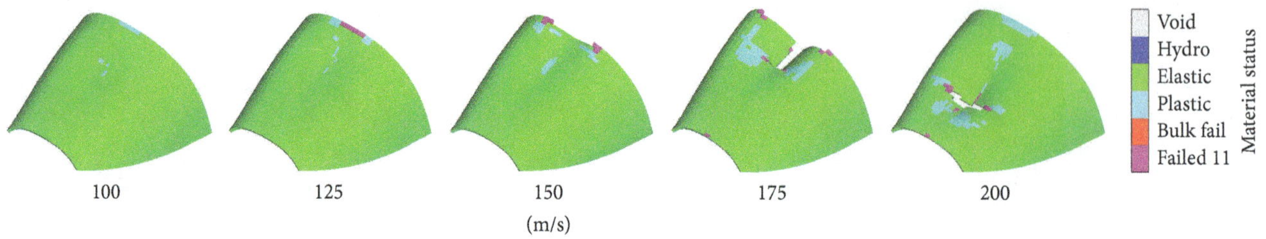

FIGURE 16: Deformation modes of windshield at different impact velocities.

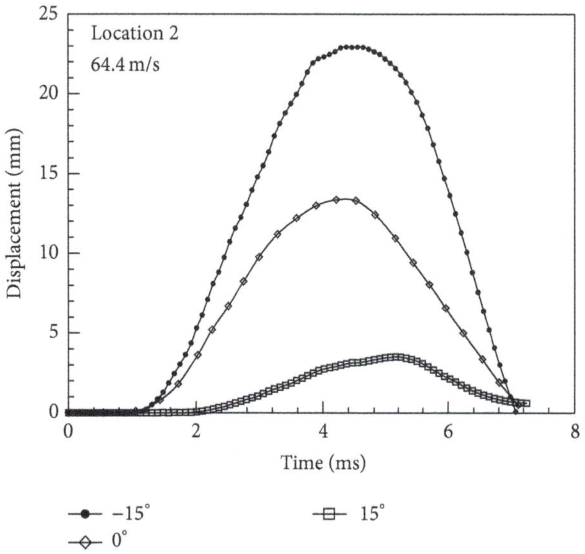

FIGURE 17: Displacement time plot for different impact angles.

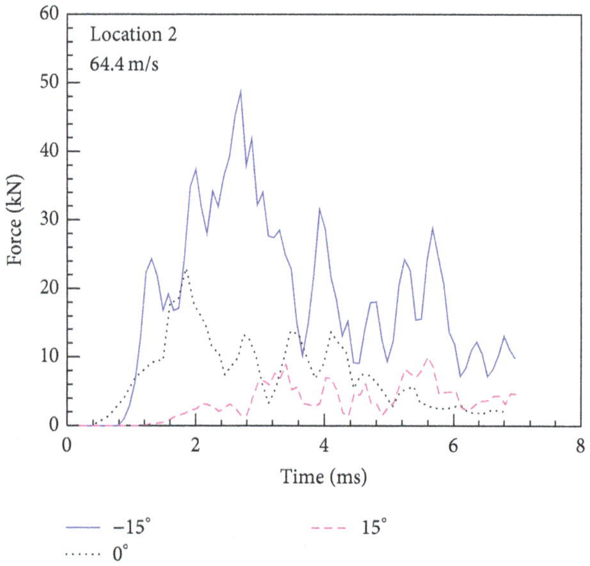

FIGURE 18: Impact force history for different impact angles.

in the state of high stresses which also defines the critical impact velocity of the particular windshield. The modes of deformation and initiation of plasticity are shown in Figure 16. It can be seen that when the velocity is 100 m/s, sign of plastic deformation appears around the point of impact and upper end of windshield. The plastic deformation increases and windshield starts to fail at its upper end when the velocity is increased to 150 m/s. The further increase in velocity leads to major windshield failure at upper end and point of impact.

The influence of three different bird angles at the point of impact was examined. Figure 17 shows the plots for normal displacement at 64.4 m/s impact velocity from which it can be observed that at 15° angle of impact the maximum normal displacement is 3.5 mm which increases to 13 mm for 0° and 23 mm for −15° angle. The impact force also increases sharply with the change of impact angle as shown in Figure 18. At 15° angle, the maximum impact force of 9.9 kN is recorded at 5.6 ms which rises to 48.6 kN at 2.7 ms for −15° impact angle. Almost 5 times increase in peak impact force was observed due to change in impact angle from 15° to −15°.

The effect of impact angle on stress and strain history is shown in Figure 19. The maximum stress amplitude of 11.8 MPa and corresponding strain value of 0.00352 occur at 5.1 ms for 15° angle of impact. When the impact angle changes to −15°, the peak stress and strain value rise to 67.8 MPa and 0.0201 at 3.75 ms representing most severe impact conditions at 64.4 m/s velocity and windshield approaching its yield stress limit. Hence, impact angle is a critical parameter in determining the critical impact velocity for windshield.

The effect of bird mass on impact response was very obvious and shown in Figure 20. With the increase in bird mass, the corresponding value of displacement and stress increases because more kinetic energy of bird is transferred to windshield. This also shows that the critical impact velocity will be different for different bird masses to produce same deformation in the windshield.

The response of windshield impacted by three different shape birds was studied in this work. Similar displacement trends were observed for hemispherical- and ellipsoidal-shaped birds while cylindrical-shaped bird produces higher displacement as shown in Figure 21(a). The peak values of equivalent stress for hemispherical and ellipsoidal shapes are same and higher than cylindrical shape (Figure 21(b)).

5. Conclusions

The behavior of windshield against high speed bird impact was successfully simulated and the effect of various parameters on its dynamic response was studied. Bird impact velocity

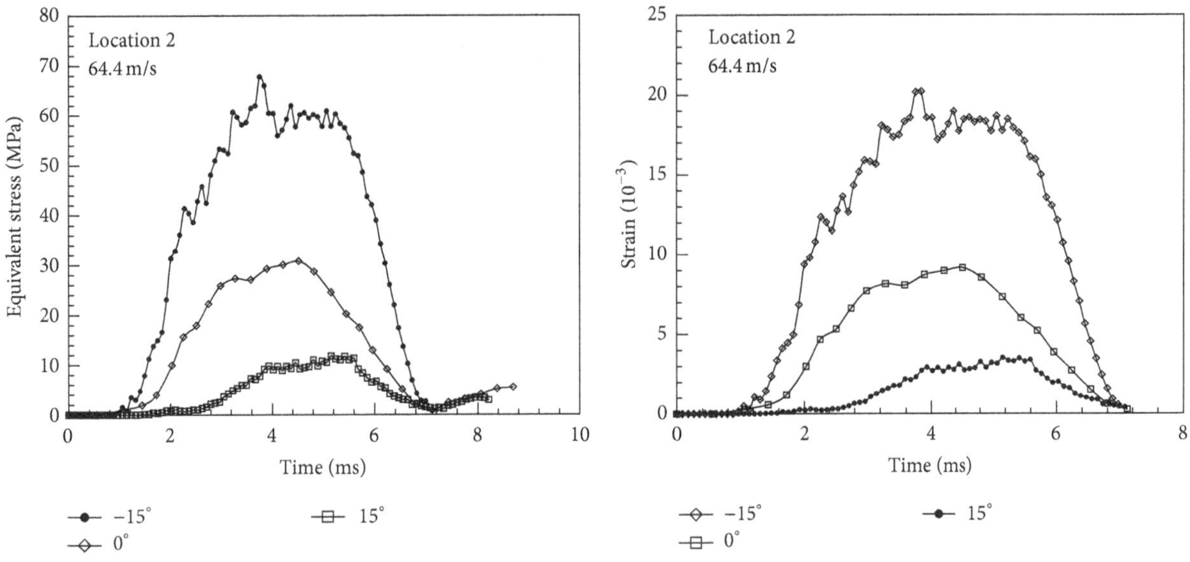

FIGURE 19: (a) Equivalent stress and (b) strain history during different angles of impact.

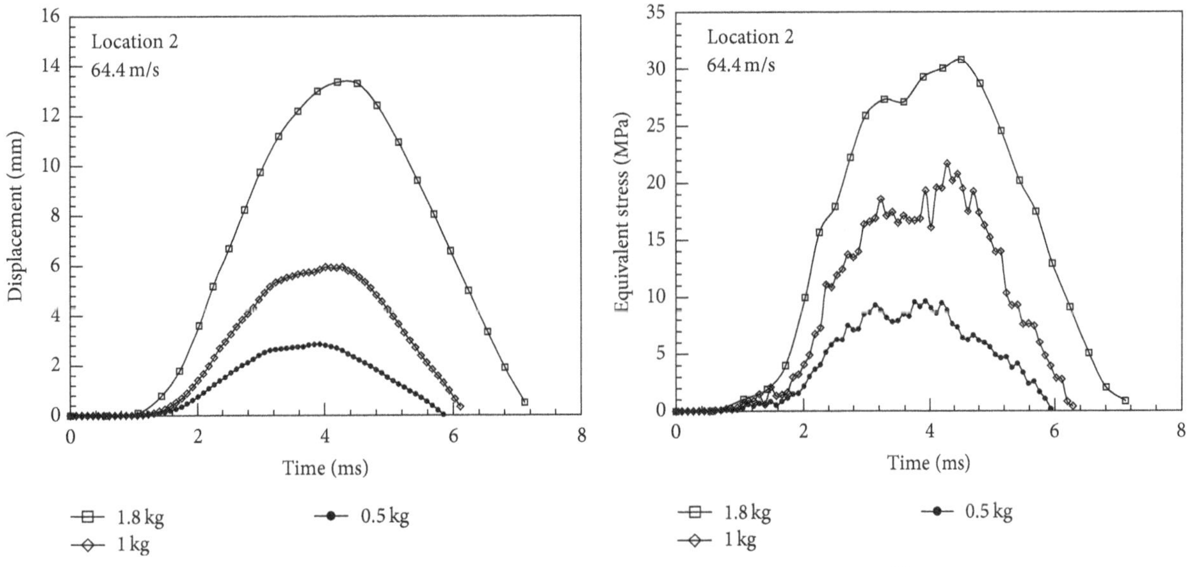

FIGURE 20: Effect of bird mass on (a) maximum normal displacement and (b) equivalent stress.

was among the most critical parameter having a strong influence on dynamic response. With the increase in bird velocity, windshield tends to deform plastically beyond its yield strength which finally leads to its major failure. For a range of velocities simulated in this study, there was a limiting impact velocity at which windshield suffers permanent plastic deformation and vulnerable to fail at certain crucial locations. The rearward fixed part of windshield was considered weakest at the critical velocity. For different impact angles, the response of windshield differs greatly. With less steep angle, most of the bird slides along the surface of windshield causing less damage. Steeper angle on the other hand produces high deformation and a plastic dimple is observed at the point of impact. The bird with higher mass proved more fatal to the windshield as they impact more kinetic energy to the structure. Although the shape of the bird did not show significant effect in this study, however, the bird with smaller length to diameter ratio and higher instantaneous contact area can affect the shock pressure and peak stress level in the structure. These critical factors can be parameterized

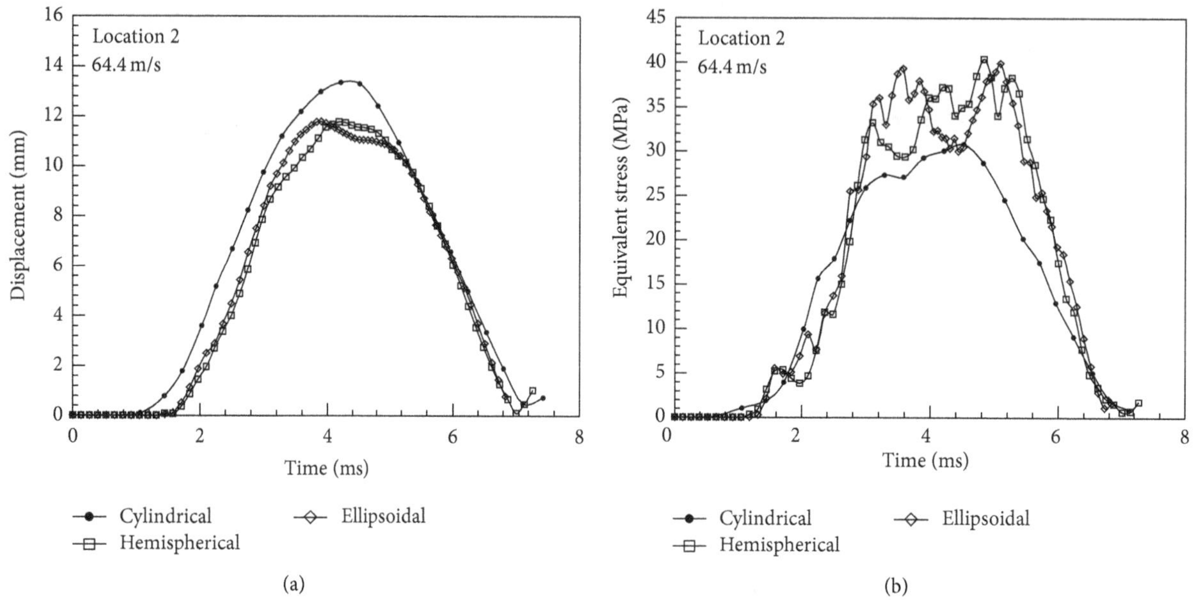

FIGURE 21: Effect of bird shape on (a) maximum normal displacement and (b) equivalent stress.

together to predict the combined effect on impact response of windshield and can provide certain guiding principles for windshield design and optimization.

Acknowledgments

This work is supported by 973 Program (2012CB025904), NPU Foundation for Fundamental Research (NPU-FFR-JC201236), and Shaanxi Provincial Natural Science Foundation (2012JQ1003).

References

[1] J. Thorpe, "Fatalities and destroyed civil aircraft due to bird strikes, 1912–2002," in *Proceedings of the 26th Meeting of the International Bird Strike Committee*, Warsaw, Poland, 2003.

[2] S. G. Zang, C. H. Wu, R. Y. Wang, and J. R. Ma, "Bird impact dynamic response analysis for windshield," *Journal of Aeronautical Materials*, vol. 20, no. 4, pp. 41–45, 2000.

[3] A. Samuelson and L. Sornas, "Failure analysis of aircraft windshields subjected to bird impact," in *Proceedings of the 15th ICAS Congress*, London, UK, 1986.

[4] R. R. Boroughs, "High speed bird impact analysis of the Learjet 45 windshield using DYNA3D," in *Proceedings of the 39th AIAA/ASME/ASCE/AHS/ASC Structures, Structural Dynamics, and Materials Conference and Exhibit and AIAA/ASME/AHS Adaptive Structures Forum*, pp. 49–59, Long Beach, Calif, USA, April 1998.

[5] R. E. McCarty, "Finite element analysis of a bird-resistant monolithic stretched acrylic canopy design for the F-16A aircraft," in *Proceedings of the American Institute of Aeronautics and Astronautics, Aircraft Systems and Technology Conference*, Dayton, Ohio, USA, 1981.

[6] R. E. McCarty, M. G. Gran, and M. J. Baruch, "MAGNA nonlinear finite element analysis of T-46 aircraft windshield bird impact," in *Proceedings of the AIAA/AHS/ASEE Aircraft System Design and Technology Meeting*, AIAA Paper 86-2732, Dayton, Ohio, USA, 1986.

[7] F. S. Wang and Z. F. Yue, "Numerical simulation of damage and failure in aircraft windshield structure against bird strike," *Materials and Design*, vol. 31, no. 2, pp. 687–695, 2010.

[8] F. S. Wang, Z. F. Yue, and W. Z. Yan, "Factors study influencing on numerical simulation of aircraft windshield against bird strike," *Shock and Vibration*, vol. 18, no. 3, pp. 407–424, 2011.

[9] X. Wang, Z. Feng, F. Wang, and Z. Yue, "Dynamic response analysis of bird strike on aircraft windshield based on damage-modified nonlinear viscoelastic constitutive relation," *Chinese Journal of Aeronautics*, vol. 20, no. 6, pp. 511–517, 2007.

[10] M. Guida, A. Grimaldi, F. Marulo, and A. Sollo, "FE study of windshield subjected to high speed bird impact," in *Proceedings of the 26th International Congress of the Aeronautical Sciences (ICAS '08)*, 2008.

[11] J. Liu, Y. L. Li, and F. Xu, "The numerical simulation of a bird-impact on an aircraft windshield by using the SPH method," *Advanced Materials Research*, vol. 33–37, pp. 851–856, 2008.

[12] S. Zhu, M. Tong, and Y. Wang, "Experiment and numerical simulation of a full-scale aircraft windshield subjected to bird impact," in *Proceedings of the 50th AIAA/ASME/ASCE/AHS/ASC Structures, Structural Dynamics, and Materials Conference*, Palm Springs, Calif, USA, 2009.

[13] J. Yang, X. Cai, and C. Wu, "Experimental and FEM study of windshield subjected to high speed bird impact," *Acta Mechanica Sinica*, vol. 19, no. 6, pp. 543–550, 2003.

[14] W. Lili, Z. Xixiong, S. Shaoqiu, G. Su, and B. Hesheng, "Impact dynamics investigation on some problems in bird strike on windshields of high speed aircrafts," *Acta Aeronautica et Astronautica Sinica*, vol. 12, no. 2, pp. B27–B33, 1991.

[15] F. Zhou, L. Wang, and S. Hu, "A damage-modified nonlinear visco-elastic constitutive relation and failure criterion of PMMA at high strain-rates," *Explosion and Shock Waves*, vol. 12, no. 4, pp. 333–342, 1992.

[16] A. Wang, X. Qiao, and L. Li, "Finite element method numerical simulation of bird striking multilayer windshield," *Acta Aeronautica et Astronautica Sinica, Series A and B*, vol. 19, pp. 446–450, 1998.

[17] Z. Zhi-lin, Z. Qi-qiao, and L. Ming-xing, "Bird impact dynamic response analysis for aircraft arc windshield," *Acta Aeronautica et Astronautica Sinica*, vol. 9, article 018, 1992.

[18] R. Doubrava and V. Strnad, "Bird strike analyses on the parts of aircraft structure," in *Proceedings of the 27th Congress of the International Council of the Aeronautical Sciences*, France, 2010.

[19] J. Bai and Q. Sun, "On the integrated design technique of windshield against bird strike," *Mechanics and Engineering*, vol. 27, no. 1, pp. 14–18, 2005.

[20] Y. Zhang and Y. Li, "Analysis of the anti-bird impact performance of typical beam-edge structure based on ANSYS/LS-DYNA," *Advanced Materials Research*, vol. 33–37, pp. 395–400, 2008.

[21] A. F. Johnson and M. Holzapfel, "Modelling soft body impact on composite structures," *Composite Structures*, vol. 61, no. 1-2, pp. 103–113, 2003.

[22] J. Cheng and W. K. Binienda, "Simulation of soft projectiles impacting composite targets using an arbitrary Lagrangian-Eulerian formulation," *Journal of Aircraft*, vol. 43, no. 6, pp. 1726–1731, 2006.

[23] F. Stoll and R. A. Brockman, "Finite element simulation of high-speed soft-body impacts," in *Proceedings of the 38th AIAA/ASME/ASCE/AHS/ASC Structures, Structural Dynamics, and Materials Conference*, pp. 334–344, April 1997.

[24] R. Hedayati and S. Ziaei-Rad, "A new bird model and the effect of bird geometry in impacts from various orientations," *Aerospace Science and Technology*, 2012.

[25] AUTODYN Theory manual Rev. 4.3. Century Dynamics, a subsidiary of ANSYS Inc, 2005.

[26] J. Wilbeck, "Impact behavior of low strength projectiles," Report AFML-TR- 77-134, Air Force Materials Laboratory, 1977.

[27] C. J. Welsh and V. Centonze, "Aircraft transparency testing artificial birds," Report AEDC-TR-86-2, US Air Force, 1986.

6

Multisized Inert Particle Loading for Solid Rocket Axial Combustion Instability Suppression

David R. Greatrix

Department of Aerospace Engineering, Ryerson University, 350 Victoria Street, Toronto, ON, Canada M5B 2K3

Correspondence should be addressed to David R. Greatrix, greatrix@ryerson.ca

Academic Editor: Valsalayam Sanal Kumar

In the present investigation, various factors and trends, related to the usage of two or more sets of inert particles comprised of the same material (nominally aluminum) but at different diameters for the suppression of axial shock wave development, are numerically predicted for a composite-propellant cylindrical-grain solid rocket motor. The limit pressure wave magnitudes at a later reference time in a given pulsed firing simulation run are collected for a series of runs at different particle sizes and loading distributions and mapped onto corresponding attenuation trend charts. The inert particles' presence in the central core flow is demonstrated to be an effective means of instability symptom suppression, in correlating with past experimental successes in the usage of particles. However, the predicted results of this study suggest that one needs to be careful when selecting more than one size of particle for a given motor application.

1. Introduction

Over the last number of decades, a multitude of research efforts have been directed towards understanding the physical mechanisms, or at least the surrounding factors, behind the appearance of symptoms typically associated with nonlinear axial combustion instability in solid-propellant rocket motors (SRMs). The principal symptoms are the presence within the motor chamber of stronger finite-amplitude traveling axial pressure waves that may be shock fronted, commonly (although not always) accompanied by some degree of base chamber pressure rise (dc shift). Note that low-magnitude pressure waves due to vortex shedding from segmented/gapped components in the motor chamber are not included (here) in this more traditional category of nonlinear axial instability. Studies of nonlinear axial combustion instability have ranged from numerous experimental test firing series on the one hand [1–3], and linear/nonlinear acoustic theory modeling on the other (largely, the analysis producing frequency-based standing wave solutions for a given chamber geometry, but without some useful quantitative information) [4–7]. On occasion, researchers have employed a numerical modeling approach, to work towards a more comprehensive quantitative understanding of the physics involved (the numerical model producing a traveling wave solution to a limit wave amplitude and corresponding small or larger dc shift, typically a time-based result evolving from an initial pulse disturbance introduced into the chamber flow) [8, 9]. Available computational power and associated result turnaround times commonly forced some simplifications in the given numerical model.

The motivation for the experimental, analytical, and numerical studies noted above was and is of course to bring this better understanding to bear in more precisely suppressing, if not eliminating, these axial instability symptoms. For example, it has been long known that inert (nonreactive) or reactive particles in the internal core flow can help to suppress axial combustion instability symptoms [10–12]. As pointed out by Blomshield [13], in his wide-ranging review of a number of cases of different motors experiencing combustion instability over the years, it is not always clear as to the quantity of particle loading (and corresponding particle size) that is needed to adequately suppress the given symptoms, if that is the suppression technique being exploited. An additional potential complication, pointed out by Waesche [14], is that it is not always clear that the effectiveness of particle or additive loading is due entirely to particle/drag effects within the central core flow, or in fact in part or in whole due to altered combustion response of the propellant, as a result of the presence of the particle/additive at the burning surface before entering the core flow region. Waesche suggests that this effect may be more readily observed for reactive particles, rather than inert ones, considering the heat transfer effects in the solid phase [14]. Given this background from past experimental observations, it would be advantageous to have a predictive numerical simulation model that would help establish the particle loading/sizing requirement for a given SRM, in this case for using inert particles of differing sizes.

An effective numerical model combines the effects of the unsteady one- or two-phase flow, the transient combustion process, and the structural dynamics of the surrounding propellant/casing structure. A case study reported by Blomshield [13], where the changing of a heavyweight static-test motor casing to a flightweight casing structure led to the appearance of combustion instability symptoms, provides one motivation for inclusion of structural effects in the numerical model. Experimental observations by Dotson and Sako [15] on in-flight fluid-structure interaction effects lend further weight in this regard.

In the present investigation, an updated numerical model incorporating the above attributes is used in the prediction of the unsteady instability-related behavior in a cylindrical-grain motor and allows for an evaluation of the corresponding effectiveness of using two or more sets of inert spherical particles (same material (nominally aluminum), differing diameters) in suppressing instability symptoms. While aluminum as a common solid propellant fuel addition is reactive (noninert) in practice (and its burning and other behavior at and away from the propellant surface in the central flow may have a significant influence on the given SRM's combustion stability), the properties of aluminum are assumed for the inert particles in this study so as to allow for comparison to the results of future studies where the aluminum particles are modelled as reactive. In practice, one can note that inert particles composed of such materials as aluminum oxide (which forms from the combustion of aluminum and oxygen) or zirconium carbide do see usage for combustion stabilization purposes. The present study is a followon to the study reported in [16] (where the use of a single set of inert spherical particles is examined). In practice, one might see the use of two or more particle sizes in a given motor. This is sometimes done to target two or more different pressure wave frequencies that have been identified as problematic (e.g., one longitudinal and two transverse if using three particle sets). The Dobbins-Temkin correlation [17] indicates that the best particle diameter d_{opt} for suppression is a function of the inverse square root of the target frequency; that is, likely a smaller diameter particle is more effective at a higher transverse frequency, versus a lower axial frequency, everything else being equal

$$d_{\text{opt}} = \sqrt{\frac{9\mu}{\pi f \rho_m}}. \quad (1)$$

In the present paper, the focus for presented results will largely be on those cases where the transient burning response of the propellant is the primary mechanism for sustaining appreciable traveling pressure waves in the combustion chamber. A few additional results will illustrate the effect of normal acceleration (through radial vibration) as a complementary mechanism acting on the transient combustion process.

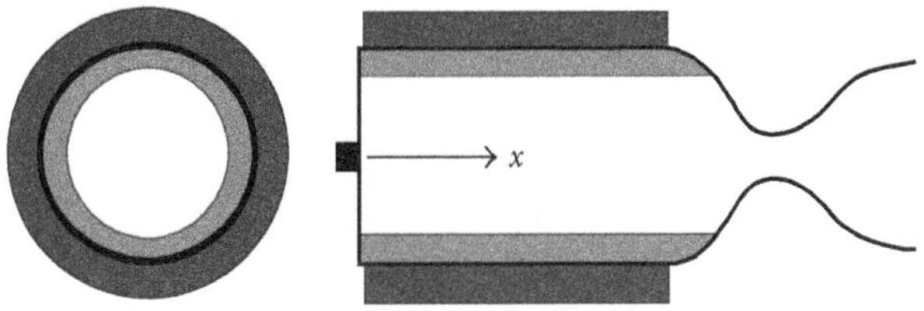

Figure 1: Schematic diagram of sleeved cylindrical-grain SRM, showing reference x-direction.

2. Numerical Model

A simplified schematic diagram of the physical system of an SRM, that is placed on a static test stand, is provided in Figure 1. In this case, the cylindrical-grain motor is free to vibrate radially without any external constraint (i.e., only constrained as indicated by the thick steel static-test sleeve surrounding the aluminum flightweight motor casing), while axial motion is constrained to a large degree by a thrust-measuring load cell at the lefthand boundary. Under normal (nominal) quasiequilibrium operating conditions, the internal gas flow (or gas-particle flow, if two-phase) moves smoothly from the burning propellant surface into the central core flow, heading downstream to eventually pass through and beyond the exhaust nozzle.

2.1. Equations Relevant to Two-Phase Flow in SRM. One defines b as the nonequilibrium sound speed of a 2-phase mixture, which for a single set of monodisperse particles within the gas can be estimated via [18]

$$b = a\left[\frac{1+\beta_g\left(C_m/C_p\right)\left(T_p/T\right)}{\mathfrak{C}+\alpha_g\beta_g(\gamma-1)\left(u_p/u\right)\left(1-\left(u_p/u\right)\right)}\right]^{1/2}, \quad (2)$$

where $\mathfrak{C}$ denotes $\alpha_g^2(1+\beta_g(u_p/u))(1+\gamma\beta_g(C_m/C_p)(T_p/T))$. The gas phase void fraction α_g is defined by

$$\alpha_g = \frac{\forall - \forall_p}{\forall} = 1 - \frac{\forall_p}{\forall}. \quad (3)$$

For a lower particle loading, one can assume that α_g is close to unity in value ($\forall_p$ is the volume occupied by the particles in an elemental total volume of $\forall$). The particle-gas mass flux ratio β_g is stipulated as

$$\beta_g = \frac{\rho_p u_p}{\rho u \alpha_g}. \quad (4)$$

One can define an average two-phase density in a given volume as

$$\rho_{2\text{ph}} = \frac{m}{\forall} \approx \frac{N \cdot m_p + \rho\forall}{\forall} = \frac{\rho_p\forall + \rho\forall}{\forall} = \alpha_p\rho_{2\text{ph}} + \rho, \quad (5)$$

where α_p is the particle-loading fraction in the flow, and N is the number of particles of average mass m_p in the elemental volume above, then one can show the correlation between particle density ρ_p and gas density ρ

$$\rho_p = \frac{N \cdot m_p}{V\!\!\!\!-} = \alpha_p \rho_{2\text{ph}} = \frac{\alpha_p}{1-\alpha_p} \cdot \rho. \tag{6}$$

When one has two or more sets of particles of differing sizes (comprised of the same material; thus, the same solid specific heat C_m), one can use the following correlations:

$$\rho_p = \frac{\sum_{i=1}^{N_{\text{set}}} N_i m_{p,i}}{V\!\!\!\!-} = \sum_{i=1}^{N_{\text{set}}} \rho_{p,i},$$

$$\alpha_p = \sum_{i=1}^{N_{\text{set}}} \alpha_{p,i},$$

$$\beta_g = \frac{\dot{m}_p}{\dot{m}_g} = \frac{\sum_{i=1}^{N_{\text{set}}} \rho_{p,i} u_{p,i}}{\rho u \alpha_g},$$

$$\beta_g \frac{C_m}{C_p} \frac{T_p}{T} = \frac{C_m}{C_p T} \cdot \frac{\sum_{i=1}^{N_{\text{set}}} \rho_{p,i} u_{p,i} T_{p,i}}{\rho u \alpha_g} = \Sigma_1, \tag{7}$$

$$\beta_g \frac{u_p}{u} = \frac{\sum_{i=1}^{N_{\text{set}}} \rho_{p,i} u_{p,i}^2}{\rho u^2 \alpha_g} = \Sigma_2,$$

$$\beta_g \frac{u_p^2}{u^2} = \frac{\sum_{i=1}^{N_{\text{set}}} \rho_{p,i} u_{p,i}^3}{\rho u^3 \alpha_g} = \Sigma_3,$$

so that the nonequilibrium sound speed can be estimated via

$$b = a \left[\frac{1+\Sigma_1}{\alpha_g^2 (1+\Sigma_2)(1+\gamma\Sigma_1) + \alpha_g (\gamma-1)(\Sigma_2 - \Sigma_3)} \right]^{1/2}. \tag{8}$$

Under nominal flow conditions, u/b would be unity at the nozzle throat.

The effect of particle mass loading ΔM_p into a solid propellant, that originally was of solid density $\rho_{s,o}$, on the loading mass fraction into a solid volume $V\!\!\!\!-$ is given by

$$\alpha_{p,s} = \frac{\Delta M_p}{\Delta M_{s,o} + \Delta M_p} = \frac{\rho_m V\!\!\!\!-_m}{\rho_{s,o}(V\!\!\!\!- - V\!\!\!\!-_m) + \rho_m V\!\!\!\!-_m} = \frac{1}{(\rho_{s,o}/\rho_m)(V\!\!\!\!-/V\!\!\!\!-_m - 1) + 1}. \tag{9}$$

The new effective solid propellant overall density becomes

$$\rho_{s,\text{new}} = \frac{\Delta M_p + \Delta M_{s,o}}{V\!\!\!\!-} = \frac{\rho_{s,o}(V\!\!\!\!- - V\!\!\!\!-_m) + \rho_m V\!\!\!\!-_m}{V\!\!\!\!-} = \rho_{s,o}\left(1 - \frac{V\!\!\!\!-_m}{V\!\!\!\!-}\right) + \rho_m \frac{V\!\!\!\!-_m}{V\!\!\!\!-}. \tag{10}$$

By substitution, one can show that

$$\rho_{s,\text{new}} = \frac{\rho_{s,o}}{1 - \alpha_{p,s} + \alpha_{p,s}(\rho_{s,o}/\rho_m)}. \tag{11}$$

2.2. Equations of Motion-Governing Two-Phase Flow. The equations of motion describing the nonsteady core flow within the SRM must be solved in conjunction with the local pyrolysis rate r_b of the solid propellant, and the surrounding structure's instantaneous geometric deformation. As pertains to the present study of a small motor having a larger length-to-diameter ratio, the quasi-one-dimensional hydrodynamic conservation equations for the axial gas flow are given below

$$\begin{aligned}
&\frac{\partial \rho}{\partial t} + \frac{\partial(\rho u)}{\partial x} = -\frac{1}{A}\frac{\partial A}{\partial x}\rho u + \left(1-\alpha_p\right)\rho_s \frac{4r_b}{d} - \left(\frac{4r_b}{d} + \kappa\right)\rho, \\
&\frac{\partial(\rho u)}{\partial t} + \frac{\partial}{\partial x}(\rho u^2 + p) = -\frac{1}{A}\frac{\partial A}{\partial x}\rho u^2 - \left(\frac{4r_b}{d} + \kappa\right)\rho u - \rho a_\ell - \sum_{i=1}^{N_{\text{set}}} \frac{\rho_{pi}}{m_{pi}} D_i, \\
&\frac{\partial(\rho E)}{\partial t} + \frac{\partial}{\partial x}(\rho u E + u p) = -\frac{1}{A}\frac{\partial A}{\partial x}(\rho u E + u p) - \left(\frac{4r_b}{d} + \kappa\right)\rho E \\
&\quad + \left(1-\alpha_p\right)\rho_s \frac{4r_b}{d}\left(C_p T_f + \frac{v_f^2}{2}\right) - \rho u a_\ell - \sum_{i=1}^{N_{\text{set}}} \frac{\rho_{pi}}{m_{pi}}\left(u_{pi} D_i + Q_i\right).
\end{aligned} \tag{12}$$

Here, the total specific energy of the gas is defined for an ideal gas as $E = p/[(\gamma-1)\rho] + u^2/2$. The corresponding equations of motion for an ith inert (nonburning) particle set within the axial flow may be found from

$$\begin{aligned}
&\frac{\partial \rho_{pi}}{\partial t} + \frac{\partial(\rho_{pi} u_{pi})}{\partial x} = -\frac{1}{A}\frac{\partial A}{\partial x}\rho_{pi} u_{pi} + \alpha_{pi}\rho_s \frac{4r_b}{d} - \left(\frac{4r_b}{d} + \kappa\right)\rho_{pi}, \\
&\frac{\partial(\rho_{pi} u_{pi})}{\partial t} + \frac{\partial(\rho_{pi} u_{pi}^2)}{\partial x} = -\frac{1}{A}\frac{\partial A}{\partial x}\rho_{pi} u_{pi}^2 - \left(\frac{4r_b}{d} + \kappa\right)\rho_{pi} u_{pi} - \rho_{pi} a_\ell + \frac{\rho_{pi}}{m_{pi}} D_i, \\
&\frac{\partial(\rho_{pi} E_{pi})}{\partial t} + \frac{\partial(\rho_{pi} u_{pi} E_{pi})}{\partial x} = -\frac{1}{A}\frac{\partial A}{\partial x}(\rho_{pi} u_{pi} E_{pi}) - \left(\frac{4r_b}{d} + \kappa\right)\rho_{pi} E_{pi} \\
&\quad + \alpha_{pi}\rho_s \frac{4r_b}{d}\left(C_m T_f + \frac{v_f^2}{2}\right) - \rho_{pi} u_{pi} a_\ell + \frac{\rho_{pi}}{m_{pi}} \times \left(u_{pi} D_i + Q_i\right).
\end{aligned} \tag{13}$$

Here, the total specific energy of a local grouping of particles from an ith set is given by $E_{pi} = C_m T_{pi} + u_{pi}^2/2$, where T_{pi} is the mean temperature of that group. As outlined in [19], the viscous interaction between the gas and a particle from the ith particle set is represented by the drag force D_i, and the heat transfer from the core flow to a particle from the ith set is defined by Q_i. In the case of drag between the gas and a representative spherical particle at a given axial location, one notes that

$$D_i = \frac{\pi d_{mi}^2}{8} C_d \rho \left(u - u_{pi}\right) \left| u - u_{pi} \right|, \tag{14}$$

where C_d is the drag coefficient for a sphere in a steady flow with low-flow turbulence (determined as function of relative Reynolds number, relative flow Mach number, and temperature difference between the particle and the gas). In the case of heat transfer from the core flow to a representative particle at a given axial location, the following applies:

$$Q_i = \pi d_{mi} k \cdot \mathrm{Nu} \cdot \left(T - T_{pi}\right), \tag{15}$$

where the Nusselt number Nu can be found as a function of Prandtl and relative Reynolds number for a sphere of mean diameter d_{mi}. One will need to solve (13) for each of N_{set} particle sets as part of the calculation process, where for dual or triple particle set loading, N_{set} has a value of 2 or 3 in the present study.

Longitudinal acceleration a_l appears in the gas and particle momentum and energy equations as a body force contribution within a fixed Eulerian reference (fixing of $x = 0$ to motor head end, x positive moving right on structure as per Figure 1; acceleration of local surrounding structure rightward is designated positive a_l) and may vary both spatially along the length of the motor and with time. The effects of such factors as turbulence can be included through one or more additional equations that employ the information from the bulk flow properties arising from the solution of the above-one-dimensional equations of motion. The principal differential equations themselves can be solved via a higher-order, explicit, and finite-volume random-choice method (RCM) approach [19, 20]. The RCM solver employs a Riemann-solution technique noted for low artificial dispersion with time of wave activity in tubes, and so forth. The equations of motion of the gas and particles will be solved over a given time step Δt (on the order of 1×10^{-7} s for the present study, given the motor solution node allocation in the axial direction from head end to nozzle exit plane), in sequence with additional equations for structural motion and propellant burning rate as described below.

2.3. Equations for Structural Motion. Structural vibration can play a significant role in nonsteady SRM internal ballistic behavior, as evidenced by observed changes in combustion instability symptoms as allied to changes in the structure surrounding the internal flow (e.g., propellant grain configuration, wall thickness, and material properties) [13, 15, 21]. The level of sophistication required for modeling the motor structure (propellant, casing, static-test sleeve, and nozzle) and applicable boundary conditions (load cell on static test stand) can vary, depending on the particular application and motor design. Loncaric et al. [22] and Montesano et al. [23] employed a finite-element approach towards the structural modeling of the given motor configuration. In the present study, a cylindrical-grain configuration allows for a simpler finite-difference approach via thick-wall theory, as reported in [20, 24]. The radial deformation dynamics of the propellant/casing/sleeve are modeled by a series of independent ring elements along the length of the motor. Axial motion along the length of the structure is modeled via beam theory, and bounded by the spring/damper load cell at the motor's head end. Viscous damping is applied in the radial and axial directions. Reference structural properties are assumed for an ammonium-perchlorate/hydroxyl-terminated polybutadiene (AP/HTPB) composite propellant surrounded by an aluminum casing and steel sleeve. For greater accuracy, some properties like the propellant/casing/sleeve assembly's natural radial frequency may be predetermined via a finite-element numerical solution, rather than via theoretical approximations [24].

2.4. Equations for Propellant Burning Rate. With respect to transient, frequency-dependent burning rate modeling, the Z-N (Zeldovich-Novozhilov) solid-phase energy conservation approach used in the present simulation program may be represented by the following time-dependent temperature-based relationship [25]:

$$r_b^* = r_{b,\mathrm{qs}} - \frac{1}{\left(T_s - T_i - \Delta H_s/C_s\right)} \frac{\partial}{\partial t} \int_{-\infty}^{0} \Delta T dy, \tag{16}$$

where $r_{b,\mathrm{qs}}$ is the quasisteady burning rate (value for burning rate as estimated from steady-state information for a given set of local flow conditions), T_i is the initial propellant temperature, and in this context, $\Delta T = T(y,t) - T_i$ is the temperature distribution in moving from the burning propellant surface at $y = 0$ (and $T = T_s$) to that spatial location in the propellant where the temperature reaches T_i. One may note at this juncture the inclusion of a net surface heat release term, ΔH_s, in the calculations. The transient heat conduction in the solid phase can be solved by an appropriate finite-difference scheme. One needs to take care in setting the solid-phase spatial increment Δy, to be in accordance with the Fourier stability limit, $\Delta y_{\mathrm{Fo}} = (2\alpha_s \Delta t)^{1/2}$, which is a function of the chosen time increment Δt [25]. The time increment itself must be coordinated between the flow and structural model solution systems [23].

In (16), r_b^* is the nominal (unconstrained) instantaneous burning rate, and its value at a given propellant grain location is solved at each time increment via numerical integration of the temperature distribution through the heat penetration zone of the solid phase. The actual instantaneous burning rate r_b may be found as a function of r_b^* through the empirical rate-limiting equation [25]

$$\frac{dr_b}{dt} = K_b\left(r_b^* - r_b\right). \tag{17}$$

The rate-limiting coefficient K_b effectively damps the unconstrained burning rate r_b^* when for a finite time increment Δt

$$K_b < \frac{1}{\Delta t}. \quad (18)$$

In the present approach, the surface-thermal gradient is free to find its own value at a given instant. One can argue that the use of (17) or some comparable damping function, while empirical, parallels the approach taken by past researchers in using a stipulated surface-thermal gradient; both approaches act to constrain the exchange of energy through the burning surface interface, allow for some variability in better comparing to a given set of experimental data, and prevent so-called burning-rate "runaway" (unstable divergence of r_b with time) [26]. As discussed in [25], the use of K_b at a set value does allow for a converged solution that is independent of the increment size for Δt and Δy, as long as one respects the Fourier stability requirement noted earlier.

The quasisteady burning rate $r_{b,\text{qs}}$ may be ascertained as a function of various parameters; in this study, as a function of local static pressure p, core flow velocity u (erosive burning component), and normal/lateral/longitudinal acceleration, such that:

$$r_{b,\text{qs}} = r_p + r_u + r_a. \quad (19)$$

The pressure-based burning component may be found through de St. Robert's law

$$r_p = Cp^n. \quad (20)$$

The flow-based erosive burning component (negative and positive) is established through the following expression [27]:

$$r_b = \left.\frac{r_b}{r_o}\right|_{\delta_r} \cdot r_o + r_e, \quad (21)$$

where at lower flow speeds, the negative component resulting from a stretched combustion zone thickness ($\sigma_r > \sigma_o$) may cause an appreciable drop in the base burn rate r_o, while at higher flow speeds, the positive erosive burning component r_e, established from a convective heat feedback premise [27], should dominate:

$$r_e = \frac{h\left(T_f - T_s\right)}{\rho_s\left[C_s(T_s - T_i) - \Delta H_s\right]}. \quad (22)$$

For the above case, where the base burning rate r_o is a function of the other mechanisms (pressure and acceleration), one finds the velocity-based component of burn rate from (21) via $r_u = r_b - r_o$. At higher flow speeds, r_u becomes equivalent to r_e. The effect of normal acceleration a_n resulting from radial propellant/casing/sleeve vibration may be determined via [28]

$$r_b = \left[\frac{C_p\left(T_f - T_s\right)}{C_s(T_s - T_i) - \Delta H_s}\right]\frac{(r_b + G_a/\rho_s)}{\exp\left[C_p\delta_o(\rho_s r_b + G_a)/k\right] - 1}, \quad (23)$$

where the compressive effect of normal acceleration and the dissipative effect of steady or oscillatory longitudinal (or lateral, if say for a star grain configuration) acceleration a_l are stipulated through the accelerative mass flux G_a

$$G_a = \left\{\frac{a_n p}{r_b}\frac{\delta_o}{RT_f}\frac{r_o}{r_b}\right\}_{\phi=0^\circ} \cos^2\phi_d. \quad (24)$$

Note that the longitudinal/lateral-acceleration-based displacement orientation angle ϕ_d is greater than the nominal acceleration vector orientation angle (ϕ; zero when only normal acceleration a_n relative to the burning propellant surface is present) [28]. One should also note that a_n is negative when acting to compress the combustion zone, and treated as zero when directed away from the zone. For the above case, where the base burning rate r_o is a function of the other flow mechanisms (pressure and core flow), one finds the acceleration-based component of burning rate from (23) via $r_a = r_b - r_o$.

With respect to the burning surface temperature T_s, one has the option of treating it as constant, or allowing for its variation, depending on the phenomenological approach being taken for estimating the burning rate [25]. While in the past a number of estimation models might have used a constant value for T_s, more recently the usage of a variable T_s has become prevalent. However, based on good comparisons in general to experimental data as reported in [25], a constant T_s was employed in the present Z-N-based phenomenological numerical combustion model, for the present investigation.

3. Results and Discussion

The characteristics of the reference motor for this study are listed in Table 1. The motor, based in large measure on a similar experimental motor [20] is a smaller cylindrical-grain design with an aluminum casing and static-test steel sleeve, with a relatively large length-to-diameter ratio. The motor at the time of pulsing has a moderate port-to-throat area ratio, with a considerable propellant web thickness remaining. The predicted frequency response for the AP/HTPB propellant at three different settings for the net surface heat release value may be viewed in Figure 2 (positive value, exothermic heat release). The general response is given in terms of the nondimensional limit magnitude M_l, defined by

$$M_\ell = \frac{r_{b,\text{peak}} - r_{b,o}}{r_{b,\text{qs,peak}} - r_{b,o}}, \quad (25)$$

where the reference burning rate $r_{b,o}$ in this case is the motor's approximate mean burn rate at the point of pulsing (1.27 cm/s). The propellant's resonant frequency f_r is set via the value of K_b (20000 s^{-1}) to be on the order of 1 kHz (a value within the range of what might be expected for this type of composite propellant at that base burning rate). This

TABLE 1: Reference motor characteristics.

Parameter	Value
Propellant grain length, L_p	52 cm
Initial port diameter, d_i	3.6 cm
Nozzle throat diameter, d_t	1.6 cm
Grain/nozzle-conv. length ratio, L_p/L_c	16 : 1
Propellant specific heat, C_s	1500 J/kg-K
Propellant density, ρ_s	1730 kg/m^3
Propellant thermal conductivity, k_s	0.4 W/m-K
Propellant thermal diffusivity, α_s	1.54×10^{-7} m^2/s
Propellant flame temperature, T_f	3000 K
Propellant surface temperature, T_s	1000 K
Propellant initial temperature, T_i	294 K
Gas specific heat, C_p	1920 J/kg-K
Specific gas constant, R	320 J/kg-K
Gas thermal conductivity, k	0.2 W/m-K
Gas absolute viscosity, μ	8.07×10^{-5} kg/m-s
Gas specific heat ratio, γ	1.2
De St. Robert exponent, n	0.35
De St. Robert coefficient, C	0.05 cm/s-(kPa)^n
Particle solid density, ρ_m	2700 kg/m^3
Particle specific heat, J/kg-K	900 J/kg-K
Particle mass fraction, α_p	0%
Propellant elastic modulus, E_A	45 MPa
Propellant Poisson's ratio, ν_A	0.497
Casing inner wall radius, r_m	3.24 cm
Casing wall thickness, h_B	0.127 cm
Casing material density, ρ_B	2700 kg/m^3
Casing elastic modulus, E_B	80 GPa
Casing material Poisson's ratio, ν_B	0.33
Sleeve wall thickness, h_C	0.47 cm
Sleeve material density, ρ_C	7850 kg/m^3
Sleeve elastic modulus, E_C	200 GPa
Sleeve material Poisson's ratio, ν_C	0.30
Casing/prop. rad. damping ratio, ξ_R	0.35
Casing/prop. long. damping ratio, ξ_L	0.10

value for f_r is in fact relatively close to the fundamental longitudinal acoustic frequency f_{1L} of the combustion chamber, in providing examples later in this paper that are close to the worst-case scenario for susceptibility to axial combustion instability symptoms.

An initial pulsed-firing simulation run was completed as a starting reference for this study, in which no particles are present or any other suppression technique being applied. In Figure 3 for head-end pressure p_c as a function of time, one can see that, at some point, the principal compression wave reaches its quasiequilibrium strength from an initial disturbance pressure Δp_d of 2 atm, the sustained compression wave front arriving about every 1 ms, oscillating at the fundamental frequency f_{1L} of 1 kHz. The base pressure is not appreciably elevated over the nominal operating chamber pressure. The effect of normal acceleration on the burning process (related to the radial vibration of the motor

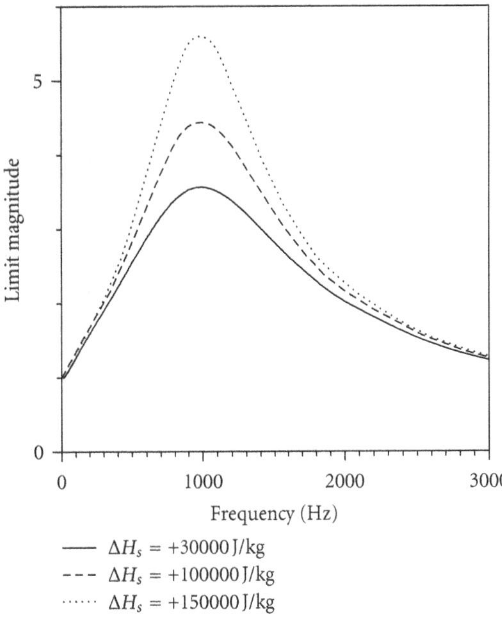

FIGURE 2: Frequency response of reference propellant ($r_{b,o}$ = 1.27 cm/s, K_b = 20000 s^{-1}, differing ΔH_s) in terms of nondimensional limit magnitude.

propellant/casing; see [16, 20, 23]) has been nullified for this simulation (in order to isolate frequency-dependent Z-N combustion response as the predominant instability symptom driver), a factor in reducing the development of a dc shift. One can note that the limit pressure wave magnitude (Δp_w, peak to trough) is decreasing gradually with time after first reaching its quasiequilibrium level, as the cylindrical grain burns back and the base pressure rises.

One can refer to Figure 4 for the pressure-time profile for the same motor, but now with 5% particle loading (by mass) of inert spherical aluminum particles having a mean 10-μm diameter. Of course, in practice, the aluminum particles would in fact be reactive (burning, if sufficient reactants like oxygen or chlorine are present in the surrounding gas), and as a result, in general continually decreasing in diameter with particle surface regression as they move aft towards the nozzle. There is also the possibility of particle agglomeration, or the coming together of two or more particles. Given the scope of the present investigation, calculations for particle regression or agglomeration were not to be done; one can consider the mean inert aluminum particle diameter for a given set as a reference size, providing results which may prove useful as a guideline when one does move to inclusion of particle burning in the computational model. Observing the results of Figure 4, suppression of axial wave development after an initial 2-atm pulse is near-complete (limit magnitude of the sustained pressure wave, at 0.26 s, is about 0.045 MPa [Δp_w], as compared to 1.42 MPa for the 0% loading case noted earlier [$\Delta p_{w,\text{peak}}$], giving a nondimensional attenuation M_a, defined by

$$M_a = \frac{\Delta p_{w,\text{peak}} - \Delta p_w}{\Delta p_{w,\text{peak}}}, \tag{26}$$

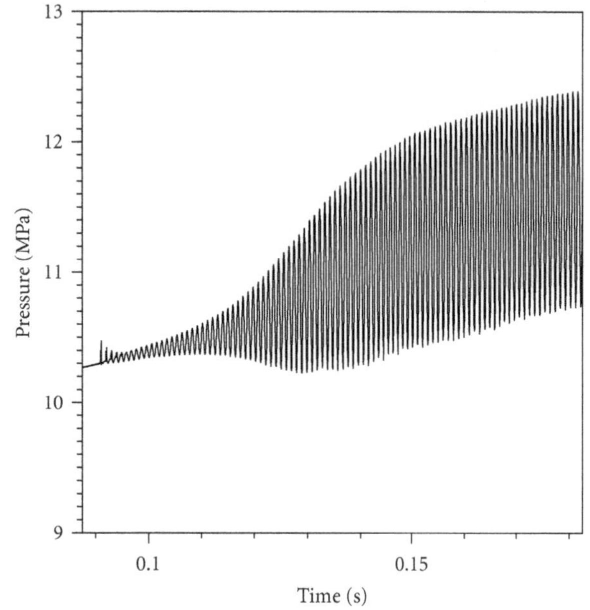

FIGURE 3: Predicted head-end pressure-time profile, reference motor ($K_b = 20000\,\text{s}^{-1}$, $\Delta H_s = 150000\,\text{J/kg}$, $\Delta p_d = 2\,\text{atm}$, $\alpha_p = 0\%$), acceleration nullified, no particles.

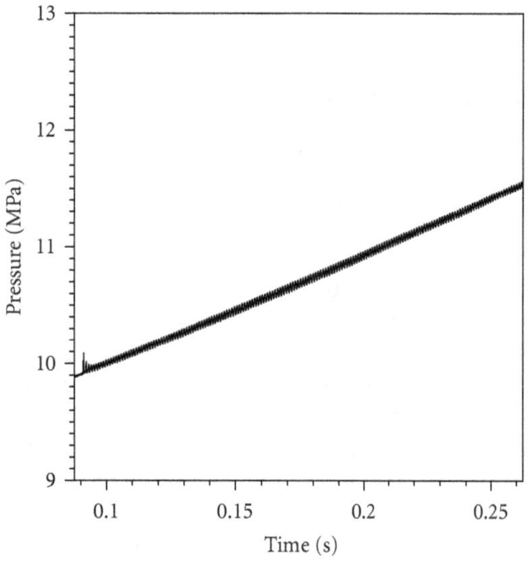

FIGURE 4: Predicted head-end pressure wave profile, reference motor ($K_b = 20000\,\text{s}^{-1}$, $\Delta H_s = 150000\,\text{J/kg}$, $\Delta p_d = 2\,\text{atm}$), acceleration nullified, 5% 10-μm Al particle loading by mass, single particle set.

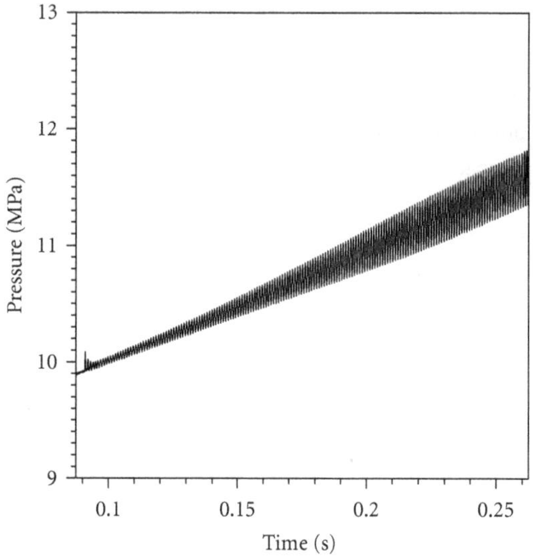

FIGURE 5: Predicted head-end pressure wave profile, reference motor ($K_b = 20000\,\text{s}^{-1}$, $\Delta H_s = 150000\,\text{J/kg}$, $\Delta p_d = 2\,\text{atm}$, $\alpha_p = 5\%$, $d_{m1} = 10\,\mu\text{m}$, $d_{m2} = 5\,\mu\text{m}$, $\alpha_{p1} = 2.5\%$, $\alpha_{p2} = 2.5\%$), two particle sets, acceleration nullified.

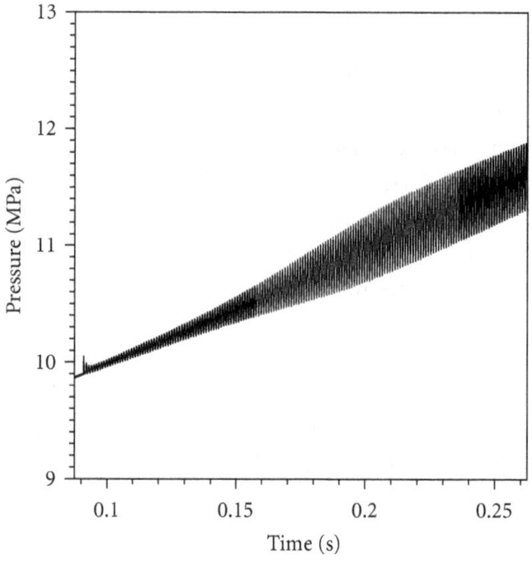

FIGURE 6: Predicted head-end pressure wave profile, reference motor ($K_b = 20000\,\text{s}^{-1}$, $\Delta H_s = 150000\,\text{J/kg}$, $\Delta p_d = 2\,\text{atm}$, $\alpha_p = 5\%$, $d_{m1} = 10\,\mu\text{m}$, $d_{m2} = 5\,\mu\text{m}$, $d_{m3} = 40\,\mu\text{m}$, $\alpha_{p1} = 1.67\%$, $\alpha_{p2} = 1.67\%$, $\alpha_{p3} = 1.66\%$), three particle sets, acceleration nullified.

a value of 0.97, noting that a value of unity is complete suppression). Historically, suppression of high-frequency tangential and radial pressure waves in SRMs by the use of particles in the range of 1 to 3% loading by mass has been in general largely successful. In the case of axial pressure waves, the effectiveness of particles from 1% to over 20% loading in suppressing wave development has been less consistent, relative to the previously mentioned transverse cases. In the case of Figure 4, remembering that acceleration as a factor has been nullified in the combustion process, a loading of 5% at 10 μm does appear to effectively suppress axial wave development in this particular motor, at this point in its firing.

In considering an example of two particle sets being used (5 and 10 μm diameters for particles of the same material), in Figure 5 one sees the result of an evenly split 1 : 1 distribution of the two sets at an overall loading of 5%. The limit = magnitude of the sustained pressure wave, at 0.26 s, is about 0.47 MPa, as compared to about ten times less for the uniform 10-μm-loading case noted for Figure 4, giving a nondimensional attenuation M_a value of 0.67 as compared to

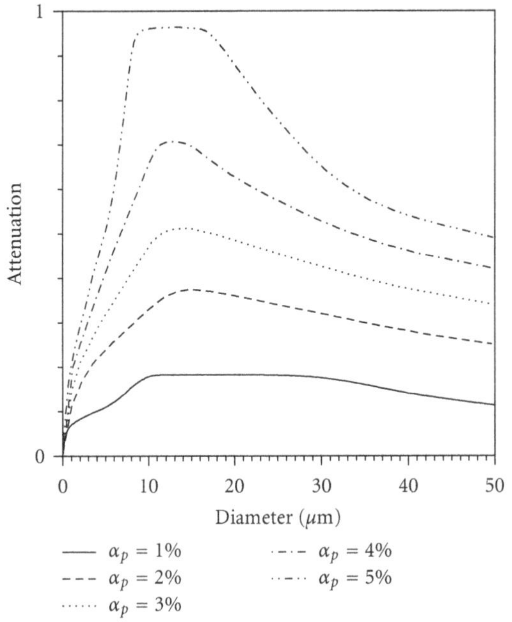

FIGURE 7: Nondimensional attenuation as function of particle diameter and loading of a single particle set, reference motor, acceleration nullified.

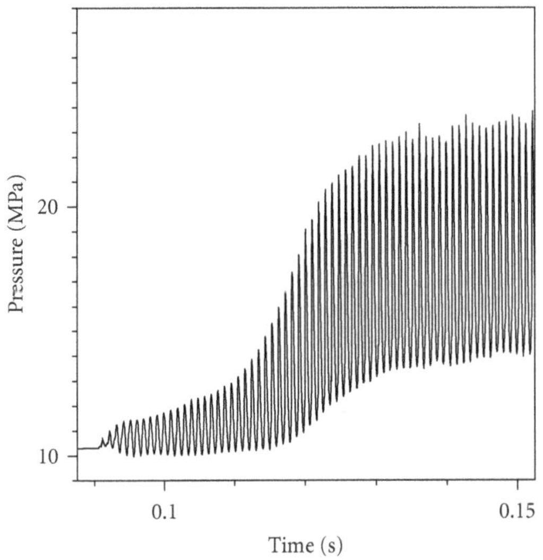

FIGURE 8: Predicted head-end pressure-time profile, reference motor (K_b = 20000 s^{-1}, ΔH_s = 150000 J/kg, Δp_d = 2 atm, α_p = 0%), acceleration active, no particles.

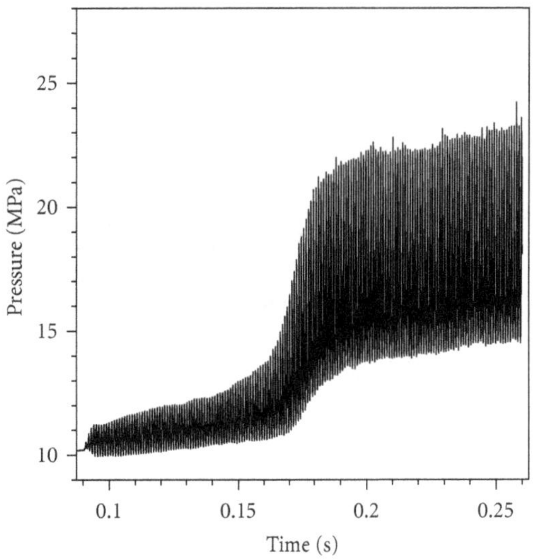

FIGURE 9: Predicted head-end pressure wave profile, reference motor (K_b = 20000 s^{-1}, ΔH_s = 150000 J/kg, Δp_d = 2 atm, α_p = 1.5%, d_{m1} = 10 μm, d_{m2} = 5 μm, d_{m3} = 60 μm, α_{p1} = 0.5%, α_{p2} = 0.5%, α_{p3} = 0.5%), three particle sets, acceleration active.

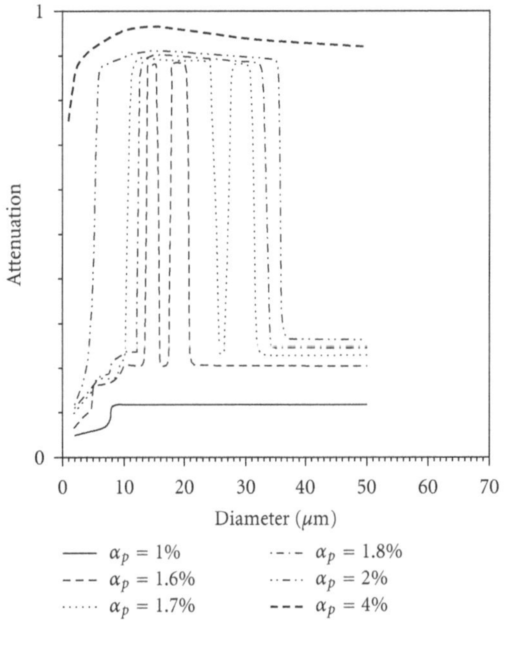

FIGURE 10: Nondimensional attenuation as function of particle diameter and loading of a single particle set, reference motor, acceleration active.

0.97. In considering an example of three particle sets being used (5, 10, and 40 μm diameters for particles of the same material), in Figure 6 one can observe the result of an evenly split 1 : 1 : 1 distribution of the three sets at an overall loading of 5%. The limit magnitude of the sustained pressure wave, at 0.26 s, is about 0.57 MPa, giving an M_a value of 0.6, or a bit less than the previous two-particle example (0.67). Referring to Figure 7 [16], one can observe that, in the limit of just using 5 μm particles at a 5% overall loading, the value for M_a is around 0.52, while, for 40 μm particles at a 5% overall loading, the value for M_a is around 0.55.

Let's consider the case when vibration-induced acceleration is active as a mechanism working in conjunction with the transient response of the burning solid propellant. Referring to Figure 8, allowing for the effect of vibration-induced acceleration on combustion, one has a much more active motor in the absence of particles in the flow, with a substantially bigger limit magnitude (around 10.2 MPa at

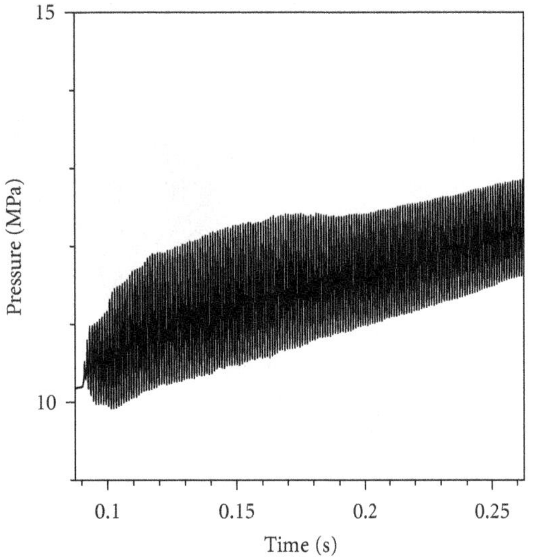

Figure 11: Predicted head-end pressure wave profile, reference motor ($K_b = 20000\,s^{-1}$, $\Delta H_s = 150000$ J/kg, $\Delta p_d = 2$ atm, $\alpha_p = 1.5\%$, $d_{m1} = 10\,\mu m$, $d_{m2} = 5\,\mu m$, $d_{m3} = 40\,\mu m$, $\alpha_{p1} = 0.5\%$, $\alpha_{p2} = 0.5\%$, $\alpha_{p3} = 0.5\%$), three particle sets, acceleration active.

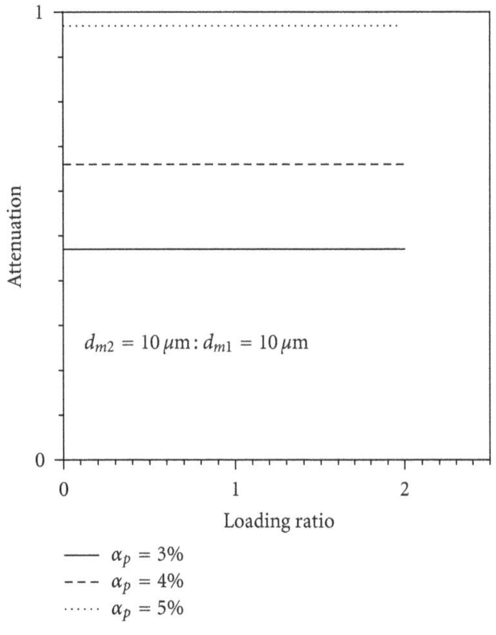

Figure 12: Nondimensional attenuation as function of particle diameter and loading distribution [$\alpha_{p2} : \alpha_{p1}$] for two sets of particles (10 μm [d_{m1}] and 10 μm [d_{m2}] particles), reference motor, acceleration nullified.

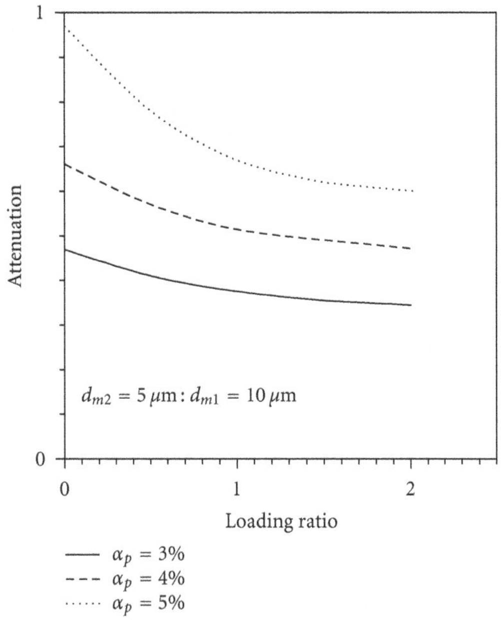

Figure 13: Nondimensional attenuation as function of particle diameter and loading distribution [$\alpha_{p2} : \alpha_{p1}$] for two sets of particles (10 μm [d_{m1}] and 5 μm [d_{m2}] particles), reference motor, acceleration nullified.

0.26 s) for the now shock-fronted axial pressure wave moving back and forth within the chamber than that seen in Figure 3, and a much more visible dc shift is present. In considering an example of three particle sets being used (5, 10, and 60 μm diameters for particles of the same material), in Figure 9 one can observe the result of an evenly split 1 : 1 : 1 distribution of the three sets at an overall loading of 1.5%. The limit magnitude of the sustained pressure wave, at 0.26 s, is about 8.97 MPa, giving an M_a value of 0.121. Referring to the single-set chart of Figure 10 [16], one can observe that in the limit of just using 5 μm particles at a 1.5% overall loading, the value for M_a is less than 0.15, using 10 μm particles at a 1.5% overall loading produces a value for M_a is less than 0.2, and for 60 μm particles at a 1.5% overall loading, the value for M_a is less than 0.2. In qualitative terms, one can state that the low-suppression attenuation observed in Figure 9 is reasonably consistent with what one would expect from the aggregate of the three individual diameter results of Figure 10 although, quantitatively, at least in this case, the 3-set attenuation seen in Figure 9 is a bit worse (lower) than any of the three in isolation.

One adjustment in particle size produces a significant change in the result of Figure 9, as evidenced by Figure 11. Using 40 μm particles in place of the 60-μm particles at a 0.5% loading as part of the overall 1.5% loading, the limit pressure wave magnitude is significantly decreased, down to 1.22 MPa from the earlier value of 8.97, or producing an M_a of around 0.88. Referring to the single-set trend chart of Figure 10, this indicates that the system has decided to shift to the high-suppression domain by this one adjustment, moving from the low-suppression domain that the system preferred in Figure 9.

Along the lines of Figures 7 and 10, one can produce charts showing trends as relates to using two sets of particles of the same material, but differing diameters. As the baseline case, Figure 12 shows an attenuation chart at three different overall particle mass loadings (α_p) for various loading distributions ($\alpha_{p_2} : \alpha_{p_1}$) of 10-$\mu$m particles. Acceleration effects on combustion are nullified for these examples. The

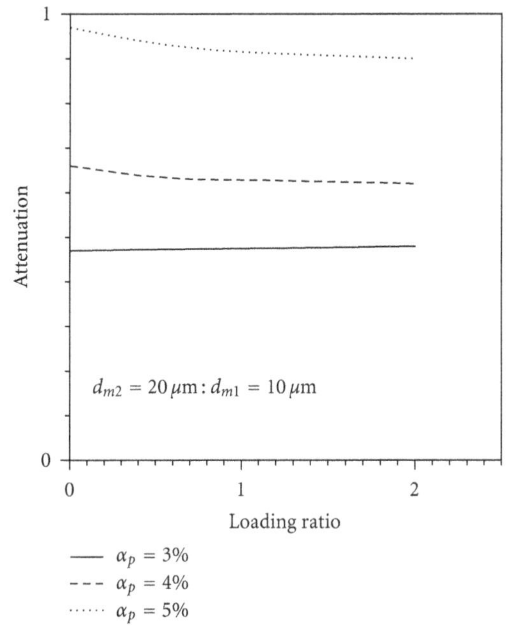

Figure 14: Nondimensional attenuation as function of particle diameter and loading distribution [α_{p2} : α_{p1}] for two sets of particles (10 μm [d_{m1}] and 20 μm [d_{m2}] particles), reference motor, acceleration nullified.

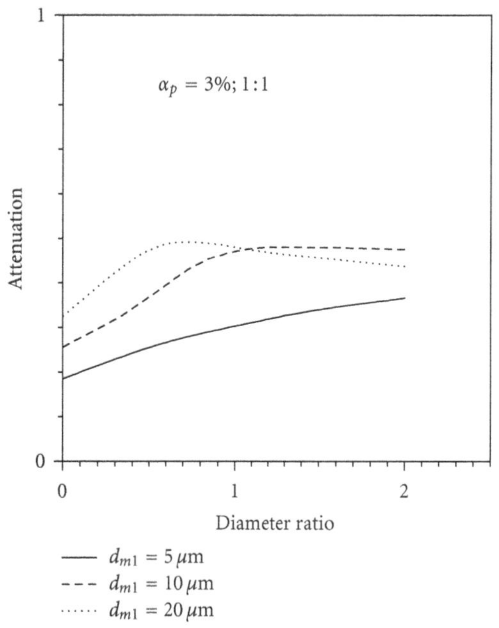

Figure 15: Nondimensional attenuation as function of particle diameter (d_{m2} : d_{m1}), for a 1 : 1 loading distribution for two sets of particles at total loading of 3%, reference motor, acceleration nullified.

loading ratio in these examples is referenced to the 10 μm size (d_{m1}), since as illustrated in Figure 7, it is nominally around the best size for attenuation of pressure-wave development in the single-size case. Because the particles of the two sets are the same size, there would not be any change in attenuation, as reflected by the resulting curves being horizontal lines from left to right. Figure 13 by contrast shows downward sloping lines when using a distribution of 5 and 10 μm particles at different overall loadings. Each of the 3 curves would continue rightward to their respective asymptotic limit, whereby only 5 μm particles are ultimately present (see Figure 7 for limit values of attenuation at the three loadings of the 5-μm size). A chart such as that shown by Figure 13 would also be instructive towards the case of using reactive particles that would reduce in size (due to burnback) during their lifetime in the motor chamber. The level of attenuation may not be as effective as one might expect from the baseline nominal starting value for the particle diameter. Figure 14 shows downward sloping lines when using a distribution of 20 and 10 μm particles at different overall loadings, but of a less severe nature (relative to that seen in Figure 13), given the similar effectiveness of the two particle sizes below an overall loading of 5% as per Figure 7.

An alternative format for illustrating trends associated with using two sets of particles is provided in Figures 15, 16, and 17, for overall particle mass loading percentages of 3, 4, and 5%. Here, the three graphs are restricted to evaluating an even 1 : 1 split in the distribution of the two sets, but with the particle diameters varied relative to each other. A more dramatic change in attenuation effectiveness as a function of varying the particle diameters is seen in the curves of Figure 17, for the 5% loading case, which again (referring to the previous paragraph), tends to correlate with the trends associated with Figure 7.

4. Concluding Remarks

A numerical evaluation of the use of two or three sets of different-sized nonburning particles within the flow as a means for suppressing axial pressure wave development has been completed for a reference cylindrical-grain composite-propellant motor, in cases where the transient burning response of the propellant is the primary mechanism for driving the instability symptoms, and in cases where both the transient burning response in conjunction with vibration-induced acceleration is playing a role. The ability of the particles to suppress axial wave development is evident, at relatively low loading percentages, results that are consistent with experimental experience. This is clearly reflected by the respective attenuation maps. If this or a comparable numerical model proves to be suitably accurate, such maps could prove a useful tool for motor designers evaluating their own motor configurations for instability behavior.

The adverse effect of loading a second (or a third) particle set that has a size that is less effective in suppressing pressure wave motion is also made evident in the present results. This study also gives some indication of what might be expected with reactive particles, where due to particle size reduction with time (under burning) the suppression effectiveness may not be quite what one would have ideally expected. In a similar vein, the present results are an indicator of the potential adverse effects of the agglomeration of particles in producing particle sizes bigger than what one would ideally find desirable for suppressing wave activity. These and other

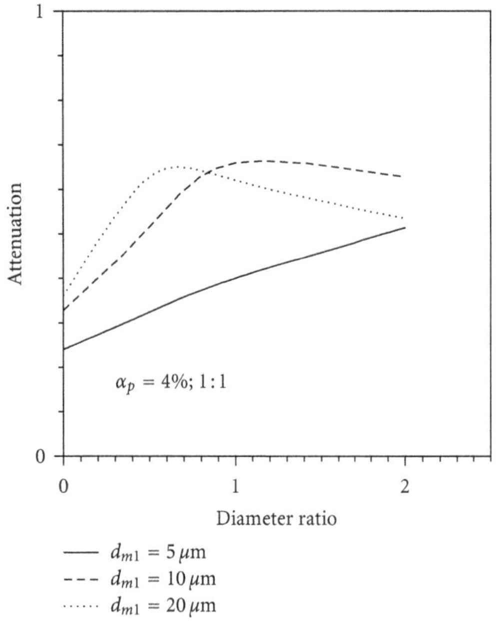

FIGURE 16: Nondimensional attenuation as function of particle diameter (d_{m2} : d_{m1}), for a 1 : 1 loading distribution for two sets of particles at total loading of 4%, reference motor, acceleration nullified.

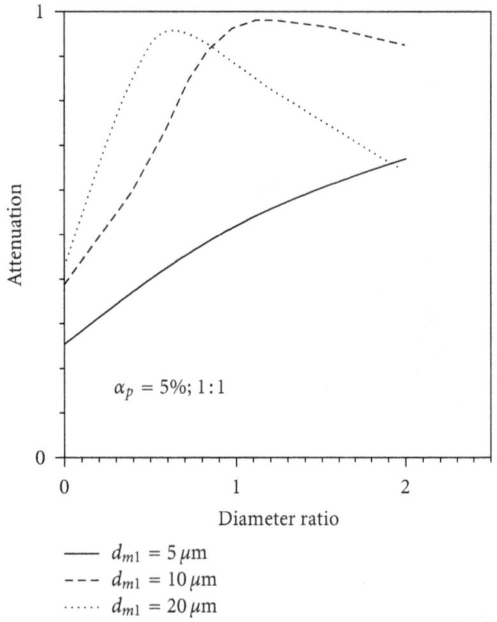

FIGURE 17: Nondimensional attenuation as function of particle diameter (d_{m2} : d_{m1}), for a 1 : 1 loading distribution for two sets of particles at total loading of 5%, reference motor, acceleration nullified.

issues, some of them potentially quite complex, remain to be explored in regards to the use and modelling of reactive particles for suppression of combustion instability in SRMs.

Nomenclature

A: local core cross-sectional area, m^2
a: gas sound speed, m/s
a_l: longitudinal (or lateral) acceleration, m/s^2
a_n: normal acceleration, m/s^2
b: nonequilibrium sound speed of two-phase mixture
C: de St. Robert coefficient, m/s-Pa^n
C_m: particle specific heat, J/kg-K
C_p: gas specific heat, J/kg-K
C_s: specific heat, solid phase, J/kg-K
D_i: drag of gas on a particle from ith particle set, N
d: local core hydraulic diameter, m
d_{mi}: mean particle diameter for ith particle set, m
E: local total specific energy of gas in core flow, J/kg
E_{pi}: local total specific energy, ith particle set in flow, J/kg
f: frequency, Hz, or Darcy-Weisbach friction factor
G_a: accelerative mass flux, kg/m^2-s
h: convective heat transfer coefficient, W/m^2-K
ΔH_s: net surface heat of reaction, J/kg
K_b: burn rate limiting coefficient, s^{-1}
k: gas thermal conductivity, W/m-K
k_s: thermal conductivity, solid phase, W/m-K
M_a: magnitude of attenuation
M_ℓ: limit magnitude, cyclic input
m_{pi}: mean mass of a particle from ith particle set, kg
N_i: number of particles from the ith set in a given volume
N_{set} : total number of particle sets
n: exponent, de St. Robert's law
p: local gas static pressure, Pa
Δp_d: initial pulse disturbance step pressure, Pa
Q_i: heat transfer from gas to a particle from ith particle set, W
R: specific gas constant, J/kg-K
r_b: instantaneous burning rate, m/s
$r_{b,o}$: reference burning rate, m/s
$r_{b,\text{qs}}$: quasisteady burning rate, m/s
r_b^*: unconstrained burning rate, m/s
r_o: base burning rate, m/s
T_f: flame temperature, gas phase, K
T_i: initial temperature, solid phase, K
T_{pi}: temperature of particle from ith set, K
T_s: burning surface temperature, K
Δt: time increment, s
u: core axial gas velocity, m/s
u_{pi}: core axial particle velocity for particle from ith set, m/s
v_f: nominal flamefront velocity, m/s
$\not{V}$: elemental volume, m^3
x: distance from head end, m
Δx: spatial increment in axial direction, m
y: radial distance from burning surface, m
Δy: spatial increment in radial direction, solid phase, m
Δy_{Fo}: Fourier limit spatial increment, m
α_g: gas phase void fraction

α_p: total particle mass fraction of overall core flow
α_{pi}: particle mass fraction of ith particle set in overall core flow
α_s: thermal diffusivity, solid phase, m^2/s
β_g: particle-gas mass flux ratio
γ: gas ratio of specific heats
δ_o: reference combustion zone thickness, m
δ_r: resultant combustion zone thickness, m
k: vibration-based wall dilatation term $(1/A \cdot \delta A/\delta t)$, s^{-1}
μ: absolute gas viscosity, kg/m-s
ρ: gas density, kg/m^3
ρ_p: total density of particles in core flow, kg/m^3
ρ_{pi}: density, ith particle set in core flow, kg/m^3
ρ_m: solid density of particle, kg/m^3
ρ_s: solid density of propellant, kg/m^3
ρ_{so}: reference solid density of propellant (no particle loading), kg/m^3
ϕ: acceleration orientation angle, rad
ϕ_d: longitudinal/lateral-acceleration-based displacement orientation angle, rad.

References

[1] W. G. Brownlee, "Nonlinear axial combustion instability in solid propellant motors," *AIAA Journal*, vol. 2, no. 2, pp. 275–284, 1964.

[2] F. S. Blomshield, J. E. Crump, H. B. Mathes, R. A. Stalnaker, and M. W. Beckstead, "Stability testing of full-scale tactical motors," *Journal of Propulsion and Power*, vol. 13, no. 3, pp. 349–355, 1997.

[3] F. S. Blomshield, H. B. Mathes, J. E. Crump, C. A. Beiter, and M. W. Beckstead, "Nonlinear stability testing of full-scale tactical motors," *Journal of Propulsion and Power*, vol. 13, no. 3, pp. 356–366, 1997.

[4] E. W. Price, "Solid rocket combustion instability–an American historical account," in *Nonsteady Burning and Combustion Stability of Solid Propellants*, L. De Luca, E. W. Price, and M. Summerfield, Eds., vol. 143 of *Progress in Astronautics & Aeronautics Series*, pp. 1–16, AIAA Publications, Washington, DC, USA, 1992.

[5] M. Barrère, "Introduction to nonsteady burning and combustion instability," in *Nonsteady Burning and Combustion Stability of Solid Propellants*, L. De Luca, E. W. Price, and M. Summerfield, Eds., vol. 143 of *Progress in Astronautics & Aeronautics Series*, pp. 17–58, AIAA Publications, Washington, DC, USA, 1992.

[6] F. E. C. Culick, "Prediction of the stability of unsteady motions in solid-propellant rocket motors," in *Nonsteady Burning and Combustion Stability of Solid Propellants*, L. De Luca, E. W. Price, and M. Summerfield, Eds., vol. 143 of *Progress in Astronautics & Aeronautics Series*, pp. 719–780, AIAA Publications, Washington, DC, USA, 1992.

[7] S. R. Fischbach, J. Majdalani, and G. A. Flandro, "Acoustic instability of the slab rocket motor," *Journal of Propulsion and Power*, vol. 23, no. 1, pp. 146–157, 2007.

[8] D. E. Kooker and B. T. Zinn, "Triggering axial instabilities in solid rockets: numerical predictions," in *Proceedings of the 9th AIAA//SAE Joint Propulsion Conference*, Las Vegas, Nev, USA, November 1973, AIAA Paper. No. 73-1298.

[9] J. D. Baum and J. N. Levine, "Modeling of nonlinear longitudinal instability in solid rocket motors," *Acta Astronautica*, vol. 13, no. 6-7, pp. 339–348, 1986.

[10] K. Ramohalli, "Technologies and techniques for instability suppression in motors," in *Nonsteady Burning and Combustion Stability of Solid Propellants*, L. De Luca, E. W. Price, and M. Summerfield, Eds., vol. 143 of *Progress in Astronautics & Aeronautics Series*, pp. 805–848, AIAA Publications, Washington, DC, USA, 1992.

[11] E. W. Price, "Experimental observations of combustion instability," in *Fundamentals of Solid-Propellant Combustion*, K. K. Kuo and M. Summerfield, Eds., vol. 90 of *Progress in Astronautics & Aeronautics Series*, pp. 733–790, AIAA Publications, Washington, DC, USA, 1984.

[12] F. S. Blomshield, K. J. Kraeutle, R. A. Stalnaker, M. W. Beckstead, and B. Stokes, "Aluminum combustion effects on combustion instability of high burn rate propellants," in *the 28th JANNAF Combustion Meeting*, vol. 3, pp. 419–438, October 1991.

[13] F. S. Blomshield, "Historical perspective of combustion instability in motors: case studies," in *Proceedings of the 37th AIAA/ASME/SAE/ASEE Joint Propulsion Conference*, Salt Lake City, Utah, USA, July 2001, AIAA Paper No. 2001-3875.

[14] R. H. W. Waesche, "Mechanisms and methods of suppression of combustion instability by metallic additives," *Journal of Propulsion and Power*, vol. 15, no. 6, pp. 919–922, 1999.

[15] K. W. Dotson and B. H. Sako, "Interaction between solid rocket motor internal flow and structure during flight," *Journal of Propulsion and Power*, vol. 23, no. 1, pp. 140–145, 2007.

[16] D. R. Greatrix, "Inert particles for axial-combustion-instability suppression in a solid rocket motor," *Journal of Propulsion and Power*, vol. 24, no. 6, pp. 1347–1354, 2008.

[17] R. A. Dobbins and S. Temkin, "Measurements of particulate acoustic attenuation," *AIAA Journal*, vol. 2, no. 6, pp. 1106–1111, 1964.

[18] M. Forde, "Quasi-one-dimensional gas/particle nozzle flows with shock," *AIAA Journal*, vol. 24, no. 7, pp. 1196–1199, 1986.

[19] J. J. Gottlieb and D. R. Greatrix, "Numerical study of the effects of longitudinal acceleration on solid rocket motor internal ballistics," *Journal of Fluids Engineering*, vol. 114, no. 3, pp. 404–410, 1992.

[20] D. R. Greatrix and P. G. Harris, "Structural vibration considerations for solid rocket internal ballistics modeling," in *Proceedings of the 36th AIAA/ASME/SAE/ASEE Joint Propulsion Conference*, Huntsville, Ala, USA, July 2000, AIAA Paper No. 2000-3804.

[21] H. Krier, S. T. Surzhikov, and R. L. Glick, "Prediction of the effects of acceleration on the burning of AP/HTPB solid propellants," in *Proceedings of the 39th AIAA Aerospace Sciences Meeting*, Reno, Nev, USA, January 2001, AIAA Paper No. 2001-0343.

[22] S. Loncaric, D. R. Greatrix, and Z. Fawaz, "Star-grain rocket motor—nonsteady internal ballistics," *Aerospace Science and Technology*, vol. 8, no. 1, pp. 47–55, 2004.

[23] J. Montesano, K. Behdinan, D. R. Greatrix, and Z. Fawaz, "Internal chamber modeling of a solid rocket motor: effects of coupled structural and acoustic oscillations on combustion," *Journal of Sound and Vibration*, vol. 311, no. 1-2, pp. 20–38, 2008.

[24] C. Baczynski and D. R. Greatrix, "Steepness of grain geometry transitions on instability symptom suppression in solid rocket motor," in *Proceedings of the 45th AIAA/ASME/SAE/ASEE Joint

Propulsion Conference, Denver, Colo, USA, August 2009, AIAA Paper No. 2009-5177.

[25] D. R. Greatrix, "Transient burning rate model for solid rocket motor internal ballistic simulations," *International Journal of Aerospace Engineering*, vol. 2008, Article ID 826070, 10 pages, 2008.

[26] D. E. Kooker and C. W. Nelson, "Numerical solution of solid propellant transient combustion," *Journal of Heat Transfer*, vol. 101, no. 2, pp. 359–364, 1979.

[27] D. R. Greatrix, "Model for prediction of negative and positive erosive burning," *Canadian Aeronautics and Space Journal*, vol. 53, no. 1, pp. 13–21, 2007.

[28] D. R. Greatrix, "Parametric analysis of combined acceleration effects on solid-propellant combustion," *Canadian Aeronautics and Space Journal*, vol. 40, no. 2, pp. 68–73, 1994.

Parachute-Payload System Flight Dynamics and Trajectory Simulation

Giorgio Guglieri

Dipartimento di Ingegneria Meccanica e Aerospaziale, Politecnico di Torino, Corso Duca degli Abruzzi 24, 10129 Torino, Italy

Correspondence should be addressed to Giorgio Guglieri, giorgio.guglieri@polito.it

Academic Editor: C. B. Allen

The work traces a general procedure for the design of a flight simulation tool still representative of the major flight physics of a parachute-payload system along decelerated trajectories. An example of limited complexity simulation models for a payload decelerated by one or more parachutes is given, including details and implementation features usually omitted as the focus of the research in this field is typically on the investigation of mission design issues, rather than addressing general implementation guidelines for the development of a reconfigurable simulation tool. The dynamics of the system are modeled through a simple multibody model that represents the expected behavior of an entry vehicle during the terminal deceleration phase. The simulators are designed according to a comprehensive vision that enforces the simplification of the coupling mechanism between the payload and the parachute, with an adequate level of physical insight still available. The results presented for a realistic case study define the sensitivity of the simulation outputs to the functional complexity of the mathematical model. Far from being an absolute address for the software designer, this paper tries to contribute to the area of interest with some technical considerations and clarifications.

1. Introduction

The purpose of a parachute is to decelerate and provide stability to a payload in flight. The aerodynamic and stability characteristics of the parachute system are governed by the geometry of the parachute as such careful consideration is paid to this in the design process. The effects of deployment and opening force are critical in the safe operation of the parachute and the integrity of the payload. The opening characteristics also feature heavily in the selection of geometry and other parameters in the design process.

Parachutes for aerospace applications [1–4] are in general symmetric about the canopy axis. This axis passes through the center of the canopy and the confluence point of the suspension lines. The canopy is the cloth surface that inflates to provide the desired lift, drag, and stability. The suspension lines transmit the retarding force from the canopy to the payload either directly or through a riser attached below the confluence point of the suspension lines. The deceleration force may be distributed on the payload over more than one mechanical joint linked to the riser by a set of short hardly extensible strips (bridles).

There are a number of different kinds of parachutes that have been designed for various applications. The different applications parachutes are typically used for pilot, drogue, deceleration, descent, extraction, supersonic drogue and stabilization, flight termination, and landing.

The dynamics of parachutes are complex and difficult to model accurately. During both the inflation process and the terminal descent stage, the dynamics of a parachute are governed by a coupling between the structural dynamics of the parachute system and the surrounding fluid flow. Both of these dynamic systems must be addressed as a coupled system to gain a proper representation of the dynamic system as a whole.

When the parachute is in a steady state, the air flowing around the decelerator will separate at some location on the canopy. The shedding of the vortices from the canopy can affect the stability and cause a periodic motion of both parachute and payload. The wake from a porous parachute consists of air that flowed around the canopy and air that flowed through the canopy. A payload body in the speed range of parachute usage sheds a very turbulent wake. Part of the flow that is entering the parachute is therefore of a disturbed nature and should be considered regarding the aerodynamic performance of the parachute. For many types of parachutes, this change in oncoming airflow can be quite significant during the time required for the parachute to inflate. The implication of a rapid deceleration is that second-order effects are likely to be present.

To summarize, calculation of parachute deployment, inflation, and deceleration requires the numerical solution to the equations of motion for a viscous, turbulent, separated airflow. The parachute is also a flexible body having dynamic behavior coupled with the behavior of the flow, which passes through and around it. From the above description it is obvious that a full-time dependent solution of this system is far from being easily feasible. To make a mathematical model that is feasible, simplifications must be made, as long as the model can be validated satisfactorily by experiment or by comparison with reference data.

The overall behavior of parachutes is related to various parameters: added masses, filling time, parachute shape (inflated canopy elongation), porosity, suspension line length, reefing, clustering, snatch loads at deployment, and aero-mechanical and inflation instability. In the past, most of these effects could be generally modeled in an imprecise way by simulation tools. A comprehensive computational technique is presented in [5–7] for carrying out three-dimensional simulations of parachute fluid-structure interactions, and this technique is applied to simulations of airdrop performance and control phenomena in terminal descent. The technique uses a stabilized space-time formulation of the time-dependent, three-dimensional Navier-Stokes equations of incompressible flows for the fluid dynamics part. A finite-element formulation derived from the principle of virtual work is used for the parachute structural dynamics. The parachute is represented as a cable-membrane tension structure. Coupling of the fluid dynamics with the structural dynamics is implemented over the fluid-structure interface, which is the parachute canopy surface.

According to the different missions, several types of payloads have been used in combination with aerodynamic decelerators: paratroops, equipment, hardware, materiel, weapons, missiles, aircraft, unmanned aerial vehicles, aerospace lifting, and nonlifting spacecraft. The present analysis is focused on aerospace applications for planetary and atmospheric entry vehicles, where the payload is typically a blunt body. The purpose of a blunt body is primarily to provide a large source of drag to facilitate a deceleration. Applications of blunt bodies can be seen with both manned reentry and planetary exploration missions. An outline of aerodynamic decelerators for robotic planetary exploration missions is given in [8]. In the development programs of such blunt bodies and also in subsequent studies, it has been shown that they may be dynamically unstable in all or part of the sub-, trans-, super-, and in some cases hypersonic speed regimes.

Blunt bodies for which there exist examples can be classified into two categories, large angle cones and capsules [3]. An example of large angle cone that has performed actual missions is the Viking probe. It should be noted that in the two Viking missions confirmed the dynamic instability upon entry into the Martian atmosphere. The examples of capsules are well-known ones due to the intense space activities from the former USSR and the USA. The capsules that were developed for manned flights in this period were Mercury, Gemini, Apollo and Soyuz, which are still considered a reference for performing missions to this day. As was the case with the large angle cones, there were found to be speed ranges over which the capsules were unstable [9].

The flow field that is associated with the capsule configuration is highly complex. For almost the entire speed range that the capsules operate over, the flow remains attached on the forward face. At the point of maximum diameter, the flow is accelerated such that the boundary layer rapidly grows to the point of separation. After the maximum diameter, the flow then remains separated and turbulent. This flow is unstable and coupled with an unsteady near wake the dynamic instability associated with capsules is produced [11].

As dynamic instability exists for blunt bodies over various speed ranges, to complete missions successfully, there is then a requirement for some kind of accurate stability assessment with a potential impact on both stability augmentation (if any reaction control system is implemented) and mission design (parachute deployment sequence). This issue applies for probes and capsules as thrusters and parachutes are the unique available sources of additional damping.

The measurement of stability derivatives for probes and capsules was a concern of designers since the origin of space flight [9], as analytical methods did not provide (at least in the past) adequate estimation of these relevant parameters. The wind tunnel experimental techniques considered are free oscillation, forced oscillation, free flight tests, and even ballistic range experiments. It should be noted that wind tunnel experiments are not without limitations, so there remains a strong desire to make available numerical data, for the purposes of both finding solutions and comparing with experimental results.

2. Background on Parachute-Payload Modeling and Simulation

Several studies analyze the descent and landing trajectories of parachute-payload systems. Generally, these analyses consist of performing a simulation (typically a 3–6 DOFs rigid body model is adopted) of the atmospheric entry phase to predict deceleration peaks, descent attitudes, and terminal conditions. In addition, a stochastic dispersion analysis (Monte Carlo or similar) is usually performed to assess the impact of off-nominal conditions that may arise to determine the robustness of the mission design.

A two-dimensional parachute model is presented in [12] to compute the various characteristics of the steady descent of a parachute system. A three degree-of-freedom analysis is presented and validated in [13] giving the longitudinal motion of a typical vehicle during the recovery phase. The parachute and the payload are supposed to be rigid and interconnected by an elastic riser. Aerodynamic loads acting on the two subsystems are considered. Computer results showed good agreement with test results in terms of oscillation amplitude and frequency, riser force, and parachute wrapup about the vehicle for the simulation of a pad-abort situation. The three-dimensional motion of a freely descending parachute is studied in [14] with a five degree-of-freedom analysis (roll motion is neglected). Exact expressions are given for the longitudinal and lateral small disturbance stability of the gliding motion of parachutes. The analysis confirms that large longitudinal disturbance of most parachutes will result in a large pitching motion, whereas a large lateral disturbance will usually cause a large angle vertical coning motion (coning mode). The longitudinal mode damps out very quickly in the stable case. The three-dimensional motion of a nonrigid parachute and payload system is studied in [15]. Both the parachute and the payload are assumed to have five degrees of freedom (roll about axis of symmetry is again neglected). They are coupled together by a fixed-length connector. The general nonlinear equations are linearized using small perturbation theory. The evaluation of the stability of an unstable payload decelerated by a parachute is performed. The authors observed that increasing riser length and parachute weight promotes system instability. A nine degree-of-freedom computer program was developed in [16] for the simulation of the trajectory and the dynamic behavior of a rotating parachute system. An accurate mathematical model of the joint between the load and the parachute was found to be necessary to predict the dynamic behavior of a rotating decelerated system. A computer model based on a six degree-of-freedom analysis is described in [17] and compared with drop test data. The payload is rigidly connected, the aerodynamic forces on canopy and payload are determined by the instantaneous angle of attack of the impinging airflow, the apparent masses are constant, but they depend on the direction of the acceleration. Full nonlinear equations of motion for the axisymmetric parachute have been obtained in [18]. In particular, the correct form of the added mass tensor for a rigid axisymmetric parachute in ideal flow has been implemented in a six degree-of-freedom computer model [19], and the results indicate that added mass effects are significant. In particular, the component of added mass along the axis of symmetry has a strong effect on parachute dynamic stability. However, design and testing experience shows that dynamic stability of the parachute is a second order design problem [3] for high-performance decelerators that usually have both high static and dynamic stability due to the porosity of the canopy. Many works, including recent models developed and validated by means of drop tests, consider the bridles as rigid elements [20, 21].

3. The Mathematical Model of the Payload-Parachute System

A first mathematical representation of a parachute-payload system (designated as SM1) is defined according to a very comprehensive approach that enforces the simplification of the coupling mechanism between the payload and the parachute, with an adequate level of physical insight still available in terms of sensitivity of system performance metrics to design parametric changes. In modeling a payload-parachute system, two bodies and one device have been used. They are the parachute canopy and the payload, which are then connected by a single riser. In reality the parachute canopy is connected to the payload by suspension lines, a riser section, and then a set of bridles. In the simulation only the riser and the bridles are modeled in terms of a dynamic response (suspension lines are neglected). However, in terms of connection points only the riser has been modeled. The riser is assumed to have flexible connections at both ends and provides stiffness above a certain threshold distance and zero stiffness below this distance (slack conditions).

The payload is considered as a rigid body with six degrees of freedom. The forces and the moments that act on the body are provided by its aerodynamics, by the weight, by the inertial actions, and, finally, by the force applied by the riser in the suspension point. The parachute canopy acts as a rigid body and, by the action of aerodynamic drag, strains the riser which then transmits a force to the payload. Figure 1 shows the general philosophy used in modeling the system.

In terms of a dynamic response, the riser is modeled as having linear stiffness and damping, where the force at any given point in time is given by

$$F_R = kl_r\varepsilon + c\dot{\varepsilon}. \tag{1}$$

Note that k is stiffness, l_r is riser length, ε and $\dot{\varepsilon}$ are strain and strain rate, respectively, and c is the damping coefficient. The above equation is implemented in the code to calculate the force present in the riser at any point. The equivalent stiffness of the bridles and riser is given by the following equation:

$$k = \frac{n_b \cdot k_r k_b}{k_r + n_b \cdot k_b}, \tag{2}$$

where n_b is the number of bridles, k_r is riser stiffness and k_b is single bridle stiffness. After calculation of the equivalent stiffness, the damping coefficient c can be calculated by the following formulation (m_P is the parachute mass and ξ is the damping ratio):

$$c = 2\xi m_P\sqrt{\frac{k}{m_P}}. \tag{3}$$

The values for strain and strain rate are computed by the difference in displacement and velocity of each end of the riser, that is, as the displacement and the velocity of the payload and that of the parachute canopy. Also note that where this calculation occurs a check is in place for the condition where the riser is slack:

$$\varepsilon = \frac{s_c - s_p - l_0}{l_r}, \qquad \dot{\varepsilon} = V_c - V_p, \tag{4}$$

where ε and $\dot{\varepsilon}$ are, respectively, strain and strain rate, s_c and s_p are, respectively, payload and parachute displacement, l_r is the riser length, and V_c and V_p are, respectively, the payload and parachute velocity. From these two calculations, the force in the riser can be derived by application of (1).

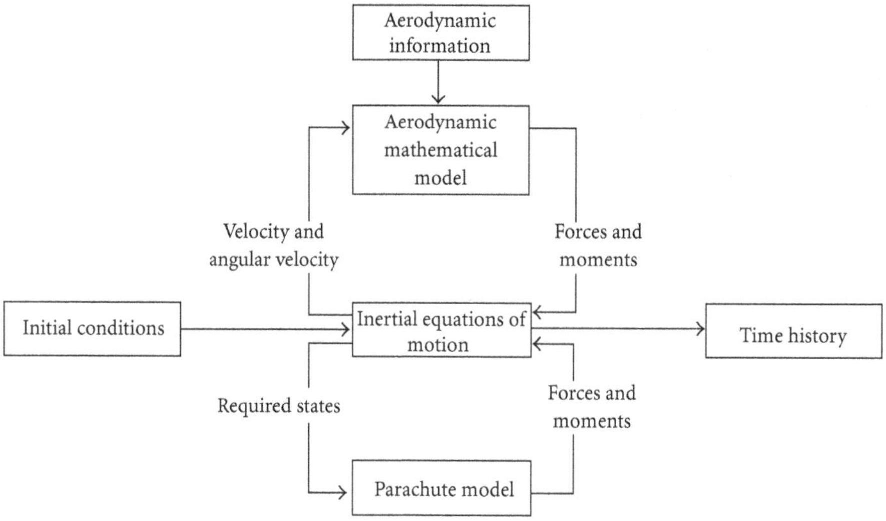

FIGURE 1: Flow chart of the simulation model.

The performance of the parachute has been modeled using an approximation for the added mass m_a. When a parachute passes through the air, it drags some of the surrounding with it this means that this air must be accelerated from rest up to the speed of the parachute. The parachute will also accelerate some of the air immediately in front and behind. The energy that is required to facilitate this acceleration is taken from the kinetic energy of the parachute-payload system. These fluid inertia effects are taken into account in the equation of motion. The idea is that the mass has a component added to take into account the extra mass associated with the accelerated air. This can be seen below, where the change in momentum of the parachute system is equal to the sum of the drag force and gravitational force component:

$$\frac{d}{dt}[M^* V] = -\frac{1}{2}\rho V^2 C_D S_P + m_P \sin\gamma, \quad (5)$$

$$M^* = m_a + m_P. \quad (6)$$

After combining (5) and (6) and then rearranging, the following expression for system acceleration (or deceleration) can be used:

$$(m_P + m_a)\frac{dV}{dt} = -\frac{1}{2}\rho V^2 C_D S_P - V\frac{dm_a}{dt} + m_P g \cdot \sin\gamma. \quad (7)$$

However, for (7) to be useful, an expression or knowledge of $m_a(t)$ must be available. Shown below is an expression for added mass, similar to that for apparent mass, that has been used widely:

$$m_a = k_a \rho \frac{4}{3}\pi R^3. \quad (8)$$

The value of k_a, a constant, can be found in various ways [1, 3]. Klimas calculated values of k_a for porous hemispherical ribbon parachutes. In the present case, added mass is calculated by using a slight variation. In the previously presented method, the volume of a hemisphere is used, whereas in the model used in the simulation the volume of an ellipsoid is used. Hence, added mass has been calculated using the following expression:

$$m_a = n_P k_a \rho \frac{4}{3}\pi R^2 h, \quad (9)$$

where n_P is the number of parachutes, k_a is the added mass coefficient, R is the inflated radius, and h is the inflated height. The integer n_P for the number of parachutes is necessary because a cluster of multiple parachutes is being modeled as one parachute of equivalent size. The added mass remains constant throughout inflation, so the values for parachute radius and height are constant, whereas in reality they are of course changing quite significantly through the inflation process. The added mass coefficient is given by:

$$k_a = 1.068(1 - 1.465p - 0.25975p^2 + 1.2626p^3), \quad (10)$$

where p is the parachute porosity (see [3, 22] for further details concerning the effect of porosity on added masses). This coefficient remains constant through inflation. The porosity for both drogue and main parachutes is set at a value of 20% ($p = 0.2$).

The significance of fluid inertial effects is relevant. As an example, opening times and loads increase with altitude. The reason for this can be seen when it is considered that at increasing altitude the true airspeed will increase for constant indicated airspeed, as such the inertial effects of the air are greater. After comparison with flight test data, the added mass approximation tends to overestimate the deceleration that will occur. From a design point of view, this means that results obtained from the added mass approximation can be viewed as conservative. So while the concept of added mass is only an approximation, the effects of fluid inertia in transient parachute aerodynamics are significant enough to be included in modeling. At present this method of accounting for the inertial effects of air is the most practical tool available to be used in the transient phase of parachute operation.

The parachute as previously stated is modeled as a separate body (rigid canopy). The deceleration of this body is therefore calculated by extension of (7) including the riser tension force:

$$\frac{dV}{dt} = \dot{V}_P = \frac{F_r - D_P - m_P g \sin\gamma}{(m_P + m_a)}, \tag{11}$$

where a is acceleration, F_r is the riser force, D_P is the parachute drag, m_P is the mass of the parachute, m_a is the added mass, and γ is the flight path angle. The drag force provided by the parachute is calculated using the following simple expression:

$$D_P = \frac{1}{2}\rho V_P^2 C_D S_P \cdot \eta, \tag{12}$$

where ρ is air density, V_P is parachute velocity, C_D is the canopy drag coefficient, S_P is the canopy projected area, and η is an efficiency factor used to account for payload wake effects ($\eta = 0.7 \div 0.75$ for the drogue parachute and $\eta = 1$ for the main parachute). The product $C_D S_P$ (drag area) of a parachute depends on its type, inflated shape, size, Reynolds' number, Mach's number, Froude's number, material elasticity, and porosity.

In (11) the third term on the numerator is the gravitation force component. The denominator is then the addition of the mass of the parachute and the added air mass.

Due to the complexity involved with the filling process, filling time cannot be calculated by a purely analytical method. The alternative once again lies with an empirical approach that is validated by flight test. Knacke defined the following empirical relation for filling time:

$$t_f = \frac{8D_0}{V_S^{0.9}}, \tag{13}$$

where D_0 is nominal diameter and V_S is velocity at line stretch. It should be noted here that (13) is dimensionally incorrect. This means that the actual physics of the system are not properly represented. While (13) has been shown to provide reasonable answers, care should always be taken in using such an equation so that it is not being used outside the intended range of conditions.

An underlying assumption used in the modeling of the parachute canopy is that it remains aligned with the velocity vector at all times. This corresponds to neglecting the lift and moment coefficients of the parachute.

The second major assumption made used in the parachute canopy model is that the added mass remains constant throughout inflation and parachute operation.

Whilst in the model the value of added mass is kept constant at the value for the fully inflated parachute (conservative approach for the estimation of deceleration peak), the value for drag area is ramped up as the parachute inflates. It is this feature of the model that provides a simulation of the inflation process. Inflation modeling is done by ramping the value of the drag area, $C_D S_P$, this being the multiplication of the drag coefficient and the projected area. The effective

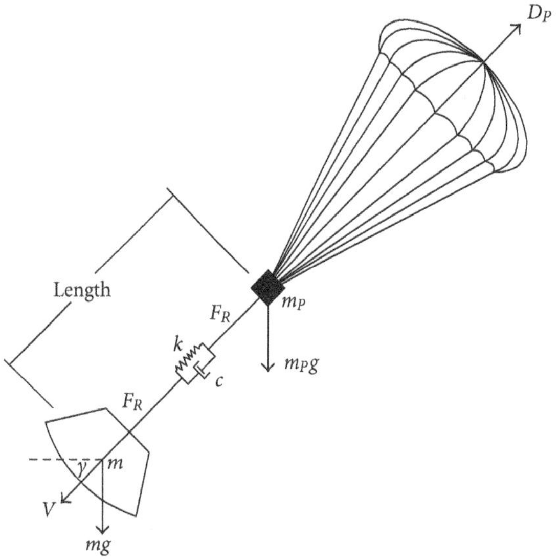

FIGURE 2: Layout of the suspension system (parachute-payload configuration).

drag area is time-scheduled according to the planned reefing stages and can be described by the following expressions:

$$(C_D S_P)_t = (C_D S_P)_0 \cdot \left[\sum_{k=1}^{i-1} \tau_k + \tau_i \left(\frac{t - t_i}{t_{fi}}\right)^{n_i}\right], \quad \text{for } t \in \left[t_i \div t_{fi}\right],$$
$$(C_D S_P)_t = (C_D S_P)_0 \cdot \sum_{k=1}^{i} \tau_k, \quad \text{for } t \in \left[t_{fi} \div t_{i+1}\right], \tag{14}$$

where reefing starting time is t_i, duration t_{fi}, and ratio τ_k. The power for the growth, n_i, has been set at 2 for the initial stages and at 2.5 for the final stage, for both drogue and main parachutes (the exponents are selected to fit the available flight test data). Selection of these values is justified by a better prediction of peak loads. This value of $C_D S_P$ is then substituted into the equation for parachute drag.

The final assumption made is with the clustered parachutes. In the simulation one parachute of equivalent size has been used, with a cluster efficiency factor applied.

In Figure 2, the forces acting on both the parachute and payload can be seen (with the exception of the body aerodynamic forces and moments).

The forces and moments acting on the payload are calculated as the relevant contributions are added. The equations are

$$X = \frac{1}{2}\rho V S\left(V C_X + C_{X_q} q d\right) - F_R \cos\alpha \cos\beta,$$

$$Y = \frac{1}{2}\rho V S(V C_Y + C_{Y_r} r d) - F_R \sin\beta,$$

$$
\begin{aligned}
Z &= \frac{1}{2}\rho VS\left(VC_Z + C_{Zq}qd\right) - F_R \sin\alpha \cos\beta,\\
L &= \frac{1}{2}\rho VSd\left(VC_l + C_{lp}pd\right) - F_R \sin\beta\left(z_{cg} - z_p\right)d\\
&\quad - F_R \sin\alpha \cos\beta\left(y_{cg} - y_p\right)d,\\
M &= \frac{1}{2}\rho VSd\left(VC_M + C_{Mq}qd\right) + F_R \cos\alpha \cos\beta\left(z_{cg} - z_p\right)d\\
&\quad + F_R \sin\alpha \cos\beta\left(x_{cg} - x_p\right)d,\\
N &= \frac{1}{2}\rho VSd(VC_N + C_{Nr}rd) - F_R \sin\beta\left(x_{cg} - x_p\right)d\\
&\quad - F_R \cos\alpha \cos\beta\left(y_{cg} - y_p\right)d,
\end{aligned}
\tag{15}
$$

where X, Y, and Z are the body forces, F_R is the riser force (tension), L, M, and N are the body moments, V is velocity, d is payload diameter, S is reference surface area, p is roll rate, q is pitch rate, r is yaw rate, α is angle of attack, β is angle of sideslip, and x_p, y_p, and z_p are the parachute-payload attachment coordinates. All coordinates are normalized with a reference length.

In the above-presented model, there are a number of featured that need to be discussed. The aerodynamic data for the payload—which the simulation uses as input—are given in the longitudinal plane only. In order to calculate the correct coefficient, the center of gravity must be shifted accordingly. This is done using the following equations:

$$
\begin{aligned}
C_M &= C_M - C_Z\left(x_{cg} - x_{\text{ref}}\right) - C_X\left(z_{cg} - z_{\text{ref}}\right),\\
C_{Mq} &= C_{Mq} - C_{Zq}\left(x_{cg} - x_{\text{ref}}\right) - C_{Xq}\left(z_{cg} - z_{\text{ref}}\right),
\end{aligned}
\tag{16}
$$

where C_M is the moment coefficient, C_Z is the force coefficient in the Z direction, C_X is the force coefficient in the X direction, C_{Mq} is the moment coefficient due to pitch rate, C_{Zq} is the force coefficient in the Z direction due to pitch rate, C_{Xq} is the force coefficient in the X direction due to pitch rate, x_{cg} and z_{cg} are the x and z locations of the payload center of gravity, and finally x_{ref} and z_{ref} are the x and z locations of the reference data center of gravity. The coordinates are normalized with a reference length, that is, for axis-symmetric bodies the payload diameter d.

The lateral-directional coefficients are obtained assuming the geometric symmetry of the payload. In taking this approach, the aerodynamic response to angle of attack and angle of sideslip have been decoupled. The implications of this decoupling are that there are no aerodynamic roll moments ($L = 0$), no sideslip effects in the longitudinal plane, and finally no angle of attack effects in the lateral plane. It should also be noted that, since the body is axis-symmetric, total angle of attack would need to be considered for aerodynamic coefficients interpolation, as implemented in [23].

The body axis system has been adopted, and this system can be seen in Figure 3. The system of differential equations that describes the parachute-payload system can be broken

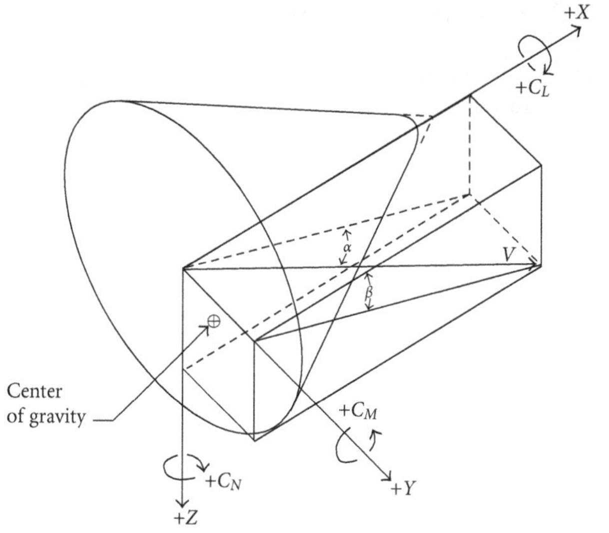

FIGURE 3: The reference frame and the aerodynamic angles for the payload.

into sections, six rigid body equations of motion, four quaternion equations, three position equations, and the velocity and displacement of the parachute.

The system written in residual form can be seen below:

$$
\begin{aligned}
r_1 &= \dot{u} - \left(\frac{(X + mgC_{BE}(1,3))}{m} - qw + rv\right),\\
r_2 &= \dot{v} - \left(\frac{(Y + mgC_{BE}(2,3))}{m} - ru + pw\right),\\
r_3 &= \dot{w} - \left(\frac{(Z + mgC_{BE}(3,3))}{m} + qu - pv\right),\\
r_4 &= I_{XX}\dot{p} - I_{XY}\dot{q} - I_{XZ}\dot{r} - I_{XZ}pq + (I_{ZZ} - I_{YY})qr\\
&\quad + I_{XY}pr + (r^2 - q^2)I_{YZ} - L,\\
r_5 &= I_{YY}\dot{q} - I_{XY}\dot{p} - I_{YZ}\dot{r} - I_{YZ}pq + (I_{XX} - I_{ZZ})pr\\
&\quad - I_{XY}qr + (p^2 - r^2)I_{XZ} - M,\\
r_6 &= I_{ZZ}\dot{r} - I_{XZ}\dot{p} - I_{YZ}\dot{q} + I_{XZ}qr + (I_{YY} - I_{XX})pq\\
&\quad - I_{YZ}pr + (q^2 - p^2)I_{XY} - N,\\
r_7 &= \dot{e}_0 + \frac{1}{2}(e_1p + e_2q + e_3r),\\
r_8 &= \dot{e}_1 - \frac{1}{2}(e_0p - e_3q + e_2r),\\
r_9 &= \dot{e}_2 - \frac{1}{2}(e_3p + e_0q - e_1r),\\
r_{10} &= \dot{e}_3 + \frac{1}{2}(e_2p - e_1q - e_0r),\\
r_{11} &= \dot{x} - C_{BE}(1,1)u - C_{BE}(1,2)v - C_{BE}(1,3)w,\\
r_{12} &= \dot{y} - C_{BE}(2,1)u - C_{BE}(2,2)v - C_{BE}(2,3)w,\\
r_{13} &= \dot{h} + C_{BE}(3,1)u + C_{BE}(3,2)v + C_{BE}(3,3)w,\\
r_{14} &= \dot{s}_p - V_p\\
r_{15} &= \dot{V}_p - \frac{\left(F_r - D_p - m_pg\sin\gamma\right)}{\left(m_p + m_a\right)},
\end{aligned}
\tag{17}
$$

where u, v, and w are the body axis velocities in the X, Y, and Z directions, respectively, p, q, and r are the roll rate, pitch rate, and yaw rate in body axis, e_0 through to e_3 are the quaternion values for orientation, γ is the flight path angle, x, y, and h give the position in the Earth fixed axis, s_p is the absolute displacement of the parachute canopy, and finally V_p is the parachute canopy total velocity.

The magnitude of the velocity vector, that is, the airspeed, is found from the three body-axis velocities:

$$V = \sqrt{u^2 + v^2 + w^2}. \tag{18}$$

The derivative of the velocity vector is then calculated by

$$\dot{V} = \frac{(u\dot{u} + v\dot{v} + w\dot{w})}{V}. \tag{19}$$

The angle of attack is calculated by

$$\alpha = \tan^{-1}\left(\left|\frac{w}{u}\right|\right). \tag{20}$$

The angle of sideslip is calculated by

$$\beta = \sin^{-1}\left(\left|\frac{v}{V}\right|\right). \tag{21}$$

Note that the angles of attack and sideslip are put into the correct phase after this calculation, since the inverse trigonometric function will only ever return values between $\pm 90°$.

The conversion from the Euler angles to quaternion values is computed by the following set of equations:

$$\begin{aligned} e_0 &= \cos\left(\frac{\psi}{2}\right)\cdot\cos\left(\frac{\theta}{2}\right)\cdot\cos\left(\frac{\phi}{2}\right) + \sin\left(\frac{\psi}{2}\right)\cdot\sin\left(\frac{\theta}{2}\right)\cdot\sin\left(\frac{\phi}{2}\right),\\ e_1 &= \cos\left(\frac{\psi}{2}\right)\cdot\cos\left(\frac{\theta}{2}\right)\cdot\sin\left(\frac{\phi}{2}\right) - \sin\left(\frac{\psi}{2}\right)\cdot\sin\left(\frac{\theta}{2}\right)\cdot\cos\left(\frac{\phi}{2}\right),\\ e_2 &= \cos\left(\frac{\psi}{2}\right)\cdot\sin\left(\frac{\theta}{2}\right)\cdot\cos\left(\frac{\phi}{2}\right) + \sin\left(\frac{\psi}{2}\right)\cdot\cos\left(\frac{\theta}{2}\right)\cdot\sin\left(\frac{\phi}{2}\right),\\ e_3 &= \sin\left(\frac{\psi}{2}\right)\cdot\cos\left(\frac{\theta}{2}\right)\cdot\cos\left(\frac{\phi}{2}\right) + \cos\left(\frac{\psi}{2}\right)\cdot\sin\left(\frac{\theta}{2}\right)\cdot\sin\left(\frac{\phi}{2}\right). \end{aligned} \tag{22}$$

The transformation matrix that relates from Earth fixed to body axis is calculated using quaternion values:

$$C_{BE} = \begin{bmatrix} e_0^2 + e_1^2 - e_2^2 - e_3^2 & 2(e_1e_2 + e_0e_3) & 2(e_1e_3 - e_0e_2) \\ 2(e_1e_2 - e_0e_3) & e_0^2 - e_1^2 + e_2^2 - e_3^2 & 2(e_2e_3 + e_0e_1) \\ 2(e_0e_2 + e_1e_3) & 2(e_2e_3 - e_0e_1) & e_0^2 - e_1^2 - e_2^2 + e_3^2 \end{bmatrix}. \tag{23}$$

The flight path angle is calculated using a sequence of C transformation matrices. There exists a problem in extracting the flight path angle as it approaches 90°, in that there exists a singularity in the inverse sine function. To get around this problem, a third axis system has been introduced that is rotated 90° in pitch with respect to the wind axis. The flight path angle is then given by the following:

$$\begin{aligned} C_{WB} &= C_1(0)\cdot C_2(\alpha)\cdot C_3(-\beta),\\ C_{BW} &= C_{WB}^{-1},\\ C_{GW} &= C_1(0)\cdot C_2(90°)\cdot C_3(0),\\ C_{WE} &= C_{WB}\cdot C_{BE},\\ C_{GE} &= C_{GW}\cdot G_{WE},\\ \gamma &= \sin^{-1}(-C_{GE}(1,3)) - 90°. \end{aligned} \tag{24}$$

The coordinates of the payload along the trajectory are calculated from

$$s = \sqrt{x^2 + y^2 + (h - h_0)^2}, \tag{25}$$

where h_0 is the initial altitude.

The atmospheric profile (density and speed of sound) is approximated with a cubic curve fit of the reference atmosphere.

A second simulation model (designated as SM2) was defined, implemented, and validated in [23]. The major differences with respect to the less complex SM1 model are

(i) the aerodynamics of the bodies are defined in terms of total angle of attack due to the geometrical symmetry of payload and parachute;

(ii) the dynamics of the parachute are modeled with a full-state rigid body six degree-of-freedom representation;

(iii) the added masses of the parachute are defined by a tensor including rotational inertial properties [18, 19];

(iv) the aerodynamics of the parachute include lift and damping coefficients;

(v) the suspension system is represented by a more realistic layout with distributed elements and suspension links whose strain is estimated with an iterative numerical method;

(vi) the effects of atmospheric turbulence and asymmetries (either geometrical or inertial unbalance) can be included.

4. Software Implementation

The simulation software is written in Fortran language. This program simulates the dynamics of the parachute-payload system (time domain integration). The user is offered predefined initial conditions from reference flight conditions or the option of trimming the payload at any desired altitude and using those generated initial conditions. The inertial properties of the payload (mass, moments

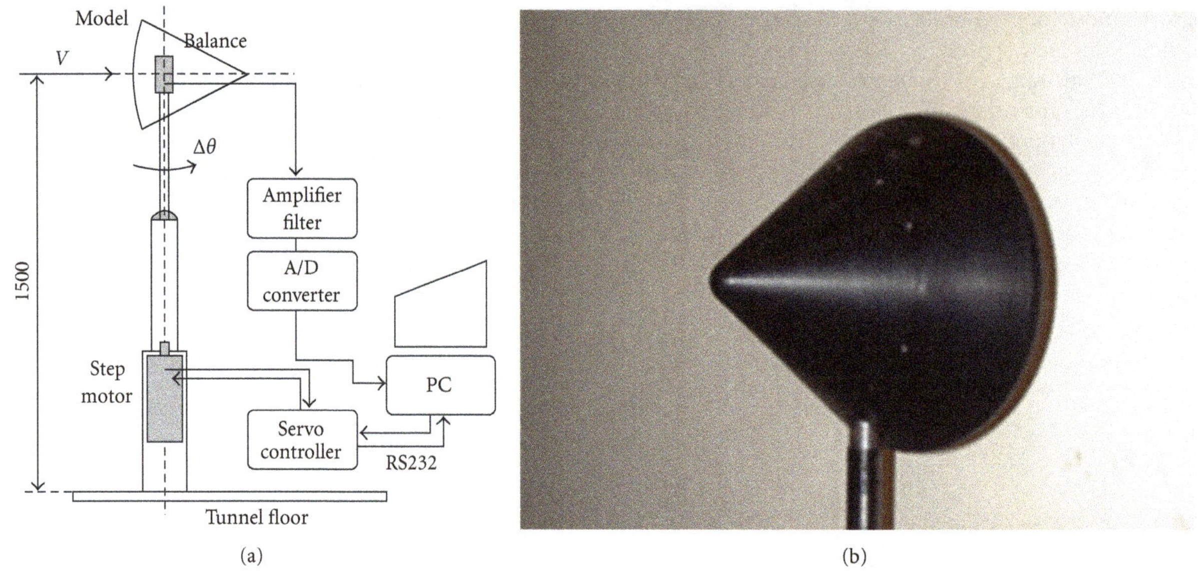

FIGURE 4: The experimental setup [10].

of inertia, and position of the center of gravity) and the parachute characteristics (size, mass, opening, and staging sequences) are defined for a selected altitude ranges (mission table lookup).

The solver used in the simulation program is DASSL (differential algebraic system solver) originally developed by Petzold [24–26] for the solution of systems of differential algebraic equations (DAEs). This routine is based on backward differentiation formulas (BDFs). The solution algorithm then attempts to make the residual $r(t)$ equal to zero at each time step:

$$r(t) = f(\dot{x}, x, t) = \dot{x} - f'(x, t). \tag{26}$$

Equation (26) also shows that it is very simple to modify existing explicit ODE (ordinary differential equation) systems to be used with an implicit formulation. With DAE solvers [27], the equations of motion can be implemented directly in the form of residuals. Therefore, no symbolic expansions are needed to identify acceleration terms, and no ad-hoc algorithms need to be used to determine the vector of derivatives at every time step. The DAE solver may automatically adjust the order of the integration formula and the integration step size to achieve the desired accuracy, whereas typical solvers have fixed order, fixed step size, and no automatic accuracy control.

The initial conditions for the payload (with or without decelerator deployed) may be computed for a given initial altitude h_0, returning the quasi-trim values for V velocity, α angle of attack, β angle of sideslip, θ pitch angle, and γ flight path angle (stabilized fall). Note that it has been decided to leave roll as the untrimmed equation. The purpose is to solve a system of n simultaneous nonlinear equations in n unknowns. It solves the problem $f(x) = 0$ where x is a vector with components $x(1),\ldots,x(n)$ and f is a vector of nonlinear functions. Each equation is of the form:

$$f_k(x(1),\ldots,x(n)) = 0, \quad \text{for } k = 1,\ldots,n. \tag{27}$$

The algorithm used for the solution is based on an iterative method, which is a variation of Newton's method using Gaussian elimination in a manner similar to the Gauss-Seidel process. Convergence is roughly quadratic. All partial derivatives required by the algorithm are approximated by first difference quotients. The convergence behavior of this code is affected by the ordering of the equations, and it is advantageous to place linear and mildly nonlinear equations first in the sequence. The convergence is started from statically stable conditions (initial payload tumbling is avoided). Numerical integration was performed assuming a time step of 0.001 s with an absolute and relative error tolerance of 10^{-7}. Longitudinal and directional planes are considered separately. The system of equations for the longitudinal plane is defined as follows:

$$\begin{gathered} f_1 = M, \quad f_2 = X - mg\sin\theta, \quad f_3 = Z + mg\cos\theta, \\ f_4 = \theta - \gamma - \alpha, \end{gathered} \tag{28}$$

where

$$\begin{aligned} M &= \frac{1}{2}\rho V^2 S d C_M + F_R \cos\alpha\left(z_{cg} - z_p\right)d \\ &\quad + F_R \sin\alpha\left(x_{cg} - x_p\right)d, \\ X &= \frac{1}{2}\rho V^2 S C_X - F_R\cos\alpha, \\ Z &= \frac{1}{2}\rho V^2 S C_Z - F_R \cdot \sin\alpha. \end{aligned} \tag{29}$$

5. Results

The reference mission described in [23] was used to evaluate the two simulation models. The data available from the

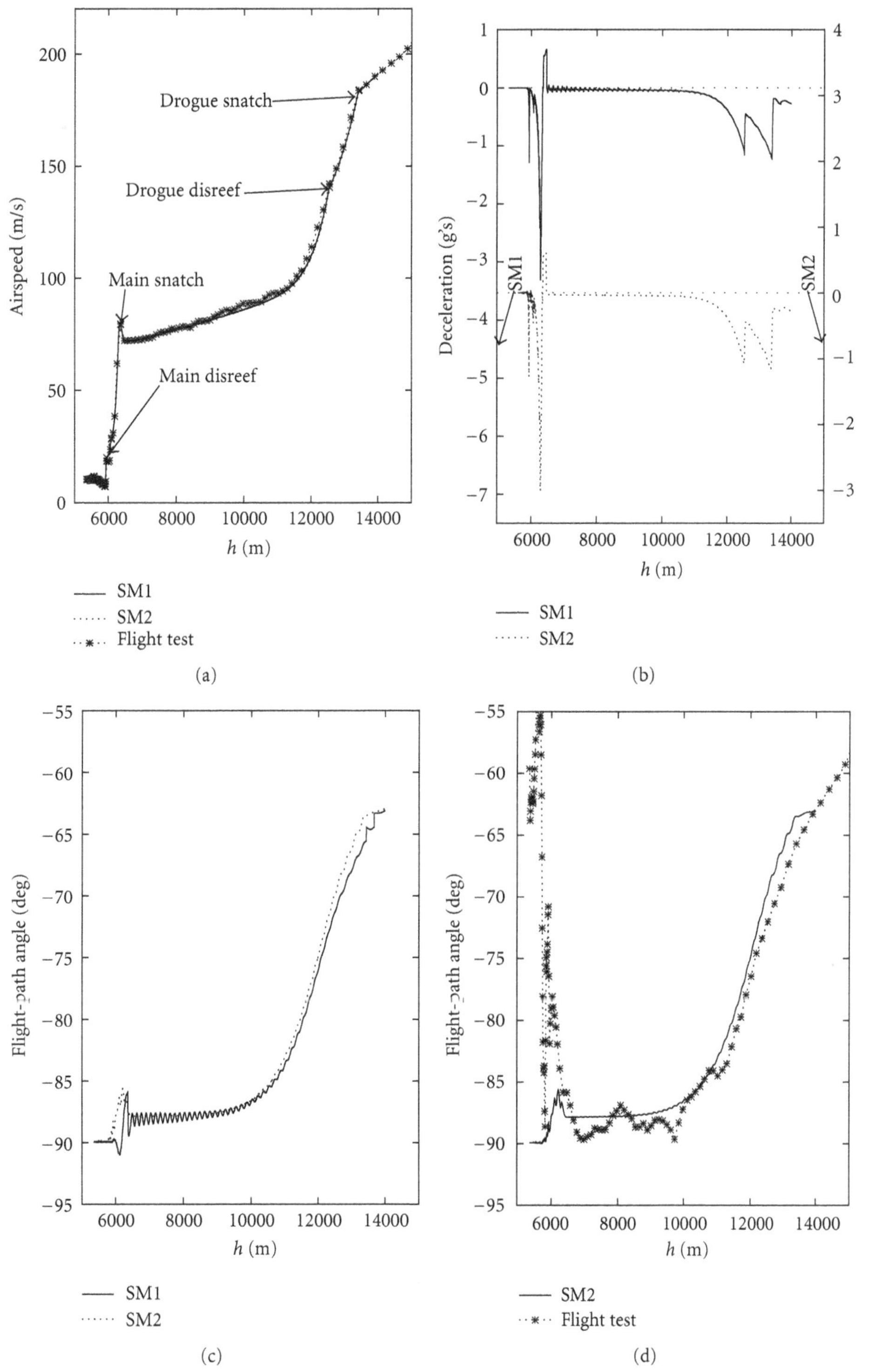

Figure 5: The simulation of the trajectory profile for the parachute-payload system.

reference flight tests [28, 29] are used for the validation of the simulation of capsule terminal reentry dynamics. The capsule is substantially a 70%-scaled version (diameter is 2.8 m) of the original Apollo Command Module decelerated by a single conical ribbon drogue parachute (nominal diameter 5.8 m) inflated in two reefing stages and a cluster of three main polyconical slotted parachutes (nominal diameter 22.9 m) inflated in three reefing stages. The sequence of inflation (see Table 1) and the parachute drag profile are time-scheduled according to the planned reefing stages and inflation times. Note that staging is obtained through parachute reefing (i.e., through restricted

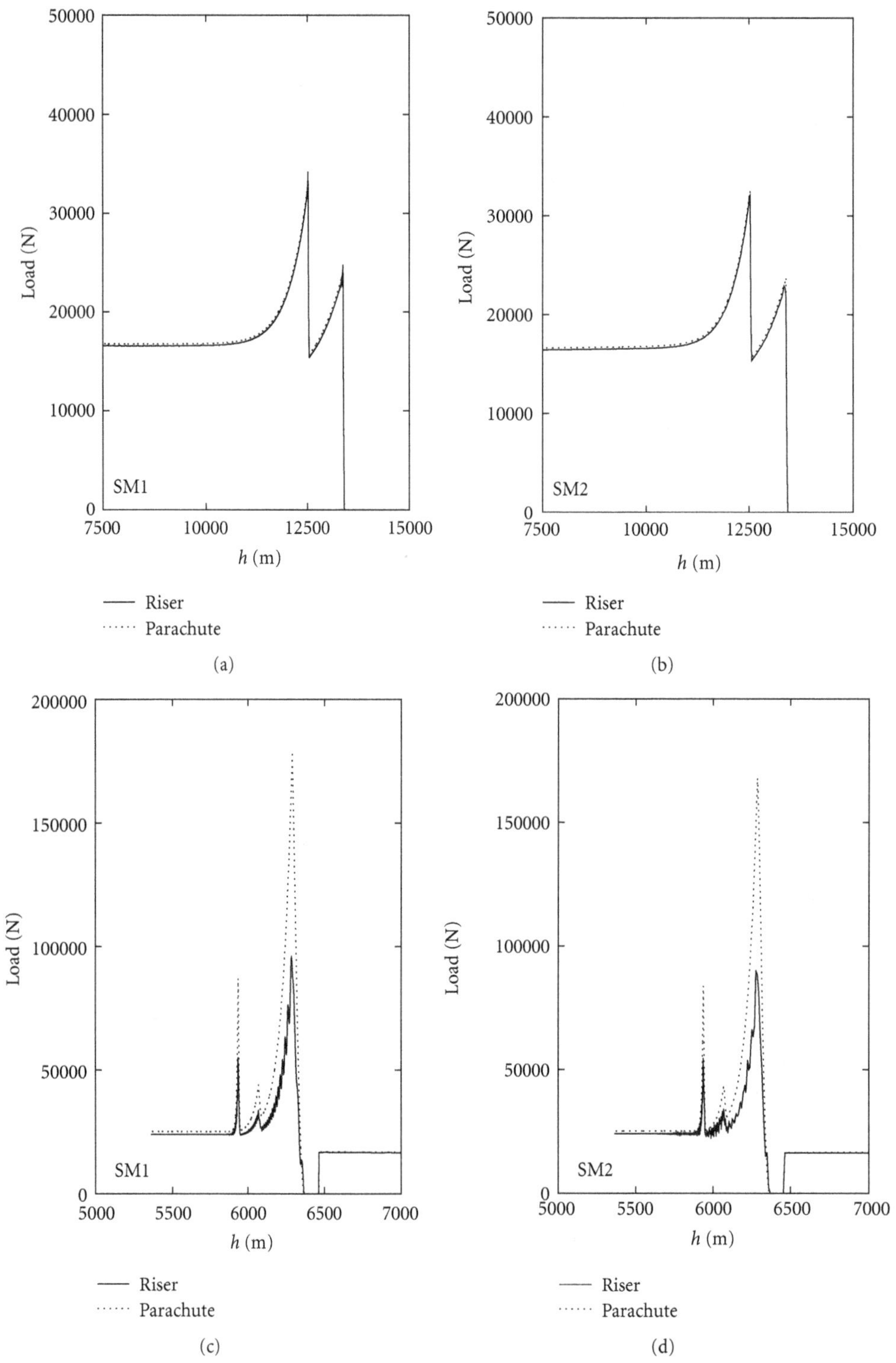

Figure 6: The loads acting on the riser and the decelerator (drogue and main parachute).

canopy deployment) with the purpose of limiting peak loads and deployment shocks. A detailed description of parachute deployment sequences for aerospace applications is given in [2, 3].

The aerodynamic data for the payload—which the simulation uses as input—were obtained with wind tunnel static and forced oscillation tests performed at Politecnico di Torino [10]. The wind tunnel is a closed-loop tunnel with a cylindrical working section of 3 m diameter by 5 m length. The model reproduces the scaled geometry of the NASA Apollo Command Module (model diameter is 340 mm). The longitudinal loads on the scaled vehicle (axial force, normal force, and pitching moment) are measured by an internal 3-component balance fit within the capsule model. Body

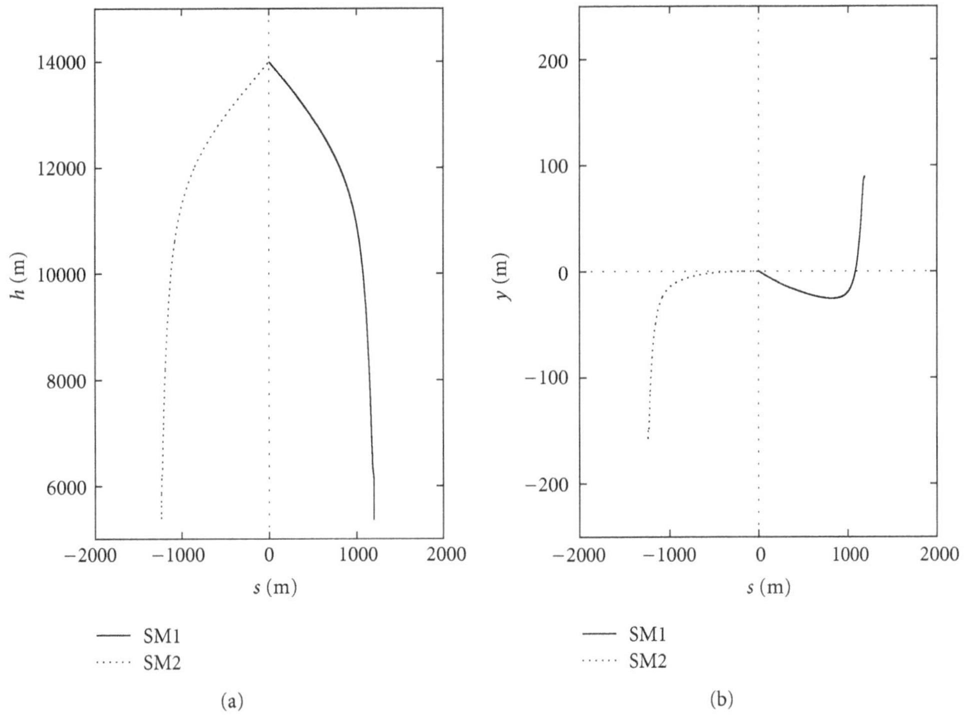

FIGURE 7: The simulation of the trajectory profile for the parachute-payload system.

TABLE 1: The parachute deployment sequence.

Sequence	Time (s)	Altitude (m)
Mortar firing (Pilot Jettison)	0.00	13994
Back cover separation	—	13666
Drogue snatch/1st stage inflation	3.34	13430
Drogue 2nd stage inflation/disreef	9.34	12546
Drogue bridles cut	—	6461
Main snatch/1st stage inflation	82.30	6364
Main 2nd stage inflation	88.30	6092
Main 3rd stage inflation/disreef	94.55	5950
Main bridle 1 cut	—	5360

axes are adopted for the reduction of experimental results centered in the actual reference center of gravity. The angle of attack is set to 180° when the spherical base (thermal shield) is exposed to the wind. The model is supported (see Figure 4) by a vertical strut that can rotate over the complete range of angle of attack ($\alpha = \pm 180°$). Positioning is performed by a step-motor that is controlled by a driving unit interfaced to the data acquisition PC. The stability derivatives are evaluated according to the small amplitude direct forced oscillation technique [30]. The oscillation is generated by the step-motor ($\Delta\theta = 1° - 5°$) and $f = 0.5$–5 Hz. The position of the driving shaft is acquired by means of a digital encoder. The measurement repeatability for the averaged static coefficients is $\sigma \approx \pm 0.1\%$ estimated over the full range while the measurement repeatability for the longitudinal stability derivatives (measured from frequency response) is $\sigma \approx \pm 1\%$.

The reentry profile is correctly reproduced by the two simulators—as presented in Figure 5—and the trajectory, shaped by the parachute opening sequence, is matched by both SM1 and SM2 models. The trend of the load factor exhibits a set of marked peaks given by the staging of both main and drogue parachute. The SM1 results exhibit an oscillatory behavior in the second part of the parachute deployment phase, induced by the oscillatory behavior of the payload, not found for SM2 model results. Note that low-altitude flight test data are affected by the asymmetry of the suspension system induced by the cut of one of the bridles of the main parachutes (see Table 1).

The ability of the SM1 model to estimate the riser and parachute loading is outlined in Figure 6. The trend of forces is coherent with the deceleration profile presented in Figure 5. The offset between the force acting on the canopy and the riser is well marked for the main clustered parachute, modeled as an equivalent single decelerator. This offset is due to the mutual influence of the added masses (inertia-induced delays) and the elongation of the riser. The effect of the elasticity of the main cable is also visible in terms of damped oscillatory strain, triggered by the bouncing of the payload (and by its attitude dynamics) after each inflation phase (staged reefing).

The projected trajectory is compared in Figure 7. Accurate trajectory coordinates were not available from flight test data. The profiles (mirrored for the purpose of checking their symmetry) are very similar in the vertical plane. Nevertheless, a slightly different path is found comparing the

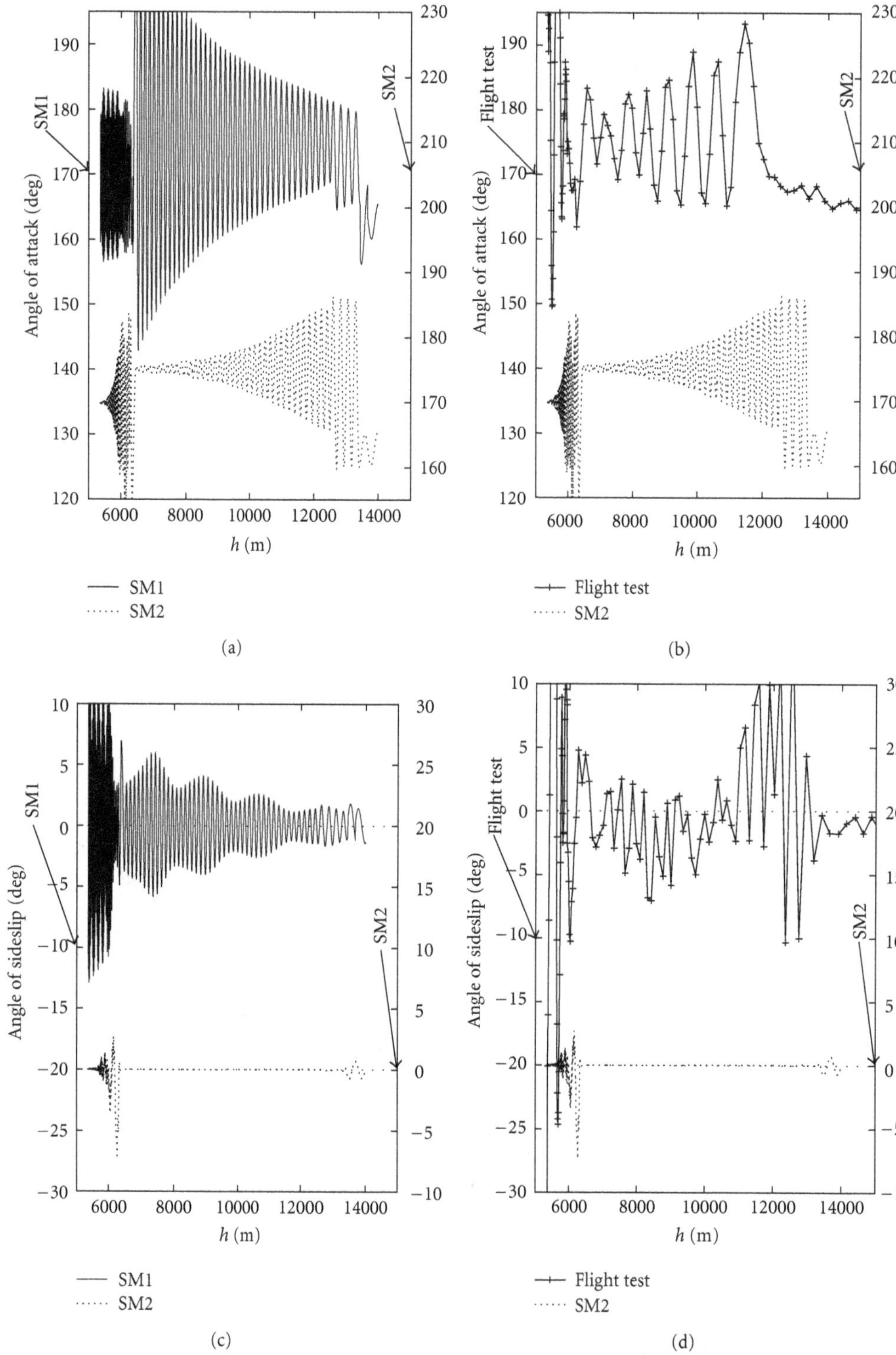

Figure 8: The simulation of the angles of attack and sideslip for the payload.

traces for the lateral-directional plane. This is the impact of the aerodynamic model adopted for the payload, which for the SM1 case decouples and superimposes the effects for the longitudinal and the lateral-directional planes. The use of total angle of attack for the interpolation and reconstruction of payload aerodynamic coefficients (as in SM2 model) provides a more accurate fit.

The aerodynamic angles of the payload are plotted in Figure 8. The major discrepancy between the two models is the level of dynamic stability of the modal response (short-period dynamics). This can be explained with the extreme sensitivity of the aerodynamic coefficients to center of gravity location and angle of attack (see Figures 9 and 10). Remind that the center of gravity location is updated

Table 2: The index of similarity for the pitch rate spectral response.

Flight test ⇒ SM1			Flight test ⇒ SM2		
Drogue parachute			Drogue parachute		
I1	I2	I3	I1	I2	I3
1.392	1.005	0.989	0.645	1.145	1.110
Main parachute			Main parachute		
I1	I2	I3	I1	I2	I3
0.578	1.022	1.029	0.463	1.088	1.098

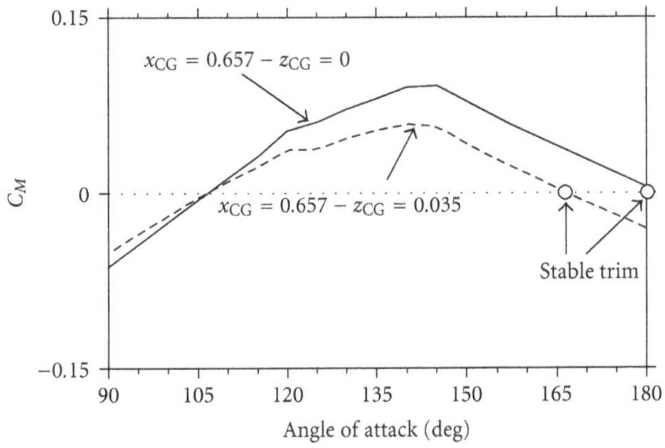

Figure 9: The aerodynamic coefficients of the payload (pitching moment static stability) [10].

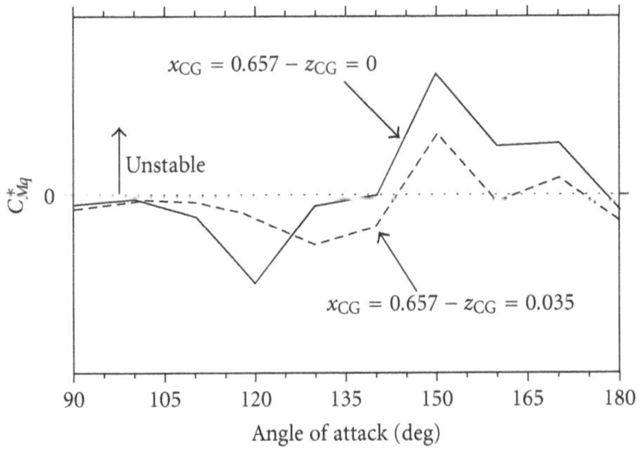

Figure 10: The aerodynamic coefficients of the payload (pitching moment dynamic stability) [10].

during the descent, accounting for covers, deployment bags, and parachutes release. As a matter of fact, minor parametric changes (α, x_{cg}, and z_{cg}) shift the attitude for the rotational equilibrium of the payload and alter its dynamic stability. This point is also a concern for the validation of simulation models by comparison with flight test data [23], in which a transition between different equilibrium states and levels of dynamic stability can be externally forced by the atmospheric perturbations, exciting the combined parachute-payload system. Another contributing element is the dynamics of the parachute that in the higher fidelity model (SM2) allows for attitude changes decoupled from the suspended body, providing a more realistic behavior of the parachute-payload system. Differently, the parachute trajectory in the SM1 model is aligned instantly with the velocity vector (flight path) of the payload. Furthermore, a multilink suspension system (as modeled in SM2 implementation) induces a stronger steering effect on the payload.

In order to verify the modal response, at least for the longitudinal short-period natural frequency, the pitch rate spectral response for the two simulation models is compared with the elaboration of available flight logs in Figures 11 and 12 (note that from flight test reports gyro signal measurements are very noisy after drogue release). The comparison of the spectral data is based on the crosscheck of the index of similarity (see Table 2), defined according to (30), where A_i is the amplitude for the given frequency ω_i:

$$
\begin{aligned}
s_0 &= \sum_{i=1}^{N} A_i, \\
s_1 &= \sum_{i=1}^{N} A_i \cdot \omega_i \longrightarrow \quad
\begin{aligned}
\sigma_1 &= \frac{s_0}{N} \\
\sigma_2 &= \frac{s_1}{s_0} \\
\sigma_3 &= \sqrt{\frac{s_2}{s_0}}
\end{aligned}
\quad \longrightarrow \quad
\underbrace{
\begin{aligned}
I_1 &= \frac{\sigma_1^a}{\sigma_1^b} \\
I_2 &= \frac{\sigma_2^a}{\sigma_2^b} \\
I_3 &= \frac{\sigma_3^a}{\sigma_3^b}
\end{aligned}
}_{a \to b} \\
s_2 &= \sum_{i=1}^{N} A_i \cdot \omega_i^2.
\end{aligned}
\tag{30}
$$

The results show that the range for natural frequency of the short period response is matched by both simulation models as demonstrated by the fact that I_2 and I_3 are close to unity. Other than that, the averaged amplitude of the spectral response is less precisely reproduced by the simulation models, as experimental data represent a modal response that is overexcited by atmospheric disturbances and suspension system asymmetries, mainly at parachute deployment.

6. Conclusions

The present work outlines a comparative analysis of two simulation models (SM1 and SM2) of a parachute-payload system with different levels of complexity. An example of limited complexity reconfigurable simulation models for a

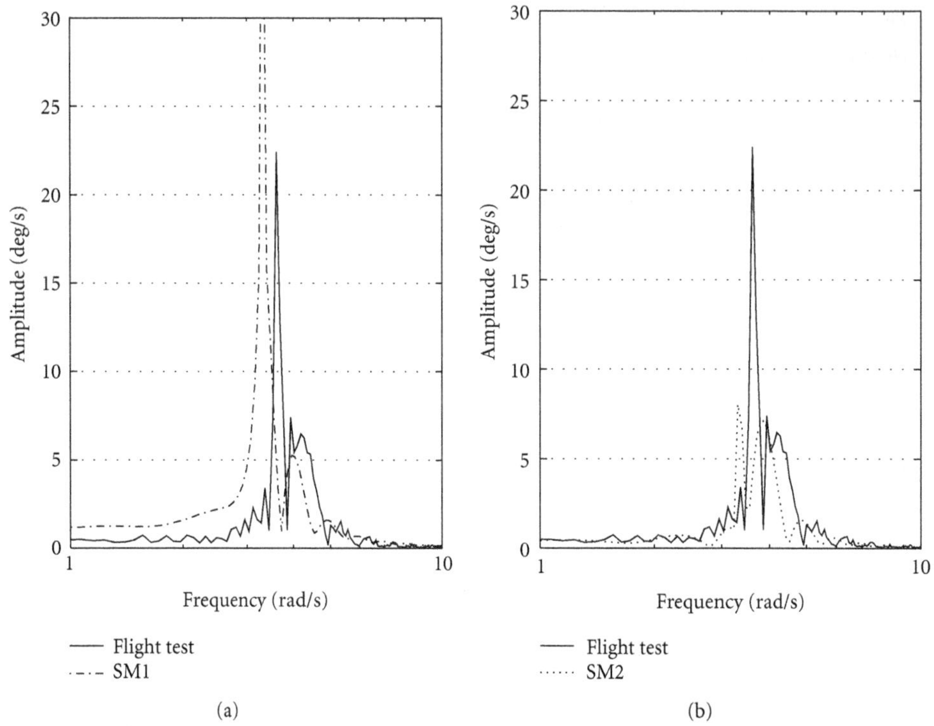

Figure 11: The effect of model complexity on pitch rate spectral response (drogue parachute).

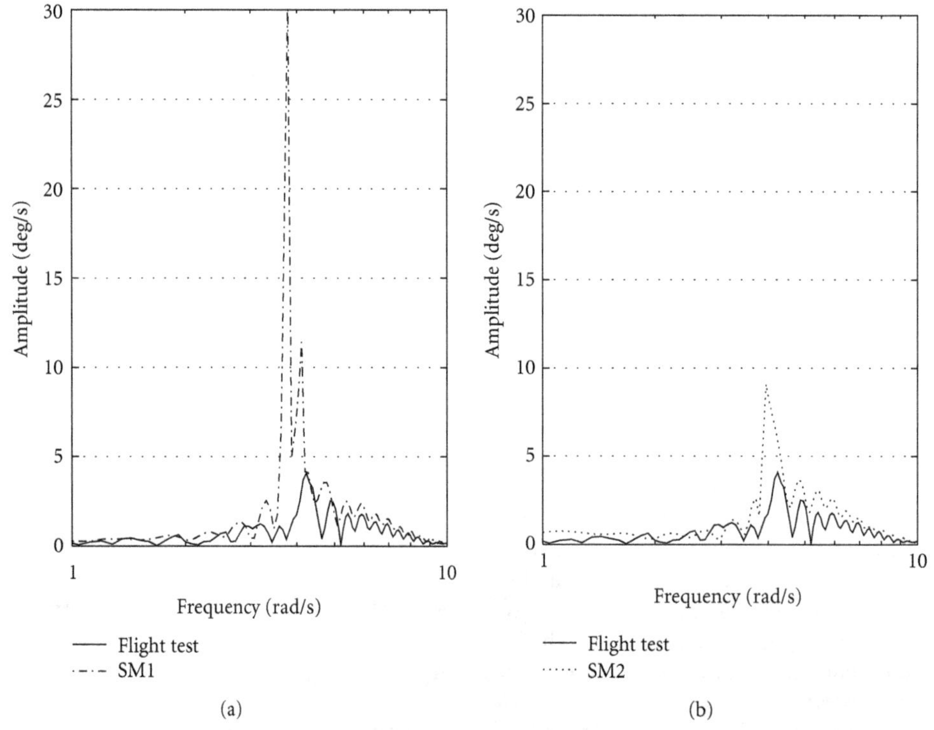

Figure 12: The effect of model complexity on pitch rate spectral response (main parachute).

payload decelerated by one or more parachutes is given, including details and implementation features usually omitted as the focus of the research in this field is typically on the investigation of mission design issues, rather than addressing general implementation guidelines for the development of a reconfigurable simulation tool. The dynamics of the system are modeled through a simple multibody model that represents the expected behavior of an entry vehicle during the final deceleration phase by means of a system of aerodynamic decelerators. The simulators are designed according to a comprehensive vision that enforces the simplification of the coupling mechanism between the payload and the parachute, with an adequate level of physical insight still available in terms of sensitivity of system performance metrics to design parametric changes.

A reference mission described was used to evaluate the two simulation models. The aerodynamic data for the payload—which the simulation uses as input—were obtained with wind tunnel static and forced oscillation tests performed at the Politecnico di Torino.

The trajectory is reproduced by the two simulators as the profiles, mainly shaped by the parachute opening sequence, are matched by both SM1 and SM2 models, even if the use of total angle of attack for the interpolation and reconstruction of payload aerodynamic coefficients (as in SM2 model) provides a more accurate fit. The riser and parachute loading are estimated in a way that is coherent with the deceleration profiles.

The major discrepancy between the two models is the level of dynamic stability of the modal response (short-period dynamics). This can be explained with the extreme sensitivity of the aerodynamic coefficients to center of gravity location and angle of attack. Another contributing element is the dynamics of the parachute which in the higher fidelity model (SM2) allows for attitude changes decoupled from the suspended body, providing a more realistic behavior of the parachute-payload system. Differently, the parachute trajectory in the SM1 model is aligned instantly with the velocity vector of the payload, probably a too crude approximation if the purpose of the analysis is the attitude dynamics of the suspended vehicle. Furthermore, a multilink suspension system (as modeled in SM2 implementation) induces a steering effect on the payload, neglected by the single-point suspension of SM1 model.

The spectral results show that the range for the natural frequency of the short-period response—as measured in flight—is matched by both simulation models as demonstrated by the fact that the indices of similarity I_2 and I_3 are close to unity. Differently, the averaged amplitude of the spectral response is less precisely reproduced by the simulation models, as the experimental data represent a modal response that is over-excited by atmospheric disturbances and suspension system asymmetries, at parachute deployment mainly.

As a general conclusion, the lower fidelity simulation tool SM1 exhibits a limited advantage in terms of computational workload, mainly providing a simplified approach for preliminary trajectory estimation and parachute sizing.

Abbreviations

CG: Center of gravity
SM1: Simulation model 1
SM2: Simulation model 2.

Symbols

c: Damping coefficient
C_{BE}: Body-to-Earth axis transformation matrix
C_{BW}: Body-to-wind axis transformation matrix
C_D: Parachute drag coefficient
C_{GW}: Auxiliary transformation matrix
C_l: Roll moment coefficient
C_{lp}: Roll moment coefficient due to roll rate
C_M: Pitching moment coefficient
C_{Mq}: Pitching moment coefficient due to pitch rate
C_N: Yaw moment coefficient
C_{Nr}: Yaw moment coefficient due to yaw rate
C_{WB}: Wind-to-Body axis transformation matrix
C_X: Force coefficient in x-axis direction
C_{Xq}: Force coefficient in x-axis direction due to pitch rate
C_Y: Force coefficient in y-axis direction
C_{Yr}: Force coefficient in y-axis direction due to yaw rate
C_Z: Force coefficient in z-axis direction
C_{Zq}: Force coefficient in z-axis direction due to pitch rate
d: Payload reference length
D_0: Canopy nominal diameter
D_P: Parachute drag
$e_0 \cdots e_3$: Quaternion parameters
F_R: Riser force (tension)
g: Acceleration of gravity
h: Canopy height
h: Altitude
I_i: Index of similarity
I_{XX}: Moment of inertia
I_{XY}: Product of inertia
I_{XZ}: Product of inertia
I_{YY}: Moment of inertia
I_{YZ}: Product of inertia
I_{ZZ}: Moment of inertia
k: Stiffness
k_a: Added mass coefficient
k_b: Bridle stiffness
k_r: Riser stiffness
L: Rolling moment
l_0: Offset length
l_r: Riser length
m: Payload mass
M: Pitching moment
M^*: Parachute total mass

m_a: Parachute added mass
m_P: Parachute mass
N: Yawing moment
p: Canopy porosity
p: Roll rate
q: Pitch rate
r: Yaw rate
R: Canopy radius
r_k: Numerical residual
s: Payload displacement
s_P: Parachute displacement
S_P: Parachute reference area
t: Time
u: Velocity in x body direction
v: Velocity in y body direction
V: Payload-free stream velocity
V_c: Payload velocity
V_P: Parachute velocity
V_s: Parachute velocity (at line stretch)
w: Velocity in z body direction
x: North displacement (trajectory)
X: Force in x body direction
x_{cg}: x-wise CG location
x_p: x-wise suspension point location
x_{ref}: x-wise CG location (reference)
y: East displacement (trajectory)
Y: Force in y body direction
y_{cg}: y-wise CG location
y_p: y-wise suspension point location
z: Down displacement (trajectory)
Z: Force in z body direction
z_{cg}: z-wise CG location
z_p: z-wise suspension point location
z_{ref}: z-wise CG location (reference)
α: Angle of attack
β: Angle of sideslip
γ: Flight path angle
ε: Strain
$\dot{\varepsilon}$: Strain rate
η: Wake penalty factor
θ: Pitch angle
ξ: Damping ratio
ρ: Air density
σ: Measurement error
ϕ: Roll angle
ψ: Yaw angle
ω_i: Frequency.

Acknowledgment

The author wishes to acknowledge the support given by Mr. Michael Gordon.

References

[1] D. J. Cockrell and A. D. Young, "The aerodynamics of parachutes," AGARD AG-295, 1987.
[2] T. W. Knacke, *Parachute Recovery Systems Design Manual*, Para Publishing, 1st edition, 1992.
[3] R. C. Maydew and C. W. Peterson, "Design and testing of high performance parachutes," AGARD AG-319, 1991.
[4] NASA Langley Research Center, "Deployable aerodynamic deceleration systems," NASA SP-8066, 1971.
[5] M. L. Accorsi, J. W. Leonard, R. J. Benney, and K. R. Stein, "Structural modeling of parachute dynamics," *AIAA Journal*, vol. 38, no. 1, pp. 139–146, 2000.
[6] M. L. Accorsi, R. J. Benney, V. Kalro, J. W. Leonard, K. R. Stein, and T. E. Tezduyar, "Parachute fluid-structure interactions: 3-D computation," *Computer Methods in Applied Mechanics and Engineering*, vol. 190, no. 3-4, pp. 373–386, 2000.
[7] M. L. Accorsi, R. J. Benney, J. W. Leonard, K. R. Stein, and T. E. Tezduyar, "Fluid-structure interactions of a round parachute: modelling and simulation techniques," *Journal of Aircraft*, vol. 38, no. 5, pp. 800–808, 2001.
[8] J. R. Cruz and J. S. Lingard, "Aerodynamic decelerators for planetary exploration: past, present, and future," in *Proceedings of the AIAA Guidance, Navigation, and Control Conference*, pp. 5342–5361, August 2006.
[9] M. Ballion, "Blunt bodies dynamic derivatives," AGARD R808, 1995.
[10] G. Guglieri and F. Quagliotti, "Low speed dynamic tests on a capsule configuration," *Aerospace Science and Technology*, vol. 4, no. 6, pp. 383–390, 2000.
[11] L. E. Ericsson and J. P. Reding, "Re-entry capsule dynamics," *Journal of Spacecraft and Rockets*, vol. 8, no. 6, pp. 579–586, 1971.
[12] R. G. Hume, *A Two Dimensional Mathematical Model of a Parachute in Steady Descent*, Aeronautical Research Council, 1973, C.P. No. 1260.
[13] M. Neustadt, R. E. Ericksen, J. J. Guiteras, and J. A. Larrivee, "A parachute recovery system dynamic analysis," *Journal of Spacecraft and Rockets*, vol. 4, no. 3, pp. 321–326, 1967.
[14] F. M. White and D. F. Wolf, "A theory of three-dimensional parachute dynamic stability," *Journal of Aircraft*, vol. 5, no. 1, pp. 86–92, 1968.
[15] D. Wolf, "Dynamic stability of a nonrigid parachute and payload system," *Journal of Aircraft*, vol. 8, no. 8, pp. 603–609, 1971.
[16] K. F. Doherr and H. Schilling, "Nine degree-of-freedom simulation of rotating parachute systems," *Journal of Aircraft*, vol. 29, no. 5, pp. 774–781, 1992.
[17] C. Tory and R. Ayres, "Computer model of a fully deployed parachute," *Journal of Aircraft*, vol. 14, no. 7, pp. 675–679, 1977.
[18] J. A. Eaton, "Added masses and dynamic stability of parachutes," *Journal of Aircraft*, vol. 19, no. 5, pp. 414–416, 1982.
[19] J. A. Eaton, "Added fluid mass and the equations of motion of a parachute," *Aeronautical Quarterly*, vol. 34, no. 3, pp. 226–242, 1983.
[20] P. A. Cuthbert, "A software simulation of cargo drop test," in *Proceedings of the 17th AIAA Aerodynamic Decelerator Systems Technology Conference*, Monterey, Calif, USA, 2003, AIAA Paper 2003-2132.
[21] P. A. Cuthbert and K. J. Desabrais, "Validation of a cargo airdrop simulator," in *Proceedings of the 17th AIAA Aerodynamic Decelerator Systems Technology Conference*, Monterey, Calif, USA, 2003, AIAA Paper 2003-2133.
[22] P. C. Klimas, "Fluid mass associated with an axisymmetric canopy," *Journal of Aircraft*, vol. 14, no. 6, pp. 577–580, 1977.
[23] G. Guglieri and F. Quagliotti, "Validation of a simulation model for a planetary entry capsule," *Journal of Aircraft*, vol. 40, no. 1, pp. 127–136, 2003.

[24] U. M. Ascher and L. R. Petzold, *Computer Methods for Ordinary Differential Equations and Differential Algebraic Equations*, SIAM, Philadelphia, Pa, USA, 1999.
[25] K. E. Brenan, S. L. Campbell, and L. R. Petzold, *Numerical Solution of Initial Value Problems in Differential Algebraic Equations*, SIAM, Philadelphia, Pa, USA, 1996.
[26] L. R. Petzold, "Description of DASSL: a differential algebraic system solver," in *Proceedings of the 10th International Mathematics and Computers Simulation Congress on Systems Simulation and Scientific Computation*, Montreal, Canada, 1982.
[27] R. Celi, "Solution of rotary-wing aeromechanical problems using differential-algebraic equation solvers," *Journal of the American Helicopter Society*, vol. 45, no. 4, pp. 253–263, 2000.
[28] S. Portigliotti, J. C. Paulat, and Rives J., "ARD descent and recovery subsystem dynamics and trajectory flight data evaluation," in *Proceedings of the 1st AAAF International Symposium on Atmospheric Reentry Vehicles and Systems*, Arcachon, France, 1999.
[29] S. Portigliotti, "Parachute flight dynamics modeling and simulation: code validation through ARD post-flight analysis," in *Proceedings of the 2nd AAAF International Symposium on Atmospheric Reentry Vehicles and Systems*, Arcachon, France, 2001.
[30] E. S. Hanff, "Direct forced oscillation techniques for the determination of stability derivatives in wind tunnel," AGARD LS-114, 1981.

Exfoliation Corrosion and Pitting Corrosion and Their Role in Fatigue Predictive Modeling: State-of-the-Art Review

David W. Hoeppner and Carlos A. Arriscorreta

Department of Mechanical Engineering, University of Utah, 50 South Central Campus Drive, Room 2010, Salt Lake City, UT 84112, USA

Correspondence should be addressed to David W. Hoeppner, dhoeppner@comcast.net

Academic Editor: Christopher J. Damaren

Intergranular attack (IG) and exfoliation corrosion (EC) have a detrimental impact on the structural integrity of aircraft structures of all types. Understanding the mechanisms and methods for dealing with these processes and with corrosion in general has been and is critical to the safety of critical components of aircraft. Discussion of cases where IG attack and exfoliation caused issues in structural integrity in aircraft in operational fleets is presented herein along with a much more detailed presentation of the issues involved in dealing with corrosion of aircraft. Issues of corrosion and fatigue related to the structural integrity of aging aircraft are introduced herein. Mechanisms of pitting nucleation are discussed which include adsorption-induced, ion migration-penetration, and chemicomechanical film breakdown theories. In addition, pitting corrosion (PC) fatigue models are presented as well as a critical assessment of their application to aircraft structures and materials. Finally environmental effects on short crack behavior of materials are discussed, and a compilation of definitions related to corrosion and fatigue are presented.

1. Introduction

This paper deals with the effects of intergranular attack and exfoliation corrosion on structural integrity of aircraft structures and materials with emphasis on aluminum alloys used over many decades for airframe components of military, commercial, and general aviation aircraft. Aluminum alloys have been the material of choice for many components of airframes in the past and remain so even though some aircraft are using more titanium alloys and resin-based composites in many airframe components. The general background on phases of life and methods for dealing with corrosion in general and aspects of HOLSIP (Holistic Structural Integrity Processes) paradigm are presented to some extent. (See http://www.holsip.com/). This is followed by a discussion of corrosion effects on SI (Structural Integrity) with some details provided on significant effects of corrosion on maintainability and reliability of structures with extensive background material. Subsequently a section that describes intergranular attack and exfoliation in general terms follows which then is followed by a discussion of cases where IG attack and exfoliation caused significant structural integrity issues in aircraft in operational fleets. Studies oriented toward evaluating the effects of IG and exfoliation on fatigue behavior with emphasis on the long crack aspects are presented. The final section then presents recommended studies in order to develop and validate models to allow prediction and management of IG attack and exfoliation as part of a Holistic Structural Integrity Processes paradigm [1–67], (numbers in parentheses refer to the references in order of appearance).

1.1. Phases of Life and Modeling. The phases of life of a structure may be classified according to the division in the Table 1. Thus, the total life ($\mathbf{L}_T$) of a structure is $\mathbf{L}_T = \mathbf{L}_1 + \mathbf{L}_2 + \mathbf{L}_3 + \mathbf{L}_4$. Figure 1 presents a depiction of the degradation process from a holistic perspective. The regions shown in Figure 1, for example, 1, 2, 3, and 4, illustrate the portion of life, on the abscissa, and the corresponding growth in discontinuity size plotted schematically on the ordinate. This paper concentrates on the phases of life $\mathbf{L}_1$ and $\mathbf{L}_2$. that is, the corrosion process or processes that results in the formation

TABLE 1: Phases of Life. See Figure 1. from Hoeppner, 1972 [67], 1981 [38], 1985 [39].

(i) Formation or nucleation of degradation/damage by a specificphysical or corrosion process interacting with the fatigue processif appropriate. Corrosion and other processes may act alone toform/nucleate the damage. A transition from theformation/nucleation stage to the next phase must occur. Phase L_1 to some other phase.
(ii) Microstructurally dominated crack linkup and propagation ("short" or "small" crack regime). Phase L_2.
(iii) Crack propagation in the regime where LEFM, EPFM, or FPFM may be applied both for analysis and material characterization (the "long" crack regime). Phase L_3.
(iv) Final instability. Phase L_4.

NOTE: In some cases in practice not all the phases cited above occur.

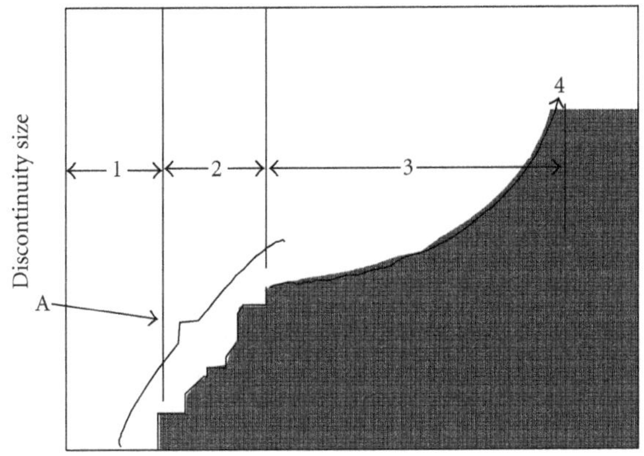

A = "first" detectable crack
1. Nucleation phase, "no crack"
2. "small crack" phase steps related to local structure (anisotropy)
3. Stress-dominated crack growth, LEFM, EPFM
4. Crack at length to produce instability

FIGURE 1: A depiction of the degradation process after Hoeppner-1972 [67], 1981 [38], 1985 [39].

or nucleation of a specific form of corrosion generating a specific form of discontinuity that is not necessarily a crack-like discontinuity (EDS or MDS—see list of definitions in the appendix) and the development of short cracks and their propagation from the initial discontinuity state or from the evolved or modified discontinuity state (IDS—see the list of definitions in the appendix) formed by the mechanism in question. The requirement of the community to come up with *design methods* to deal with corrosion or other time-based degradation, that is, fatigue, creep, and wear is essential and some of the elements are depicted in Figure 2. This figure illustrates that most of the quantitative methods that have been developed used the concepts of mechanics of materials with an incorporation of fracture mechanics.

The sections of this paper that follow will discuss the following major areas:

(i) general effects of corrosion on structural integrity;

(ii) intergranular attack and exfoliation corrosion (EC) in aircraft structural aluminum alloys;

(iii) efforts to date on modeling effects of exfoliation corrosion in aircraft structure with emphasis on fatigue and fatigue crack propagation behavior.

The issue of the effects of corrosion on structural integrity of aircraft has been a question of concern for some time [1–36]. The potential effects are many and they can be categorized as follows.

(An attempt has been made to provide as simple a statement of each potential problem as possible. In the discussion below the use of the terms global and local refers to the likely extent of the corrosion on the surface of a component. Global means the corrosion would be found on much of the component whereas local means the corrosion may be localized to only small, local areas.)

(1) Reduction of section with a concomitant increase in stress (e.g., thickness change, etc.). Global or local.

(2) Production of stress concentration. Local.

(3) Nucleation of cracks. Local, possibly global. Source of Multiple-site cracking.

(4) Production of corrosion debris. This may result in surface pillowing by various means, which may significantly change the stress state and structural behavior. Local and global.

(5) Creation of a situation that causes the surfaces to malfunction. Local and global.

(6) Cause environmentally assisted crack growth (EACG) under cyclic (corrosion fatigue or corrosion-fatigue) or sustained loading (SCC) conditions. Local.

(7) Create a damage state that is missed in inspection when the inspection plan was not developed for corrosion or when corrosion is missed. Local and global.

(8) Change the structurally significant item due to the creation of a damage state not envisioned in the structural damage analysis or fatigue and strength analysis. If the SSI is specified, for example, by location of maximum stress or strain, then the corrosion may cause another area(s) to become significant. Local or global.

(9) Create an embrittlement condition in the material that subsequently affects behavior. Local or global.

(10) Create a general aesthetic change from corrosion that creates maintenance to be done and does damage to the structure. Local or global.

(11) Corrosion maintenance does not eliminate all the corrosion damage and cracking or the repair is specified improperly or executed improperly thus creating a damage state not accounted for in the design. Local or global.

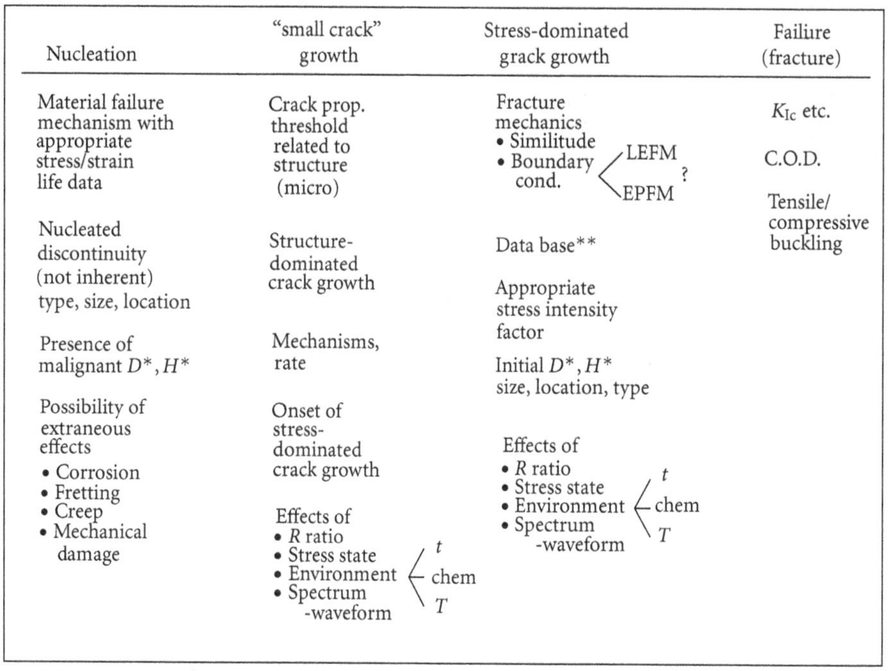
Nucleation
"small crack" growth
Stress-dominated grack growth
Failure (fracture)
Material failure mechanism with appropriate stress/strain life data
Nucleated discontinuity (not inherent) type, size, location
Presence of malignant D*, H*
Possibility of extraneous effects
• Corrosion
• Fretting
• Creep
• Mechanical damage
Crack prop. threshold related to structure (micro)
Structure-dominated crack growth
Mechanisms, rate
Onset of stress-dominated crack growth
Effects of
• R ratio
• Stress state
• Environment
• Spectrum -waveform
t
chem
T
Fracture mechanics
• Similitude
• Boundary cond.
LEFM
EPFM
?
Data base**
Appropriate stress intensity factor
Initial D*, H* size, location, type
Effects of
• R ratio
• Stress state
• Environment
• Spectrum -waveform
t
chem
T
KIc etc.
C.O.D.
Tensile/compressive buckling

Figure 2: Methods for each life phase after Hoeppner-1972 [67], 1981 [38], 1985 [39].

(12) Generation of a damage state that alters either the durability phase of life or the damage tolerant assessment of the structure or both.

(13) Creation of a widespread corrosion damage (WCD) state or a state of corrosion that impacts the occurrence of widespread fatigue damage (WFD) and its concomitant effects. [1, 3, 4, 13, 15, 25–27, 31–36].

(14) Produce a condition that may cause a loss of fail safety in conjunction with one or more issues noted above.

The question of whether corrosion, corrosion fatigue, corrosion/fatigue, and or stress corrosion cracking (see appendix for definitions used herein) are safety concerns or just maintenance/economic concerns has been a point of discussion related to aircraft structural integrity for over 50 years. Nonetheless, a great deal of the aircraft structural integrity community believes that corrosion-related degradation is just an economic or maintenance concern. The issue of type of corrosion and its effects on structural integrity has been addressed in other summaries. This brief introduction gives a summary of some of the compilations of information related to corrosion in general. The major section that follows presents more information on the studies to date that have focused or are focusing on intergranular attack and exfoliation corrosion.

It was with the issue of safety or economic concerns that led Campbell and Lehay [12] and Wallace et al. [13] to pursue the presentation of technical facts and knowledge to illustrate the potential for a safety issue as well as maintenance and/or economic issue. Finally, Hoeppner et al. [27] reviewed failure data obtained from USAF, USN, USA, FAA, and the NTSB related to aircraft incidents and accidents in the USA from 1975–1994 to evaluate further the potential for corrosion and fretting-related degradation to be significant safety issues. A quote from the introduction to the paper [27] follows:

"On July 25, 1990, a pilot and crew were killed when the right wing outboard of the engine nacelle separated from their Aero Commander (now Twin Commander) 680 while performing a geological survey. The aircraft entered an uncontrolled decent and crashed into a field near Hassela, Sweden. Investigations revealed that the wing failed due to corrosion pits, which nucleated fatigue cracks in the lower spar cap" [27]. Although the accident occurred in Sweden, this accident sparked inspections of other Twin Commander aircraft worldwide. In November of 1991, Twin Commander released a service report detailing extensive cracking problems found in the lower spar cap of a US registry airplane. The Australian Civil Airworthiness Authority (CAA), on behalf of the Federal Aviation Administration (FAA), conducted fractographic analyses on ten cracks found in the component. The CAA determined that the cracks formed by intergranular attack, pits and resulted in stress corrosion cracks and that further extension occurred by fatigue. These failures will be referred to later in the section on IG and exfoliation and more detail will be provided. The mechanism overlap (two or more corrosion or degradation mechanisms being involved in change of damage state) frequently has been observed as documented by [13] as well.

The above example illustrates how corrosion pits and IG attack can severely jeopardize the structural integrity and safety of aircraft. In addition to corrosion, fretting and fretting fatigue have proven, on occasion, to be significant safety hazards. This paper will not deal with fretting and fretting fatigue as the first author has written extensively about this elsewhere and these mechanisms of degradation were not to be included in this brief summary.

Although the aircraft industry directs a great deal of attention to safety concerns, for many years it has relegated corrosion and fretting to maintenance, economic, and inspection issues. While the industry has developed some corrosion/fretting prevention programs, it has not done what it possibly should to quantitatively evaluate the effects of corrosion/fretting on structural integrity. What attempts have been made in this area appear to be sporadic and limited in number [27].

Walter Schütz addressed this issue further in the Plantema lecture at ICAF [25]. Furthermore, anyone that doubts the potential catastrophic consequences of corrosion-related degradation of aircraft structure would be assisted by reading Steve Swift's insightful presentation related to "The Aero Commander Chronicle" [26]. As a part of a technical paper delivered by Hoeppner et al. at ICAF-1999, they found the following with regard to pitting corrosion and pitting corrosion fatigue as listed in Table 2.

The examples shown in the table, taken with the general information cited in the references, clearly show that corrosion-related degradation is a significant safety issue in the assurance of structural integrity of aircraft. No such compilation has been done for exfoliation alone but needs to be done in the authors' opinion.

In recent years more emphasis has been placed on this issue of corrosion effects on structural integrity-especially after the fleet surveys subsequent to the Aloha Airlines accident (AA243) in 1988 [16]. Even though the NATO-AGARD community authorized the production of a manual on corrosion case studies and a great deal of information was presented in the manual published by AGARD [13], it is essential that the RMS deficiencies that may arise before accidents occur be recognized. This clearly has not been the case in all major fleets of aircraft whether they are military or commercial [12, 16, 19–22, 25–27]. Another issue that is clear is that deficiencies in the analysis of failures and the databases exist [27, 28].

The potential regrettable occurrence of accidents from corrosion-related crack formation/nucleation is a constant threat to aircraft safety. The following quote from the recent NATO RTO conference on fatigue in the presence of corrosion adds some understanding to the need for greater effort to understand the potential role of effects of corrosion on structural integrity.

> *Some of the workshop papers discussed the significance of corrosion-fatigue as a safety issue or an economic issue. There is ample data to support the contention that it is definitely an economic issue. There is also ample data to support the contention that it has not been a significant safety problem. However, the problem is certainly a potential safety concern if maintenance does not perform their task diligently. In addition, management must continuously update established maintenance and inspection practices to address additional real-time degradation threats for aircraft operated well beyond their initial design certification life. The economic issue alone is sufficient to motivate the support of research and development that can reduce the maintenance burden. This research will also reduce the threat of catastrophic failure from the corrosion damage.*

(*Lincoln, J., Simpson, D., Introduction to* [36]).

Another quote from a different reference sheds further light on this issue [33, page 1-1].

> *At the present time, structural life assessments, inspection requirements, and inspection intervals, are determined by Durability and Damage Tolerance Assessments (DADTAs) using fracture mechanics crack growth techniques in accordance with the Aircraft Structural Integrity Program (ASIP). These techniques do not normally consider the effects of corrosion damage on crack initiation or crack growth rate behavior. Also, these techniques do not account for multiple fatigue cracks in the DADTAs of the structural components susceptible to WFD. For aircraft that are not expected to have significant fatigue damage for many years, such as the C/KC-135, this approach has severe limitations since it does not account for corrosion damage or WFD. The impact of corrosion damage and WFD on stress, fatigue life, and residual strength must be understood to ensure maintenance inspections and repair actions are developed and initiated before serious degradation of aircrew/aircraft safety occurs.*

Thus, the community clearly now recognizes the potential impact of corrosion-related degradation on structural integrity of aircraft. The need to understand the potential for the occurrence of corrosion on aircraft components is critical. Thus, to even begin the assessment of this potential the community needs to know the following:

(i) the chemical environment likely to be encountered on the structure of interest at the location of interest,

(ii) the material from which the component is manufactured,

(iii) the orientation of the critical forces (loads) applied externally and internally with respect to the critical directions in the material,

(iv) the susceptibility of the material to occurrence of a given type of corrosion,

(v) the temperature of exposure of the component,

(vi) the type of forces applied (i.e., sustained force or cyclic force-constant amplitude loading or variable amplitude loading),

(vii) the type of exposure to the chemical environment (i.e., constant, intermittent), concomitant with the forces (corrosion fatigue or stress corrosion cracking) or sequentially with force (corrosion/fatigue or corrosion-fatigue),

(viii) the rates of corrosion attack,

TABLE 2: Incidents from pitting corrosion and corrosion fatigue.

Aircraft	Location of failure	Cause	Incident severity	Place	Year	From
Bell Helicopter	Fuselage, longeron	Fatigue, corrosion and pitting present	Serious	AR	1997	NTSB
DC-6	Engine, master connecting rod	Corrosion pitting	Fatal	AK	1996	NTSB
Piper PA-23	Engine, cylinder	Corrosion pitting	Fatal	AL	1996	NTSB
Boeing 75	Rudder control	Corrosion pitting	Substantial damage to plane	WI	1996	NTSB
Embraer 120	Propeller blade	Corrosion pitting	Fatal and serious, loss of plane	GA	1995	NTSB
Gulfstream GA-681	Hydraulic line	Corrosion pitting	Loss of plane, no injuries	AZ	1994	NTSB
L-1011	Engine, compressor assembly disk	Corrosion pitting	Loss of plane, no injuries	AK	1994	NTSB
Embraer 120	Propeller blade	Corrosion pitting	Damage to plane, no injuries	Canada	1994	NTSB
Embraer 120	Propeller blade	Corrosion pitting	Damage to plane, no injuries	Brazil	1994	NTSB
Mooney Mooney 20	Engine, interior	Corrosion pitting, improper approach	Minor injuries	TX	1993	NTSB
C-130	Bulkhead "Pork chop" fitting	Fatigue, corrosion pitting	Pressurization leaks	—	1995	LMAS
C-141	FS998 main frame	Corrosion pitting, stress corrosion cracking	Found crack during inspection	—	1991	LMAS

(ix) the potential influence of the effects of corrosion on fatigue crack nucleation and propagation,

(x) the impact of any related corrosion degradation to residual strength,

(xi) the potential for widespread corrosion damage to occur (WCD),

(xii) the potential impact of corrosion on the occurrence of widespread fatigue damage (WFD) and its impact on structural integrity.

Obviously this is a formidable list but the assessment of these items is possible to some degree to make the estimation of the effects of corrosion more accurate than they have been to date.

This paper deals with the identification of the issues to be dealt with in establishing methods of estimating (predicting) the effects of corrosion. To do this, various *models* are employed to be able to identify methods of establishing those components most susceptible to the ravages of corrosion.

2. Corrosion in Aircraft Structural Aluminum Alloys

2.1. General. Corrosion is an electrochemical reaction process between a metal or metal alloy and its environment [37]. For corrosion to occur, four conditions must exist, namely, an anode, a cathode, an electrolyte, and an electrical path (flow of electrons). The anode and the cathode could be of two dissimilar metals or anodic and cathodic cells could be formed in the same metal alloy because of the potential difference in the constituent chemical elements or grain interior and grain or phase boundaries. Moreover, depending on the availability of oxygen (differential aeration cells) and electrolyte (differential concentration cells) on the surface of the metal alloy, special types of localized corrosion could occur. 2xxx (Al-Cu alloys) and 7xxx (Al-Zn alloys) series aluminum alloys are most commonly used in manufacturing aircraft structural components. This is currently true and has been true for some time. Depending upon strength and toughness requirements, different types of aluminum alloys such as 2024, 7075, 7178, and many others are used for commercial and military aircraft fuselage skins, wing skins, and other extrusions and forging such as stringers and fuselage frames. In general, 2024-T3 is used for skins and 7075-T6 for stringers and frames although many applications of these and other alloys in the 2xxx and 7xxx families exist. Lap or butt splices are the common configuration for longitudinal joints whereas butt joints are for circumferential joints. A common joining method is riveting and in some cases it is in combination with adhesive bonding. In older aircraft, spot welding also can be found. As Wallace and Hoeppner mentioned in their AGARD report on "Aircraft Corrosion: Causes and Case Histories," in the initial stages, corrosion is in the form of filiform or pitting in the interior and exterior of fuselage skins [38]. Moreover, as noted in their report, crevice corrosion between the riveted sheets in fuselage joints is a significant issue and it is usually associated with the trapped small "stagnant solution." Furthermore, depending upon the chemical conditions this could lead to a combination of pitting, galvanic, or exfoliation corrosion. As well, it is recognized that fretting corrosion/wear in faying surfaces and within fastener holes plays a role in the corrosion mechanisms within aircraft joints [38]. The process of corrosion may start early in the process of manufacturing and continues when the aircraft enters its service. Therefore, it has been realized that the corrosion prevention and control program (CPCP) should be planned concurrently from the initial design until the aircraft is out of service. Furthermore design allowables should be established as with other major integrity issues.

Many types of corrosion mechanisms such as intergranular, exfoliation, pitting, crevice, fretting, microbiologically influenced corrosion, stress corrosion cracking, and hydrogen embrittlement have been found to occur in aircraft structural aluminum alloys [38]. Moreover, the synergistic effects of corrosion and the loading conditions have been found to initiate the corrosion fatigue failure process and the stress corrosion cracking failure process of aluminum alloy aircraft structural components. As identified, recently, in a report by the National Research Council's National Materials Advisory Board [39], corrosion in aircraft structural joints would result in the following: (i) significant changes in the applied stress because of material loss as well as corrosion product buildup that may cause "pillowing" or bulging of aluminum alloy sheet, (ii) hydrogen embrittlement that may result in reduced toughness, strength, and ductility of the material, and (iii) increase in fatigue crack growth rates that may severely hamper the planned inspection intervals. These issues have been discussed in workshops presented for the US-FAA and UCLA as well as FASIDE Int. Inc. workshops since 1971. In addition, the first author has frequently discussed the following other potential effects of corrosion on structural integrity:

(i) production of localized stress concentrations that act as crack nucleation sites,

(ii) change of the structurally significant item (SSI),

(iii) modification of the fail safety by any of the above.

Moreover, recently, an attempt has been made to model loss of thickness due to crevice corrosion growth in a corroded lap joint.

Several metallurgical, mechanical, and environmental factors influence the corrosion process in aluminum alloys [40]. Metallurgically induced factors include heat treatment, chemical composition of alloying element, material discontinuities such as the presence of voids, inclusions, precipitates, second-phase particles, and grain boundaries as well as grain orientation. Environmental factors include temperature, moisture content, pH, type of electrolyte, and the time of exposure. Aircraft often are exposed to both external and internal environments. External surfaces of the aircraft are exposed to a variety of environments including rain, humidity, acid rain, deicing fluid, industrial pollutants, hot and cold temperatures, dust, high content of deposits of exhaust gases from engines, and salts. In addition, the inside of the aircraft is affected by condensed moisture, spilled beverages, cargo leak, deicing fluid, lavatory seepage, and accumulated water in the fuel tank, and others. Moreover, aircraft are exposed to wide ranges of environment depending upon their route and geographical location, namely, tropical, marine, industrial, and rural [38, 51].

In both military and commercial aircraft, internal and external wing structures as well as the fuselage bilge areas and flight control surfaces are found to be most affected by corrosion in a marine and tropical environment [41]. The major causes of corrosion in aging aircraft as observed in Indonesian aging aircraft were found to be due to spillage of toilet liquid, contamination due to spillage or evaporation from the cargo compartment, and contamination due to high humidity [42]. In addition, in these aircraft, corrosion was often found in the area surrounding the cargo compartment, wing structure, and landing gear. The types of corrosion found in these aircraft were of exfoliation, galvanic, filiform, and stress corrosion, and among these exfoliation corrosion was found in most cases [42].

Several "structural issues" such as exfoliation, pitting, stress corrosion cracking, fatigue cracking, fastener corrosion, wear, fatigue and corrosion, delamination and disbonds have been observed in the US Air Force aging aircraft as shown in Table 3 [39]. For example, in C/KC 135 fleet, crevice corrosion in the spot welded lap joint/doubler and corrosion around the steel fasteners on the upper wing skin have been recognized as significant corrosion issues [43]. In the later case, as was noted, there was a possibility of moisture from condensation or deicing solution trapped around fastener heads forming a galvanic couple. This was observed to result in intergranular attack of the grain boundaries leading to exfoliation [43].

Examination of C/KC 135 fuselage lap splices (stiffened aluminum lap joint) revealed that outer skin corrosion was predominantly intergranular and exfoliation [44]. Moreover, extensive cracking was noted at these sites in the outer skin. In addition, extensive "pillowing" with more than 300% change in volume due to corrosion products along the faying surfaces was observed. In the rivet/shank region, severe localized corrosion and intergranular corrosion were observed. The fracture of rivet heads was attributed to high local stress due to environmentally assisted cracking at the junction. As well, in this study, solution samples were collected from selected areas of lap splice joints and the solution analysis showed the presence of several cations such as Al^{3+}, Ca^{2+}, Na^{+}, K^{+}, and Ni^{2+} and also anions Cl^{-}, SO_4^{2-}, and NO_3^{-}. Subsequent potentiodynamic tests using solution containing these ions led to the belief that dissolution rates

Table 3: Corrosion and fatigue issues in the US Air Force aging aircraft.

	Type of aircraft	Issues
1	C/KC 135 (Tanker aircraft)	*Corrosion* between fuselage lap joints and spot welded double layers, around fasteners in the 7178-T6 aluminum upper wing skins, between wing skins and spars, between bottom wing skin and main landing gear trunnions, between fuselage skin and steel doublers around pilot windows, *Stress Corrosion Cracking* (SCC) of large 7075-T6 aluminum forging (fuselage station 620, 820, and 960), *corrosion* and *SCC* of fuselage station 880 and 890 floor beams, wing station 733 closure rib, and corrosion in the E model engine struts.
2	C-141B (Transport aircraft)	*Widespread Fatigue Damage* (WFD) in the fuel drain holes in the lower surfaces of the wings, corrosion and *SCC* in the upper surface of the center wing, fatigue cracking and *SCC* around the wind shield, *fatigue* cracking in the stiffeners in the aft pressure door, SCC in the fuselage main frames, and *corrosion* in the empennage.
3	C-5 (Airlifter)	*SCC* of the 7075-T6 aluminum mainframes, keel beam, and fittings in the fuselage, 7079-T6 fuselage lower lobe and aft upper crown.
4	B-52H (Bomber)	*Cracking* in the bulkhead at body station 694, *fatigue cracking* in flap tracks and in the thrust brace lug of the forward engine support bulkhead, *cracking* in the side skin of the pressure cabin, aft body skins, and upper surface of the wing.
5	F-15 (Fighter aircraft)	*Low-cycle fatigue cracking* in the upper wing surface runouts, upper wing spar cap seal grooves, front wing spar conduit hole, upper in-board longeron splice plate holes, *corrosion* in nonhoneycomb structure including fuselage fuel tank, the outboard leading-edge structure of the wings, and the flap hinge beam.
6	F-16 (Fighter aircraft)	*Cracking* of the vertical tail attachment bulkhead at fuselage station 479, fuel vent holes of the lower wing skin, the wing attach bulkhead at fuselage station 341, the upper wing skin, fastener problems on the horizontal tail support box beam, and the ventral fin.
7	A-10 (Attack aircraft)	*Fatigue cracking in the wing* auxiliary spar cutout of the center section rib at wing station 90, outer panel front spar web at wing station 118 to 126, outer panel upper skin at leading edge. Fatigue cracking in the center fuselage forward fuel cell floor at the boost pump, forward fuselage gun bay compartment, forward fuselage lower longeron and skin at fuselage station 254, and center fuselage overwing lower floor panel stiffeners. Fatigue cracking in the aft nacelle hanger frame, thrust fitting and the engine inlet ring assembly skin/frame. Fatigue cracking in the main landing gear shock strut outer cylinder. *Exfoliation corrosion* in the 2024-T351 aluminum lower wing skin, 7075-T6 aluminum upper wing at the leading edge, 2024-T3511 aluminum lower front spar cap, 7075-T6 aluminum fuselage bottom skin 2024-T3/7075-T6 aluminum fuselage side skin and beaded pan, and 2024-T3511 aluminum horizontal stabilizer upper spar caps. *Pitting corrosion* in the 9Ni-4Co-0.3C steel wing attach fitting bushing and lug bore, main landing gear fitting attach bolts, 7075-T6 aluminum aft fuel cell aft bulkhead, and 2024-T351 center fuselage upper longeron. SCC in the wing attach bushing flange, and the main landing gear attach bolts.
8	E-3A (Airborne Warning and Control System)	*Fatigue and corrosion* in the 7178-T6 rudder skins and spoiler actuator clevis. *Exfoliation corrosion* in the 7178-T6 upper wing skin, leading edge slats, main landing gear door, fillet flap, fuselage stringer 23, and magnesium parts. *Delamination* and disbonds in the windows, floor panels, and nose radome core. *Wear* in the antenna pedestal turntable bearings.
9	E-8 (Joint Surveillance and Attack Radar System)	*"Small" fatigue cracks* in fastener holes in the 7075-T6 aluminum stringers, in the 2024-T3 aluminum skins.
10	T-38 (Air training command aircraft)	*Fatigue cracking* in the lower surface of the wing, lower wing skin fastener holes, wing skin access panel holes, milled pockets on the lower wing skin, and the fuselage upper cockpit longerons. *SCC* in the fuselage cockpit upper and lower longerons, fuselage forgings. Honeycomb *corrosion* in the horizontal stabilizer (due to water intrusion), and the landing gear strut door.

could completely penetrate the fuselage outer skin during service life [44].

Thus, in addition to fatigue cracking, different corrosion mechanisms occur in aircraft structures depending upon their location, geometry, exposure to environment, and loading conditions. Research studies conducted within the Quality and Integrity Design Engineering Center (QIDEC) at the University of Utah as well as other related studies are briefly discussed below.

2.2. Intergranular and Exfoliation Corrosion. Exfoliation corrosion is believed to be a manifestation of intergranular corrosion. Intergranular corrosion results from either the segregation of reactive impurities or from the depletion of passivating elements at the grain boundaries. This makes the regions at or surrounding the grain boundaries less resistant to corrosion resulting in preferential corrosion. The high strength aluminum alloys such as 2xxx and 7xxx series are highly susceptible to intergranular corrosion [37].

Exfoliation corrosion is a form of intergranular attack that occurs at the boundaries of grains elongated in the rolling direction. The 7xxx series aluminum alloys are particularly less resistant to exfoliation corrosion because during heat treatment (to achieve maximum desirable strength) their constituent elements copper and zinc accumulate at grain boundaries leaving the adjacent region free of precipitates. As aluminum and aluminum intermetallic compounds are highly reactive in the EMF series and aluminum is anodic to copper in the galvanic series, the resulting galvanic couples cause the grain boundaries to preferentially corrode (intergranular attack). McIntyre and Dow have related the localized corrosion problems in the 7075-T7352 fuel tanks of underwater weapon systems to intergranular corrosion [45]. In their study, aluminum alloys 7075 and 6061 were exposed to artificial seawater containing nitrate ions. It was observed that accelerated intergranular corrosion occurred in 7075 alloys. From the test results, they hypothesized that refueling the improperly cleaned fuel tank may cause the propellant in contact with the small quantity of sea water remaining in the fuel tank resulting in the release of nitrate ions from a hydrolysis process leading to reduced pH that may cause the dissolution of the oxide film (localized corrosion). They further hypothesized that corrosion eventually propagates to the bulk regions of the alloy due to intergranular attack by the preferential corrosion of reactive $MgZn_2$ intermetallic compounds located at grain boundaries. This was found to be true for 7075 aluminum alloy but not for 6061 aluminum alloy because the latter does not contain either Cu or Zn as alloying element [45].

Reducing the impurities such as iron and silicon as well as heat treatment modifications in aluminum alloys have resulted in an increase in the resistance to exfoliation corrosion [46]. For example, overaged 7075-T7 alloys are more resistant to exfoliation corrosion when compared to 7075-T6 alloys. In addition, Rinnovatore showed that in the T6 temper, exfoliation corrosion resistance was found to be greater for forgings produced from rolled bar stock than forgings from extruded bar stock [47]. Moreover, it was shown that rapid quenching from the solution temperature in cold water increased exfoliation corrosion resistance of forgings tempered to T6.

Fatigue and exfoliation interactions have been studied. Mills reports that most of the studies have been performed during the last five years on this issue although Shaffer in 1968 reported significant reduction in the fatigue life of exfoliated extruded 7075-T6 spar caps [48]. Moreover, multiple crack nucleation sites were observed in 7075-T651 [49] and 2024-T3 [50] aluminum alloy specimens when the specimens were subjected to exfoliation corrosion and then fatigue tested under positive R values with constant amplitude loading. Mills found an 88% decrease in the fatigue life of the specimens with prior exfoliation corrosion damage when compared to specimens tested without prior-corrosion damage. Chubb et al. showed in their study using panels containing fastener holes that the end grains exposed in the rivet holes would be the potential corrosion sites that could eventually result in multiple site damage.

In a recent study [48], experiments were performed to determine the effect of exfoliation on the fatigue crack growth behavior of 7075-T651 aluminum alloy. First the specimens were subjected to prior-corrosion damage using ASTM standard EXCO corrosive solution and then fatigue tested in corrosion fatigue environments of dry air, humid air, and artificial acid rain. Test results indicated that prior-corrosion damage resulted in higher crack growth rates than when tested in dry air as well as in acid-rain environments when compared to uncorroded specimens. Fractographic analysis showed quasi-cleavage fracture close to the exfoliated edge of the specimens tested in all the three environments indicating embrittlement by prior corrosion. Thus, embrittlement by prior corrosion was stated to "result in accelerated crack nucleation, faster short crack growth, and earlier onset of fatigue phenomena such as multiple site damage."

2.3. Corrosion Fatigue. Corrosion fatigue is defined as "*the process in which a metal fractures prematurely under conditions of simultaneous corrosion and repeated cyclic loading at lower stress levels or fewer cycles than would be required in the absence of the corrosive environment*" [40]. Corrosion acting conjointly with fatigue can have major effects on materials in structures of aircraft. First, corrosion can create discontinuities (pits, cracks, etc.) that act as origins of fatigue cracks with significant reductions in life at *all* stress levels. In crack propagation, corrosion effects are well known to produce accelerated fatigue crack propagation. The combination of aggressive environment and cyclic loading conditions has been observed to accelerate crack growth rates in aluminum alloys. Several mechanisms were proposed to explain the corrosion fatigue process [37]. They are (i) dissolution of material at the crack tip in corrosive environment, (ii) hydrogen embrittlement in which diffusion of hydrogen (a byproduct of corrosion process) into the lattice space could weaken the atomic bonds thereby reducing the fracture energy, (iii) theory of adsorbed ions in which the transport of critical species to the crack tip results in lowering of the energy required for fracture, and (iv) film-induced cleavage in which it is hypothesized that crack speed would increase at the film-substrate interface when the crack grows through the low-toughness oxide layer leading to the rupture of the film.

In general, corrosion fatigue effects on crack propagation are more pronounced at lower stress intensities whereas at higher stress intensities the crack propagates at such a high rate that the effects of chemical dissolution or localized embrittlement will be negligible. Several parameters affect corrosion fatigue crack propagation rates. For example, crack growth rates increase with increase in the stress intensity range. Also, at lower frequency corrosion fatigue effects will be more severe than at higher frequency because of the time-dependent nature of the process. Increase in R value has been found to generally increase corrosion fatigue crack propagation rates. As well, increasing the concentration of corrosive species, lowering the pH, increasing the moisture content, and temperature usually result in more severe effects [40].

The most common corrosion fatigue environment that is simulated in laboratory testing is 3.5% NaCl as it is believed to result in severe general corrosion rates and it represents roughly the salinity of sea water. In addition, other environments such as humid air, salt sprays, and artificial acid rain (to simulate industrial pollutants) also are used to characterize corrosion fatigue crack growth behavior of aluminum alloys. As aircraft are exposed to several complex chemical environments both inside and outside, no single environment could simulate the actual condition. Therefore, a few studies used sump tank water that was considered close to a "realistic chemical environment" [7]. The quest for realistic corrosion fatigue environment led Swartz et al. [51] to collect and analyze solution samples from bilge areas, external galley, and lavatories of five different airplanes. As a result, a new chemical environment was developed to perform corrosion fatigue crack growth experiments on 2024-T351, 2324-T39, 7075-T651, and 7150-T651 aluminum alloys. For all the alloys studied the fatigue crack propagation rates in synthetic bilge solution were found to be between the dry air and the 3.5% NaCl data. In another study [52], cyclic wet and dry environment was simulated in characterizing the corrosion fatigue crack growth rates in 2024-T351 aluminum alloy. It was hypothesized that during the dry cycle the partial evaporation of the aqueous solution may allow some chemical species to get deposited at the crack tip, and then in the wet cycle when the rehydration occurs, corrosion could occur at a greater rate than before.

To simulate aircraft service corrosion, fatigue crack growth studies were conducted on service corroded 2024-T3 aluminum panels extracted from a C/KC-135 aircraft [53]. Test results showed that in some cases fatigue crack growth rates were two or three times greater in the corroded material, however, in other cases, there was little difference. It was observed that "the difference in the crack growth rates was due to high variability in the amount of corrosion damage between specimens."

2.4. Corrosion Pillowing and Its Effect on Structural Integrity of Aircraft Lap Joints. Recently, some studies have shown that the increase in stress levels is not only because of the thickness loss due to corrosion but also due to the volume of the corrosion product buildup in a joint [54]. Also, evidence shows that lap joints contain "faying cracks" under the rivet heads in the corroded areas. The complexity of this issue as explained by Komorowski et al. is that "the majority of the cracks had not penetrated the outer skin surface and appeared to grow more rapidly along the faying surface creating a high aspect ratio semi-elliptical crack and it is difficult to detect and affects the structural integrity of the joint" [54]. As reported by Komorowski et al. [54], the major corrosion product in the lap splices is found to be aluminum oxide trihydrate, an "oxide mix" which has a high molecular volume ratio to the alloy. As the oxide is insoluble, it is found to remain within the joint and in turn is responsible to deform the skins in the joint which usually gives a bulging appearance, commonly termed as "pillowing." Moreover, finite element analysis revealed that for a two-layer joint, the stresses due to 6% thinning due to corrosion resulted in stress more than the yield strength of 2024-T3 aluminum alloy [55]. In addition, "pillowing-induced deformation" was observed on the corroded joints after removal of the rivets and the separation of the skin. Moreover, multiple cracks were found to nucleate from rivet holes. Fracture mechanics analysis has shown that as the pillowing increases, the stress intensity factor for the crack edge along the faying surface increases [55]. On the other hand, the stress intensity factor decreases for the crack edge along the outer surface. Therefore, it was hypothesized that pillowing produces compressive stresses in the rivet area on the outer surface because of the resultant bending stresses. At the same time, high tensile stress is produced on the faying surface resulting in more rapid growth of faying surface cracks in the direction of the row of rivets than through the skin towards the outer surface [55].

2.5. Pitting Nucleation Theories. Pitting corrosion is defined as "localized corrosion of a metal surface, confined to a point or small area that takes the form of cavities" [40]. Pitting is a deleterious form of localized corrosion, and it occurs mainly on metal surfaces which owe their corrosion resistance to passivity. The major consequence of pitting is the breakdown of passivity, that is, pitting, in general, occurs when there is breakdown of surface films when exposed to pitting environment. Pitting corrosion is so complicated in nature because "oxide films formed on different metals vary one from another in electronic conduction, porosity, thickness, and state of hydration" [56]. The empirical models that have been developed to understand the pitting process are closely related to the integrity of the metal oxide film. The salient features of the empirical theories related to pit nucleation mechanisms are mentioned in Table 4.

Therefore, nucleation of pits generally involves certain localized changes in the structure and properties of the oxide film. However, propagation of pits is related to the dissolution of the underlying bulk metal. Further discussion on this subject is presented later in this paper.

3. Pitting Corrosion

3.1. Overview. Pitting is classified as a localized attack that results in rapid penetration and removal of metal at "small" discrete areas [68]. An electrolyte should be present for pitting to occur. The electrolyte could be a film of condensed moisture or a bulk liquid. How and when pitting occurs on a metal depends on numerous factors, such as type of alloy, its composition, integrity of its oxide film, presence of any material or manufacturing-induced discontinuities, and chemical and loading environment, to name a few. Many metals and their alloys are subject to pitting in different environments. These include alloys of carbon steels, stainless steels, titanium, nickel, copper, and aluminum [69].

In passivated metals or alloys that are exposed to solutions containing aggressive anions, primarily chloride,

Table 4: Pit nucleation theories.

Proposed by	Theory
Evans [57] (1929-30)	Proposed penetration theory. Ability of a chloride ion to penetrate the film was linked to the occurrence of pitting. Halide ions are assumed to be transported from the film-solution interface to the metal-oxide interface either by the application of electric field or exchange of anions.
Hoar [58] (1967), Hoar and Wood [59] (1960s)	Assumed the adsorption of anions on the oxide surface as the key aspect in the pit nucleation process. Proposed "ion-migration" model that involves activating anions that enter the oxide film lattice without exchange thereby increasing the ionic conductivity of the film resulting in local high anodic dissolution rates and pitting. Proposed "mechanical" model in which it was assumed that adsorption of anions at the oxide-solution interface lowers the interfacial energy resulting in the formation of cracks in the protective oxide film under the influence of the "electrostatic repulsion" of the adsorbed anions. Suggested a concept of local acidification of pit as a critical factor in pit growth.
Bohni and Uhlig [60] and Kolotyrkin and Ya [61] (1961–1967)	Proposed adsorption theory in which at a certain value of the potential (pitting potential) the adsorption of aggressive anions on the metal surface displaces the passivating species such as oxygen. Kolotyrkin suggested that adsorption of anions at preferred sites forming soluble complexes with metal ions from the oxide. Once such species leave the oxide, thinning of the film starts locally increasing the electric field strength which accelerates the dissolution of the oxide.
Sato [62–64] (1971, 1982)	Proposed that at a critical potential an internal film pressure exceeds the critical compressive stress for film fracture. Considered thinning of film at local sites and suggested that pitting occurs only when a critical concentration of aggressive anions and a critical acidity is locally built up.
Lin et al. [65] (1981)	Proposed that metal vacancies may accumulate as a result of the diffusion of metal cations from the metal/film to the film/solution interface, forming voids at the metal/film interface. When the voids grow to a critical size the passive film will collapse leading to pit growth.

pitting corrosion results in local dissolution leading to the formation of cavities or "holes." The shape of the pits or cavities can vary from shallow to cylindrical holes, and the cavity is approximately hemispherical [70]. The pit morphology depends on the metallurgy of the alloy and chemistry of the environment and the loading conditions. As observed first by McAdam and Gell in 1928 [71], these pits may cause local increase in stress concentration and cracks may nucleate from them [71].

According to Foley [72], pitting corrosion of aluminum occurs in four steps: (1) adsorption of anions on the aluminum oxide film, (2) chemical reaction of the adsorbed anion with the aluminum ion in the aluminum oxide lattice, (3) penetration of the oxide film by the aggressive anion resulting in the thinning of the oxide film by dissolution, and (4) direct attack of the exposed metal by the anion.

The susceptibility of a metal to pitting corrosion as well as the rate at which pitting occurs on its surface depends on the integrity of its oxide film. Therefore, a brief overview of the mechanisms of the formation of passive film is discussed below.

3.2. Formation of Passive Films and Their Growth. The following discussion on the oxide film formation and its growth is extracted from [73].

Early investigators examined the effects of natural waters on metals by placing them outside. One investigator, Liversidge, in 1895, observed that an aluminum specimen,

> ... *"lost its brilliancy, and became somewhat rough and speckled with grey spots mixed with larger light grey patches; it also became rough to the feel, the grey parts could be seen to distinctly project above the surface, and under the microscope they presented a blistered appearance. This encrustation is held tenaciously, and does not wash off, neither is it removed on rubbing with a cloth"* [74].

Liversidge proposed that a hydrated aluminum oxide had formed, but did not confirm this with further testing of the layer. He did, however, note that when weighed, the aluminum specimens gained weight with exposure, rather than losing weight [75]. It was later confirmed that the weight gain was due to formation of an oxide film [76]. Although Liversidge suggested the formation of an aluminum oxide film, subsequent investigators proposed other theories to explain the passive behavior of aluminum. Some of these were changes in the state of electric charge on the surface, changes in valence at the surface, and a condensed oxygen layer [77]. Dunstan and Hill proved the presence of the oxide film on the surface of the metals in 1911. Through experiments with iron, they determined that the passive film was reduced at 250°F, the temperature at which magnetic iron oxide is reduced. Similar films were found on other metals [77]. Barnes and Shearer attempted to determine the constitution of passive films on aluminum and magnesium in 1908. They determined that aluminum formed hydrogen peroxide when reacting with water and that the passive film consisted of $Al_2(OH)_6$ [76]. This was later determined to be incorrect [78].

3.3. Structure of the Passive Film in Aluminum. It later was determined that this film on aluminum consists of an aluminum oxide created when the aluminum comes in contact with an environment. Generally, this film is amorphous; however, under certain circumstances it will develop one of seven crystalline structures:

(1) gibbsite (also called hydrargillite): (α-$Al_2O_3 \cdot 3H_2O$),

(2) bayerite: (β-$Al_2O_3 \cdot 3H_2O$),

(3) boehmite: (α-$Al_2O_3 \cdot H_2O$ or $AlO \cdot OH$),

(4) diaspore: (β-$Al_2O_3 \cdot H2O$),

(5) gamma alumina: (γ-Al_2O_3),

(6) corundum: (α-Al_2O_3),

(7) combinations of aluminum oxides with inhibitors, for example, ($2Al_2O_3 \cdot P_2O_5 \cdot 3H_2O$).

Gibbsite and diaspore structures are not found during corrosion of aluminum, but are frequently found in bauxite ores. Boehmite, bayerite, gamma alumina, and corundum are sometimes found in the passive layers of aluminum under certain conditions. Additionally, bayerite is frequently found as a corrosion product during pitting of aluminum. Combinations of aluminum oxides with inhibitors are not understood very well in the literature, but it is known that they will combine with oxide layer to form improved corrosion resistance through changing the passive film structure. Several researchers have studied changes in the amorphous structure of the oxide film. In one investigation, the passive film formed on the pure aluminum sheet revealed changes in structure with an increase in temperature and oxygen content. Prior to heating, the structure was reported to be amorphous oxide. As the temperature was increased, the amorphous film thickened, formed boehmite, and bayerite. The rate of film formation increased with temperature, and with an increase in oxygen content, intergranular attack began. The researcher suggested the following sequence of events in the formation: boehmite is nucleated at dislocation centers that are at the surface of the amorphous film; it then grows by a diffusion mechanism. During thickening of the boehmite, a process occurs that allows aluminum ions to escape into the solution, which results in bayerite growth [79].

Other investigations revealed that aluminum in the molten state would develop an oxide film of gamma alumina which will convert to corundum when exposed to dry air. Aluminum sheet in water at temperatures below 70 to 85°C after long aging will develop a passive film consisting of bayerite. Boehmite is found on aluminum exposed to water at high temperatures (above 70 to 85°C) [78]. More recently, researchers have found small regions of crystallized γ-alumina within the amorphous layers created during anodizing [80].

During exposure to air and water, alumina will form a passive film with a duplex structure. The film will consist of two layers, a permeable outer layer and a protective, nonporous layer next to the metal's surface. In the case of an air environment, the protective layer is thicker and the permeable layer is comparatively thin. In the case of an immersion in water, the permeable layer is thicker and the protective layer is thinner. In both cases, the total thickness of the duplex film is the same [78].

The protective layer will quickly reach maximum thickness, with the permeable layer growing slower. The growth rate of each layer depends on a few parameters. In air, it is dependent on temperature; in water, it is dependent on temperature, oxygen content, pH, and the type of ions present in the electrolyte; in anodization procedures, it depends on electrolyte and applied potential. The film is typically formed on pure aluminum when the pH of the solution is between 4.5 to 8.5 [78].

Other researchers have suggested that the permeable outer layer consists of hexagonal close-packed pores in pure aluminum. The size of these pores will depend on conditions of formation. Sealing processes in an attempt to improve the characteristics of the passive film sometimes control these conditions of formation. In sealing processes, the pores are blocked or made smaller by boehmite or gamma alumina formation, nickel acetate is added to obstruct the pores, and dichromates or chromates can be added to create pores of a different structure [59].

The passive film formed on metals will differ according to the environment in which it forms. Studies done by Seligman and Williams in the 1920s illustrate this difference. In experiments with tap water, the presence or absence of certain impurities caused either the passive film to breakdown and the metal to corrode or the film will become thick and less susceptible to corrosion. They determined that nitrates and chromates would combine with the passive film and serve to increase resistance of the passive film to localized corrosion [81]. Later studies emphasized this conclusion. One researcher found a film of 55,000 angstroms in distilled water and another found a film of only 4,800 angstroms for the same alloy (AA-1099) immersed in tap water [78]. Additionally, experiments performed by Bengough and Hudson on aluminum in sea water showed that the passive film varied with corroding liquid and with different alloying elements [75].

In a more recent paper, researchers determined that the reaction between aluminum and water takes place in three steps: formation of the amorphous oxide, dissolution of the oxide, and deposition of the dissolved products as hydrous oxide. In the first step, the amorphous oxide layer is formed and grows by the anodic and cathodic reactions present at the water/metal interface. The second step involves a hydrolysis reaction with the surface which depends on temperature, pH and aluminum concentration, and the last step is accomplished when the resulting hydroxide is deposited on the surface. The rate at which the film will grow is controlled by the diffusion of water molecules through the existing layers. At temperatures between 50 and 100°C, pseudoboehmite grows on the amorphous oxide. At 40°C, however, bayerite crystallization occurs and with time will overcome the pseudoboehmite [82].

Upon exposure of an air-formed film to water, the air-formed film will break down and another film will form that is thicker and contains more water. The rate at which the film is reformed depends on the anions present and the temperature [78]. In more recent work, the water in the aluminum passive film has been stated to be a medium for the mobilization for aluminum cations and deposited anions [83].

In air, the thickness of the passive film is dependent on humidity. In higher humidity, the oxide layer is thicker.

The growth rate of the film, however, does not depend on humidity. Rosenfeld et al. found that in high purity air, the growth rate was not changed. However, when small amounts of impurities were added, growth was accelerated in humid air [84]. In addition to impurities, the growth of the film is highly dependent on temperature. Below 200°C, the film will grow only to a few hundred angstroms, above 300 to 400°C, the rate gradually increases, between 400 and 600°C, the film will grow to a thickness of 400 angstroms, at 450°C, the film will crystallize to gamma alumina [78].

3.4. Pitting Potential and Induction Time. According to Szklarska [69], the susceptibility of a metal or alloy to pitting can be estimated by determination of one of the following criteria:

(i) characteristic pitting potential,

(ii) critical temperature of pitting,

(iii) number of pits per unit area or weight loss,

(iv) The lowest concentration of chloride ions that may cause pitting.

One of the most important criteria to determine an alloy's susceptibility to pitting corrosion is to find the pitting potential, that is, the potential at which the passive film starts to break down locally. The potential above which pits nucleate is denoted by E_p and the potential below, which pitting does not occur and above which the nucleated pits can grow, is often indicated by E_{pp}. Once the passive film begins to breakdown, the time it takes to form pits on a passive metal exposed to a solution containing aggressive anions, for example, Cl^-, is called the induction time or incubation time [69]. The induction time is meaningful in a statistical sense as it represents the average rate of reaction over the whole surface to produce a measurable increase in current. It should not be considered as the time to form the first pit. This is because "micro" pits have been observed to form during the induction time [72]. The induction time is usually denoted by τ. It is measured as the time required producing an appreciable anodic current at a given anodic potential. It is expressed as $1/\tau = k'(E - E_p)$, where E is the applied potential and K' is a function of Cl^- ion concentration [85]. In general, pitting potential decreases with increasing Cl^- ion concentration.

The most commonly used relation for estimating t is based on an exponential relationship between time and activation energy, that is, $1/\tau = Ae^{-Ea/RT}$; the activation energy needed for pit nucleation can be obtained from an Arrhenius plot of log $(1/\tau)$ versus 1/temperature [72]. As well, Hoar [58] has proposed a relationship $1/\tau = K(\mathrm{Me})^m(\mathrm{X}^-)^n$ to estimate the induction time. Where Me is the metal ion concentration, X^- is the halide ion concentration, and m and n are orders of reaction which are determined experimentally. Subsequent to the nucleation of pits it has been observed that they grow. The following subsection presents a discussion of pit growth.

3.5. Pit Growth Rate and Pit Morphology. Godard [86] developed a simple but effective relation based on the experimental data to estimate the rate at which pits grow. The empirical relation he developed was $d = K(t^{1/3})$. Even though he found this relation when tested using aluminum, it was observed to be true for other materials in different types of water environments. In general, the rate of pit growth depends on several factors such as temperature, pH, properties of passive films, chloride ion concentration, presence of anions and cations in solution, and the orientation of the material [5]. The pit growth can be viewed as a direct interaction of the exposed metal with the environment.

Upon observing the geometry of the pits formed on 7075 aluminum alloy in halide solutions, Dallek and Foley [87] proposed a pit growth rate expression $i - i_p = a(t - t_i)^b$ in which current was expressed as a function of time. In this expression, i is the dissolution current, i_p is the passive current, t is the time, t_i is the induction time, a is the constant depending on the halide, and b is the constant depending on the geometry of the pit. From this expression, a plot of $\log(i - i_p)$ versus $\log(t - t_i)$ will give the slope b. Dallek predominantly observed pits of hemispherical shape. However, Nguyen and Foley [88] have observed hemispherical pits at low potential on 1199 aluminum alloy in chloride solutions and at high potential they observed a porous layer film covered on the pit mouth with orifice at the center. This study indicated the effect of potential on the morphology of pits.

Chloride ion concentration also was found to affect the pit morphology. Baumgartner and Kaesche [89] observed that in dilute to medium concentrated solutions, pit morphology was "rough" whereas at high concentration, pits were found to be "smooth and rounded." In addition, a recent study by Grimes [73] showed clearly the effect of loading conditions on the morphology of pits. This study was conducted on 7075-T6 aluminum alloy in 3.5% salt water under three different loading conditions, namely, zero, sustained, and cyclic. It was found that the pits propagated under cyclic loads were three times larger in cross-sectional area when compared to those grown under sustained or zero load conditions. Also, it was found that most of the pits originated from the grain boundaries. This study concludes that the effect of both mechanical and chemical environment must be considered in pitting corrosion studies. However, when studying the effect of pitting on the fatigue life of aluminum alloy 7075-T6 in 3.5% NaCl solution, Ma [90] found that although the test frequency (5 and 20 Hz) had a pronounced effect on the total corrosion fatigue life, the fatigue test frequency did not have any effect on the pit morphology. On the other hand, Chen et al. [91] have found that the size of the pit from which a crack nucleated was comparatively larger at the lower frequencies and stresses than at higher frequencies and stresses when fatigue tested using 2024-T3 aluminum alloy.

3.6. Mechanisms of Pit Nucleation. In general, pit nucleation mechanisms are classified into three categories: (i) adsorption-induced mechanisms, (ii) ion-migration and penetration models, and (iii) mechanical film breakdown theories.

3.6.1. Adsorption-Induced Mechanisms. In this section mechanisms of pit nucleation based on the adsorption of aggressive anions at energetically favored sites are discussed. Many researchers including Uhlig et al. [60, 92, 93], Hoar [58], Hoar and Jacob [94], and Kolotyrkin [95] have suggested mechanisms related to the ion-adsorption concepts (see Table 5). Many of the mechanisms proposed in the literature consider this as a necessary step in the pit nucleation process. Uhlig [60, 92, 93] and Kolotyrkin [95] independently proposed that both oxygen and chlorine anions can be adsorbed onto the metal surfaces. When the metal is exposed in air, oxygen is adsorbed by the metal resulting in the formation of passive oxide film. Consequently, a chemical bond is established between the oxygen anion and the metal cation. This process is known in corrosion terminology as "chemisorption." Chemisorption results in the formation of a metal compound that covers the surface of a metal. If aluminum is exposed in oxygen, the resulting compound is aluminum oxide, that is, Al_2O_3. However, the type of compound that is formed on the metal surface depends on the environment in which the metal is exposed. For example, in the case of salt water, Cl^- ions in addition to oxygen are present. When oxygen is adsorbed, passivation of metal occurs whereas if chlorine anion is adsorbed, it does not result in passivation but breakdown of passivity occurs.

As proposed by Kolotyrkin [95], below the pitting potential, metals may prefer to adsorb oxygen, and above this critical potential metals may adsorb halides, such as Cl^-. This mechanism is termed "competitive adsorption" as the presence of different anions will compete with the oxygen to be chemisorbed by the metal. Therefore, at or above the pitting potential, chlorides and other aggressive anions if present combine with the metal and then diffuse from the metal's surface into the solution. Subsequently, it combines with water in solution to form metallic oxides, hydrogen and chloride ions. These chloride ions are attracted to the surface of the metal and the process begins again. It was hypothesized by many researchers that the chloride ions might diffuse to regions of high energy such as inclusion, dislocations, and other form of discontinuities.

Hoar [58] and Hoar and Jacob [94] originally proposed a "complex ion formation theory" which stated that the formation of Cl^- containing complexes on the film-solution interface might lead to a locally thinned passive layer. This was proposed because Cl^- -containing complexes are more soluble when compared to complexes formed in the absence of halides. They assumed that a high-energy complex is formed when a small number of Cl^- ions jointly adsorb around a cation in the film surface, which can readily dissolve into solution. This creates a stronger anodic field at this site that will result in the rapid transfer of another cation to the surface where it will meet more Cl^- and enter into solution. Experimental support was provided for this concept by Strehblow et al. [96] by conducting an investigation on the attack of passive iron by hydrogen fluoride. They found that the breakdown process occurred with complete removal of the passivated oxide layer. It was observed that hydrogen fluoride catalyzed the transfer of Fe^{3+} and Ni^{2+} ions from oxide into the electrolyte. As mentioned in a paper by Bohni [97], similar observation was made in another study by Heusler et al. regarding the influence of chloride containing borate and phthalate solutions on the passive film breakdown of iron. Different behavior of Cl^- and F^- ions in the pit nucleation process was proposed in a model by Heusler et al. Cl^- ions were suggested to form only two dimensional "clusters" leading to the localized thinning of the passive layer. However, it was proposed that F^- ions adsorb homogeneously on the oxide surface thereby promoting a general attack. It should be noted that the proposed models did not take into account material discontinuities such as point "defects," dislocations, inclusions, voids, and others. Also, another model based on the concept of an increased probability of "electrocapillary film breakdown" was proposed by Sato [63, 64] (see Table 3). Although Sato includes the effect of dislocations in this purely theoretical approach, no experimental evidence was found in the literature to support his model. However, Sato's theoretical model proposed that n-type passive oxide films are more stable than p-type films because of the difference in the band structure of electron levels.

From these studies it can be concluded that in addition to chloride anions, other anions such as chromate and sulphate also get adsorbed changing the nature of the compound. In addition, as observed by Richardson and Wood [98], enhanced adsorption takes place at the "imperfections or flaws" in the oxide film. These discontinuities in the film usually become the sites of anion adsorption. Nilsen and Bardal [99] have observed by measuring the pitting potential of four aluminum alloys (99% pure Al, Al-2.7Mg, Al-4.5Mg-Mn, and Al-1Si-Mg) and found that the pitting potential values for the four alloys were within only 25 mv. From this study, they concluded that alloy composition does not directly depend on the adsorption step of the process.

3.6.2. Ion-Migration and Penetration Models. A few models (see Table 6) were proposed based on either penetration of anions from the oxide/electrolyte interface to the metal/oxide interface or migration of cations or their respective vacancies. This theory is based on the concept that Cl^- ions migrate through the passive film and results in breakdown of the film once they reach the metal/film interface. Hoar [58] explained that when a critical potential is reached, smaller ions, like Cl^-, may penetrate the film under the influence of an electrostatic field which exists across the film. These aggressive anions prefer the high energy regions like grain boundaries and impurities as sites for migration because these regions produce thinner passive films locally. During the migration, the ions either pass through the film completely or they may combine with the metal cation in the midst of the film resulting in the formation of what is called a "contaminated" film, which is a better conductor than the "uncontaminated" film. This process results in

Table 5: Adsorption-induced mechanisms.

	Proposed by	Summary	Description	Limitations
1	Uhlig et al. [92] 1950–69, Kolotyrkin 1961 [95], Hoar 1967 [58]	(i) Proposed concepts based on either competitive adsorption or surface complex ion formation.	(i) In competitive adsorption mechanism Cl^- anions and passivating agents are simultaneously adsorbed. Above a critical potential Cl^- adsorption is favored resulting in the breakdown of passivity. (ii) Kolotyrkin suggested that there were critical Cl^-/inhibitor concentration ratios, depending on the potential above which pitting would occur.	Occurrence of induction times varying with passive film thickness cannot be explained.
2	Sato 1982 [63, 64]	(i) Proposed a theoretical concept based on the potential dependent transpassive dissolution which depends on the electronic properties of the passive film. (ii) The electrochemical stability of a passive film depends strongly on the "electron energy band structure" in the film.	(i) Stated that the critical potential above which potential-dependent dissolution of the film occurs will be less noble at the sites of chloride ion adsorption. (ii) As a result of the increased dissolution rate above the critical potential, local thinning of the passive films occurs until a steady state is reached. (iii) Proposed that the local thinning of the oxide film as a mechanism of pit "initiation". (iv) Included the effect of dislocations similar to the influences of Cl^- ions.	Knowledge of the electronic properties of passive films has not been fully understood. Experimental evidence for this mechanism is lacking.

an autocatalytic reaction, which encourages more ions to penetrate the film. This hypothesis is supported by some researchers as they have observed a higher concentration of Cl^- ions over thin films on the surface of iron and that the time to breakdown the film increases with the thickness of the film [94]. It was further hypothesized that Cl^- ions first fill anion vacancies on the surface of the passive film and then migrate to the metal/oxide interface. However, other works revealed that the time required for Cl^- to penetrate through the film is much longer than the induction time measured experimentally [97].

Later, Chao et al. [100] proposed a model in which the growth of the passive film was explained by the transport of both anions (e.g., oxygen ion) and cations (e.g., metal ion). Diffusion of anion from film-solution interface to metal-film interface results in thickening of the film. Cation diffusion from the metal-film interface to the film-solution interface results in the creation of metal vacancies at the metal/film interface. These metal vacancies usually "submerge" into the metal itself. However, if the cation diffusion rate is higher than the rate of vacancy submergence into the bulk metal, the metal vacancies will increase leading to the formation of voids at the metal/film interface. This process is known as "pit incubation." Subsequently, when the void reaches a critical size, the pit incubation period ends leading to the local rupture of the passive film. This eventually results in pit growth at that local site. Based on this theory, Chao et al. expressed a criterion for pit "initiation" as stated below.

$$(J_{ca} - J_m) \times (t - \tau) = \xi, \tag{1}$$

where, J_{ca} is the cation diffusion rate in the film, J_m is the rate of submergence of the metal vacancies into the bulk metal, t is the time required for metal vacancies to accumulate to a critical amount x, τ is a constant.

Also, in this model, the role of the halide ion in accelerating the film breakdown by increasing J_{ca} was suggested.

The ion penetration and migration theories do not include the effect of mechanical breakdown of the oxide film that may result because of the scratches from which pits can nucleate, nor is the mechanical breakdown of the oxide film included that results from strain and local cracking of the oxide film.

In addition Lin et al. [65] have proposed a "point defect" model for anodic films to calculate J_{ca} for "thin" films on the order of 10–40 A. Also, the "point defect" model could be used to calculate incubation times. Although, the "point defect" model was one of the most detailed models proposed, this model has some limitations as mentioned in Table 4.

3.6.3. Mechanical Film Breakdown Theories—Chemicomechanical Breakdown Theories. Pit nucleation

TABLE 6: Ion-migration and penetration models.

	Proposed by	Summary	Description	Limitations
1	Hoar et al. 1967 [58, 59]	(i) Presented that when the electrostatic field across the film/solution interface reaches a critical value corresponding to the critical breakdown potential, the anions adsorbed on the oxide film enter and penetrate the film.	(i) Favored sites for ion migration are suggested to be high-energy regions like grain boundaries and impurities where thinner passive films are produced. (ii) If the aggressive ions meet a metal cation, contaminated film is produced that encourages further ions to penetrate the film. Then, this process continues as an autocatalytic reaction.	(i) Did not explain the observation that pits often form from mechanical breaks in the oxide film or from scratches.
2	Lin et al. [65] 1981	(i) Presented a theoretical model to explain the chemical breakdown of passive film.	(i) Proposed that metal vacancies may accumulate as a result of the diffusion of metal cations from the metal/film to the film/solution interface, forming voids at the metal/film interface. When the voids grow to a critical size the passive film will collapse leading to pit growth.	(i) Surface discontinuities such as grain boundaries and so forth were not considered in developing the model. (ii) No direct observation of void formation was made. (iii) As the measured induction times usually show a large scatter, definite quantitative agreement is difficult to obtain.

models proposed so far based on the concepts of the "chemicomechanical" breakdown of films have not included the effect of externally applied stresses (see Table 7). Sato [63, 64] showed that a significant film pressure always acted on "thin" films that he attributed to "electrostriction." Sato expressed a relation between the film pressure, thickness, and surface tension of the film as follows:

$$p = p_o + \left[\frac{(\delta(\delta - 1)\xi^2)}{8\pi}\right] - \frac{\gamma}{L}, \qquad (2)$$

where p is the film pressure, p_o is the atmospheric pressure, δ is the film dielectric constant, ξ is the electric field, γ is the surface tension, L is the film thickness.

According to his hypothesis, both γ and L have significant influence on film pressure p. Based on this relation, Sato suggested that the adsorption of chloride ion significantly reduces the surface tension γ thereby increasing p. Also, he proposed that when p is above the critical value, the film might break down. In addition, Sato proposed that breakdown of the film occurs when it attains a thickness at which mechanical stresses caused by "electrostriction" become critical. Therefore, building up of critical stresses in the film could cause pitting.

In addition to the aforementioned theory, some researchers have observed the influence of mechanically produced discontinuities (such as scratches in the passive film) on the formation of pits along those scratches [98]. If there is a scratch in the passive film that sets up a local anodic site, which will, eventually, be the preferred site for pit to form, this smaller anode/cathode ratio results in higher local potential leading to the nucleation of pits. Other researchers proposed a similar theory that is related to the value of product of the length of the discontinuity and the current density. Assuming a unidirectionally growing pit, if this value exceeds a critical value, the discontinuity such as "fissures" in the oxide film may form a local area of low pH leading to the formation of pits from them. This happens due to the difference in the pH at the local site (fissure) when compared to the bulk solution. It was proposed that a fissure of size in the order of 10^{-6} cm could be a limiting condition for this to happen [101].

Hoar and Jacob [94] also assumed that the presence of pores or "flaws" could mechanically stress and damage the passive films in contact with an aggressive solution. Moreover, Hoar assumed that aggressive anions would replace water and reduce surface tension at the solution-film interface by repulsive forces between particles, producing cracks.

In conclusion, there is no full agreement among the researchers regarding the mechanisms of pit nucleation. However, as the pitting process itself is a complex one, the commonly accepted view is that the first step in the pit nucleation process is the localized adsorption of aggressive anions on the surface of the passivated metal. Several experimental studies also have indicated that the preferred sites for the passage of anions through the oxide film are the discontinuities present in an alloy. Such discontinuities are nonmetallic inclusions; second-phase precipitates, pores or voids, grain or phase boundaries, and other mechanical damages [69]. These discontinuities eventually may become pit nucleation sites. The aforementioned theories on pit nucleation are based purely on electrochemical concepts. However, the breakdown of surface film is dependent not only on the solution conditions (e.g., pH) and the electrochemical state at the metal/solution interface, but also on the nature of the material as well as the stress state. In addition, the aforementioned pit nucleation mechanisms did not take into account the material parameters such

Table 7: Chemicomechanical breakdown theories.

	Proposed by	Summary	Description	Limitations
1	Sato 1971 [62]	(i) Proposed a breakdown mechanism for anodic films from thermodynamic considerations.	(i) Showed that thin films always contain film pressure due to "electrostriction." (ii) Hypothesized that both the surface tension of the film and the film thickness have a significant effect on film pressure. (iii) Proposed that adsorption of chloride ions, depending on their concentration, greatly reduces surface tension.	Experimental proof is not found.
2	Sato 1982 [63, 64]	(i) Derived an equation for the work required to form a cylindrical breakthrough pore in the passive film.	(i) Proposed that for a pit nucleus to grow to macroscopic size a critical radius corresponding to critical pore formation energy must be exceeded.	Experimental proof is not found. Microstructural parameters such as grain boundaries, inclusions that may influence pitting "initiation" were not considered.

as the microstructural effects, inherent discontinuities such as voids, inclusions, second-phase particles as well as the externally applied stress. Moreover, localized corrosion also may take place at slip bands during fatigue loading [102].

Once the pit is formed, the rate of pit growth is dependent mainly on the material, local solution conditions, and the state of stress. Cracks have been observed to form from pits under cyclic loading conditions. Therefore, to estimate the total corrosion fatigue life of an alloy, it is of great importance to develop some realistic models to establish the relationship between pit propagation rate and the stress state. Furthermore, pitting corrosion in conjunction with externally applied mechanical stresses, for example, cyclic stresses, has been shown to severely affect the integrity of the oxide film as well as the fatigue life of a metal or an alloy. Therefore, to understand these phenomena, some models based on pitting corrosion fatigue mechanisms have been proposed as discussed below.

3.7. Pitting Corrosion Fatigue. Linear Elastic Fracture Mechanics (LEFM) concepts are widely used to characterize the crack growth behavior of materials under cyclic stresses in different environmental conditions. It is important to note that both pitting theory and crack growth theory have been used in model development as follows. Pit growth rate theory proposed by Godard is combined with fatigue crack growth concepts. The time to form/nucleate a Mode I crack from the pit (under cyclic loading) could be modeled using LEFM concepts. Based on this idea, a few models [103–106] were proposed since 1971 (see Table 8). All of the models assume hemispherical geometry for the pit shape, and the corresponding stress intensity relation is used to determine the critical pit depth using the crack growth threshold (ΔK_{th}) that is found empirically. For hemispherical pit geometry, these models provide a reasonable estimate for the total corrosion fatigue life. However, it is well known that corrosion pit morphology varies widely. Thus, this aspect must eventually be dealt with in LEFM models that attempt to deal with pit growth and the ultimate nucleation of crack(s) from pit(s).

As mentioned before, the combined effect of corrosion and the applied cyclic loading have been shown to produce cracks from corrosion pits. In addition, pits have frequently been the source of cracks on aircraft that are operating in fleets. Depending upon the fatigue loading and corrosion conditions, some studies have shown that the crack formation/nucleation site may change from slip bands to corrosion pits [107]. This observation was made when fatigue was tested at reduced strain rates in Al-Li-Cu alloy. Another study also showed an anodic dissolution in slip bands in Al-Li-Zr alloy at high stress levels whereas at low stress levels fatigue cracks nucleated from corrosion pits [108]. Therefore, it was hypothesized that at higher stress levels, conditions are favorable to form cracks from slip bands before the corrosion pit reaches the critical condition to favor the nucleation of crack from it. In addition, a recent study also showed that larger pit was formed at lower stress and frequency. It also was observed in 2024-T3 (bare) aluminum alloy in NaCl solution that once pits formed from the constituent particles, because of the applied cyclic stresses, the pits coalesced laterally and in depth to form larger pits from which crack was observed to nucleate [109]. Therefore, modeling the transition of a pit first to a "short" crack and then to a "long" crack is considered to be important in characterizing the total corrosion fatigue life of a material as discussed in the next section [66, 106, 110].

4. Environmental Effects on "Short" Crack Behavior of Materials

A few "small" crack studies under corrosion fatigue conditions have been performed to characterize the transition of a pit to a "small" crack. In 2024 aluminum alloy, Piascik and Willard have shown a three times increase in crack growth rates of "small" cracks in salt water environment when compared to air. Moreover, their studies clearly have observed

TABLE 8: Pitting corrosion fatigue models background-references [112, 113, 122–217].

	Proposed by	Summary	Description	Advantages/limitations
1	Hoeppner [67] (1972 - current)	(i) Proposed a model to determine critical pit depth to nucleate a Mode I crack under pitting corrosion fatigue conditions. (ii) Combined with the pit growth rate theory as well as the fatigue crack growth curve fit in a corrosive environment, the cycles needed to develop a critical pit size that will form a Mode I fatigue crack can be estimated.	(i) Using a four-parameter Weibull fit, fatigue crack growth threshold (ΔK_{th}) was found from corrosion fatigue experiments for the particular environment, material, frequency, and load spectrum. (ii) The stress intensity relation for surface discontinuity (half penny-shaped crack) was used to simulate hemispherical pit. i.e. $K=1.1\sigma\sqrt{\pi(a/Q)}$, where, σ is the applied stress, a is the pit length, and Q is the function of $a/2c$, S_{ty}. (iii) Using the threshold determined empirically, critical pit depth was found from the stress intensity relation mentioned above. (iv) Then, the time to attain the pit depth for the corresponding threshold value was found using $t=(d/c)^3$, where, t is the time, d is the pit depth, and c is a material/environment parameter.	(i) This model provides a reasonable estimate for hemispherical geometry of the pits. (ii) This model is useful to estimate the total corrosion fatigue life with knowledge of the kinetics of pitting corrosion and fatigue crack growth. (iii) This model did not attempt to propose mechanisms of crack nucleation from corrosion pits. (iv) Quantitative studies of pitting corrosion fatigue behavior of materials can be made using this model. (v) This model is valid only for the conditions in which LEFM concepts are applicable. (vi) Material dependent.
2	Lindley et al. [104]	(i) Similar to Hoeppner's model, a method for determining the threshold at which fatigue cracks would grow from the pits was proposed.	(i) Pits were considered as semielliptical-shaped sharp cracks (ii) Used Irwin's stress intensity solution for an elliptical crack in an infinite plate and came up with the relationship to estimate threshold stress intensity values related to fatigue crack nucleation at corrosion pits. i.e. $\Delta K_{th} = \frac{\Delta\sigma\sqrt{(\pi a)[1.13-0.07(a/c)^{1/2}]}}{[1+1.47(a/c)^{1.64}]^{1/2}}$, where, $\Delta\sigma$ is the stress range, a is the minor axis, and c is the major axis of a semi-elliptical crack. (iii) From the observed pit geometry that is, for a/c ratio, threshold stress intensity can be calculated. (iv) For the corresponding a/c ratio, critical pit depth can be estimated.	(i) The proposed stress intensity relation can be used in tension-tension loading situations where stress intensity for pits and cracks is similar. (ii) Critical pit depths for cracked specimens can be estimated using the existing threshold stress intensity values. (iii) This model is valid only for the conditions in which LEFM concepts are applicable. (iv) Material dependent.
3	Kawai and Kasai [105]	(i) Proposed a model based on estimation of allowable stresses under corrosion fatigue conditions with emphasis on pitting. (ii) As corrosion is not usually considered in developing S-N fatigue curves, a model for allowable stress intensity threshold involving corrosion fatigue conditions was proposed.	(i) Considered corrosion pit as an elliptical crack. (ii) Based on experimental data generated on stainless steel, new allowable stresses based on allowable stress intensity threshold were proposed. i.e. $\Delta\sigma_{all} = \Delta K_{all}/F\sqrt{\pi h_{max}}$, where ΔK_{all} can be determined from a da/dN versus ΔK plot for a material, h_{max} is the maximum pit depth, and F is a geometric factor.	(i) Using this model, allowable stress in relation to corrosion fatigue threshold as a function of time can be estimated. (ii) Material dependent. (iii) This model is valid only for the conditions in which LEFM concepts are applicable.

Table 8: Continued.

	Proposed by	Summary	Description	Advantages/limitations
4	Kondo [111]	(i) Corrosion fatigue life of a material could be determined by estimating the critical pit condition using stress intensity factor relation as well as the pit growth rate relation.	(i) Pit diameter was measured intermittently during corrosion fatigue tests. (ii) From test results, corrosion pit growth law was expressed as $2c\alpha C_p t^{1/3}$, where $2c$ is the pit diameter, t is the time, and C_p is an environment/material parameter. Then, critical pit condition (ΔK_p) in terms of stress intensity factor was proposed by assuming pit as a crack. $\Delta K_p = 2.24\sigma_a\sqrt{\pi c\alpha/Q}$, where σ_a is the stress amplitude, a is the aspect ratio, and Q is the shape factor. (iii) Critical pit condition was determined by the relationship between the pit growth rate theory and fatigue crack growth rates: $c = c_p(N/f)^{1/3}$, where N is the number of stress cycles, f is the frequency, and $2c$ is the pit diameter. (iv) The pit growth rate dc/dN was developed using ΔK relation as given below $dc/dN = (1/3)C_p^3 f^{-1}\alpha^2\pi^2 Q^{-2}(2.24\sigma_a)^4\Delta K^{-4}$ dc/dN was determined using experimental parameter Cp. (v) Finally, the critical pit size $2C_{cr}$ was calculated from the stress intensity factor relation. i.e., $2C_{cr} = (2Q/\pi\alpha)(\Delta K_p/2.24\sigma_a)^2$.	(i) The aspect ratio was assumed as constant. (ii) Material and environment dependent.

the transition of pits formed at the constituent particles to intergranular "microcracks" and then to transgranular fracture path once the crack reaches the depth of 100 mm. In addition, the increase in "small" crack growth rates was observed even at very low mode $I\Delta K$ (<1 MPa $\sqrt{m}$). As well, Kondo [111] also observed in two low alloy steels that "short" cracks from pits propagated at ΔK that is well below the threshold value of a long crack for these materials.

In a recent in situ fatigue study, prior pitted 2024-T351 and 7075-T651 aluminum alloy specimens exhibited faster crack growth rates in the "short" crack regime when compared to specimens without prior corrosion damage (Hoeppner [now Taylor] [112]). This study showed that prior corrosion damage did influence the "small" crack growth rates. It also was observed that the 7075 aluminum alloy specimen had faster crack growth rates compared to the 2024 aluminum alloy specimens. Also, in this study cracks were observed to form from pits on the prior corroded specimens whereas on the specimens without any prior corrosion damage, cracks formed from constituent particles.

In addition to a few previous studies (Hoeppner, 1979, [103]) in which pitting was modeled statistically with different materials and specimen types, recently, as discussed before in this paper, there was a study demonstrating that corrosion fatigue induced "short" crack formation from pits [113]. Also, recent studies [73, 110, 112] have shown that pits form in different shapes depending upon environment and loading conditions in contradiction to general assumption that pits have hemispherical shape. Although this assumption simplifies the modeling part of research [111], further studies to characterize the formation of cracks from pits in the "short" crack regime must be evaluated as indicated by A. Hoeppner [112]. Apart from these studies the literature search has not found any "short" crack studies to evaluate the formation of cracks from pits and their crack morphologies and paths. Moreover, fretting mechanism(s) in conjunction with fatigue and corrosion may further aggravate this.

5. Conclusions and Recommendations

The review of the literature clearly shows that much progress has been made on modeling the effects of corrosion on material behavior and structural integrity. It is clear that to date the models have centered on characterizing the corrosion and modeling the effects of the corrosion as one or more of the following:

(i) section change that affects the area/volume that modifies the stress,

(ii) formation/nucleation of localized debris that may modify the stress (part of pillowing) that modifies the stress or stress intensity;

(iii) nucleation of intergranular corrosion that is involved in pillowing that modifies the stress or stress intensity;

(iv) nucleation of localized corrosion (pitting, fretting, etc.) that modifies the local stress and may ultimately nucleate cracks;

(v) production of products of corrosion that produce localized embrittlement effects that may alter the material behavior and produce accelerated crack propagation.

All of the above have been reviewed in the preceding sections and lead to the recognition that one of the most pressing issues to be resolved is the actual quantitative characterization of the corrosion in relation to the physical damage state that is underway. Some of this has been accomplished in the past with the efforts of the past at the University of Utah as discussed in the earlier sections of this paper. From the work of L. Grimes at Utah as well as additional efforts at the University of Utah, the use of the confocal microscope will be of great assistance in characterizing the three-dimensional (3D) surface "damage" that results from corrosion of various forms.

Within the last few years interest in corrosion and the effects of corrosion has picked up in part due to numerous failures in many industries including nuclear power plants, gas and oil pipelines, and aircraft to name a few. Roberge [114–116] has introduced excellent reference books on aspects of corrosion and also a web page (http://www.corrosion-doctors.org/) that contain a wealth of information related to many of the topics covered in this paper. A recent issue of business week [117] states that the USA DOD spends "22.9 billion a year fighting rust". There is little doubt that this number will become much larger and more of the structures in use in aircraft and many other applications age and it is unlikely that more funds will be appropriated to replace many aging aircraft components. Thus, many of the issues covered herein will become more important in both the design, operational, and maintenance strategies to combat the issue of corrosion. This also is clear from the fact that the USA DOD has established a Corrosion Policy and Oversight Office Congress in the pentagon as was mandated by the US Congress in 2003. It remains to be seen whether this will result in significant cost savings to combat corrosion and reduce the number of accidents from corrosion-related issues.

Even though fracture mechanics-based modeling has been extremely useful in modeling the effects of corrosion on structural integrity it has taken many simplifications and, depending on the manner in which the fracture mechanics is used in the model, has resulted in downgrading the real corrosion characterization issue and understanding the 3D nature of the corrosion degradation process. New tools and models will have to be brought to bear on the formation/nucleation and growth of the corrosion with or without load of either sustained (SCC) or cyclic nature (EANC/F) (Environmentally assisted nucleation and cracking with fatigue loading). Furthermore the transitions of corrosion to actual cracks will have to be understood to improve the models that currently exist and any new ones that may be developed. Aspects of this were discussed by Hoeppner [118] and Swift [119] in recent ICAF meetings. No doubt more attention will be focused on this in the future.

The characterization of chemically dependent short crack propagation and modeling of it will have to be much better understood. One area not addressed in the article is the effect of either prior corrosion and/or concomitant corrosion on either fatigue crack propagation or stress-corrosion cracking. Both of these issues are extremely important to the overall area of model development and consideration should be given to expanding at many laboratories in the future.

The importance of corrosion to DOD activities within the USA has recently be noted

Appendix

List of definitions related to corrosion fatigue and stress corrosion cracking. Corrosion-related Definitions of terms for use in CFSD Phase II, CMI and CP programs. Prepared by David W. Hoeppner, P.E., Ph.D. 1999–2011. Significant input into the preparation of this document has been made by the following: Nick Bellinger, Graeme Eastaugh, and Jerzy Komorowski- All of NRC, Ottawa, Ontario, Canada. Mr. Craig Brooks, APES Inc. Dr. Charles Elliott, Dr. Paul Clark, Ms. Amy Taylor-University of Utah and FASIDE International Inc. This document is still undergoing change. Please submit recommendations to the corresponding author.

Age Degradation Process (ADP). Any one of or combination of physical or chemical degradation such as fatigue, environmental effects (corrosion on metals and joints), creep, wear, and synergisms of these.

Crack-like discontinuity (CLD). A discontinuity that meets the criteria for a crack. A stress singularity exists near or at the tip of the discontinuity; no "traction forces" exist on the surfaces of the discontinuity.

Durability and Damage Tolerance Analysis (DADTA). The procedure of performing a durability and damage tolerance analysis. Analysis of the ability of the airframe or component to resist damage (including fatigue cracking, environmentally assisted cracking, hydrogen-induced cracking, corrosion, thermal degradation, delamination, wear, and the effects of foreign objects) and failure due to the presence of damage, for a specified period of unrepaired usage. From JSSG-2006.

Defect (Various Definitions Exist). The most common definition is any feature that is outside the boundary conditions of a given component/product design that will make the component/product incapable of meeting its requirements when it is needed. Defects are also defined related to product manufacturing and also related to representation of the product.

Discontinuity Evolution Process (DEP). The specific process by which a population of discontinuities evolves.

Discontinuity Nucleation Process (DNP). Any one or more or specific physical or chemical processes that may form discontinuities not inherent to a material. Example: in some materials fatigue deformation occurs by dislocation movement and the production of slip bands on external or internal surfaces. The slip band is thus a nucleated discontinuity. Example two. In some aluminum alloys intrinsic particles are known to nucleate corrosion pits if the pitting potential for nucleation is achieved. The pit is formed by a DNP.

Discontinuity State (DS). See IDS, EDS, and MDS below.

Discontinuity State Evolution Process (DSEP). The specific physical or chemical processes by which the discontinuity state evolves. The major forms of time-dependent or-related phenomena by which the state is changed are corrosion (more generally environmental degradation), creep, fatigue, wear, and sequential combinations and synergisms of them. SEE EDS below.

Discontinuity State Evolution Response (DSER). Any change in state of an IDS population.

Environmentally Assisted Cracking (EAC). It may occur under sustained load from either applied load or "residual stresses" EAC_{sl}. And may also occur under either constant amplitude cyclic forces or variable amplitude cyclic forces EAC_{fatigue}.

Equivalent Corrosion Damage (ECD). A modified discontinuity state (MDS) at some specific time that is made equivalent to a crack size often referred to as a "flaw" to start a residual life analysis by subcritical crack growth analysis.

Evolving Discontinuity State (EDS). The description of the evolution of the discontinuity and the progression of changes to the discontinuity or population of discontinuities over time and cyclic load exposures. (Subsequent to either the nucleation of a discontinuity or the activation of an IDS by a specific physical or chemical process acting alone or conjointly the resultant discontinuity or population of them may evolve in state with time or cyclic load exposure. Various metrics are used to describe the EDS).

Equivalent Initial Flaw Size (EIFS). A term used to describe a discontinuity size usually determined by extrapolation from a set of fatigue data. The EIFS has no direct relationship to any specific IDS.

Fatigue Crack Propagation (FCP). Extension of a crack under cyclic or repeated loads. The stages of crack propagation are divided into four phases, namely, (1) small or short crack propagation, long crack propagation in the linear elastic regime, and long crack propagation in either the elastic-plastic or fully plastic regime.

Initial Discontinuity State (IDS). The initial (intrinsic) population of discontinuities that are in a structure made of a given material as it was manufactured in a given geometric form. The IDS is a geometric and material characteristic that is a function of composition, microstructure, phases and phase morphology, and the manufacturing process used to process the material. The geometric and material discontinuities can be modeled separately.

Examples of material IDS types include constitutive particles, inclusions, grain boundaries, segregated phases, phase boundaries, voids (vacancies, microporosity, and porosity), intrinsic cracks, and so forth.

Manufacturing processes such as machining and assembly can introduce additional discontinuities at fasteners, fillets, and so forth, that extend the tail (larger discontinuity sizes) of the IDS distribution.

Initial Material Discontinuity (IDS_{ms}). The initial population of intrinsic material discontinuities. See IDS.

Initial Manufacturing Discontinuity (IDS_{mfg}). The resultant effect on the population of discontinuities from a given manufacturing process or sequence of manufacturing processes including joining of the three major types (namely, mechanical joining, thermal joining, and adhesive bonding).

Initial Geometric Discontinuity (IDS_{geo}). The initial geometric discontinuities in a product. These often are generally referred to as a "notch".

Modified Discontinuity State (MDS). The physical state of a discontinuity or damage state at any given time in its evolution. Various metrics may be used to describe the state. Example: a crack has grown to a given size, and it is an MDS at a specific time and thus size. Example2: a corrosion pit has grown to state at some point in time. The IDS may progress (EDS) to various MDS values through the mechanisms of corrosion, creep, fatigue, wear, or combinations of these over time.

Principal Structural Element (PSE)

Safe Life. A term usually taken to mean structural design based on ideal continuum mechanics assumptions and practices without consideration of cracks or crack-like discontinuities based on the assumptions of homogeneity and continuity. In traditional safe life design toughness, subcritical crack growth, directed inspection, and inspection intervals are not dealt with for fatigue, corrosion, and related items.

Structurally Significant Location (SSL). The significant locations on a structure determined by the potential behavior and changes in state that may occur in the structure related to its use under conditions of interest.

List of definitions related to Corrosion Fatigue and Stress Corrosion Cracking. *The following definitions are taken from* [120].

Standard G15-97a-Standard Terminology Relating to Corrosion and Corrosion Testing

Corrosion Fatigue. the process in which a metal fractures prematurely under conditions of simultaneous corrosion and repeated cyclic loading at lower levels or fewer cycles than would be required in the absence of the corrosive environment. G15-99b, p69.

Corrosion Fatigue Strength. the maximum repeated stress that can be endured by a metal without failure under definite conditions of corrosion and fatigue and for a specific number of stress cycles and a specified period of time. G15-99b, p69.

Exfoliation Corrosion. corrosion that proceeds laterally from the sites of initiation along planes parallel to the surface, generally at grain boundaries, forming corrosion products that force metal away from the body of the material, giving rise to a layered appearance. (G15-99B).

Pitting. —corrosion of a metal surface, confined to a point or small area, that takes the form of small cavities.

Stress—Corrosion Cracking. A cracking process that requires the simultaneous action of a corrodent and sustained tensile stress. (This excludes corrosion-reduced sections which fail by fast fracture. It also excludes intercrystalline or transcrystalline corrosion which can disintegrate an alloy without either applied or residual stress.)

The following definitions are taken from [121].

E1823-96-Standard Terminology Relating to Fatigue and Fracture Testing

Corrosion Fatigue. The process by which fracture occurs prematurely under conditions of simultaneous corrosion and repeated cyclic loading at lower stress levels or fewer cycles than would be required in the absence of the corrosive environment. E1823-96, p1016. (Note slight word differences between this definition of corrosion fatigue and the one above. It is possible that these differences have been eliminated in the newer version of G15. I am checking into this.)

Environment-Assisted Cracking (EAC). A cracking process in which the environment promotes crack growth or higher crack growth rates that would occur without the presence of the environment. E1823-96, p1028. Same definition in E 1681-95, p944 (see below).

Fatigue. The process of progressive localized permanent structural change occurring in a material subjected to conditions that produce fluctuating stresses and strains at some point or points and that may culminate in cracks or complete fracture after a sufficient number of fluctuations. E1823-96, p1019.

The following definition is taken from [121].

E7-97a- Standard Terminology Relating to Metallography

Stress-Corrosion Crack. A crack which may be intergranular or transgranular depending on the material, resulting from the combined action of corrosion and stress, either external (applied) or internal (residual). E7-97a, p52.

The following definitions are taken from [121].

E 1681-95-Standard Test Method for Determining a Threshold Stress Intensity Factor for Environment-Assisted Cracking of Metallic Materials under Constant Load

Stress-Corrosion Cracking (SCC). A cracking process that requires the simultaneous action of a corrodent and sustained tensile stress. E1681-95, p943.

Environment-Assisted Cracking (EAC). Same as above in E1823-96. E1681-95, p944.

We have found no standard definitions for either corrosion-fatigue or corrosion/fatigue. Thus, unless someone can find a standard or suggest one for our work it is suggested we stick with only standard terminology. We have added some other definitions in the following appendix. These terms all are some I have heard used at conferences and our various team meetings. Thus, I have added them.

Additional Definitions. The definitions in this section are not standard definitions.

Corrosion+Fatigue. Fatigue occurs in a material/structure that has undergone corrosion. The fatigue may occur as either pure fatigue or corrosion fatigue. See ASTM definitions previously supplied.

Corrosion-Fatigue: Fatigue occurs in a material/structure that has undergone corrosion. The fatigue may occur as either pure fatigue or corrosion fatigue. See ASTM definitions previously supplied.

Corrosion/Fatigue. Fatigue occurs in a material/structure that has undergone corrosion. The fatigue may occur as either pure fatigue or corrosion fatigue. See ASTM definitions previously supplied.

Fretting Fatigue. Fatigue occurs in the presence of fretting. Thus, the action is concomitant. This situation occurs in many holes with fasteners moving in the holes or on faying surfaces in splice joints.

Fretting+Fatigue. Fatigue occurs on a material/structure that has undergone fretting. The fatigue may occur as either pure fatigue or corrosion fatigue.

Fretting/Fatigue. Fatigue occurs on a material/structure that has undergone fretting. The fatigue may occur as either pure fatigue or corrosion fatigue.

(Prior Corrosion)+Fatigue. Fatigue occurs in a material/structure that has undergone corrosion. The fatigue may occur as either pure fatigue or corrosion fatigue. See ASTM definitions previously supplied.

(Prior Corrosion)/Fatigue. Fatigue occurs in a material/structure that has undergone corrosion. The fatigue may occur as either pure fatigue or corrosion fatigue. See ASTM definitions previously supplied.

Mechanism Overlap. The interaction of more than one degradation mechanism in generation of the degradation condition in a material/structure.

Missed Corrosion+Fatigue. Fatigue occurs in a material/structure that has undergone corrosion. The fatigue may occur as either pure fatigue or corrosion fatigue. See ASTM definitions previously supplied.

Missed Corrosion/Fatigue. Fatigue occurs in a material/structure that has undergone corrosion. The fatigue may occur as either pure fatigue or corrosion fatigue. See ASTM definitions previously supplied.

SSI/Corrosion. A structurally significant item designated by its propensity to become a critical item based on the potential for corrosion degradation of any type.

SSI/Corrosion Fatigue. A structurally significant item designated by its propensity to become a critical item based on the potential for corrosion fatigue degradation.

SSI/Fatigue. A structurally significant item designated by its propensity to become a critical item based on the potential for fatigue degradation. These sites are usually determined by durability and/or damage tolerance assessment.

SSI/Fatigue/Durability. A structurally significant item designated by its propensity to become a critical item based on the potential for fatigue degradation as determined by the durability assessment.

SSI/Fatigue/Damage Tolerance. A structurally significant item designated by its propensity to become a critical item based on the potential for fatigue degradation as determined by the damage tolerance assessment.

SSI/Fretting Fatigue. A structurally significant item designated by its propensity to become a critical item based on the potential for fretting fatigue degradation.

SSI/SCC. A structurally significant item based on its propensity to undergo the degradation mechanism of stress corrosion cracking. See the ASTM standard previously supplied on stress corrosion cracking.

SSL/Corrosion. A structurally significant location designated by its propensity to become a critical location based on the potential for corrosion degradation of any type.

SSL/Corrosion Fatigue. A structurally significant location designated by its propensity to become a critical location based on the potential for corrosion fatigue degradation.

SSL/Fatigue/Durability. A structurally significant location designated by its propensity to become a critical location based on the potential for fatigue degradation as determined by the durability assessment.

SSL/Fatigue/Damage Tolerance. A structurally significant item designated by its propensity to become a critical location based on the potential for fatigue degradation as determined by the damage tolerance assessment.

SSL/Fretting Fatigue. A structurally significant item designated by its propensity to become a critical location based on the potential for fretting fatigue degradation.

SSL/SCC. A structurally significant location based on its propensity to undergo the degradation mechanism of stress corrosion cracking. See the ASTM standard previously supplied on stress corrosion cracking.

Local Corrosion. Corrosion of a skin or web (wing, fuselage, empennage, or strut) not exceeding one frame, stinger, or stiffener bay) or corrosion of a single frame, chord, stringer, or stiffener, or corrosion of more than one frame, chord, stringer, or stiffener but, no corrosion on two adjacent members on each side of the corroded member.

Widespread Corrosion. Corrosion of two or more adjacent skin or web bays defined by frame, stringer, or stiffener spacing. Or corrosion of two or more adjacent frames, chords, stringers, or stiffeners.

Level 1 Corrosion. Corrosion damage occurring between successive inspections that is *local* and can be reworked/blended out within allowable limits as defined by the manufacturer. Or corrosion damage that is *local* but exceeds allowable limits and can be attributed to an event *not typical* of the operator's usage of other airplane's in the same fleet. Or operator experience over several years has demonstrated only light corrosion between successive inspections but latest inspection and cumulative blend-out now exceed allowable limits.

Level 2 Corrosion. Corrosion occurring between successive inspections that requires rework/blendout which exceeds allowable limits, requiring a repair or complete or partial replacement of a principal structural element as defined by the original equipment manufacturer's structural repair manual. Or corrosion occurring between successive inspections that is widespread and requires blendout approaching the allowable rework limits.

Level 3 Corrosion. Corrosion found during the first or subsequent inspections, which is determined (normally by the operator) to be a potential urgent airworthiness concern requiring expeditious action.

The above are taken from Boeing Commercial Airplane Company and FAA documents.

Acknowledgments

The authors wish to express appreciation to the former Dr. John DeLuccia who provided valuable comment and insights for the preparation of this paper. In addition, they express our deep appreciation to FASIDE International Inc. and the University of Utah for provision of facilities and a magnificent library system. Ms. Amy Taylor and Dr. Chandrasekaran provided much valuable input over a period of years. They appreciate all their efforts. The following three individuals provided extensive discussions on the topics covered herein: Mr. Nick Bellinger and Mr. Jerzy Komorowski (both of NRC-IAR-Canada), and Mr. Craig Brooks (AP/ES-USA). The authors are grateful for the many discussions and interactions with them over many years.

References

[1] "Stress corrosion cracking in aircraft structural materials," in *Proceedings of the Symposium held by the Structures and Materials Panel of AGARD*, vol. 18 of *AGARD Conference Proceedings Series*, NATO-AGARD, Turin, Italy, April, 1967.

[2] "Fundamental aspects of stress corrosion cracking," in *Proceedings of the conference held at the Ohio State University*, R. W Staehle, A. J. Forty, and D. van Rooyen, Eds., National Association of Corrosion Engineers, Ohio, USA, September, 1967.

[3] "Effects of environment and complex load history on fatigue life, ASTM STP 462," in *Proceedings of the Symposium on Effects of Environment and Complex Load History on Fatigue Life*, M. Rosenfeld, D. W. Hoeppner, and R. I. Stephens, Eds., ASTM, Atlanta, Ga, USA, September, 1986.

[4] "Corrosion fatigue: chemistry, mechanics, and microstructure," in *Proceedings of the Conference held at the University of Connecticut*, O. Devereux, A. J. McEvily, and R. W. Staehle, Eds., National Association of Corrosion Engineers, Connecticut, Conn, USA, June, 1971.

[5] B. F. Brown, *Stress-Corrosion Cracking in High Strength Steels and in Titanium and Aluminum Alloys*, Naval Research Laboratory, Washington, DC, USA, 1972.

[6] L. R. Hall, R. W. Finger, and W. F. Spurr, "Corrosion fatigue crack growth in aircraft structural materials," Tech. Rep. number AFML-TR-73-204, Boeing Company, 1973.

[7] D. Pettit, J. Ryder, W. Krupp, and D. Hoeppner, "Investigation of the effects of stress and chemical environments on the prediction of fracture in aircraft structural materials," Tech. Rep. number AFML-TR-74-183, Lockheed California Company, 1974.

[8] "Corrosion-fatigue technology, ASTM STP 642," in *Proceedings of the Symposium held in Denver, CO*, H. L. Craig Jr., T. W. Crooker, and D. W. Hoeppner, Eds., ASTM, Denver, Colo, USA, November, 1976.

[9] "Aircraft corrosion," in *Proceedings of the 52nd Meeting of the AGARD Structures and Materials Panel*, no. 315, NATO-AGARD, Cesme, Turkey, April, 1981.

[10] "Corrosion fatigue," in *Proceedings of the 52nd Meeting of the AGARD Structures and Materials Panel*, no. 316, NATO-AGARD, Cesme, Turkey, April, 1981.

[11] "Corrosion fatigue, STM STP 801," in *Proceedings of the Symposium on Corrosion Fatigue: Mechanics, Metallurgy, Electrochemistry, and Engineering*, T. W. Crooker and B. N. Leis, Eds., ASTM, St. Louis, Mo, USA, October, 1981.

[12] D. W. Hoeppner and V. Chandrasekaran, "Corrosion and corrosion fatigue predictive modeling-state of the art review," Tech. Rep., FASIDE International, 1998.

[13] W. Wallace, D. Hoeppner, and P. V. Kandachar, "Aircraft corrosion: causes and case histories," in *AGARD Corrosion Handbook*, vol. 1, North Atlantic Treaty Organization, 1985.

[14] *ASM Handbook*, vol. 13, ASM International, Metals Park, Ohio, USA, 1987.

[15] The Boeing Company, "Corrosion prevention and control, manual for training operators of Boeing commercial," Aircraft, Seattle, Wash, USA, 1988.

[16] NTSB Metallurgist's, "Aloha airlines flight 243," Tech. Rep. number 88–85, Materials Laboratory, 1988.

[17] "Environment induced cracking of metals," in *Proceedings of the International Conferencen*, R. P. Gangloff and M. B. Ives, Eds., National Association of Corrosion Engineers NACE-10, Kohler, Wash, USA, October, 1988.

[18] *ASM Handbook*, vol. 18, Friction Lubrication and Wear Technology ASM International, Metals Park, Ohio, USA, 1992.

[19] "Naval Aviation Safety Program," US Navy, OPNAVINST 3750.6Q CH-1, OP-05F, 1991.

[20] Federal Aviation Administration, "Aircraft accident and incident synopses related to corrosion, fretting, and fatigue for the period 1976–1993," 1994.

[21] National Transportation Safety Board, "Aircraft accident and incident synopses related to corrosion, fretting, and fatigue for the period 1975–1993," 1994.

[22] F. Karpala and O. L. Hageniers, "Characterization of corrosion and development of a broadboard model of a D sight aircraft inspection system," Report to DOT phase 1, Diffracto, 1994.

[23] United States Air Force, "Navy, and army, aircraft accident and incident synopses related to corrosion, fretting, and fatigue," 1994.

[24] G. Cooke, P. J. Vore, C. Gumienny, and G. Cooke Jr., "A study to determine the annual direct cost of corrosion maintenance for weapon systems and equipment in the United States Air Force," Final report contract number F09603-89-C-3016, SEPT., 1990.

[25] W. Schütz, "Corrosion fatigue-the forgotten factor in assessing durability," in *Proceedings of the 18th Symposium on the International Committee of Aeronautical Fatigue, (ICAF '95)*, J. M. Grandage and G. S. Jost, Eds., vol. 1 of *Estimation, Enhancement and Control of Aircraft Fatigue Performance*, pp. 1–52, EMAS, Melbourne, Australia, May, 1995.

[26] S. J. Swift, "The aero commander chronicle," in *Proceedings of the 18th Symposium on the International Committee of Aeronautical Fatigue, (ICAF '95)*, J. M. Grandage and G. S. Jost, Eds., vol. 1 of *Estimation, Enhancement and Control of Aircraft Fatigue Performance*, pp. 507–530, EMAS, Melbourne, Australia, May, 1995.

[27] D. W. Hoeppner, L. Grimes, A. Hoeppner, J. Ledesma, T. Mills, and A. Shah, "Corrosion and fretting as critical aviation safety issues," in *Proceedings of the 18th Symposium on the International Committee of Aeronautical Fatigue, (ICAF '95)*, J.M. Grandage and G.S. Jost, Eds., vol. 1 of *Estimation, Enhancement and Control of Aircraft Fatigue Performance*, pp. 87–106, EMAS, Melbourne, Australia, May, 1995.

[28] C. L. Brooks, K. Liu, and R.G. Eastin, "Understanding fatigue failure analyses under random loading using a C-17 test report," in *Proceedings of the 18th Symposium on the International Committee of Aeronautical Fatigue, (ICAF '95)*, J. M. Grandage and G. S. Jost, Eds., vol. 1 of *Estimation, Enhancement and Control of Aircraft Fatigue Performance*, pp. 449–468, EMAS, Melbourne, Australia, May, 1995.

[29] *ASM Handbook*, vol. 19, Fatigue and Fracture ASM International, Metals Park, Ohio, USA, 1996.

[30] C. G. Schmidt, J. E. Crocker, J. H. Giovanola, C. H. Kanazawa, and Schockey, "Characterization of early stages of corrosion fatigue in aircraft skin," Final report, contract No. 93-G-065, SRI International Final Report to DOT, Menlo Park, Calif, USA, Report No. DOT/FAA/AR-95/108, 1996.

[31] G. K. Cole, G. Clark, and P. K. Sharp, "Implications of corrosion with respect to aircraft structural integrity," Tech. Rep. number DSTO-RR-0102, AMRL, Melbourne, Australia, 1997.

[32] *Aging of U.S. Air Force Aircraft*, Report of the committee on aging of U.S. Air Force Aircraft, NMAB (National Materials Advisory Board), Commission on Engineering and Technical Systems, National Research Council, Publication NMAB-488-2, 1997.

[33] Boeing, "Corrosion damage assessment framework, corrosion/fatigue effects on structural integrity," Tech. Rep. number D500-13008-1, USAF contract No. F9603-97-C-0349, 1998.

[34] C. Paul and T. Mills, "Corrosion/fatigue," in *Proceedings of the Aerospace Materials Conference*, 1998.

[35] "A study to determine the cost of corrosion maintenance for weapon systems and equipment in the United StatesAir Force," Final Report Contract number F09603-95-D-0053, 1998.

[36] "Fatigue in the presence of corrosion," in *Proceedings of the Workshop of the RTO Applied Technology (AVT) Panel*, vol. 18 of *RTO Proceedings*, no. AC/323(AVT) TP/8, NATO, Research and Technology Organization, Corfu, Greece, October, 1998.

[37] D. Jones, *Principles and Prevention of Corrosion*, Macmillan Publishing, New York, NY, USA, 1992.

[38] D. W. Hoeppner, "Estimation of component life by application of fatigue crack growth threshold knowledge," in *Fatigue, Creep and Pressure Vessels for Elevated Temperature Service*, C. W. Lawton and R. R. Seeley, Eds., pp. 1–83, The American Society of Mechanical Engineers, New York, NY, USA, 1981.

[39] D. W. Hoeppner, "Parameters that input to application of damage tolerance concepts to critical engine components," in *Proceedings of the AGARD Conference Damage Tolerance Concepts for Critical Engine Components*, no. AGARD-CP 393, pp. 4.1–4.16, NATO-AGARD, San Antonio, Tex, USA, April, 1985.

[40] American Society for Metals, *Metals Handbook*, vol. 13 of *Corrosion*, American Society for Metals (ASM), Metals Park, Ohio, USA, 9th edition, 1987.

[41] A. Alvarez, "Corrosion on aircraft in marine-tropical environments: a technical analysis," *Material Performance*, vol. 36, no. 5, pp. 33–38, 1997.

[42] B. M. Suyitno and T. Sutarmadji, "Corrosion control assessment for Indonesian aging aircraft," *Anti-corrosion Methods and Materials*, vol. 44, no. 2, pp. 115–122, 1997.

[43] D. J. Groner, "US Air Force aging aircraft corrosion," in *Current Awareness Bulletin, Structures Division*, Wright Laboratory, Spring, 1997.

[44] R. S. Piascik, R. G. Kelly, M. E. Inman, and S. A. Willard, "Fuselage lap splice corrosion," Tech. Rep. number WL-TR-96-4094, ASIP, 1996.

[45] J. F. McIntyre and T. S. Dow, "Intergranular corrosion behavior of aluminum alloys exposed to artificial seawater in the presence of nitrate anion," *Corrosion*, vol. 48, no. 4, pp. 309–319, 1992.

[46] J. J. Thompson, E. S. Tankins, and V. S. Agarwala, "A heat treatment for reducing corrosion and stress corrosion cracking susceptibilities in 7xxx aluminum alloys," *Materials Performance*, pp. 45–52, 1987.

[47] J. V. Rinnovatore, K. F. Lukens, and J. D. Corrie, "Exfoliation corrosion of 7075 aluminum die forgings," *Corrosion*, vol. 29, no. 9, pp. 364–372, 1973.

[48] T. B. Mills, *The combined effects of prior-corrosion and aggressive chemical environments on fatigue crack growth behavior in aluminum alloy 7075—T651*, Ph.D. dissertation, University of Utah, 1997.

[49] T. B. Mills, *The effects of exfoliation corrosion on the fatigue response of 7075-T651 aluminum plate*, M.S. thesis, University of Utah, 1994.

[50] J. P. Chubb, T. A. Morad, B. S. Hockenhull, and J.W. Bristow, "The effect of exfoliation corrosion on the fatigue behavior of structural aluminum alloys," in *Structural Integrity of Aging Airplanes*, S. N. Atluri, S. G. Sampath, and P. Tong, Eds., pp. 87–97, Springer, New York, NY, USA, 1991.

[51] D. D. Swartz, M. Miller, and D. W. Hoeppner, "Chemical environments in commercial transport aircraft and their effect on corrosion fatigue crack propagation," in *Proceedings of the 14th symposium of the International Committee on Aeronautical Fatigue*, J.M. Grandage and G. S. Jost, Eds., vol. 1 of *Estimation, Enhancement and Control of Aircraft Fatigue Performance*, pp. 353–364, Melbourne, Australia, May, 1995.

[52] J. Kramer and D. W. Hoeppner, "Effects of cyclic immersion in 3.5% NaCl solution on fatigue crack propagation rates in aluminum 2024-T351," in *Proceedings of the USAF Structural Integrity Program Conference*, vol. 2, no. WL-TR-96-4093, pp. 1089–1112, 1995.

[53] T. B. Mills, D. J. Magda, S. E. Kinyon, and D. W. Hoeppner, "Fatigue crack growth and residual strength analyses of service corroded 2024-T3 aluminum fuselage panels," Tech. Rep., Oklahoma City Air Logistics Center and Boeing Defense and Space Group, University of Utah, 1995.

[54] J. P. Komorowski, N. C. Bellinger, and R. W. Gould, "The role of corrosion pillowing in NDI and in the structural integrity of fuselage joints," in *Proceedings of the 19th Symposium of the Industrial College of the Armed Forces, (ICAF '97)*, Fatigue in New and Aging Aircraft, 1997.

[55] N. C. Bellinger, J. P. Komorowski, and R. W. Gould, "Damage tolerance implications of corrosion pillowing on fuselage lap joints," *American Institute of Aeronautics and Astronautics Journal*, vol. 3, pp. 317–320, 1997.

[56] M. Jayalakshmi and Muralidharan Muralidharan, "Empirical and deterministic models of pitting corrosion—an overview," *Corrosion Reviews*, vol. 14, no. 3-4, pp. 375–402, 1996.

[57] V. R. Evans, " XVI.—the passivity of metals. Part II. The breakdown of the protective film and the origin of corrosion currents," *Journal of the Chemical Society*, pp. 92–110, 1929.

[58] T. P. Hoar, "The production and breakdown of the passivity of metals," *Corrosion Science*, vol. 7, pp. 341–355, 1967.

[59] T. P. Hoar and G. C. Wood, "The sealing of porous anodic oxide films on aluminum," *Electrochemica Acta*, vol. 7, pp. 333–353, 1962.

[60] H. Bohni and H. H. Uhlig, "Environmental factors affecting the critical pitting potential of aluminum," *Journal of the Electrochemical Society*, vol. 116, pp. 906–910, 1969.

[61] Ja. M. Kolotyrkin and M. Ya, "Effects of anions on the dissolution kinetics of metals," *Journal of the Electrochemical Society*, vol. 108, no. 3, pp. 209–216, 1961.

[62] N. Sato, "A theory for breakdown of anodic oxide films on metals," *Electrochemica Acta*, vol. 16, no. 10, pp. 1683–1692, 1971.

[63] N. Sato, "Anodic breakdown of passive films on metals," *Journal of the Electrochemical Society*, vol. 129, no. 2, pp. 255–260, 1982.

[64] N. Sato, "The stability of pitting dissolution of metals in aqueous solution," *Journal of the Electrochemical Society*, vol. 129, no. 2, pp. 260–264, 1982.

[65] L. F. Lin, C. Y. Chao, and D. D. Macdonald, "A point defect model for anodic passive films II. Chemical breakdown and pit initiation," *Journal of the Electrochemical Society*, vol. 128, no. 6, pp. 1194–1198, 1981.

[66] D. W. Hoeppner, D. Mann, and J. Weekes, "Fracture mechanics based modeling of corrosion fatigue process," in *Proceedings of the 52nd Meeting of the AGARD Structural and Materials Panel*, Corrosion Fatigue, Cesme, Turkey, April, 1981.

[67] D. W. Hoeppner, "Corrosion fatigue considerations in materials selections and engineering design," in *Corrosion Fatigue: Chemistry, Mechanics, and Microstructure*, pp. 3–11, NACE, 1972.

[68] D. Jones, *Principles and Prevention of Corrosion*, Macmillan Publishing, New York, NY, USA, 1992.

[69] S. Szklarska, *Pitting Corrosion of Metals*, National Association of Corrosion Engineers (NACE), Houston, Tex, USA, 1986.

[70] J. E. Hatch, Ed., *Aluminum Properties and Physical Metallurgy*, American Society for Metals (ASM), Metals Park, Ohio, USA, 1984.

[71] D. J. McAdam and G. W. Gell, "Pitting and its effect on the fatigue limit of steels corroded under various conditions," *Journal of the Proceedings of the American Society for Testing Materials*, vol. 41, pp. 696–732, 1928.

[72] R. T. Foley, "Localized corrosion of aluminum alloys—a review," *Corrosion*, vol. 428, no. 5, pp. 277–288, 1986.

[73] L. Grimes, *A comparative study of corrosion pit morphology in 7075-T6 aluminum alloy*, M.S. thesis, University of Utah, 1996.

[74] Liversidge, "On the corrosion of aluminum," in *Chemical News, LXXI*, 1895.

[75] G. D. Bengough and O. F. Hudson, "Aluminum, fourth report to the corrosion committee," *Journal of the Institute of Metals*, vol. 21, no. 1, p. 105, 1919.

[76] H. T. Barnes and G . W. Shearer, "A hydrogen peroxide cell," *Journal of Physical Chemistry*, vol. 12, p. 155, 1908.

[77] W. R. Dunstan and J. R. Hill, "The passivity of iron and certain other metals," *Journal of the Chemical Society, Transactions*, vol. 99, pp. 1853–1866, 1911.

[78] H. P. Godard, *The Corrosion of Light Metals*, John Wiley and Sons, New York, NY, USA, 1967.

[79] R. K. Hart, "The oxidation of aluminum in dry and humid oxygen atmospheres," *The Proceedings of the Royal Society*, vol. A236, p. 68, 1956.

[80] G. E. Thompson, K. Shimizu, and G. C. Wood, "Observation of flaws in anodic films on aluminum," *Nature*, vol. 286, pp. 471–472, 1980.

[81] R. Seligman and P. Williams, "The action on aluminum of hard industrial waters," *Journal of the Institute of Metals*, vol. 23, pp. 159–184, 1920.

[82] R. S. Alwitt, "The growth of hydrous oxide films on aluminum," *Journal of the Electrochemical Society*, vol. 121, no. 10, pp. 1322–1328, 1974.

[83] T. E. Graedel, "Corrosion mechanisms for aluminum exposed to the atmosphere," *Journal of the Electrochemical Society*, vol. 136, no. 4, p. 204C, 1989.

[84] I. L. Rosenfield and I. S. Danilov, "Electrochemical aspects of pitting corrosion," *Corrosion Science*, vol. 7, p. 129, 1967.

[85] A. Broli and H. Holtan, "Use of potentiokinetic methods for the determination of characteristic potentials for pitting corrosion of aluminum in a deaerated solution of 3%NaCl," *Corrosion Science*, vol. 13, no. 4, pp. 237–246, 1973.

[86] H. P. Godard, *The Corrosion of Light Metals*, John Wiley and Sons, New York, NY, USA, 1967.

[87] S. Dallek and R. T. Foley, "Propagation of pitting on aluminum alloys," *Journal of the Electrochemical Society*, vol. 125, no. 5, pp. 731–733, 1978.

[88] T. H. Nguyen and R. T. Foley, "On the mechanism of pitting of aluminum," *Journal of the Electrochemical Society*, vol. 126, no. 11, pp. 1855–1860, 1979.

[89] M. Baumgartner and H. Kaesche, "Aluminum pitting in chloride solutions: morphology and pit growth kinetics," *Corrosion Science*, vol. 31, pp. 231–236, 1990.

[90] L. Ma, *Pitting effects on the corrosion fatigue life of 7075-T6 aluminum alloy*, dissertation, University of Utah, 1994.

[91] G. S. Chen, C. Liao, K. Wan, M. Gao, and R. P. Wei, "Pitting corrosion and fatigue crack nucleation," in *Effects of the Environment on the Initiation of Crack Growth, ASTM STP 1298*, W. A. Van Der Sluys, R. S. Piascik, and R. Zawierucha, Eds., pp. 18–33, American Society for Testing and Materials, 1997.

[92] H. H. Uhlig, "Adsorbed and reaction-product films on metals," *Journal of the Electrochemical Society*, vol. 97, no. 11, pp. 215c–220c, 1950.

[93] H. P. Leckie and H. H. Uhlig, "Environmental factors affecting the critical potential for pitting in 18-8 stainless steel," *Journal of the Electrochemical Society*, vol. 113, no. 2, pp. 1262–1267, 1966.

[94] T. P. Hoar and W. R. Jacob, "Breakdown of passivity of stainless steel by halide ions," *Nature*, vol. 216, pp. 1299–1301, 1967.

[95] Ya, M. Kolotyrkin, "Effects of anions on the dissolution kinetics of metals," *Journal of the Electrochemical Society*, vol. 108, no. 3, pp. 209–216, 1961.

[96] H. H. Strehblow, B. Titze, and B.P. Loechel, "The breakdown of passivity of iron and nickel by flouride," *Corrosion Science*, vol. 19, pp. 1047–1057, 1979.

[97] H. Bohni, "Localized corrosion," in *Corrosion Mechanisms*, F. Mansfeld, Ed., pp. 285–328, Marcel Dekker, New York, NY, USA, 1987.

[98] J. A. Richardson and G. C. Wood, "A study of the pitting corrosion of Al by scanning electron microscopy," *Corrosion Science*, vol. 10, no. 5, pp. 313–323, 1970.

[99] N. Nilsen and E. Bardal, "Short duration tests and a new criterion for characterization of pitting resistance of Al alloys," *Corrosion Science*, vol. 17, pp. 635–646, 1977.
[100] C. Y. Chao, L. F. Lin, and D. D. Macdonald, "A point defect model for anodic passive films I. film growth kinetics," *Journal of the Electrochemical Society*, vol. 128, no. 6, pp. 1187–1194, 1981.
[101] J. R. Galvele, "Transport processes and the mechanism of pitting of metals," *Journal of the Electrochemical Society*, vol. 123, no. 4, pp. 464–474, 1976.
[102] R. Akid, "The role of stress-assisted localized corrosion in the development of short fatigue cracks," in *Effects of the Environment on the Initiation of Crack Growth, ASTM STP 1298*, W. A. Van Der Sluys, R. S. Piascik, and R. Zawierucha, Eds., pp. 3–17, American Society for Testing and Materials, 1997.
[103] D. W. Hoeppner, "Model for prediction of fatigue lives based upon a pitting corrosion fatigue process," in *Proceedings of the ASTM-NBS-NSF Symposium*, J. T. Fong, Ed., Fatigue Mechanisms, no. ASTM STP 675, pp. 841–870, American Society for Testing and Materials, 1979.
[104] T. C. Lindley, P. McIntyre, and P. J. Trant, "Fatigue crack initiation at corrosion pits," *Metals Technology*, vol. 9, pp. 135–142, 1982.
[105] S. Kawai and K. Kasai, "Considerations of allowable stress of corrosion fatigue (focused on the influence of pitting)," *Fatigue Fracture of Engineering Materials Structure*, vol. 8, no. 2, pp. 115–127, 1985.
[106] D. W. Hoeppner, "Corrosion fatigue considerations in materials selections and engineering design," in *Corrosion Fatigue: Chemistry, Mechanics, and Microstructure*, O. Devereux, A. J. McEvily, and R. W. Staehle, Eds., pp. 3–11, NACE-2, National Association of Corrosion Engineers, 1972.
[107] M. Rebiere and T. Magnin, "Corrosion fatigue mechanisms of an 8090 Al Li Cu alloy," *Materials Science and Engineering: A*, vol. 128, no. 1, pp. 99–106, 1990.
[108] G. S. Chen and D. J. Duquette, "Corrosion fatigue of a precipitation hardened Al-Li-Zr Alloy in a 0.5 M sodium chloride solution," *Metallurgical Transactions*, vol. 23, no. 5, pp. 1563–1572, 1992.
[109] G. S. Chen, M. Gao, and R. P. Wei, "Microconstituent-induced pitting corrosion in aluminum," *Corrosion*, vol. 52, pp. 8–15, 1996.
[110] L. Ma and D. W. Hoeppner, "The effects of pitting on fatigue crack nucleation in 7075-T6 aluminum alloy," in *Proceedings of the FAA/NASA International Symposium on Advanced Structural Integrity Methods for Airframe Durability and Damage Tolerance*, vol. 3274, part 1, pp. 425–440, NASA Conference Publication, 1994.
[111] Y. Kondo, "Prediction of fatigue crack initiation life based on pit growth," *Corrosion Science*, vol. 45, no. 1, pp. 7–11, 1989.
[112] A. M. Hoeppner, *The effect of prior corrosion damage on the short crack growth rates of two aluminum alloys*, M.S. thesis, University of Utah, 1996.
[113] R. Akid and G. Murtaza, "Environment assisted short crack growth behavior of a high strength steel," in *Short Fatigue Cracks*, K.J. Miller and E.R. de los Rios, Eds., pp. 193–208, Mechanical Engineering Publications, 1992.
[114] P. R. Roberge, *Handbook of Corrosion Engineering*, McGraw Hill Book, New York, NY, USA, 1999.
[115] P. R. Roberge, *Corrosion Engineering: Principles and Practice*, McGraw Hill Book, New York, NY, USA, 2008.
[116] P. R. Roberge, *Corrosion Inspection and Monitoring*, Wiley-Interscience, Hoboken, NJ, USA, 2007.
[117] *Bloomberg Business Week*, page 37, June 12, 2011.
[118] D. W. Hoeppner, "A review of corrosion fatigue and corrosion/fatigue considerations in aircraft structural design," in *Proceedings of the 22nd Symposium of the International Committee of Aeronautical Fatigue*, M. Guillaume, Ed., vol. 1 of *ICAF 2003-Fatigue of Aeronautical Structures as an Engineering Challenge*, pp. 425–438, EMAS Publishing, Lucerne, Switzerland, May, 2003.
[119] S. Swift, "Rusty diamond," in *Proceedings of the 24th Symposium of the International Committee on Aeronautical Fatigue*, L. Lazzeri and A. Salvetti, Eds., ICAF 2007 Durability and Damage Tolerance of Aircraft Structures: Metals vs. Composites, Naples, Italy, May, 2007.
[120] *ASTM Standards volume 03.02-Wear and Erosion; Metal Corrosion*, ASTM, Philadelphia, Pa, USA, 1994.
[121] *ASTM Standards volume 03.01-Metals-Mechanical Testing; Elevated and Low Temperature Tests; Metallography*, ASTM, Philadelphia, Pa, USA, 1998.
[122] "Behavior of short cracks in airframe components," in *Proceedings of the 55th Meeting of the AGARD Structures and Materials Panel*, no. 328, Toronto, Canada, September, 1983.
[123] R. K. Bolinbroke and J. E. King, "The growth of short fatigue cracks in titanium alloys IMI550 and IMI318," in *Small Fatigue Cracks*, R. O. Ritchie and J. Lankford, Eds., pp. 129–144, Metallurgical Society, 1986.
[124] A. Boukerrou and R.A. Cottis, "The influence of corrosion on the growth of short fatigue cracks in structural steels," in *Short Fatigue Cracks*, K. J. Miller and E. R. de los Rios, Eds., pp. 209–217, Mechanical Engineering Publications, 1992.
[125] J. L. Breat, F. Mudry, and A. Pineau, "Short crack propagation and closure effects in A508 steel," *Fatigue Fracture of Engineering Materials Structures*, vol. 6, pp. 349–358, 1983.
[126] C. W. Brown, J. E. King, and M. A. Hicks, "Effects of microstructure on long and short crack growth in nickel base super alloys," *Metal Science*, vol. 18, pp. 374–380, 1984.
[127] C. W. Brown and D. Taylor, "The effects of texture and grain size on the short fatigue crack growth rates in Ti-6Al-4V," in *Fatigue Crack Threshold Concepts*, D.L. Davidson and S. Suresh, Eds., pp. 433–446, AIME, 1984.
[128] D. W. Cameron, *Perspectives and insights on the cyclic response of metal*, Ph.D. dissertation, University of Toronto, 1984.
[129] K. S. Chan and J. Lankford, "The role of microstructural dissimilitude in fatigue and fracture of small cracks," *Acta Metallurgica*, vol. 36, pp. 193–206, 1988.
[130] P. Clement, J. P. Angeli, and A. Pineau, "Short crack behavior in nodular cast iron," *Fatigue Fracture of Engineering Materials and Structures*, vol. 7, pp. 251–265, 1984.
[131] D. L. Davidson, "Small and large fatigue cracks in aluminum alloys," *Acta Metallurgica*, vol. 36, no. 8, pp. 2275–2282, 1988.
[132] D. L. Davidson, J. B. Campbell, and R. A. Page, "The initiation and growth of fatigue cracks in a titanium aluminide alloy," *Metallurgical Transactions: A*, vol. 22, pp. 377–391, 1991.
[133] E. R. De Los Rios, Z. Tang, and K. J. Miller, "Short crack fatigue behavior in a medium carbon steel," *Fatigue Fracture of Engineering Materials and Structures*, vol. 7, pp. 97–108, 1984.
[134] E. R. De Los Rios, H. J. Mohamed, and K. J. Miller, "A micromechanic analysis for short fatigue crack growth," *Fatigue Fracture of Engineering Materials and Structures*, vol. 8, pp. 49–63, 1985.
[135] E. R. De Los Rios, A. Navarro, and K. Hussain, "Microstructural variations in short fatigue crack propagation of a C-Mn steel," in *Short Fatigue Cracks*, K. J. Miller and E. R.

de los Rios, Eds., pp. 115–132, Mechanical Engineering Publications, 1992.
[136] V. B. Dutta, S. Suresh, and R. O. Ritchie, "Fatigue crack propagation in dual-phase steels: effects of ferritic-martensitic microstructures on crack path morphology," *Metallurgical Transactions: A*, vol. 15, pp. 1193–1207, 1984.
[137] J. N. Eastabrook, "A dislocation model for the rate of initial growth of stage I fatigue cracks," *International Journal of Fracture*, vol. 24, no. 1, pp. 43–49, 1984.
[138] M. H. El Haddad, N.E. Dowling, T. H. Topper, and K. N. Smith, "J-integral applications for short fatigue crack at notches," *International Journal of Fracture*, vol. 16, pp. 15–30, 1980.
[139] P. J. E. Forsyth, *The Physical Basis of Metal Fatigue*, Blackie and Son Limited, London, UK, 1969.
[140] R. P. Gangloff, "Crack size effects on the chemical driving force for aqueous corrosion fatigue," *Metallurgical Transactions: A*, vol. 16A, no. 5, pp. 953–969, 1985.
[141] R. P. Gangloff and R. P. Wei, "Small crack-environment interaction: the hydrogen embrittlement perspective," in *Small Fatigue Cracks*, R.O. Ritchie and J. Lankford, Eds., pp. 239–264, Metallurgical Society, 1986.
[142] M. Goto, "Scatter in small crack propagation and fatigue behavior in carbon steels," *Fatigue Fracture of Engineering Materials and Structures*, vol. 16, pp. 795–809, 1993.
[143] M. Goto, "Statistical investigation of the behavior of small cracks and fatigue life in carbon steels with different ferrite grain sizes," *Fatigue Fracture of Engineering Materials and Structures*, vol. 17, no. 6, pp. 635–649, 1994.
[144] J. C. Healy, L. Grabowski, and C. J. Beevers, "Short-fatigue-crack growth in a nickel-base superalloy at room and elevated temperature," *International Journal of Fatigue*, vol. 13, no. 2, pp. 133–138, 1991.
[145] M. A. Hicks and C. W. Brown, "A comparison of short crack growth behavior in engineering alloys," in *Fatigue 84*, C. J. Beevers, Ed., p. 1337, EMAS, Warley, UK, 1984.
[146] S. Hirose and M. E. Fine, "Fatigue crack initiation and microcrack propagation in X7091 type aluminum P/M alloys," *Metallurgical Transactions*, vol. 14A, no. 6, pp. 1189–1197, 1983.
[147] P. D. Hobson, "The formulation of a crack growth equation for short cracks," *Fatigue Fracture of Engineering Materials and Structures*, vol. 5, no. 4, pp. 323–327, 1982.
[148] D. W. Hoeppner, "The effect of grain size on fatigue crack propagation in copper," in *Fatigue Crack Propagation, ASTM 415*, J. Grosskreutz, Ed., pp. 486–504, American Society for Testing and Materials, 1967.
[149] Op. cit. 38.
[150] D. W. Hoeppner, "Application of damage tolerance concepts to "short cracks" in safety critical components," in *Proceedings of the 12th Industrial College of the Armed Forces Symposium*, Industrial Applications of Damage Tolerance (167) Concepts, pp. 2.1/1–2.2/20, ICAF document no. 1336, 1983.
[151] T. Hoshide, T. Yamada, and S. Fujimura, "Short crack growth and life prediction in low-cycle fatigue and smooth specimens," *Engineering Fracture Mechanics*, vol. 21, no. 1, pp. 85–101, 1985.
[152] P. Hyspecky and B. Stranadel, "Conversion of short fatigue cracks into a long crack," *Fatigue Fracture of Engineering Materials and Structures*, vol. 15, no. 9, pp. 845–854, 1992.
[153] S. Kawachi, K. Yamada, and T. Kunio, "Some aspects of small crack growth near threshold in dual phase steel," in *Short Fatigue Cracks*, K. J. Miller and E.R. de los Rios, Eds., pp. 101–114, Mechanical Engineering Publications, 1992.
[154] Y. H. Kim, T. Mura, and M.E. Fine, "Fatigue crack initiation and microcrack growth in 4140 steel," *Metallurgical Transactions: A*, vol. 9, no. 11, pp. 1679–1683, 1978.
[155] Y. H. Kim and M.E. Fine, "Fatigue crack initiation and strain-controlled fatigue of some high strength low alloy steels," *Metallurgical Transactions: A*, vol. 13, no. 1, pp. 59–71, 1982.
[156] S. Kumai, J. E. Kino, and J. F. Knott, "Short and long fatigue crack growth in a SiC reinforced aluminum alloy," *Fatigue Fracture of Engineering Materials and Structures*, vol. 13, no. 5, pp. 511–524, 1990.
[157] C. Y. Kung and M.E. Fine, "Fatigue crack initiation and microcrack growth in 2024-T4 and 2124-T4 aluminum alloys," *Metallurgical Transactions: A*, vol. 10, pp. 603–610, 1979.
[158] T. Kunio and K. Yamada, "Microstructural aspects of the threshold condition for non-propagating fatigue cracks in martensitic-ferritic structures," in *Fatigue Mechanisms, ASTM STP 675*, J.T. Fong, Ed., pp. 342–370, 1979.
[159] S. I. Kwun and R. A. Fournelle, "Fatigue crack initiation and propagation in a quenched and tempered niobium bearing HSLA steel," *Metallurgical Transactions: A*, vol. 13, pp. 393–399, 1982.
[160] J. Lankford, T. S. Cook, and G. P. Sheldon, "Fatigue microcrack growth in nickel-base superalloy," *International Journal of Fracture*, vol. 17, pp. 143–155, 1981.
[161] J. Lankford, "The growth of small fatigue cracks in 7075-T6 aluminum," *Fatigue Fracture of Engineering Materials and Structures*, vol. 5, pp. 233–248, 1982.
[162] J. Lankford, "The effect of environment on the growth of small fatigue cracks," *Fatigue Fracture of Engineering Materials and Structures*, vol. 6, pp. 15–31, 1983.
[163] J. Lankford and D. L. Davidson, "The role of metallurgical factors in controlling the growth of small fatigue cracks," in *Small Fatigue Cracks*, R.O. Ritchie and J. Lankford, Eds., pp. 51–72, The Metallurgical Society, 1986.
[164] Y. Mahajan and H. Margolin, "Low cycle fatigue behavior of Ti-6Al-2Sn-4Zr-6Mo: part I. The role of microstructure in low cycle crack nucleation and early crack growth," *Metallurgical Transactions: A*, vol. 13, no. 2, pp. 257–267, 1982.
[165] R. C. McClung, K. S. Chan, S. J. Hudak, and D. L. Davidson, "Analysis of small crack behavior for airframe applications," in *Proceedings of the FAA/NASA International Symposium on Advanced Structural Integrity Methods for Airframe Durability and Damage Tolerance*, 1994.
[166] A. J. McEvily and S. Minakawa, "Crack closure and the conditions for crack propagation," in *Fatigue Crack Growth Threshold Concepts*, D.L. Davidson and S. Suresh, Eds., pp. 517–530, TMS-AIME, 1984.
[167] Z. Mei and J. W. Morris, "The growth of small fatigue cracks in A286 steel," *Metallurgical Transactions: A*, vol. 24, no. 3, pp. 689–700, 1993.
[168] Z. Mei, C. R. Krenn, and J. W. Morris, "Initiation and growth of small fatigue cracks in a Ni-base superalloy," *Metallurgical Transactions: A*, vol. 26, no. 8, pp. 2063–2073, 1995.
[169] K. J. Miller, "The short crack problem," *Fatigue and Fracture of Engineering Materials and Structures*, vol. 5, no. 3, pp. 223–232, 1982.
[170] K. Minakawa, Y. Matsuo, and A. McEvily, "The influence of a duplex microstructure in steels on fatigue crack growth in

the near-threshold region," *Metallurgical Transactions: A*, vol. 13, no. 3, pp. 439–445, 1982.

[171] W. L. Morris, O. Buck, and H. L. Marcus, "Fatigue crack initiation and early propagation in Al 2219-T851," *Metallurgical Transactions: A*, vol. 7, no. 7, pp. 1161–1165, 1976.

[172] W. L. Morris and O. Buck, "Crack closure load measurements for microcracks developed during the fatigue of Al 2219-T851," *Metallurgical Transactions: A*, vol. 8, no. 4, pp. 597–601, 1977.

[173] Y. Murakami and M. Endo, "Quantitative evaluation of fatigue strength of metals containing various small defects or cracks," *Engineering Fracture Mechanics*, vol. 17, no. 1, pp. 1–15, 1983.

[174] D. J. Nicholls and J. W. Martin, "A comparison of small fatigue crack growth, low cycle fatigue and long fatigue crack growth in Al-Li alloys," *Fatigue Fracture of Engineering Materials and Structures*, vol. 14, no. 2-3, pp. 185–192, 1991.

[175] J. C. Newman and P. R. Edwards, "Short crack growth behavior in an aluminum alloy," NATO-AGARD report, An AGARD cooperative test Programme, 1988.

[176] M. Okazaki, T. Tabata, and S. Nohmi, "Intrinsic stage I crack growth of directionally solidified Ni-Base superalloys during low-cycle fatigue at elevated temperature," *Metallurgical Transactions: A*, vol. 21, no. 8, pp. 2201–2208, 1990.

[177] J. Z. Pan, E. R. De Los Rios, and K. J. Miller, "Short fatigue crack growth in plain and notched specimens of an 8090 Al-Li alloy," *Fatigue Fracture of Engineering Materials and Structures*, vol. 16, no. 12, pp. 1365–1379, 1993.

[178] S. Pearson, "Initiation of fatigue cracks in commercial aluminum alloys and the subsequent propagation of very short cracks," *Engineering Fracture Mechanics*, vol. 7, no. 2, pp. 235–247, 1975.

[179] J. Petit and A. Zeghloul, "Environmental and nicrostructural influence on fatigue propagation of small surface cracks," in *Environmentally Assisted Cracking: Science and Engineering, ASTM STP 1049*, W. B. Lisagor, T.W. Crooker, and B. N. Leis, Eds., pp. 334–346, American Society for Testing and Materials, Philadelphia, Pa, USA, 1990.

[180] J. Petit, J. Mendez, L. W. L. Berata, and C. Muller, "Influence of environment on the propagation of short fatigue cracks in a titanium alloy," in *Short Fatigue Cracks*, K. J. Miller and E. R. de los Rios, Eds., pp. 235–250, Mechanical Engineering Publications, 1992.

[181] J. Petit and A. Zeghloul, "On the effect of environment on short crack growth behavior and threshold," in *The Behavior of Short Fatigue Cracks*, K. J. Miller and E. R. de los Rios, Eds., pp. 163–178, Mechanical Engineering Publications, 1986.

[182] J. Petit and K. Kosche, "Stage I and Stage II propagation of short and long cracks in Al-Zn-Mg alloys," in *Short Fatigue Cracks*, K. J. Miller and E. R. de los Rios, Eds., pp. 135–151, Mechanical Engineering Publications, 1992.

[183] R. S. Piascik and S. A. Willard, "The growth of small corrosion fatigue cracks in alloy 2024," *Fatigue Fracture of Engineering Materials and Structures*, vol. 17, pp. 1247–1259, 1994.

[184] A. Plumtree and B. P. D. O'Connor, "Influence of microstructure on short fatigue crack growth," *Fatigue Fracture of Engineering Materials and Structures*, vol. 14, no. 2-3, pp. 171–184, 1991.

[185] J. P. Polak and T. Laskutin, "Nucleation and short crack growth in fatigued polycrystalline copper," *Fatigue Fracture of Engineering Materials and Structures*, vol. 13, no. 2, pp. 119–133, 1990.

[186] J. M. Potter and B. G. W. Yee, "Use of small crack data to bring about and quantify improvements to aircraft structural integrity," in *Behavior of Short Cracks in Airframe Components*, p. 18, North Atlantic Treaty Organization, NATO-AGARD, 1983.

[187] R .O. Ritchie and J. Lankford, "Small fatigue cracks: a statement of the problem and potential solutions," *Materials Science and Engineering*, vol. 84, pp. 11–16, 1986.

[188] R. O. Ritchie and S. Suresh, "Mechanics and physics of the growth of small cracks," Berkeley report , University of California, 1995.

[189] J. Schijve, "Fatigue crack closure, observations and technical significance," Tech. Rep. number NLR TR-679, National Aerospace Laboratory NLR, Amsterdam, The Netherland, 1986.

[190] J. Schijve, "Multiple-site-damage fatigue of riveted joints," in *Proceedings of the International Workshop On Structural Integrity of Aging Airplanes*, Atlanta, Ga, USA, March, 1992.

[191] J. K. Shang, J. L. Tzou, and K. J. Miller, "Role of crack tip shielding in the initiation and growth of long and small fatigue cracks in composite microstructures," *Metallurgical Transactions: A*, vol. 18, no. 9, pp. 1613–1627, 1987.

[192] G. P. Sheldon, T. S. Cook, J. W. Jones, and J. Lankford, "Some observations on small fatigue cracks in a superalloy," *Fatigue Fracture of Engineering Materials and Structures*, vol. 3, pp. 219–228, 1981.

[193] D. Sigler, M. C. Montpetit, and W. L. Haworth, "Metallography of fatigue crack initiation in an overaged high-strength aluminum alloy," *Metallurgical Transactions: A*, vol. 14, no. 5, pp. 931–938, 1982.

[194] R. R. Stephens, L. Grabowski, and D. W. Hoeppner, "The effect of temperature on the behavior of short fatigue cracks in Waspaloy using an in situ SEM fatigue apparatus," *International Journal of Fatigue*, vol. 15, no. 4, pp. 273–282, 1993.

[195] C. M. Suh, J. J. Lee, and Y. G. Kang, "Fatigue microcracks in type 304 stainless steel at elevated temperature," *Fatigue Fracture of Engineering Materials and Structures*, vol. 13, no. 5, pp. 487–496, 1990.

[196] S. Suresh, "Crack deflection: implications for the growth of long and short fatigue cracks," *Metallurgical Transactions: A*, vol. 14, pp. 2375–2385, 1983.

[197] S. Suresh, "Fatigue crack deflection and fracture surface contact: micromechanical model," *Metallurgical Transactions: A*, vol. 16, no. 2, pp. 249–260, 1985.

[198] S. Taira, K. Tanaka, and M. Hoshina, "Grain size effect on crack nucleation and growth in long-life fatigue of low-carbon steel," in *Fatigue Mechanisms, ASTM STP 675*, J. T. Fong, Ed., pp. 135–173, ASTM, Philadelphia, Pa, USA, 1979.

[199] D. Taylor and J. F. Knott, "Fatigue crack propagation behavior of short cracks; the effect of microstructure," *Fatigue Fracture of Engineering Materials and Structures*, vol. 4, no. 2, pp. 147–155, 1981.

[200] K. Tokaji, T. Ogawa, and Y. Harada, "The growth of small fatigue cracks in a low carbon steel; the effect of microstructure and limitations of linear elastic fracture mechanics," *Fatigue Fracture of Engineering Materials and Structures*, vol. 9, no. 3, pp. 205–217, 1986.

[201] K. Tokaji, T. Ogawa, Y. Harada, and Z. Ando, "Limitations of linear elastic fracture mechanics in respect of small fatigue

cracks and microstructure," *Fatigue Fracture of Engineering Materials and Structures*, vol. 9, no. 1, pp. 1–14, 1986.

[202] K. Tokaji, T. Ogawa, and Y. Harada, "Evaluation on limitation of linear elastic fracture mechanics for small fatigue crack growth," *Fatigue Fracture of Engineering Materials and Structures*, vol. 10, no. 4, pp. 281–289, 1987.

[203] K. Tokaji, T. Ogawa, S. Osako, and Y. Harada, "The growth behavior of small fatigue cracks; the effect of microstructure and crack closure," *Fatigue '87*, vol. 2, pp. 313–322, 1988.

[204] K. Tokaji and T. Ogawa, "The growth of microstructurally small fatigue cracks in a ferritic-pearlitic steel," *Fatigue Fracture of Engineering Materials and Structures*, vol. 11, pp. 331–342, 1988.

[205] K. Tokaji, T. Ogawa, and T. Aoki, "Small fatigue crack growth in a low carbon steel under tension-compression and pulsating-tension loading," *Fatigue Fracture of Engineering Materials and Structures*, vol. 13, no. 1, pp. 31–39, 1990.

[206] K. Tokaji and T. Ogawa, "The effects of stress ratio on the growth behavior of small fatigue cracks in an aluminum alloy 7075-T6 (with special interest in Stage I crack growth)," *Fatigue Fracture of Engineering Materials and Structures*, vol. 13, no. 4, pp. 411–421, 1990.

[207] K. Tokaji, T. Ogawa, Y. Kameyama, and Y. Kato, "Small fatigue crack growth behavior and its statistical properties in a pure titanium," *Fatigue '90*, vol. 2, pp. 1091–1096, 1990.

[208] K. Tokaji and T. Ogawa, "The growth behavior of microstructurally small fatigue cracks in metals," in *Short Fatigue Cracks*, K. J. Miller and E. R. de los Rios, Eds., pp. 85–100, Mechanical Engineering Publications, 1992.

[209] K. T. Venkateswara Rao, W. Yu, and R. O. Ritchie, "Fatigue crack propagation in aluminum-lithium alloy 2090: part II. Small crack behavior," *Metallurgical Transactions: A*, vol. 19, pp. 563–569, 1988.

[210] L. Wagner, J. K. Gregory, A. Gysler, and G. Lutjering, "Propagation behavior of short cracks in a Ti-8.6Al alloy," in *Small Fatigue Cracks*, R.O. Ritchie and J. Lankford, Eds., pp. 117–128, Metallurgical Society, 1986.

[211] H. A. Wood and J. L Rudd, "Evaluation of small cracks in airframe structures," personal communication, 1993.

[212] D. C. Wu, *An investigation into the fatigue crack growth characteristics of a single crystal nickel-base superalloy*, Ph.D. dissertation, University of Toronto, 1986.

[213] J. R. Yates, W. Zhang, and K. J. Miller, "The initiation and propagation behavior of short fatigue cracks in Waspaloy subjected to bending," *Fatigue Fracture of Engineering Materials and Structures*, vol. 16, pp. 351–362, 1993.

[214] A. Zeghloul and J. Petit, "Environmental sensitivity of small crack growth in 7075 aluminum alloy," *Fatigue Fracture of Engineering Materials and Structures*, vol. 8, no. 4, pp. 341–348, 1985.

[215] A. K. Zurek, M. R. James, and W. L. Morris, "The effect of grain size on fatigue growth of short cracks," *Metallurgical Transactions: A*, vol. 14, no. 8, pp. 1697–1705, 1982.

[216] D. W. Hoeppner et al., "Aircraft structural fatigue," Four volumes of notes, Course sponsored by the US FAA, 1979–1992.

[217] M. Creager, T. R. Brussat, D. W. Hoeppner, and T. Swift, "Structural integrity of new and aging metallic aircraft," Short course by UCLA Extension, Department of Engineering, Information Systems and Technical Management, Los Angeles, Calif, USA, 1970-present.

9

Assessment of Combustor Working Environments

Leiyong Jiang and Andrew Corber

Aerospace Research, Gas Turbine Laboratory, The National Research Council of Canada, 1200 Montreal Road, M-10, Ottawa, ON, Canada K1A 0R6

Correspondence should be addressed to Leiyong Jiang, leiyong.jiang@nrc-cnrc.gc.ca

Academic Editor: Victor Giurgiutiu

In order to assess the remaining life of gas turbine critical components, it is vital to accurately define the aerothermodynamic working environments and service histories. As a part of a major multidisciplinary collaboration program, a benchmark modeling on a practical gas turbine combustor is successfully carried out, and the two-phase, steady, turbulent, compressible, reacting flow fields at both cruise and takeoff are obtained. The results show the complicated flow features inside the combustor. The airflow over each flow element of the combustor can or liner is not evenly distributed, and considerable variations, ±25%, around the average values, are observed. It is more important to note that the temperatures at the combustor can and cooling wiggle strips vary significantly, which can significantly affect fatigue life of engine critical components. The present study suggests that to develop an adequate aerothermodynamics tool, it is necessary to carry out a further systematic study, including validation of numerical results, simulations at typical engine operating conditions, and development of simple correlations between engine operating conditions and component working environments. As an ultimate goal, the cost and time of gas turbine engine fleet management must be significantly reduced.

1. Motivation

It is a promising approach to assess the remaining life of gas turbine critical components, based on their service history (flight operating conditions and sensor readings), by applying aerothermodynamics, structural and material analysis models. It is consistent with condition-based maintenance (CBM) of gas turbine engines, that is, maintenance actions would be performed only when they are required. As a result, the expense of maintenance/repair of an engine fleet can be reduced, the engine operation can be made more reliable, and service life can be extended. This approach has been investigated by researchers from various disciplines under a major collaboration program [1].

One of the subprograms is the development of an aero-thermodynamics model. The model will provide more realistic distributions of temperature and pressure or loads for safety/cost critical components from engine operational conditions and sensor readings. Furthermore, it will also provide an environment for the assessment of current and enhanced sensor suites and the prediction of degraded operation with wear/damage or control changes. With the required thermal flow information, structural and material analyses can be performed, and the remaining life of engine components can be assessed with confidence.

For technology development, a practical gas turbine combustor is selected as a research case. Gas turbine combustors are exposed to high-temperature, high-pressure, and high-dynamic load environments, and failures occasionally occur during operation. The flow parameters around and inside the combustor vary significantly, which causes considerably uneven structural stresses. Increased localized metal temperatures and thermal gradients can both reduce the combustor fatigue life. A higher temperature reduces the fatigue strength of the material, while a larger gradient increases the strain excursion and causes higher stresses. Therefore, for reliable structural, material and life analyses, the accurate definition of the working environment for the combustor is deemed necessary.

Due to the harsh conditions, experimental measurements inside the engine are extremely difficult. An alternative is to use validated computational flow dynamics (CFD) methods

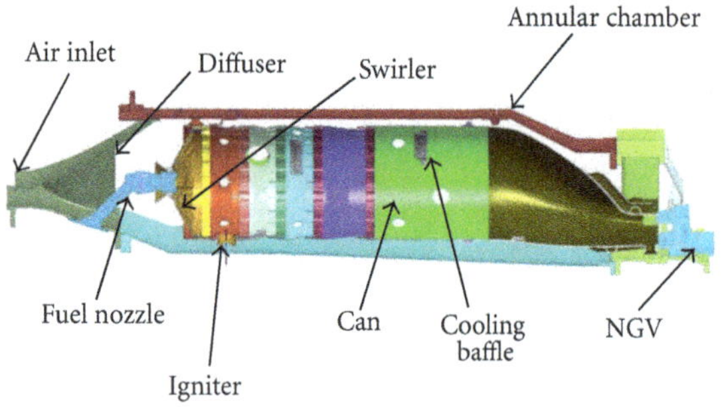

FIGURE 1: A 60-degree sector.

TABLE 1: Engine operating conditions.

	Flow parameters	
Takeoff	Air inlet total pressure (kPa)	1000
	Air mass flow rate (kg/s)	2.36
	Air inlet temperature (K)	606
	Fuel mass flow rate (kg/s)	0.053
Cruise	Air inlet total pressure (kPa)	463
	Air mass flow rate (kg/s)	1.17
	Air inlet temperature (K)	552
	Fuel mass flow rate (kg/s)	0.0207

to document typical operating conditions and then correlate these detailed numerical results with the engine operating conditions and sensor readings to form semiempirical models. These models can be used for engine component life analysis or an essential component of CBM for engine fleet management.

This paper covers a high-fidelity CFD model of the gas turbine combustor, complex flow fields inside the combustor, temperature and pressure distributions over the combustor can as well as indications or suggestions from the present study.

2. Thermal Fluid Dynamics Modeling of the Combustor

2.1. High-Fidelity CFD Model of the Combustor. Figure 1 illustrates a 60-degree sector of the gas turbine combustor. It is a can-annular design with six combustion cans and an annular air supply chamber [2]. Compressed air enters the annular chamber through a narrow annulus, decelerates in the diffuser, and then flows over and enters the combustor can or liner through air-management holes and cooling wiggle strips around the can. Inside the can, fine fuel droplets from a fuel nozzle evaporate, mix with air, and then burn. The mixture continues to react with air, cools down further downstream, and eventually reaches the air-cooled nozzle guide vanes (NGVs). For the present phase of work, the NGV is not included in the CFD model.

For traditional numerical simulations of gas turbine combustors, the computational domain is limited to the flow field inside the combustor liner, that is, the liner internal and external flow fields are decoupled. The airflow-splitting over the combustor liner is estimated based on semi-empirical discharge coefficient correlations [3, 4] and the flow rate is assumed evenly distributed over each liner airflow device (a row of holes, a wiggle strip, etc.). In the present study, both the liner internal and external flow fields are simulated simultaneously, that is, they are directly coupled.

Two engine operating conditions, takeoff and altitude cruise [2], are considered in the present study and the flow parameters for a single can are listed in Table 1.

2.2. Fuel Spray Measurements at the GTL High-Pressure Spray Rig. The temperature field in a combustor is dominated by the fuel distribution and arrangement of primary, cooling and dilution air [5]. Therefore, it is essential to have proper fuel spray parameters in order to predict the combustor flow field accurately. The fuel spray characteristics were measured in the GTL (Gas Turbine Laboratory) High-Pressure Spray Rig (HPSR) with a phase Doppler particle analyzer (PDPA), as shown in Figure 2, at the engine cruise and takeoff conditions. Figure 3 gives a picture of the fuel spray in an air box which is the domed portion of the combustor can.

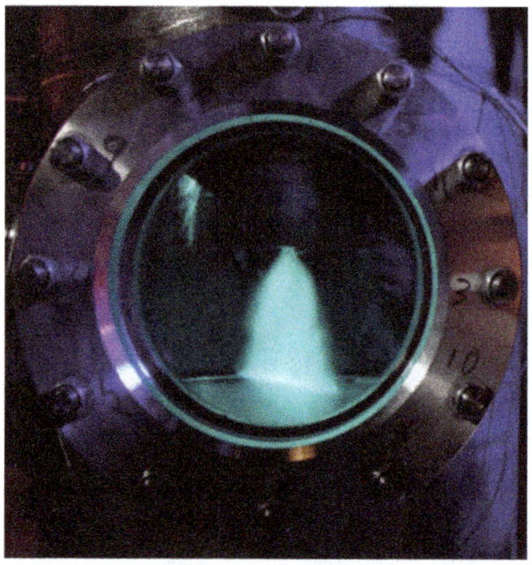

FIGURE 2: Fuel spray measurements in the GTL HPSR.

During spray measurement, the air and fuel flow rates remained the same as the flight conditions, and the pressure in the HPSR was adjusted to match the air density at flight. The measured parameters included the radial distributions of axial, tangential, and radial velocities, droplet size as well as fuel flux. All these results were used as the initial conditions of fuel spray for the predictions of the combustor thermal fields.

2.3. Mesh-Independent Studies and Combustor Mesh. Due to the geometrical complexity of practical gas turbine combustors, it is difficult to fully meet mesh-independence requirements. Researchers, such as [6], have pointed out that a gas turbine combustor simulation would need 410 million cells in order to claim mesh independence. This is understandable, for example, if the rollup of vortices along the bending jet and the structures of horseshoe and wake

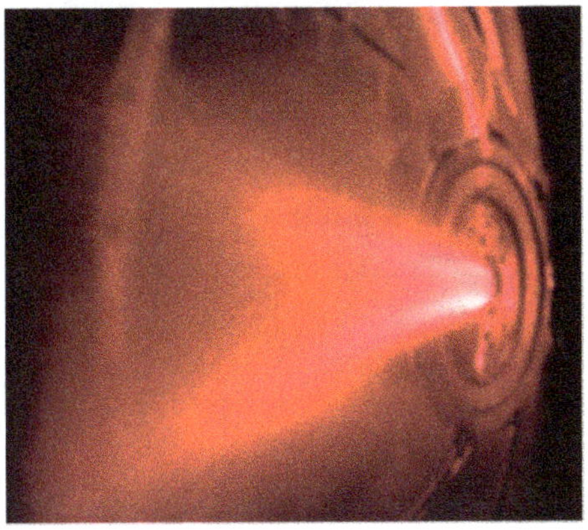

Figure 3: The fuel spray in a specifically built air box.

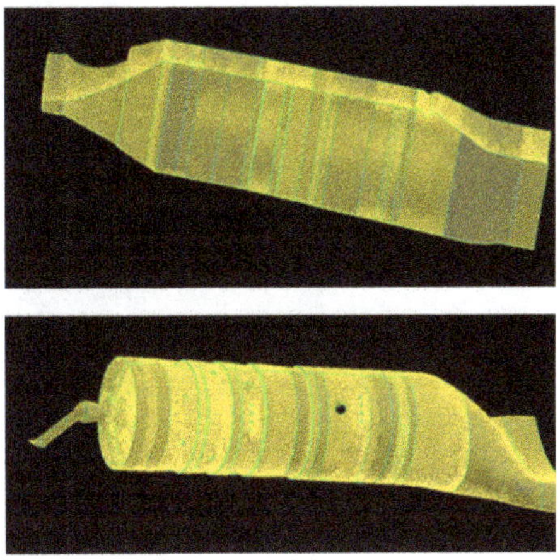

Figure 4: Meshes of 60-degree annular chamber and a combustor can.

vortices for a single cross-flow jet [7] have to be properly resolved in numerical simulations, the required mesh size can be even larger than 410 million cells. Therefore, the best way to check mesh independence would be to examine if the objectives or required flow parameters remain mesh-independent in simulations.

As mentioned before, the main objective of this work is to obtain accurate temperature and pressure distributions over the combustor can. Since the air distribution over the combustor can directly determines the combustor performance and metal wall temperature, the mesh-independence issues related to liner airflow devices (holes and wiggle strips) were studied first before meshing the whole 60-degree sector. For detail, please refer to [8, 9].

During meshing of the whole geometry, 36–84 surface elements were used for medium and large liner airflow holes. For small holes, such as the splashing holes over the dome swirler plates and cooling baffles (Figure 1), about 30 surface elements were generated. With these arrangements, the area-weighted numerical uncertainty of air distribution over the liner is less than 2%.

For the whole combustor simulation, the wiggle strip was replaced by an equivalent slot which gave the same mass flow rate as that of the wiggle-strip for similar flow conditions. For the purpose of structure/material/life analyses of wiggle strips, the flow field of a single wiggle strip element was resolved with the boundary conditions obtained from the results of whole combustor simulations.

A few meshes were created and preliminary test runs were performed to improve the quality of numerical simulations. In the end, a mesh with 13.3 million cells was used for the simulations. Figure 4 illustrates the meshes of the 60-degree annular chamber (top) and the combustor can (bottom). Efforts were made to generate hexahedral cells as far as possible. Fine nodes were laid in and around holes, cooling slots, baffles, regions, dome swirler, and dome section. The nondimensional wall boundary parameter, y+, at the combustor can walls varied from ~20 to ~250.

2.4. Flow Fields of the Combustor. A large amount of data is available to reveal complicated flow features and physical phenomena inside the combustor [8]. Only some of the results at the cruise conditions are presented here. The flow features at takeoff are similar to those at cruise. Figures 5–7 show the velocity vectors and contours of Mach number and temperature along the middle longitudinal plane of the 60-degree sector of the combustor. In the figures, the thick black lines are the cut-through surfaces of the combustor can and fuel nozzle, and the dimensions and flow parameters are normalized by their representative values.

As shown in the upper plot of Figure 5, the compressed air flows into the annular chamber through a narrow inlet, slows down in the diffuser, and then enters the can dome through the splashing holes on the dome walls. As a result, two large recirculation zones are formed immediately downstream of the dome swirler, and the shape of the lower one is distorted by the igniter (Figure 1). These swirling flows are used to start and anchor flame in the combustor (Figure 7), and at the same time cool the dome head walls. Further downstream, the air flows through cooling slots into the combustion can, and the flow velocity gradually increases towards the can exit, as shown in both upper and lower plots of Figure 5. Another recirculation zone inside the can is also observed adjacent to the two dome recirculation regions. For the annular chamber, due to the geometrical blockage, strong swirling flows are found upstream of the dome, and a mild recirculation zone is observed in the dead flow region above the can exit section.

The Mach number contours at this section are given in Figure 6. In the primary zone of the combustor can (just downstream of the dome swirler), the Mach number or flow velocity is low. This gives enough time for fuel droplets to evaporate, mix and burn with air, and provides good environments for flame stabilization. The Mach number inside the can gradually increases downstream of the dome section and reaches a high value of ~0.28 at the can exit. In Figure 6, a maximum Mach number of 0.32 is observed in the diffuser due to the narrow flow passage (Figure 1). Shown in Figure 7 are the temperature contours inside and outside the can. As illustrated in Figures 5 and 6, the flow field is not symmetric, even in the upstream region of the

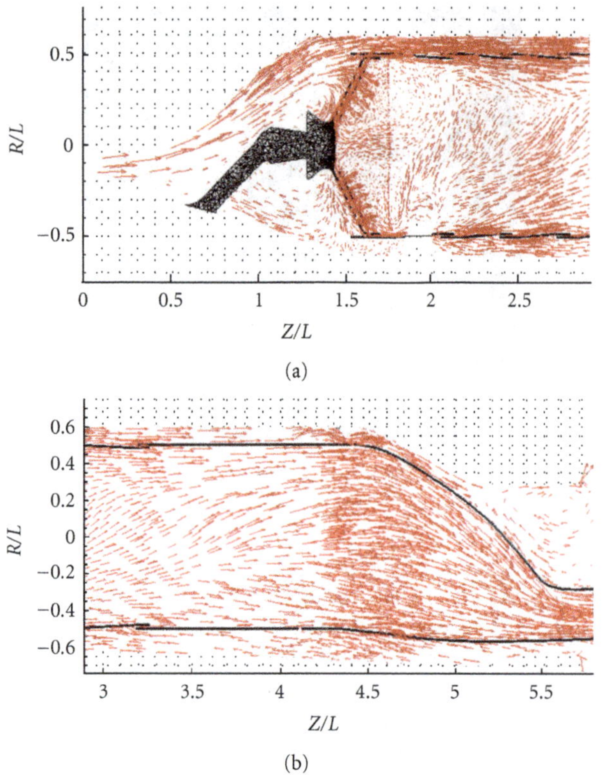

FIGURE 5: Velocity vector plots along the middle longitudinal plane.

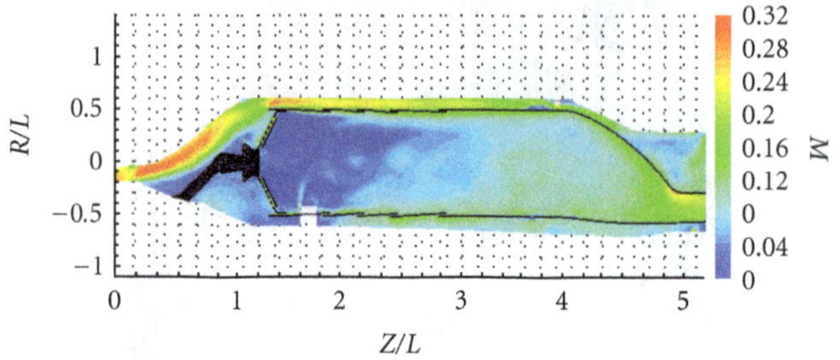

FIGURE 6: Mach number contours along the middle longitudinal plane.

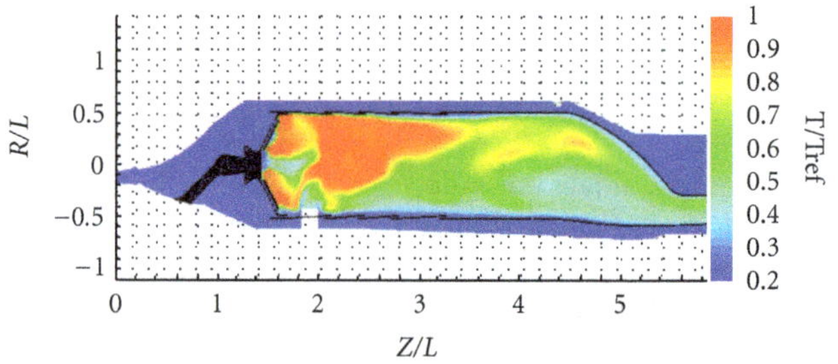

FIGURE 7: Temperature contours along the middle longitudinal plane.

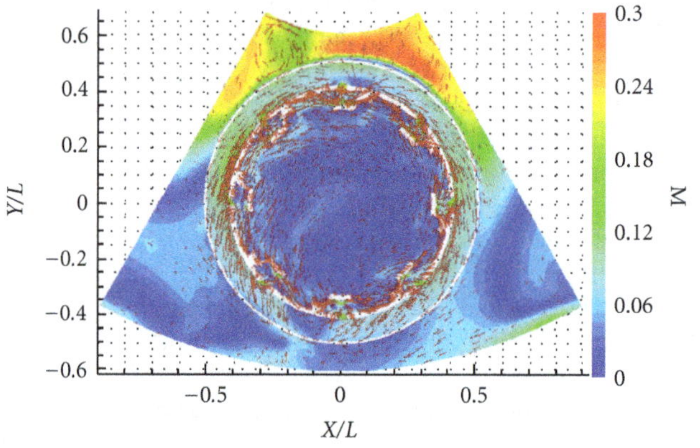

FIGURE 8: Velocity vectors and Mach numbers at a dome swirler cross-section.

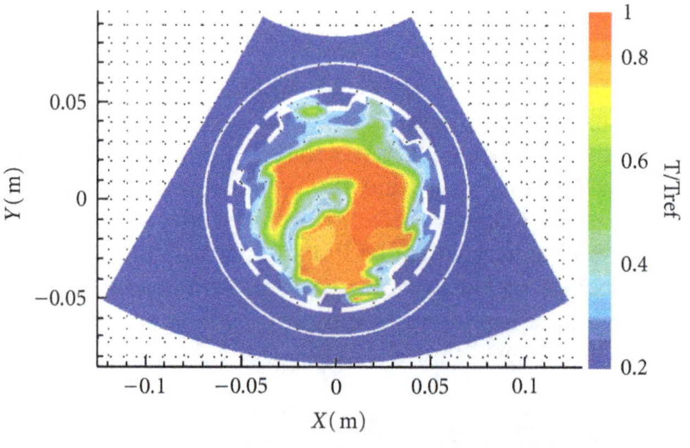

FIGURE 9: Temperature at a dome swirler cross-section.

combustor can. The high temperature region starts from the dome swirler and extends to the middle of the can with a maximum of ~1. This complies with gas turbine combustor design criteria [5], that is, the high-temperature region should be located in the primary and secondary zones. The low-temperature region in the center immediately downstream of the dome swirler is where the fuel spray is introduced into the flow field.

Figure 8 shows the velocity vector and Mach number plots, while Figure 9 illustrates temperature distribution at a dome swirler cross-section. This section cuts through the dome section, including swirler plates, splashing holes, and can liner wall, as indicated in thick white lines in Figures 8 and 9. There are eight swirler plates with 5 splashing holes for each, and only one hole for each plate is shown in the figures. As shown in Figure 8, the air enters the dome through small splashing holes and forms swirling flow at the plane perpendicular to the combustor axial axis. Referring to the two swirling regions inside the can observed along the longitudinal plane in Figure 5, it is understood that a donut-shaped vortex is formed immediately downstream of the dome swirler. Due to the geometric variation, high Mach number is found in the inner region of the annular chamber. The maximum temperature at this section (Figure 9) is close to that of the whole flow field (Figure 7), as shown in Figure 12, which could cause defects and cracks on the dome swirler [8].

The flow features at the cross-section of the second primary holes are given in Figures 10 and 11. Air enters

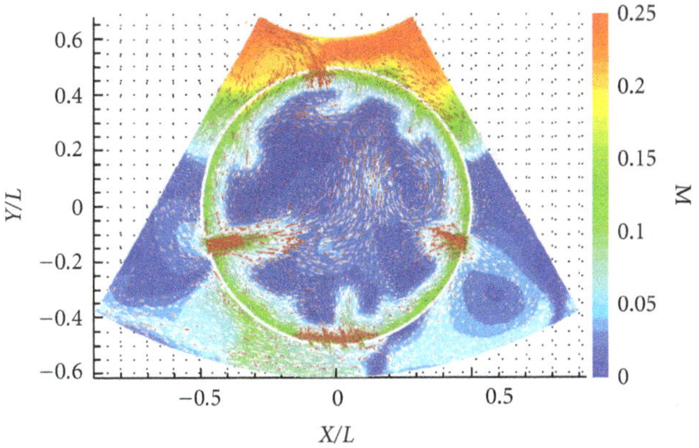

FIGURE 10: Velocity vectors and Mach numbers at the second primary hole section.

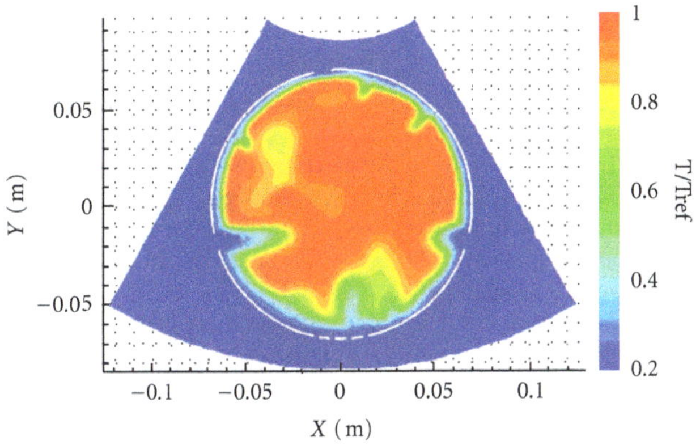

FIGURE 11: Temperature at the second primary hole cross-section.

the combustor can through 3 holes, and two large and two medium swirling regions are observed inside the can. It is important to note that although the size of the three air entry holes is the same, the airflow rate varies significantly, as shown in Figure 10. This is also observed at other liner air entry cross-sections (holes and wiggle strips), and the deviation of airflow rate can be as high as ±25%.

It is expected that the flow parameters such as flow and liner temperature distributions can be significantly different from those with the even airflow distribution over each cooling device in the traditional decoupled combustor simulations as mentioned earlier. As a result, the accuracy of combustor performance and life assessment can be considerably affected. This observation strongly suggests that for combustor simulations, the flow fields inside and outside the liner should be coupled in order to avoid potential prediction errors. At this primary zone section, the temperature is highest in the combustor flow field, as shown in Figure 11. This is because the combustion process takes place mainly in this zone. In Figures 10 and 11, a cooling baffle with four small splashing holes is displayed, and its cooling effect can be found in Figure 11.

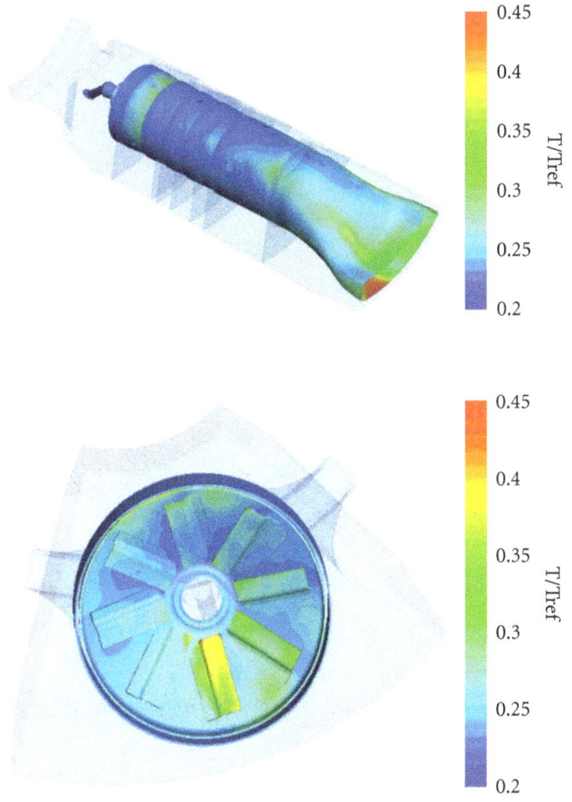

FIGURE 12: Combustor can and dome temperature contours at cruise.

2.5. Combustor Working Environment. Figures 12 and 13 present 3D temperature and absolute static pressure distributions over the combustor can and fuel nozzle at the cruise conditions. The maximum temperature occurs at the bottom left of the can exit. It reaches ~0.50, which is below but close to the allowable temperature for Hastelloy X alloy. As shown in Figure 12, the temperature over the combustor can vary significantly, which can cause considerably uneven structural stresses. As mentioned, increased localized metal temperatures and thermal gradients can both reduce the combustor fatigue life. Therefore, for reliable structural, material, and life analyses, the reliable working environments for engine components are deemed necessary.

The pressure distribution over the can walls is fairly uniform, as indicated in Figure 13. For the current combustor case, the pressure difference inside and outside the liner is minor, and therefore the pressure effect on the combustor life is probably insignificant in comparison with temperature. The detailed temperature distribution over the wiggle-strip element is illustrated in Figure 14 for cruise. The maximum temperature reaches ~0.288.

These detailed 3D temperature and pressure distributions over the combustor can and wiggle-strip element will be used for structural, material and life analyses of the combustor. It is expected that the critical or representative regions and/or parameters for the life analysis would be

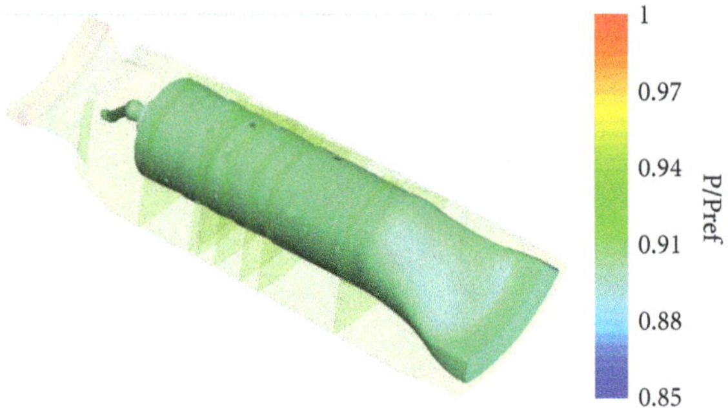

FIGURE 13: Combustor can pressure contours at cruise.

FIGURE 14: Temperature distribution over wiggle strip element at cruise.

identified, which may help the development of a simple aero-thermodynamics model. This model should be able to correlate the engine operating conditions/sensor readings with the working environments of engine critical components.

3. Conclusions and Future Work

To be able to reduce the cost of gas turbine engine fleet management, extend their service time, and perform maintenance actions only when they are required, it is vital to provide reliable aero-thermodynamic loads for the engine critical components.

As a first phase of the development of an aero-thermodynamics model, the benchmark modeling on a practical gas turbine combustor is successfully carried out, and the two-phase, steady, turbulent, compressible, reacting flow fields at the cruise and takeoff conditions are obtained. The complicated flow features inside the combustor are observed. More importantly, the present study indicates that the airflow over each liner flow device is not evenly distributed, and $\pm 25\%$ variations around the average values are observed. These findings suggest that the coupled combustor simulation should be performed.

The detailed 3D temperature and pressure distributions over the combustor can and wiggle-strip element at the engine cruise and takeoff conditions are available for structural, material, and life analyses. It is expected that the critical or representative regions and/or parameters would be identified for the development of a simple aero-thermodynamics model, which can correlate the engine operating conditions/sensor readings with the working environments of engine critical components.

The present study also indicates that it is necessary to carry out a further systematic study in order to develop an adequate aero-thermodynamics model. The on-going and future activities should include validation of numerical results, simulations at typical engine operating conditions, and development of simple correlations between engine operating conditions and component working environments.

Acknowledgments

The authors are grateful to the Department of National Defence and the Canadian Forces as well as the National Research Council Gas Turbine Laboratory for funding and supporting this collaborative research project. The technical authorities and project managers at DND and NRC, Mr. Ken McRae and Mr. Jeff Bird, respectively are gratefully acknowledged.

References

[1] J. Bird, M. Mrad, X. Wu, and C. Yang, "Propulsion System Mission and Maintenance Planning Demonstrator," LTR-GTL-2010-0092, The National Research Council of Canada.

[2] V. L. Oechsle, P. T. Ross, and H. C. Mongia, "High density fuel effects on gas turbine engines," Tech. Rep. AIAA-87-1829, San Diego, Calif, USA, 1987, AIAA/SAE/ASME/ASEE 23rd Joint Propulsion Conference.

[3] P. J. Stuttaford and P. A. Rubini, "Preliminary gas turbine combustor design using a network approach," *Journal of Engineering for Gas Turbines and Power*, vol. 119, no. 3, pp. 546–552, 1997.

[4] R. C. Adkins and D. Gueroui, "An improved method for accurate prediction of mass flows through combustor liner holes," Tech. Rep. GT86-149, Duesseldorf, Germany, 1986.

[5] A. H. Lefebvre and D. R. Ballal, *Gas Turbine Combustion*, Taylor and Francis Group, 2010.

[6] H. C. Mongia, "Recent advances in the development of combustor design tools," Tech. Rep. AIAA-2003-4495, Huntsville Alabama USA, 2003, AIAA/SAE/ASME/ASEE 39th Joint Propulsion Conference.

[7] M. Samimy, K. S. Breuer, L. G. Leal, and P. H. Steen, *A Gallery of Fluid Motion*, Cambridge University Press, Cambridge, UK, 2003.

[8] L. Y. Jiang and A. Corber, "Benchmark Modeling of T56 Gas Turbine Combustor—phase I, CFD model, flow features, air distribution and combustor can temperature distribution," LTR-GTL-2010-0088, The National Research Council of Canada, 2011.

[9] L. Jiang, "Combustor cooling wiggle strip and geometrical simplification," in *Proceedings of the International Mechanical Engineering Congress and Exposition (IMECE '08)*, pp. 177–184, November 2008.

10

Guidance Stabilization of Satellites Using the Geomagnetic Field

Francisco Miranda[1, 2]

[1] *Center for Research and Development in Mathematics and Applications, University of Aveiro, Portugal*
[2] *School of Technology and Management, Polytechnic Institute of Viana do Castelo, Avenida do Atlântico, 4900-348 Viana do Castelo, Portugal*

Correspondence should be addressed to Francisco Miranda, fmiranda@estg.ipvc.pt

Academic Editor: Kenneth M. Sobel

In the last years the small satellites have played an important role in the technological development. The attractive short period of design and low cost of them and the capacity to solve problems that are usually considered as problems to big and expensive spacecrafts lead us to study the control problem of these satellites. Active three-axis magnetic attitude stabilization of a low Earth orbit satellite is considered in this work. The control is created by interaction between the magnetic moment generated by magnetorquers mounted on the satellite body and the geomagnetic field. This problem is quite complex and difficult to solve. To overcome this difficulty guidance control is considered, where we use ε-strategies introduced by Pontryagin in the frame of differential games theory. Qualitative analysis and results of numerical simulation are presented.

1. Introduction

The problem of attitude control of spacecraft has been widely studied in the last years. If the spacecraft is equipped with three independent actuators, a complete solution to the set point and tracking control problems is available. This problem was solved by different ways (see, e.g., [1–4]). If only two independent actuators are available, as discussed in detail in [5], the problem of attitude regulation is not solvable by means of continuous time-invariant control laws, whereas a time-varying control law, achieving local asymptotic nonexponential stability, was proposed in [6]. Since magnetorquers are relatively reliable, lightweight, and energy efficient, they have become attractive for small and inexpensive satellites, but the above results are not directly applicable if the satellite is equipped with these magnetic coils as actuators. Such actuators operate on the basis of the interaction between the magnetic field generated by magnetic coils installed on the satellite body and the magnetic field of the Earth. This implies a spacecraft control significantly different. There are two classical uses of magnetic torque rods in attitude control. One is for momentum management of wheel-based systems [7]. The other is for angular momentum and nutation control of spinning [8], momentum-biased [9], and dual-spin spacecraft [10]. The present study is one of a growing number that considers active three-axis magnetic attitude stabilization of a low Earth orbit satellite considering a nontilted dipole model for the geomagnetic field. Magnetic coils have been extensively used since the early sixties as a simple and reliable technology to implement attitude control actuators in low Earth orbit satellites (see, e.g., [11–16] and the survey [17] and the references therein). The main difficulty in the implementation of this active three-axis magnetic system consists in the fact that the control torque can only be generated perpendicular to the geomagnetic field vector, which results in a time-varying nonlinear system. To solve this problem the most natural is to use the Lyapunov functions method combined with the Krasovskii-LaSalle theorem [18]. The Lyapunov function is used to construct a stabilizer depending on current time and position of the system. Then, the Krasovskii-LaSalle theorem is applied to prove the asymptotic stability of the equilibrium position. However, a rigorous verification of the Krasovskii-LaSalle theorem conditions faces serious technical difficulties which can hardly be overcome even with the help of systems for symbolic computations. Other techniques were presented (see, e.g., [19, 20]). In [19], Wiśniewski presented a simple constant gain controller (CGC) via the Linear Quadratic Regulator (LQR) method and proposed other two attitude controllers using the periodic characteristic of the Earth magnetic field: infinite and finite horizon periodic controllers. In [20], Psiaki showed us how to design a class of stabilizing attitude controllers for nadir-pointing spacecraft using only magnetic actuation. Their control laws are designed using a new type of periodic linear quadratic regulator whose Riccati equation solution is approximated by a linear time-invariant solution for an averaged system. The resulting full-state feedback controller derives its periodicity from the time-varying control influence matrix which can be derived from on-board magnetometer measurements. Both authors considered the linear time-varying system taking the periodic nature of the magnetic field as an advantage to approximate the solution of the Riccati equation. A considerable amount of work has been dedicated to the problems of analysis and design of magnetic control laws in the linear case, that is, control laws for nominal operation of a satellite near its equilibrium attitude (see, e.g., [9, 17, 19–25]). Abdelrahman et al. [26] proposed an attitude control method of a satellite with only magnetic torquers as actuators via the state-dependent Riccati equation (SDRE) technique based on pseudolinear time-varying modeling of the spacecraft nonlinear dynamics. However, limited attention has been dedicated to global formulations of the problem. In [18, 27, 28], the attitude regulation problem for Earth pointing spacecraft is studied exploiting the (quasi) periodic behavior of the system, which used the standard passivity arguments to prove local asymptotic stabilizability of open-loop equilibria. A solution to the global stabilization problem by means of full (or partial) state feedback has been studied by Lovera and Astolfi [29–32]. Conditions for almost global attitude regulation are obtained in [29, 30] for the case of full state feedback, and local results are derived for the case of output (attitude only) feedback. The paper [32] shows us how stability conditions similar to those given in [29, 30] that can be derived for control laws achieving Earth pointing for magnetically actuated spacecraft, taking also into account the effect of gravity gradient torques. For this problem, an almost global stabilization result is given for the case of full state feedback, resorting to an adaptive PD-like state feedback control law. With respect to works [29, 30, 32] the results presented in [31] that can deal with a generic magnetically actuated satellite do not rely on restrictive assumptions on the controller parameters and guarantee that there are no trajectories of the closed-loop system along which the average controllability can be lost. These results do not rely on the (frequently adopted) periodicity assumption for the geomagnetic field along the considered orbit. The proposed framework for closed-loop stability analysis of magnetically controlled spacecraft can be also exploited to predict the effect of actuator faults on the behavior of the controlled satellite.

An alternative way to solve the active three-axis global magnetic attitude stabilization problem of a low Earth orbit satellite is suggested here. The method is based on *guidance control* [33], a more general control procedure. Considering an auxiliary global stabilization problem that we know an analytic solution, the method consists to use a trajectory of the globally stabilizable auxiliary system as a guide. We construct a local stabilizer that keeps the trajectory of the system in a neighborhood of a solution of the auxiliary system. In this way, the trajectory of the system tends to the equilibrium position. According to this approach and using ε-strategies introduced by Pontryagin in the frame of differential games theory [34], the stabilizing control is constructed as a function of time defined in a small time interval and not as a feedback. From the practical point of view, ε-strategy is similar to stabilizer which depends on the time and position only, because it usually is implemented as a generator of piecewise constant controls. The numerical implementation of this method is based on the construction of multistep reachability sets [33]. The main contribution of this method is to solve a global stabilization problem applying local stabilization methods that avoid difficulties which we usually have in global stabilization problems, helping to overcome serious mathematical difficulties and being more effective in applications. This procedure also yields additional possibilities for the design of a stabilizer that eliminates the peak effect, that is, the large deviation of the solutions from the equilibrium position at the beginning of the stabilization process that we have when we construct stabilizers to obtain a high speed of damping of the control systems trajectories (studies of this effect are presented in [35–37]). The minimization problem used in this paper also eliminates this effect which represents a serious obstacle to guidance stabilization.

From all studies about the nonlinear controllability theory, the work [38] presents an application of this theory to the time-varying attitude dynamics of a magnetically actuated spacecraft in a Keplerian orbit in the geomagnetic field. The model considered in [38] does not include the gravitational torque and therefore the result can be applied to spherically symmetric satellites only. Considering a satellite subjected to gravity gradient torque, the controllability of the magnetic attitude of the satellite is studied here. Conditions of controllability when the satellite circular orbit is neither polar nor equatorial are obtained by local controllability theory.

It is assumed that the satellite moves along a circular orbit and the current value of the geomagnetic field with respect to the satellite axes is known due to three-axis magnetometer measurements. The position of the satellite mass center is also assumed to be known and the currents in the coils are control parameters.

2. Guidance Stabilization Process

Consider a control system

$$\dot{x}(t) = f(t, x(t)) + G(t, x(t))u, \quad u \in S, \qquad (1)$$

where $f : R \times R^n \to R^n$ and $G : R \times R^n \to M(n, k)$ are sufficiently smooth functions, that is, there exist all derivatives needed in our considerations. Let $0 \in \operatorname{int} S \subset R^k$ be such that $f(t, \tilde{x}) = 0$ for all t, that is, $\tilde{x} \in R^n$ is an equilibrium position. The global stabilization problem is to find a map $u : R^n \to S$ such that $u(\tilde{x}) = 0$ and the equilibrium position $\tilde{x}$ of the differential equation $\dot{x}(t) = f(t, x(t)) + G(t, x(t))u(x(t))$ is globally asymptotically stable. This problem is usually very hard to solve. One of the ways to avoid difficulties is to use a *guidance control* [33], a more general control procedure. Consider an auxiliary differential equation $\dot{\hat{x}}(t) =$

$f(t,\hat{x}(t)) + w(\hat{x}(t))$ such that $\tilde{x}$ is an globally asymptotically stable equilibrium position. The trajectories of system (1), under the guidance control, are defined as solutions to the Cauchy problem:

$$\begin{aligned} \dot{x}(t) &= f(t,x(t)) + G(t,x(t))u(t,x,\hat{x}), \\ \dot{\hat{x}}(t) &= f(t,\hat{x}(t)) + w(\hat{x}(t)), \\ x(t_0) &= x_0, \\ \hat{x}(t_0) &= \hat{x}_0, \end{aligned} \tag{2}$$

where the trajectory $\hat{x}(t)$, that is considered to be a "guide", tends to the equilibrium position $\tilde{x}$. Then we construct $u(t,x,\hat{x}) \in S$, $t \in [t_0+k\varepsilon, t_0+(k+1)\varepsilon]$, $\varepsilon > 0$, $k = 0,1,\ldots$, such that $|\overline{x}(t)| = |x(t) - \hat{x}(t)| \to 0$, $t \to \infty$, where $x(\cdot)$ is a solution to (1). To this end, we linearize system (1) along the trajectory $\hat{x}(t)$ and apply the methods developed to solve the stabilization problem for linear systems. This approach works only if $\overline{x}(t)$ is sufficiently small. For this reason, we construct the control $u(t,x,\hat{x})$ solving the following minimization problem

$$\begin{aligned} &\min_{u(t,x,\hat{x})\in S}\ \max_{t\in[t_0+k\varepsilon,t_0+(k+1)\varepsilon]} |\overline{x}(t)|, \\ &\dot{\overline{x}}(t) = A(t)\overline{x}(t) + B(t)u(t) - w(t), \\ &t \in [t_0 + k\varepsilon, t_0 + (k+1)\varepsilon], \\ &\overline{x}(t_0 + k\varepsilon) = \overline{x}_k, \\ &\overline{x}(t_0 + (k+1)\varepsilon) \in \beta|\overline{x}_k|B_n, \quad k = 0,1,\ldots, \end{aligned} \tag{3}$$

where $0 < \beta < 1$, $\varepsilon > 0$, $A(t) = \nabla_x f(t,\hat{x}(t))$, $B(t) = G(t,\hat{x}(t))$ and $w(t) = w(\hat{x}(t))$. In this way, we generate an ε-strategy, that is, a sequence of controls u defined on the intervals $[t_0 + k\varepsilon, t_0 + (k+1)\varepsilon]$, $k = 0, 1$, and so on, that minimizes the norm of the solution $\overline{x}(t)$ forcing this way the solution $\overline{x}(t)$ to be small for all t. This implies that the trajectory $x(t)$ of (1) follows the trajectory $\hat{x}(t)$ of the auxiliary differential equation. As $\hat{x}(t) \to \tilde{x}$, $t \to \infty$, then $x(t) \to \tilde{x}$, $t \to \infty$. These controls depend on t and $\overline{x}(t_0 + k\varepsilon)$, $k = 0,1,\ldots$. This problem has a solution if the control system $\dot{\overline{x}}(t) = A(t)\overline{x}(t) + B(t)u(t) - w(t)$, $u(t) \in S$, is locally controllable.

Let $u(t) \in S$, $t \geq t_0$, be a measurable bounded control. The solution of the linearized system $\dot{\overline{x}}(t) = A(t)\overline{x}(t) + B(t)u(t) - w(t)$, where $\overline{x}(t_0) = \overline{x}_0$, is denoted by $\overline{x}(\cdot, t_0, \overline{x}_0, u(\cdot))$ and the symbol $\overline{X}(t_0, \overline{x}_0)$ is used for the set of all solutions. Define the reachability set $\overline{\mathbf{R}}(t_1, t_0, \overline{x}_0) = \{\overline{x}(t_1, t_0, \overline{x}_0, u(\cdot)) \in R^n : \overline{x}(\cdot, t_0, \overline{x}_0, u(\cdot)) \in \overline{X}(t_0, \overline{x}_0)\}$. Using the following well-known result from the control theory (see [39]), we have a possibility to check if minimization problem (3) has a solution, that is, we can prove that there exists an ε-strategy such that $|\overline{x}(t)| = |x(t) - \hat{x}(t)| \to 0$, $t \to \infty$.

Theorem 1 ([39]). *The linearized system is controllable on the interval $[t_0, t_1]$, that is, $\overline{\mathbf{R}}(t_1, t_0, 0) = R^n$, if and only if, only the trivial solution $\overline{x}^*(t) \equiv 0$ to the adjoint differential equation $\dot{\overline{x}}^*(t) = -A^*(t)\overline{x}^*(t)$, $t \in [t_0, t_1]$, satisfies the orthogonality condition $B^*(t)\overline{x}^*(t) = 0$, $t \in [t_0, t_1]$. The condition $\overline{\mathbf{R}}(t_1, t_0, 0) = R^n$ implies that control system (1) is controllable around the trajectory $\hat{x}(\cdot)$ on the interval $[t_0, t_1]$.*

3. Application to the Satellite Stabilization Problem

Consider a satellite moving along a circular orbit. Introduce two Cartesian reference systems $OX_1X_2X_3$ and $ox_1x_2x_3$. The system $OX_1X_2X_3$ is the body reference system. The origin of this system, O, coincides with the satellite mass center and the axes are directed along the principal inertia axes. The system $ox_1x_2x_3$ is the orbital reference system. The origin o coincides with the center of mass of the satellite. The axis ox_3 is directed along the radius vector of the satellite mass center and the axis ox_2 is perpendicular to the orbital plane. The angular position of the satellite with respect to the orbital system is described by three positive left hand rotations defined by the Euler angles (α, β, γ). Coordinates of a vector in the satellite body system, X, and in the orbital system, x, satisfy the relation $x = BX$, where B is an orthogonal 3-1-2 rotation matrix with $\det B = 1$. The components of the matrix:

$$B = \begin{pmatrix} b_{11} & b_{12} & b_{13} \\ b_{21} & b_{22} & b_{23} \\ b_{31} & b_{32} & b_{33} \end{pmatrix} \tag{4}$$

are

$$\begin{aligned} b_{11} &= \cos\alpha\cos\beta + \sin\alpha\sin\beta\sin\gamma, \\ b_{12} &= -\cos\alpha\sin\beta + \sin\alpha\cos\beta\sin\gamma, \\ b_{13} &= \sin\alpha\cos\gamma, \\ b_{21} &= \sin\beta\cos\gamma, \\ b_{22} &= \cos\beta\cos\gamma, \\ b_{23} &= -\sin\gamma, \\ b_{31} &= -\sin\alpha\cos\beta + \cos\alpha\sin\beta\sin\gamma, \\ b_{32} &= \sin\alpha\sin\beta + \cos\alpha\cos\beta\sin\gamma, \\ b_{33} &= \cos\alpha\cos\gamma. \end{aligned} \tag{5}$$

Denote by b_i, $i = 1,2,3$, the rows of the matrix B. Obviously $b_3 = b_1 \times b_2$.

The satellite is equipped with three magnetic coils oriented along three orthogonal axes. Let Ω be the vector of angular velocity in the body axes. The rotation of the satellite subjected to gravity gradient torque is described by the following equations:

$$\begin{aligned} J\dot{\Omega} &= -\Omega \times J\Omega + 3\omega^2 b_3 \times J b_3 + U \times F, \\ \dot{b}_i &= b_i \times (\Omega - \omega b_2), \qquad i = 1,2,3, \end{aligned} \tag{6}$$

where

(i) the tensor of inertia J is a diagonal matrix with the diagonal elements J_1, J_2, J_3, such that $J_3 < J_1 \leq J_2$,

(ii) the vector F of the geomagnetic field in the body axes is given by $F = B^* f$, where $f = \mu_0 \mu_m r^{-3}(\cos\nu \sin\tau, \cos\tau, -2\sin\nu\sin\tau)$,

(iii) τ is the orbital inclination, that is, the angle between the equatorial and orbital planes,

(iv) $\nu = \omega t + \varphi_0$ is the argument of latitude,

(v) $\omega = (\mu_g/r^3)^{1/2} \approx 0.001\ \mathrm{s}^{-1}$ is the angular velocity of the orbital motion,

(vi) $r = 7.4 \times 10^6$ m is the radius of the orbit,

(vii) $\mu_m = 8.06 \times 10^{22}\ \mathrm{A\cdot m^2}$ is the Earth magnetic dipole moment,

(viii) $\mu_g = 3.986 \times 10^{14}\ \mathrm{m^3/c^2}$ is the Earth gravitational parameter,

(ix) $\mu_0 = 4\pi \times 10^{-7}$ H/m is the magnetic parameter,

(x) φ_0 describe the initial position of the mass center in the orbit,

(xi) the vector U has the components $U_i = I_i \rho L$, $i = 1, 2, 3$, where I_i stands for the current in the ith coil, ρ is the number of turns, and L is the area of a loop.

The currents I_i, $i = 1,2,3$, are control parameters. At the initial moment t_0, the satellite has an angular position that corresponds to the matrix $B(t_0) = B_0$ and the angular velocity $\Omega(t_0) = \Omega_0$. The problem is to find control laws $I_i = I_i(F)$ that drive $B(t)$ and angular velocity to I and $(0,\omega,0)$, respectively. Considering the Lyapunov function, known as the Jacobi integral,

$$V(\alpha,\beta,\gamma,\Omega) = \frac{1}{2}(\Omega - \omega b_2)^* J(\Omega - \omega b_2) + \frac{3}{2}\omega^2 b_3^* J b_3 - \frac{1}{2}\omega^2 b_2^* J b_2 - \frac{1}{2}\omega^2 (3J_3 - J_2), \tag{7}$$

we see that this is a stable equilibrium position [18, 28].

Consider the auxiliary system:

$$J\dot{\hat{\Omega}} = -\hat{\Omega} \times J\hat{\Omega} + 3\omega^2 \hat{b}_3 \times J\hat{b}_3 + W, \qquad \dot{\hat{b}}_i = \hat{b}_i \times \left(\hat{\Omega} - \omega\hat{b}_2\right), \quad i = 1,2,3, \tag{8}$$

where

$$W = \sum_{i=1}^{3} k_i r_i \times \hat{b}_i - \theta\left(\hat{\Omega} - \omega\hat{b}_2\right) + \omega\left(J\left(\hat{b}_2 \times \hat{\Omega}\right) + \hat{\Omega} \times J\hat{b}_2 + \hat{b}_2 \times J\hat{\Omega}\right) - \omega^2\left(\hat{b}_2 \times J\hat{b}_2 + 3\hat{b}_3 \times J\hat{b}_3\right), \tag{9}$$

for $k_i > 0$, $k_i \neq k_j$, $i, j = 1,2,3$, $i \neq j$, $(r_1 \ \ r_2 \ \ r_3) = I$, $\theta > 0$, $\hat{\Omega} \in R^3$ and $\hat{B} = (\hat{b}_1 \ \ \hat{b}_2 \ \ \hat{b}_3)^*$ has the same properties as B.

Theorem 2. *System (8)–(9) possesses an asymptotically stable equilibrium position* $\tilde{\Omega} = (0,\omega,0)$, $\tilde{b}_i = r_i$, $i = 1,2,3$.

Using the following lemma, the proof of Theorem 2 is trivial.

Lemma 3. *The system*

$$J\dot{\Theta} = -\Theta \times J\Theta + P, \qquad \dot{\hat{b}}_i = \hat{b}_i \times \Theta, \quad i = 1,2,3, \tag{10}$$

where $P = \sum_{i=1}^{3} k_i r_i \times \hat{b}_i - \theta\Theta$, *has the asymptotically stable equilibrium position* $\Theta = (0,0,0)$, $\hat{b}_i = r_i$, $i = 1,2,3$.

Proof. Considering the Lyapunov function $V = (\langle\Theta, J\Theta\rangle + \sum_{i=1}^{3} k_i(\hat{b}_i - r_i)^2)/2$ and following the proof of theorem in [40], we get the result. □

Remark: We can also see the proof of this lemma in [41].

Proof of Theorem 2. Using the equality:

$$\hat{\Omega} = \Theta + \omega\hat{b}_2 \tag{11}$$

in (8), we have

$$J\dot{\Theta} = -\Theta \times J\Theta + G, \qquad \dot{\hat{b}}_i = \hat{b}_i \times \Theta, \quad i = 1,2,3, \tag{12}$$

where

$$G = W - \omega\left(J\left(\hat{b}_2 \times \Theta\right) + \Theta \times J\hat{b}_2 + \hat{b}_2 \times J\Theta\right) - \omega^2\left(\hat{b}_2 \times J\hat{b}_2 - 3\hat{b}_3 \times J\hat{b}_3\right). \tag{13}$$

Considering

$$W = \sum_{i=1}^{3} k_i r_i \times \hat{b}_i - \theta\Theta + \omega\left(J\left(\hat{b}_2 \times \Theta\right) + \Theta \times J\hat{b}_2 + \hat{b}_2 \times J\Theta\right) + \omega^2\left(\hat{b}_2 \times J\hat{b}_2 - 3\hat{b}_3 \times J\hat{b}_3\right), \tag{14}$$

and using Lemma 3, we see that system (12)–(13) has the following asymptotically stable equilibrium position

$$\Theta = (0,0,0), \qquad \hat{b}_i = r_i, \quad i = 1,2,3. \tag{15}$$

Using again equality (11) in (14)-(15), we obtain the result. □

Auxiliary system (8)–(9) will be a guide for control system (6). Using the techniques presented in previous section, we numerically create a control U such that system (6) follows auxiliary system (8)–(9). Finding U, we get the currents I_i, $i = 1,2,3$. Its only necessity to prove that the linear system of minimization problem (3) is locally controllable. Considering $\lambda = (\alpha,\beta,\gamma)$, the corresponding adjoint system takes the form:

$$\dot{\overline{\Omega}}^* = A_1\overline{\Omega}^* + A_2\overline{\lambda}^*, \qquad \dot{\overline{\lambda}}^* = A_3\overline{\Omega}^* + A_4\overline{\lambda}^*, \tag{16}$$

where $\overline{\lambda}^*, \overline{\Omega}^* \in R^3$, $A_1 = (\nabla_\Omega \Phi)^*$, $A_2 = \Upsilon^*$, $A_3 = (\nabla_\lambda \Psi)^*$, $A_4 = (\nabla_\lambda \Upsilon \cdot \widehat{\Omega})^*$,

$$\Phi = \begin{pmatrix} \dfrac{J_3 - J_2}{J_1}\widehat{\Omega}_2\widehat{\Omega}_3 \\ \dfrac{J_1 - J_3}{J_2}\widehat{\Omega}_1\widehat{\Omega}_3 \\ \dfrac{J_2 - J_1}{J_3}\widehat{\Omega}_1\widehat{\Omega}_2 \end{pmatrix},$$

$$\Upsilon = -\begin{pmatrix} \dfrac{\sin\widehat{\beta}}{\cos\widehat{\gamma}} & \dfrac{\cos\widehat{\beta}}{\cos\widehat{\gamma}} & 0 \\ \sin\widehat{\beta}\tan\widehat{\gamma} & \cos\widehat{\beta}\tan\widehat{\gamma} & 1 \\ \cos\widehat{\beta} & -\sin\widehat{\beta} & 0 \end{pmatrix},$$

$$\Psi = -3\omega^2 \begin{pmatrix} \dfrac{J_3 - J_2}{J_1}\cos\widehat{\alpha}\cos\widehat{\gamma} \\ \quad\times\left(\sin\widehat{\alpha}\sin\widehat{\beta} + \cos\widehat{\alpha}\cos\widehat{\beta}\sin\widehat{\gamma}\right) \\ \dfrac{J_1 - J_3}{J_2}\cos\widehat{\alpha}\cos\widehat{\gamma} \\ \quad\times\left(\cos\widehat{\alpha}\sin\widehat{\beta}\sin\widehat{\gamma} - \sin\widehat{\alpha}\cos\widehat{\beta}\right) \\ \dfrac{J_2 - J_1}{J_3}\left(\cos\widehat{\alpha}\sin\widehat{\beta}\sin\widehat{\gamma} - \sin\widehat{\alpha}\cos\widehat{\beta}\right) \\ \quad\times\left(\sin\widehat{\alpha}\sin\widehat{\beta} + \cos\widehat{\alpha}\cos\widehat{\beta}\sin\widehat{\gamma}\right) \end{pmatrix}. \tag{17}$$

If $\overline{\Omega}^* \equiv 0$, then the first equation of the adjoint system implies $\overline{\lambda}^* \equiv 0$, because $\det A_2 \neq 0$. The orthogonality condition from Theorem 1 can be written in two equivalent forms $\overline{\Omega}^* \times F = 0$ and $\overline{\Omega}^* = \xi F$. Differentiating the latter, we obtain $\dot{\overline{\Omega}}^* = \dot{\xi}F + \xi\dot{F}$. Substituting this for $\dot{\overline{\Omega}}^*$ in the first equation of the adjoint system, we have $\overline{\lambda}^* = A_2^{-1}(\dot{\xi}F + \xi(\dot{F} - A_1F))$. Differentiating the first equation of the adjoint system and substituting the expression for $\overline{\lambda}^*$ obtained above, we get $\ddot{\overline{\Omega}}^* = (\dot{A}_2 + A_2A_4)A_2^{-1}(\dot{\xi}F + \xi(\dot{F} - A_1F)) + \xi(A_2A_3F + \dot{A}_1F + A_1\dot{F}) + \dot{\xi}A_1F$. Differentiating $\overline{\Omega}^* \times F = 0$ twice, we get $\ddot{\overline{\Omega}}^* \times F + 2\dot{\overline{\Omega}}^* \times \dot{F} + \overline{\Omega}^* \times \ddot{F} = 0$. Substituting the above formulas for $\overline{\Omega}^*$ and its derivatives, we obtain $\dot{\xi}v_1 + \xi v_2 = 0$, where

$$\begin{aligned} v_1 &= \left((\dot{A}_2 + A_2A_4)A_2^{-1}F + A_1F\right) \times F + 2F \times \dot{F}, \\ v_2 &= \left((\dot{A}_2 + A_2A_4)A_2^{-1}(\dot{F} - A_1F) + (A_2A_3 + \dot{A}_1)F + A_1\dot{F}\right) \\ &\quad \times F + F \times \ddot{F} \end{aligned} \tag{18}$$

are real analytic functions. From Theorem 1 we see that to prove the local controllability of the linearized system of (6) it suffices to show that $\overline{\Omega}^* \equiv 0$ on any interval of time. This happens, for example, if the vectors v_1 and v_2 are linearly independent. Thus we have the following result.

Lemma 4. *If $v_1 \times v_2 \neq 0$ for almost all $\nu \in [0, 2\pi]$, then the linearized system of* (6) *is controllable.*

Assume that the satellite circular orbit is neither polar nor equatorial. In this case, with the help of the symbolic computation system Mathematica, we prove that the conditions of Lemma 4 are satisfied for $J_1 = J_2 = J_3$ and $\widehat{\Omega}$ is sufficiently small and for $J_3 < J_1 \leq J_2$ and $(\widehat{\Omega}, \widehat{\lambda})$ in a neighborhood of the stable equilibrium position. See the two following examples to each one of these cases.

Example 5. Consider $J_1 = J_2 = J_3 = 1$, $\widehat{\alpha} = \widehat{\beta} = \widehat{\gamma} = 0$ and $\widehat{\Omega} \equiv 0$. Using the symbolic computation system Mathematica we obtain

$$\begin{aligned} v_1 &= \left(v_1^{(1)}, v_1^{(2)}, v_1^{(3)}\right) \\ &= \left(3\omega^2\sin 2\tau\sin\nu, 3\omega^2\sin 2\nu\sin^2\tau, 3\omega^2\sin 2\tau\cos\nu\right), \\ v_2 &= \left(v_2^{(1)}, v_2^{(2)}, v_2^{(3)}\right) \\ &= \left(-3\omega\sin 2\tau\cos\nu, 6\omega\sin^2\tau\left(\cos^2\nu + 2\sin^2\nu\right),\right. \\ &\qquad \left. 3\omega\sin 2\tau\sin\nu\right). \end{aligned} \tag{19}$$

As

$$\begin{aligned} \det\begin{pmatrix} v_1^{(1)} & v_2^{(1)} \\ v_1^{(3)} & v_2^{(3)} \end{pmatrix} &= 9\omega^3\sin^2 2\tau \cdot \det\begin{pmatrix} \sin\nu & -\cos\nu \\ \cos\nu & \sin\nu \end{pmatrix} \\ &= 9\omega^3\sin^2 2\tau \neq 0, \end{aligned} \tag{20}$$

for all $\tau \in]0, \pi/2[$, then the vectors v_1 and v_2 are linearly independent. This proves the conditions of Lemma 4.

Example 6. Now, consider any $J_3 < J_1 < J_2$, $\widehat{\alpha} = \widehat{\beta} = \widehat{\gamma} = 0$ and $\widehat{\Omega} = (0, \omega, 0)$. Using the symbolic computation system Mathematica we obtain $v_1 \times v_2 = (a, b, c)$, where

$$a = \begin{pmatrix} \dfrac{a_1\cos\tau\sin^3\tau}{J_1^2J_2J_3^2}\left(\dfrac{3}{4}(5J_1 - J_2 + J_3)J_3 + J_1(J_1 - J_2 + 2J_3)\right) \\ -\dfrac{a_2\sin^2\tau\sin 2\tau}{4J_1J_3^2} \\ \dfrac{a_1\cos\tau\sin^3\tau}{J_1^2J_2J_3^2}\left(\dfrac{1}{4}(5J_1 - J_2 + J_3)J_3 - J_1(J_1 - J_2 + 2J_3)\right) \\ +\dfrac{a_2\sin^2\tau\sin 2\tau}{4J_1J_3^2} \end{pmatrix}^* \cdot \begin{pmatrix} \cos\nu \\ \cos 3\nu \end{pmatrix},$$

$$b = \begin{pmatrix} \dfrac{b_1\cos^2\tau\sin^2\tau}{J_1^2J_2J_3} - \dfrac{b_2\sin^2 2\tau}{2J_1J_2J_3^2} \\ \dfrac{b_1\cos^2\tau\sin^2\tau}{J_1^2J_2J_3} + \dfrac{b_2\sin^2 2\tau}{2J_1J_2J_3^2} \end{pmatrix}^* \cdot \begin{pmatrix} 1 \\ \cos 2\nu \end{pmatrix},$$

$$c = \begin{pmatrix} \frac{c_1 \cos\tau \sin^3\tau}{J_1^2 J_2 J_3^2}\left(\frac{1}{4}(5J_1 - J_2 + J_3)J_3 + 3J_1(J_1 - J_2 + 2J_3)\right) \\ -\frac{c_2 \cos\tau \sin^3\tau}{4J_1^2 J_3} \\ \frac{c_1 \cos\tau \sin^3\tau}{J_1^2 J_2 J_3^2}\left(\frac{1}{4}(5J_1 - J_2 + J_3)J_3 - J_1(J_1 - J_2 + 2J_3)\right) \\ -\frac{c_2 \cos\tau \sin^3\tau}{4J_1^2 J_3} \end{pmatrix}^* \cdot \begin{pmatrix} \sin\nu \\ \sin 3\nu \end{pmatrix}, \quad (21)$$

with

$$\begin{aligned} a_1 &= -(J_1^2(2J_2 + 3J_3) + 4J_2(-J_2 + J_3)J_3 \\ &\quad + J_1(-2J_2^2 + 3J_2J_3 - 3J_3^2))\omega^3, \\ a_2 &= -(J_1 - J_2 + 2J_3)(2J_1^2 + 7J_3(-J_2 + J_3) \\ &\quad + J_1(-2J_2 + 3J_3))\omega^3, \\ b_1 &= \frac{1}{2}(-5J_1 + J_2 - J_3)(J_1^2(2J_2 + 3J_3) + 4J_2(-J_2 + J_3)J_3 \\ &\quad + J_1(-2J_2^2 + 3J_2J_3 - 3J_3^2))\omega^3, \\ b_2 &= \frac{1}{2}(J_1 - J_2 + 2J_3)(2J_1^2(J_2 + 3J_3) + J_2(-J_2 + J_3)J_3 \\ &\quad + J_1(-2J_2^2 + 3J_2J_3 - 6J_3^2))\omega^3, \\ c_1 &= (J_2(-J_2 + J_3)J_3 + 2J_1^2(J_2 + 3J_3) \\ &\quad + J_1(-2J_2^2 + 3J_2J_3 - 6J_3^2))\omega^3, \\ c_2 &= (-5J_1 + J_2 - J_3)(2J_1^2 + 7J_3(-J_2 + J_3) \\ &\quad + J_1(-2J_2 + 3J_3))\omega^3. \end{aligned} \quad (22)$$

Suppose that $b \equiv 0$. As

$$\det\begin{pmatrix} \frac{\cos^2\tau\sin^2\tau}{J_1^2 J_2 J_3} & -\frac{\sin^2 2\tau}{2J_1 J_2 J_3^2} \\ \frac{\cos^2\tau\sin^2\tau}{J_1^2 J_2 J_3} & \frac{\sin^2 2\tau}{2J_1 J_2 J_3^2} \end{pmatrix} = \frac{\cos^2\tau\sin^2\tau\sin^2 2\tau}{J_1^3 J_2^2 J_3^3} \neq 0, \quad (23)$$

for all $\tau \in]0, \pi/2[$, then we have $b_1 = b_2 = 0$. From triangular inequality for moments of inertia, we have $J_2 < J_1 + J_3$. Thus $b_1 = 0$ is equivalent to

$$2J_1J_2(J_1 - J_2) = -3J_1J_3(J_1 - J_3) - 3J_1J_2J_3 + 4J_2J_3(J_2 - J_3), \quad (24)$$

and $b_2 = 0$ is equivalent to

$$6J_1J_3(J_1 - J_3) + J_2J_3(J_3 - J_2) + 3J_1J_2J_3 = -2J_1J_2(J_1 - J_2). \quad (25)$$

From (24) and (25) we get

$$3J_1J_3(J_1 - J_3) = -3J_2J_3(J_2 - J_3). \quad (26)$$

Dividing both sides of (26) by $J_1J_2J_3$, we have

$$3\left(\frac{J_1 - J_3}{J_2}\right) = -3\left(\frac{J_2 - J_3}{J_1}\right), \quad (27)$$

that is a contradiction. Thus we have $b \equiv 0$ to a finite number of points ν. This proves the conditions of Lemma 4.

4. Stabilization Algorithm and Numerical Simulation

Consider the following stabilization algorithm based on the previous results and a numerical simulation.

4.1. Stabilization Algorithm. The stabilization algorithm presented (see Algorithm 1) is a simple illustration of the previous method that we can implement to generate a sequence of controls U defined on the intervals $[t_0 + k\varepsilon, t_0 + (k+1)\varepsilon]$, $k = 0, 1$, and so on. This sequence of controls can be used in a real situation in real time.

4.2. Numerical Simulation. To test the algorithm we consider satellite (6) and auxiliary system (8)–(9) having the same initial position in $t_0 = 0 : \Omega(0) = \widehat{\Omega}(0) = (0,0,0)$ and $B(0) = \widehat{B}(0)$, where $(\alpha, \beta, \gamma) = (\pi/3, \pi/12, -\pi/9)$ and consider the following parameters: the orbital inclination $\tau = \pi/3$, the tensor of inertia $J = \mathrm{diag}(1.7, 1.8, 1.4)$, the initial position of the mass center in the orbit $\varphi_0 = -\pi/2$, $\theta = 1$, $k_1 = 0.1$, $k_2 = 0.2$, $k_3 = 0.3$, $\varepsilon = 6$, $n = 10$ and $m = 3$.

Implementing an ε-strategy in the form of three-step reachability set construction algorithm, that is, considering $m = 3$ in the above algorithm, we obtain the results of the numerical simulation that are shown in Figures 1 and 2. This ε-strategy consists in the following. First, we divide the intervals of time $[t_0 + k\varepsilon, t_0 + (k+1)\varepsilon]$, $k = 0, 1$, and so on, on three subsets with the same size $[t_0 + (k + (j-1)/3)\varepsilon, t_0 + (k + j/3)\varepsilon]$, $j = 1, 2, 3$, and using the Euler's formula, we obtain approximations to $(\overline{\Omega}, \overline{B})_{3k+j} = (\overline{\Omega}, \overline{B})(t_0 + (k + j/3)\varepsilon)$ which depend of piecewise constant controls $U_{3k+j} \in [t_0 + (k + (j-1)/3)\varepsilon, t_0 + (k + j/3)\varepsilon]$. Finally, using an numerical algorithm to calculate the minimum of $|((\overline{\Omega}, \overline{B})_{3k+1}, (\overline{\Omega}, \overline{B})_{3k+2}, (\overline{\Omega}, \overline{B})_{3k+3})|$, we obtain these piecewise constant controls. Then we have U and therefore we find the control laws I_i, $i = 1, 2, 3$, that drive system (6) to the equilibrium position. Figure 1 represents the norm of the difference between the trajectory of satellite (6) and the trajectory of auxiliary system (8)–(9), and Figure 2 represents the norm of the difference between the trajectory of satellite (6) and the equilibrium position.

5. Conclusions

The problem of active three-axis magnetic attitude stabilization of a low Earth orbit satellite was studied. In this paper it was proposed an alternative approach to the existing

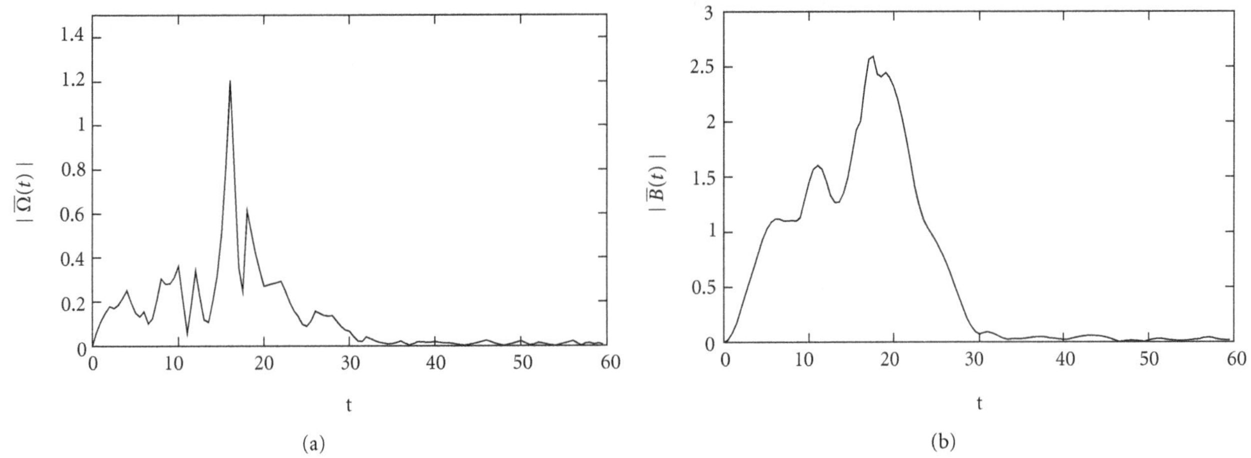

FIGURE 1: Norm of the difference between the satellite trajectory and the guide along the time (sec).

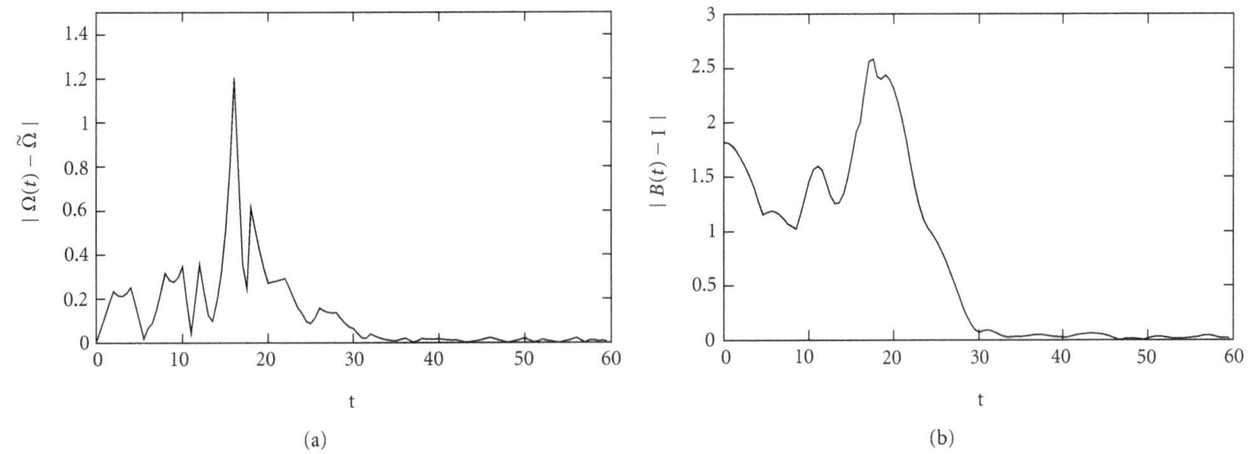

FIGURE 2: Norm of the difference between the satellite trajectory and the equilibrium position along the time (sec).

START;
CALCULATION of $(\overline{\Omega}, \overline{B})(t_0)$;
FOR $k = 0$ **TO** $n - 1$
    CALCULATION of:
    – the guide trajectory $(\widehat{\Omega}, \widehat{B})$ and the guide stabilizer W on $[t_0 + k\varepsilon, t_0 + (k+1)\varepsilon]$;
    – the satellite stabilizer
        $U = (U(t_0 + k\varepsilon), U(t_0 + (k + 1/m)\varepsilon), \ldots, U(t_0 + (k + (m-1)/m)\varepsilon))$
    using a numerical algorithm to MINIMIZE the $\max_{t \in [t_0 + k\varepsilon, t_0 + (k+1)\varepsilon]} |(\overline{\Omega}, \overline{B})(t)|$ in m STEPS, where

$$(\dot{\overline{\Omega}}, \dot{\overline{B}})(t) = A(t)(\overline{\Omega}, \overline{B})(t) + (J^{-1}(U \times F(t, \widehat{\Omega}, \widehat{B}))\ \ 0)^* - (J^{-1}W\ \ 0)^*,$$
$$A(t) = \nabla_{(\Omega, B)} f(t, \widehat{\Omega}, \widehat{B}),$$
$$f(t, \Omega, B) = \begin{pmatrix} J^{-1}(-\Omega \times J\Omega + 3\omega^2 b_3 \times J b_3) \\ b_i \times (\Omega - \omega b_2) \end{pmatrix}, \qquad i = 1, 2, 3;$$

    FOR $j = 1$ **TO** m
        CALCULATION of the satellite trajectory (Ω, B) on
            $[t_0 + (k + (j-1)/m)\varepsilon, t_0 + (k + j/m)\varepsilon]$
        using the stabilizer $U(t_0 + (k + (j-1)/m)\varepsilon)$;
    END
END

ALGORITHM 1

methods. The method used in the stabilization of satellite is based on guidance control, a more general control procedure. Conditions of controllability were also obtained in this study. According to these conditions and using ε-strategies, the stabilizing control was constructed as a function of time defined in a small time interval and not as a feedback. The simulation results obtained using the algorithm proposed prove the feasibility of the approach presented here. This simple approach using local stabilization methods solves the complex and difficult global stabilization problem.

Nomenclature

Throughout this paper we will use the following notations:

$M(n, m)$: The set of n-by-m real matrices
R: The set of real numbers
R^n: The usual n-dimensional space of vectors $x = (x_1, \dots, x_n)$, where $x_i \in R$, $i = \overline{1, n}$
$\nabla_x f$: The Jacobian matrix of the function f
$\langle x, y \rangle$: The inner product of two vectors x and y in R^n
$x \times y$: The vector product of two 3-dimensional vectors x and y
$|x|$: The Euclidean norm of a vector $x \in R^n$
$|A|$: The matrix ∞-Norm of a real matrix $A = (a_{ij})$, $i = \overline{1, m}$, $j = \overline{1, n}$
B_n: The unit ball in $R^n : B_n = \{x \in R^n : |x| \le 1\}$
I: The identity matrix
A^*: The transposed matrix of a real matrix A
A^{-1}: The inverse matrix of a real matrix A
$\operatorname{int} H$: The interior of a set $H \subset R^n$.

Acknowledgments

This work was supported by the Portuguese Foundation for Science and Technology (FCT), the Portuguese Operational Programme for Competitiveness Factors (COMPETE), the Portuguese Strategic Reference Framework (QREN), and the European Regional Development Fund (FEDER).

References

[1] J. T.-Y. Wen and K. Kreutz-Delgado, "The attitude control problem," *IEEE Transactions on Automatic Control*, vol. 36, no. 10, pp. 1148–1162, 1991.

[2] O.-E. Fjellstad and T. I. Fossen, "Comments on "The attitude control problem"," *IEEE Transactions on Automatic Control*, vol. 39, no. 3, pp. 699–700, 1994.

[3] F. Caccavale and L. Villani, "Output feedback control for attitude tracking," *Systems and Control Letters*, vol. 38, no. 2, pp. 91–98, 1999.

[4] M. R. Akella, "Rigid body attitude tracking without angular velocity feedback," *Systems and Control Letters*, vol. 42, no. 4, pp. 321–326, 2001.

[5] C. I. Byrnes and A. Isidori, "On the attitude stabilization of rigid spacecraft," *Automatica*, vol. 27, no. 1, pp. 87–95, 1991.

[6] P. Morin, C. Samson, J.-B. Pomet, and Z.-P. Jiang, "Time-varying feedback stabilization of the attitude of a rigid spacecraft with two controls," *Systems and Control Letters*, vol. 25, no. 5, pp. 375–385, 1995.

[7] H. Iida and K. Ninomiya, "A new approach to magnetic angular momentum management for large scientific satellites," *NEC Research and Development*, vol. 37, no. 1, pp. 60–77, 1996.

[8] L. D. D. Ferreira and J. J. Cruz, "Attitude and spin rate control of a spinning satellite using geomagnetic field," *Journal of Guidance, Control, and Dynamics*, vol. 14, no. 1, pp. 216–218, 1991.

[9] M. E. Pittelkau, "Optimal periodic control for spacecraft pointing and attitude determination," *Journal of Guidance, Control, and Dynamics*, vol. 16, no. 6, pp. 1078–1084, 1993.

[10] K. T. Alfriend, "Magnetic attitude control system for dual-spin satellites," *AIAA Journal*, vol. 13, no. 6, pp. 817–822, 1975.

[11] J. S. White, F. H. Shigemoto, and K. Bourquin, "Satellite attitude control utilizing the Earth's magnetic field," Tech. Rep. D-1068, NASA, Washington, DC, USA, 1961.

[12] A. C. Stickler and K. T. Alfriend, "Elementary magnetic attitude control system," *Journal of Spacecraft and Rockets*, vol. 13, no. 5, pp. 282–287, 1976.

[13] F. Martel, P. K. Pal, and M. L. Psiaki, "Active magnetic control system for gravity gradient stabilized spacecraft," in *Proceedings of the 2nd Annual AIAA/USU Conference on Small Satellites*, Logan, Utah, USA, September 1988.

[14] K. L. Musser and W. L. Ebert, "Autonomous spacecraft attitude control using magnetic torquing only," in *Proceedings of the Flight Mechanics/Estimation Theory Symposium*, pp. 23–38, NASA, May 1989.

[15] R. Wiśniewski, "Nonlinear control for satellite detumbling based on magnetic torquing," in *Proceedings of the 22nd (JSDE) Joint Service Data Exchange for Guidance, Navigation and Control*, Scottsdale, Ariz, USA, October-November 1994.

[16] R. Wiśniewski and M. Blanke, "Three-axis satellite attitude control based on magnetic torquing," in *Proceedings of the 13th IFAC World Congress*, San Francisco, Calif, USA, June 1996.

[17] E. Silani and M. Lovera, "Magnetic spacecraft attitude control: a survey and some new results," *Control Engineering Practice*, vol. 13, no. 3, pp. 357–371, 2005.

[18] R. Wiśniewski and M. Blanke, "Fully magnetic attitude control for spacecraft subject to gravity gradient," *Automatica*, vol. 35, no. 7, pp. 1201–1214, 1999.

[19] R. Wiśniewski, "Linear time-varying approach to satellite attitude control using only electromagnetic actuation," *Journal of Guidance, Control, and Dynamics*, vol. 23, no. 4, pp. 640–647, 2000.

[20] M. L. Psiaki, "Magnetic torquer attitude control via asymptotic periodic linear quadratic regulation," *Journal of Guidance, Control, and Dynamics*, vol. 24, no. 2, pp. 386–394, 2001.

[21] R. Wiśniewski and F. L. Markley, "Optimal magnetic attitude control," in *Proceedings of the 14th IFAC World Congress*, Beijing, China, July 1999.

[22] M. Lovera, E. De Marchi, and S. Bittanti, "Periodic attitude control techniques for small satellites with magnetic actuators," *IEEE Transactions on Control Systems Technology*, vol. 10, no. 1, pp. 90–95, 2002.

[23] H. Yan, I. M. Ross, and K. T. Alfriend, "Pseudospectral feedback control for three-axis magnetic attitude stabilization in elliptic orbits," *Journal of Guidance, Control, and Dynamics*, vol. 30, no. 4, pp. 1107–1115, 2007.

[24] M. Corno and M. Lovera, "Spacecraft attitude dynamics and control in the presence of large magnetic residuals," *Control Engineering Practice*, vol. 17, no. 4, pp. 456–468, 2009.

[25] T. Pulecchi, M. Lovera, and A. Varga, "Optimal discrete-time design of three-axis magnetic attitude control laws," *IEEE Transactions on Control Systems Technology*, vol. 18, no. 3, pp. 714–722, 2010.

[26] M. Abdelrahman, I. Chang, and S.-Y. Park, "Magnetic torque attitude control of a satellite using the state-dependent Riccati equation technique," *International Journal of Non-Linear Mechanics*, vol. 46, no. 5, pp. 758–771, 2011.

[27] C. Arduini and P. Baiocco, "Active magnetic damping attitude control for gravity gradient stabilized spacecraft," *Journal of Guidance, Control, and Dynamics*, vol. 20, no. 1, pp. 117–122, 1997.

[28] C. J. Damaren, "Comments on "Fully magnetic attitude control for spacecraft subject to gravity gradient"," *Automatica*, vol. 38, no. 12, p. 2189, 2002.

[29] M. Lovera and A. Astolfi, "Global attitude regulation using magnetic control," in *Proceedings of the 40th IEEE Conference on Decision and Control (CDC '01)*, pp. 4604–4609, Orlando, FL, USA, December 2001.

[30] M. Lovera and A. Astolfi, "Spacecraft attitude control using magnetic actuators," *Automatica*, vol. 40, no. 8, pp. 1405–1414, 2004.

[31] M. Lovera and A. Astolfi, "Global magnetic attitude control of inertially pointing spacecraft," *Journal of Guidance, Control, and Dynamics*, vol. 28, no. 5, pp. 1065–1067, 2005.

[32] M. Lovera and A. Astolfi, "Global magnetic attitude control of spacecraft in the presence of gravity gradient," *IEEE Transactions on Aerospace and Electronic Systems*, vol. 42, no. 3, pp. 796–805, 2006.

[33] V. Bushenkov and G. Smirnov, *Stabilization Problems with Constraints: Analysis and Computational Aspects*, Gordon and Breach Science Publishers, Amsterdam, The Netherlands, 1997.

[34] L. S. Pontryagin, "Linear differential games. I," *Soviet Mathematics. Doklady*, vol. 8, pp. 769–771, 1967.

[35] R. N. Izmailov, "The peak effect in stationary linear systems with scalar inputs and outputs," *Automation and Remote Control*, vol. 48, no. 8, pp. 1018–1024, 1987.

[36] H. J. Sussmann and P. V. Kokotovic, "The peaking phenomenon and the global stabilization of nonlinear systems," *IEEE Transactions on Automatic Control*, vol. 36, no. 4, pp. 424–440, 1991.

[37] G. Smirnov, V. Bushenkov, and F. Miranda, "Advances on the transient growth quantification in linear control systems," *International Journal of Applied Mathematics and Statistics*, vol. 14, no. J09, pp. 82–92, 2009.

[38] S. P. Bhat, "Controllability of nonlinear time-varying systems: applications to spacecraft attitude control using magnetic actuation," *IEEE Transactions on Automatic Control*, vol. 50, no. 11, pp. 1725–1735, 2005.

[39] G. Smirnov, *Introduction to the Theory of Differential Inclusions*, American Mathematical Society, Providence, RI, USA, 2002.

[40] I. V. Burkov, "Asymptotic stabilization of the position of a rigid body with fixed point without velocity measurements," *Systems and Control Letters*, vol. 25, no. 3, pp. 205–209, 1995.

[41] S. Sastry, *Nonlinear Systems: Analysis, Stability, and Control, Interdisciplinary Applied Mathematics, Vol. 10*, Springer, Berlin, Germany, 1999.

11

Development and Validation of a New Boundary Condition for Intake Analysis with Distortion

Foad Mehdi Zadeh, Jean-Yves Trépanier, and Eddy Petro

Department of Mechanichal Engineering, École Polytechnique de Montréal, 2500, Chemin de Polytechnique, Montréal, QC, Canada H3T 1J4

Correspondence should be addressed to Foad Mehdi Zadeh; foad.mehdi-zadeh@polymtl.ca

Academic Editor: Mark Price

The design of an intake for a gas turbine engine involves CFD-based investigation and experimental assessment in an intake test rig. In both cases, the engine is represented by a mass flux sink, usually positioned a few fan radii aft of the real fan face. In general, this approach is sufficient to analyze intake geometry for low distortion at the fan face, because in this case the interaction of the fan with the inlet flow can be neglected. Where there are higher levels of distortion at the fan face, the interaction could become more significant and a different approach would be preferable. One alternative that takes into account the interaction in such cases includes the fan in the analysis of the intake, using either a steady or unsteady flow model approach. However, this solution is expensive and too computationally intensive to be useful in design mode. The solution proposed in this paper is to implement a new boundary condition at the fan face which better represents the interaction of the fan with the flow in the air intake in the presence of distortion. This boundary condition includes a simplified fan model and a coupling strategy applied between the fan and the inlet. The results obtained with this new boundary condition are compared to full 3D unsteady CFD simulations that include the fan.

1. Introduction

The distortion is defined as a nonuniformity in flow properties as a function of space and time. The spatial flow distortion is usually divided into two types: radial (r), and circumferential (θ). In axial engines, the nonuniform distribution of total pressure associated with an axial velocity deficit is the type of distortion most frequently encountered. In the presence of distortion, the work done by the fan on the flow is nonuniform, and, as a result, the fan influences the distorted upstream flow in an attempt to suppress that distortion. The mass flux sink approach for the boundary condition neglects the interaction between the inlet and the fan, which could be important in operating conditions where the inlet design can result in high levels of distortion or flow recirculation at the fan face. This interaction can be captured using a 3D unsteady approach, where the full 360° fan is included in the inlet analysis. However, this requires significant computational resources. The approach proposed here is to impose a boundary condition on the engine fan face that better represents the interaction of the fan with the distorted flow. This boundary condition consists of a nonuniform static pressure plane which takes into account the effect of the nonuniformity of the work done by the fan blades. For this purpose, the fan is modeled with an Actuator Disc, and the Actuator Disc analysis is coupled with analysis of the flow in the inlet through the theory of parallel compressors. With this process, an improved boundary condition is defined on the fan face under distortion.

This paper is organized as follows: in Section 2, the pertinent literature on the interaction of fans with distortion is reviewed. Then, in Section 3, the selected simplified fan model is presented, which is a variant of the Actuator Disc model. In Section 4, the coupling methodology used to iteratively define the boundary condition on the fan face using both the Actuator Disc model and the theory of parallel compressors is described. Validation of the approach on the NASA Rotor 67 case, as described by Fidalgo et al. [1], is also presented in this section. Finally, in Section 5, an application for a nacelle at high incidence is given.

2. Fan Modeling under Distortion

The simplest approach to modeling a fan subjected to distortion forces is the theory of parallel compressors. This theory models a compressor running under circumferential total pressure distortion [2]. Two separate compressors work on the flow, one on the clean flow and one on the distorted flow, without any interaction. Their operational characteristics are the same, and they both exhaust the flow at a single static pressure, which is the same as the pressure obtained without distortion for the same total mass flow. Experimental tests [3] have confirmed this theory, and other researchers have subsequently improved and validated it [4, 5].

With the development of CFD, researchers investigate distortion using this tool using simplified models of rotating machines. Whiteld and Jameson [6] and Dang [7] described propeller/airframe interaction using analytical source terms and Actuator Disc in computational model, respectively. An affordable CFD approach for modeling turbomachinery is the Streamline Curvature method using body forces or energy sources as the effective elements of engine. In 2001, Hsiao et al. [8] implemented the body forces in an in-house code to simulate the flow through the NASA stages 35 and 22. In terms of separation due to the high angle of attack, their results were consistent with experimental tests. In 2006, Hale et al. [9] have used it to model the interaction between the fan and upstream distortion. The application involves an F-16 fighter jet. With good accuracy, their work shows a 6% decrease in the overall performance of the compressor caused by distortion.

In 2007, Yao et al. applied a URANS solver to predict the effect of inlet distortion for a two-stage compressor [10]. Their results show the formation of a static pressure distortion caused by the distortion of total pressure. This nonuniformity of static pressure induces a swirl in the flow upstream of the fan. In 2010, Fidalgo et al. did a URANS study of the aerodynamics of the NASA Rotor 67 subjected to a circumferential total pressure distortion imposed on 120° at entrance [1]. They also showed the presence of an induced swirl flow at the fan face resulting from the capacity of the fan to redistribute the mass flow in an attempt to suppress distortion.

Although the URANS approach provided detailed and very instructive information about these flows, the high CPU requirements of the approach make its use in design mode prohibitively costly. For practical purposes, it would be useful to have a simplified fan model, which could be embedded in the outflow boundary condition located on the fan face, in order to use CFD for inlet design. The main objective of this paper is to develop such a boundary condition to represent the interaction of the fan with the flow in the intake.

According to several of the documents cited above, the work of the fan is not uniform in the case of upstream distortion, and the boundary condition must take this nonuniformity into account. So, our approach will be to devise a simplified fan model and to establish a proper methodology for coupling it with a CFD analysis of the inlet. In the next section, the simplified fan model used in our methodology is presented.

3. A Simplified Fan Model

Of the many aerodynamic fan models available, the Actuator Disc is a fast and cost-effective approach that takes into account the radial distribution of flow properties. It is a mathematical model which considers the fan as a plane comprising an infinite number of blades. This plane induces a discontinuity of tangential velocity.

In 1995, Lewis [11] proposed (1) relating the upstream and downstream axial velocities to their values at the fan boundaries:

$$\begin{aligned} V_z &= V_{z1} + \frac{(\rho_2 V_{z2} - \rho_1 V_{z1})}{\rho_m} \frac{e^{kz/(r_t - r_h)}}{2}, \quad z < 0, \\ V_z &= V_{z2} - \frac{(\rho_2 V_{z2} - \rho_1 V_{z1})}{\rho_m} \frac{e^{-kz/(r_t - r_h)}}{2}, \quad z > 0. \end{aligned} \tag{1}$$

In (1), V_z is the axial velocity component, ρ is the density, ρ_m is the average density of flows that are the farthest upstream and downstream of the fan, r_t and r_h are the tip and hub radii, respectively, k is a constant related to r_t/r_h, and (1) and (2) represent locations that are the farthest upstream and downstream of the fan, respectively. Note that this version of these equations is applicable to compressible flows.

In order to obtain an equation for the radial variation of V_{z2}, a simplified form of the Navier-Stokes equation is assumed for the fluid elements farthest downstream of the fan and is given by

$$-\frac{1}{\rho}\frac{\partial P}{\partial r} = \left(V_r \frac{\partial V_r}{\partial r} - \frac{V_\theta^2}{r}\right). \tag{2}$$

Then, in (2), one can link pressure to enthalpy via the first law of thermodynamics, use the Euler equations for turbomachinery to relate enthalpy to tangential velocity changes, and relate axial and tangential velocities at the trailing edge of the blade through the velocity triangle. After some tedious manipulations, the final equation for the radial variation of V_{z2} is given by (3). More details about this development can be found in [12]. In (3), velocity components at the trailing edge are related to station 2 using (1):

$$\begin{aligned} \frac{dV_{z2}}{dr} = {} & \frac{tg^2\beta_2 (V_{z,te}/V_{z2}) \left((F/\rho_m)(d\rho_2/dr) + (d\rho_2/dr)\left(\rho_2 F/2\rho_m^2\right)\right) V_{z2}}{1 + tg^2\beta_2 (V_{z,te}/V_{z2}) \left(1 - (\rho_2/\rho_m) F\right)} \\ & + \frac{tg^2\beta_2 (V_{z,te}/V_{z,2}) (d\rho_2/dr) \left(\rho_1 F/2\rho_m^2\right) V_{z1}}{1 + tg^2\beta_2 (V_{z,te}/V_{z2}) \left(1 - (\rho_2/\rho_m) F\right)} \end{aligned}$$

$$+\frac{2\omega r\left(V_{z,te}tg\beta_2/V_{z2}r\right)-V_{z,te}tg\beta_2\left(V_{z,te}tg\beta_2/V_{z2}r\right)}{1+tg^2\beta_2\left(V_{z,te}/V_{z2}\right)\left(1-\left(\rho_2/\rho_m\right)F\right)}$$

$$-\frac{V_{z,te}r\left(d\left(tg\beta_2\right)/dr\right)\left(V_{z,te}tg\beta_2/V_{z2}r\right)}{1+tg^2\beta_2\left(V_{z,te}/V_{z2}\right)\left(1-\left(\rho_2/\rho_m\right)F\right)}, \quad (3)$$

with

$$F=\frac{1}{2}e^{-kz/(r_t-r_h)}. \quad (4)$$

In this equation, the density at the downstream location is linked to the upstream density value with an additional equation for the global fan efficiency. Integrating (3) with the constraint of respecting mass conservation enables us to obtain the radial distribution of the axial velocity at the location farthest downstream. Using these values and classical thermodynamic relations, all the other flow properties can be obtained.

With this Actuator Disc model, the flow analysis can be performed to relate the upstream and downstream locations of the fan.

The input data for this model are:

(i) blade trailing edge angle versus radial direction: $\beta_2(r)$,

(ii) fan speed: ω,

(iii) total pressure and temperature at the location farthest upstream: P_t and T_t,

(iv) flow rate: $\dot{m}$,

(v) polytropic efficiency: η.

To validate this model, an incompressible rotor presented by Lewis [13] is used. Figure 1 presents the velocity profiles along the trailing edge and at the farthest location downstream for comparison purposes. All these results are consistent with those presented by Lewis.

4. Fan and Upstream Flow Coupling

4.1. Work and Distortion. A simple analysis can be performed to illustrate the effect of distortion on the fan's work. This work on different stream tubes of a distorted flow is not uniform, mainly owing to different flow rates. To illustrate this effect, velocity triangles can be drawn for both the clean and distorted parts, as in Figure 2, where U is the rotational speed and $V_{\theta 2}$ is the swirl velocity at the exit of the blade.

One can see that a reduced mass flow results in an increase in tangential velocity at the outlet, which leads to increased work. This shifts the local operating point of the fan, and the ratio of total pressure for the distorted streamlines is higher than that for the clean part, as illustrated on the compressor map given in Figure 3.

4.2. Development of an Iterative Method to Define the Boundary Condition. A case study from Fidalgo et al. [1] is used. A sketch of their geometry is given in Figure 4. Their work is aimed at determining the nature of the interaction between NASA Rotor 67 and the flow under distortion. As shown in Figure 4, the air intake includes a cylindrical channel and a round-shaped nose in front of the fan face, identified as LE. In Figure 4, the numbers 1, 2, and 3 identify reference planes on which the flow properties are calculated for comparison purposes. The flow distortion is imposed in a circumferential zone (120°) on the inlet plane as a reduced total pressure.

The proposed boundary condition is placed on the fan face, LE. This plane is then divided into sectors, so as to impose a different boundary condition in each sector. In this study, simulations with 12 uniformly distributed radial sectors, as shown in Figure 5, are used.

4.3. Coupling of the Actuator Disc and CFD Solver. The Actuator Disc model and the intake CFD model are coupled using the theory of parallel compressors. Based on this theory, the average static pressure for all the stream tubes exhausting at the farthest location downstream of the fan has to be equal to the static pressure for the case without distortion. In order to implement this theory, an iterative approach is used, which is summarized as follows.

(i) Run a CFD simulation of the intake by imposing a uniform static pressure plane on the intake's outflow plane (corresponding to the fan face, LE). Adjust the static pressure to match the required total mass flow. For each sector, compute the average flow rate and swirl angle and send these as input to the AD program.

(ii) Run the AD program for every sector independently.

(iii) Calculate the average static pressure profile at the location farthest downstream.

(iv) Compare the average static pressure profile with the static pressure profile obtained with a uniform flow distribution (clean).

(v) Apply an iterative process to eliminate the gap between the two static pressure profiles (average and clean) by varying the static pressure for each sector.

4.4. The Iterative Process. As discussed previously, the main effect of the fan on the distorted upstream flow is to suck the regions of low mass flow more strongly. This nonuniform suction redistributes the flow on the inlet plane, an effect which is obtained by imposing a lower static pressure on the sectors having low mass flow. Our convergence criterion involves a comparison of the average static pressure profile at the farthest downstream location with the same profile

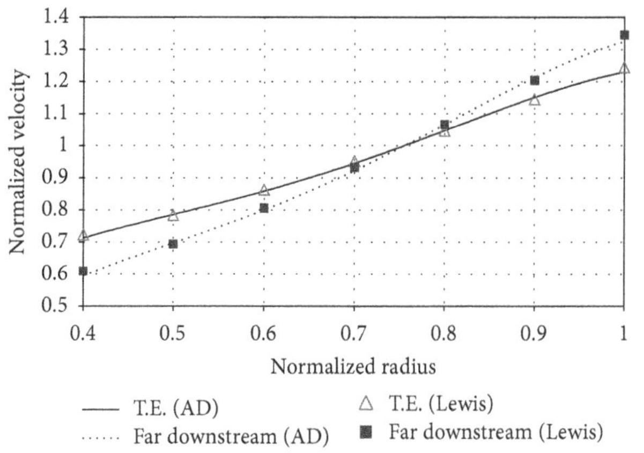

FIGURE 1: Velocity profiles.

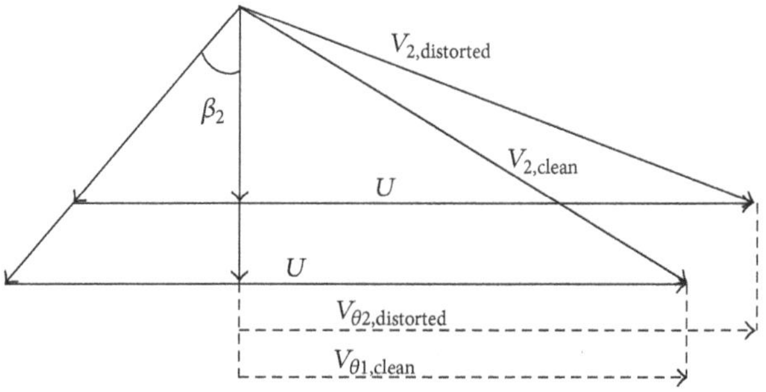

FIGURE 2: Velocity triangles.

without distortion. To achieve this, an iterative process is defined using

$$\frac{P_i^{n+1}}{\overline{P}} = 1 + \alpha^{n+1}\left(\frac{\dot{m}_i^n}{\overline{\dot{m}}} - 1\right). \quad (5)$$

In (5), the value of the static pressure for each sector (i) is redefined as a function of its flow rate at the previous iteration (n) multiplied by a factor α, which is based on the difference between the average static pressure profile and the static pressure profile obtained without distortion. According to (5), the static pressure is reduced in areas with a flow deficit and increased in areas with extra flow. This tends to balance the mass distribution. The parameter value (α) is determined using the secant method, as given by

$$\alpha^{n+1} = \alpha^n - \frac{\Delta P^n}{\Delta P'^n},$$
$$\Delta P'^n = \frac{\Delta P^{n-1} - \Delta P^n}{\alpha^{n-1} - \alpha^n}. \quad (6)$$

The mass flow rate and the static pressure distributions corresponding to the first and last iterations are reproduced in Figure 6. One can see that the iterative process has decreased the pressure in the distorted region, and, as a result, the flow rate has increased in that region.

To determine the adequate mesh size, a grid independence study was done, as presented in Figure 7. RMS

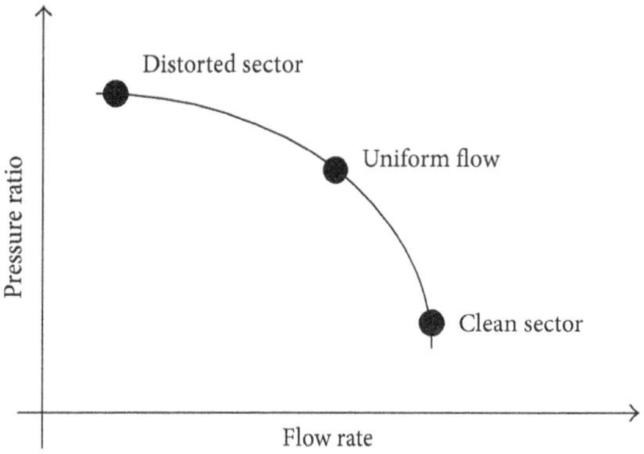

FIGURE 3: Local operation points.

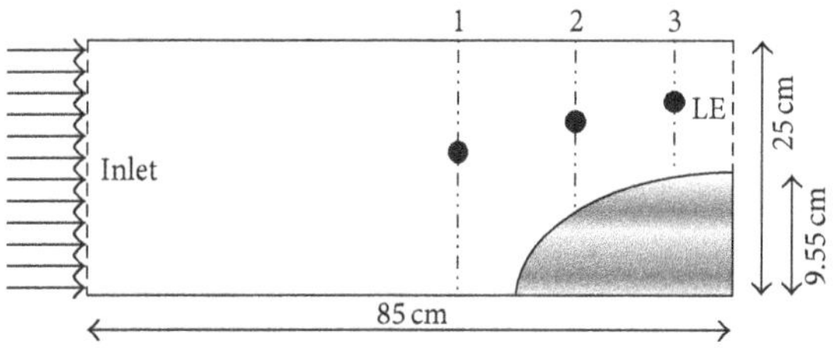

FIGURE 4: Air duct and reference planes.

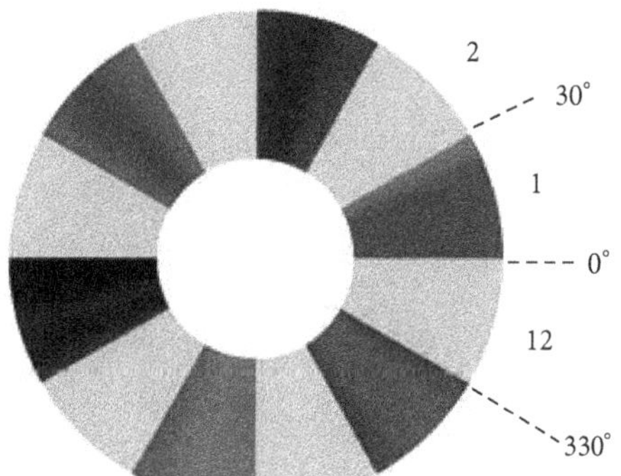

FIGURE 5: Divided sectors.

difference of the normalized variable, ρV_z, was employed as the accuracy criterion. A mesh composed of 78000 nodes is selected to perform the simulations using the standard k-epsilon turbulence model with a scalable wall treatment. As it was shown in Figure 7, above the vicinity of this node number, simulation's accuracy remains in the same order of magnitude with less than 1% of difference. It should be noted that these values are obtained in comparison with a reference grid of 167736 nodes.

4.5. Results, Validations, and Analysis. In this section, our results are compared to the URANS results from Fidalgo et al. [1]. In Figure 8, the local fan pressure ratio computed for the twelve sectors is shown. One can notice the higher pressure ratios for the distorted sectors and note the fact that the orbits are distributed around the fan speed line. However, the amplitudes of the pressure ratio around the

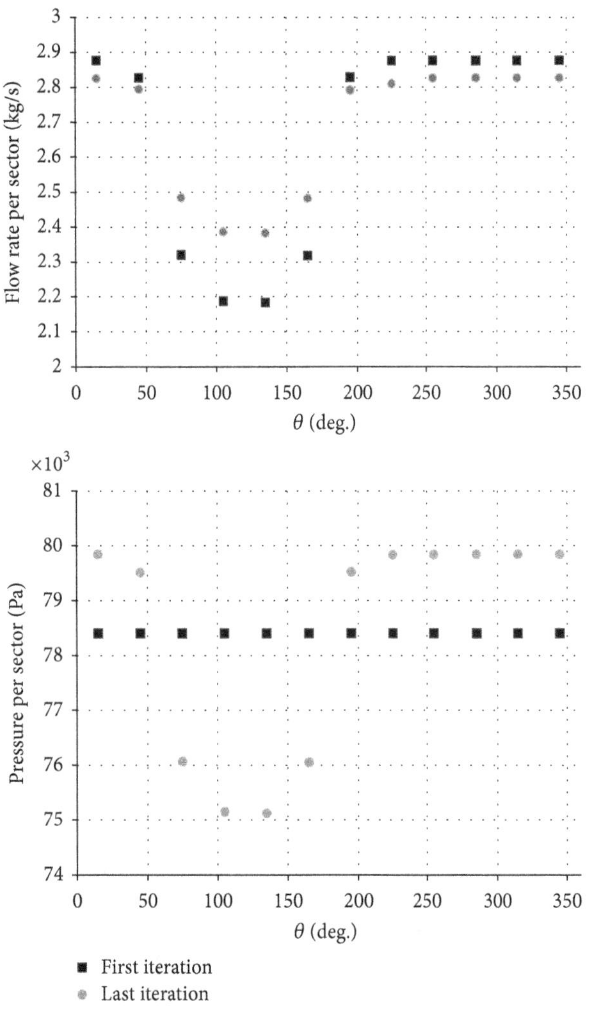

FIGURE 6: Static pressure and flow rate distribution in the first and last iterations.

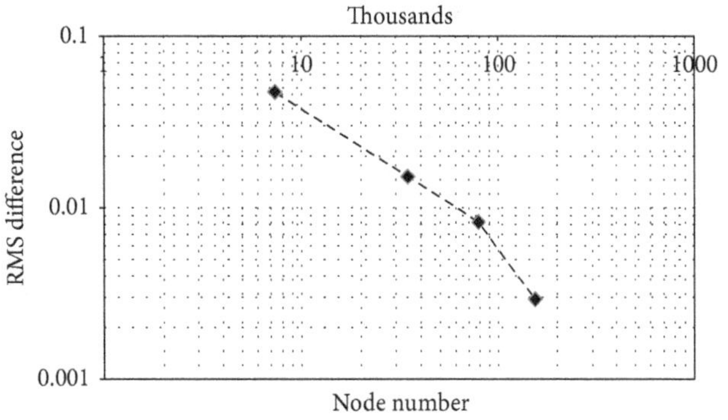

FIGURE 7: Mesh independence study.

orbits are not the same. These differences are caused by the various assumptions used in our AD-based approach, including that of a constant area flow path across the fan and a constant efficiency. Nevertheless, the qualitative behavior of the circumferential pressure ratio distribution is reproduced well, which should enable us to better represent the effect

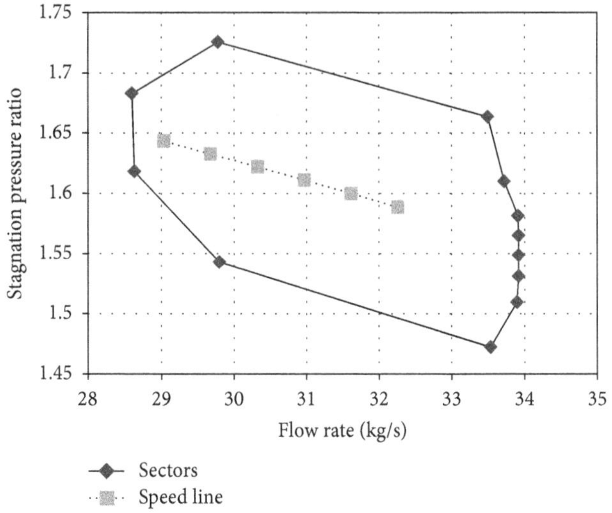

FIGURE 8: Total pressure ratio of clean and distorted sectors and the speed line.

of the fan on the upstream flow. This will be shown in the following.

Figures 9 to 11 show the circumferential distribution of various properties upstream of the fan on three planes, located at $-0.5C$, $-2.5C$, and $-5.5C$, where C is the axial chord length at the mid-point of the blade (see Figure 4).

On these three planes, three properties are compared:

(i) unit mass flow rate, normalized by the same property at the $-5.5C$ plane for a clean flow;

(ii) static pressure, normalized by the average total pressure at the inlet;

(iii) angle between the absolute velocity vector and the axial direction.

Figure 9 presents the distribution of the unit mass flow rate. According to this figure, the fan attempts to suppress the velocity deficit, and the flow velocity becomes more uniform when the flow approaches the fan face. The AD model results show slightly less distortion than those of Fidalgo et al. [1], but the prediction of the distortion suppression by the two models is similar. The same effect can be observed on the pressure field, as shown in Figure 10. One can see that, in the segments under distortion, the static pressure decreases as the flow approaches the fan. This variation is also well captured by our AD-based boundary condition. Finally, the nonuniform static pressure distribution causes secondary flows upstream of the fan. Figure 11 shows the angle between the velocity vector and the axial direction. One can see that the norm of this angle is maximal at the beginning and at the end of the distortion zone and that its amplitude increases as the flow approaches the fan face. For that property, the agreement between our approach and the URANS results is excellent. Furthermore, a comparison with the experimental results, also provided by Fidalgo et al. at the $-2.5C$ plane, also shows excellent agreement.

The conclusion from this validation is that our AD-based approach for the boundary condition at the fan face

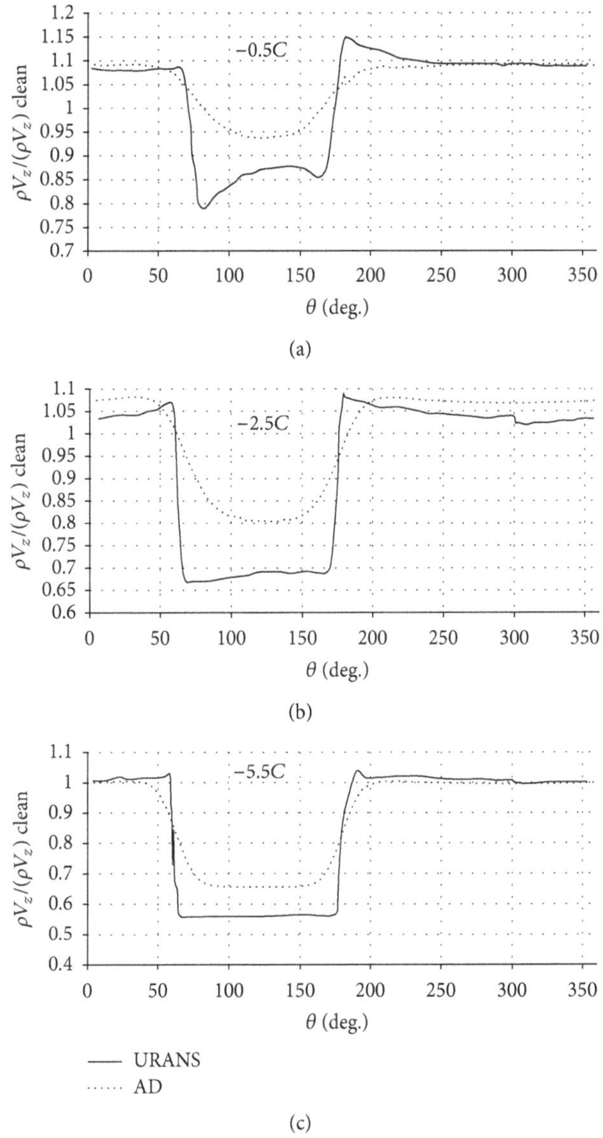

FIGURE 9: Comparison of unit mass flow rate for 3 reference planes.

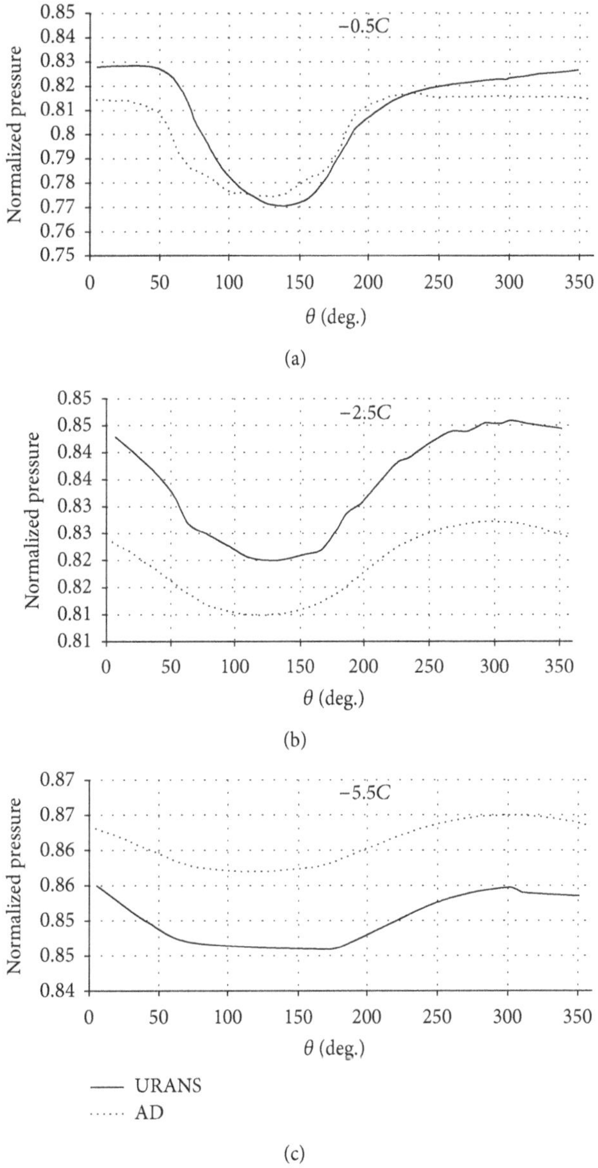

FIGURE 10: Comparison of the static pressure for 3 reference planes.

provides a good representation of the main characteristics of the interaction of the fan with the inlet flow. As such, it constitutes an improvement over the classical mass flux sink approach. The method is now ready to be applied to a more complex case.

5. A Nacelle at High Incidence

A typical situation to which the developed method could be applied is the study of a nacelle at high incidence. In this case, one would like to be able to predict the onset of flow separation in the inlet as the flow angle increases. However, the interaction of the fan and the inlet can influence the flow separation behavior. In this section, the proposed AD approach is applied and compared to various CFD approaches, obtained with different types of fan and inlet interaction modeling.

5.1. The Geometry and Meshing of the Nacelle Case. The nacelle used in our study here was designed to accommodate NASA Rotor 67. It has a throat diameter of 46.1 cm, which accommodates the rotor; see Figure 12. The nacelle is followed by an ogive-shaped surface 1 meter in length representing the jet flow at the engine exhaust. The nacelle is placed in a rectangular cubic computational domain measuring 41 m × 10 m × 20.7 m, as illustrated in Figure 13.

A partial view of the mesh of the nacelle is shown in Figure 14. The mesh contains 1714560 elements and 1774171 nodes. NASA Rotor 67 can be fitted inside this nacelle. The total mesh, including the nacelle plus the full 22 passage Rotor 67, contains 6254269 elements and 6034624 nodes.

5.2. Aerodynamic Modeling. ANSYS CFX-13.0 was used for all the CFD runs. RANS and URANS models are used

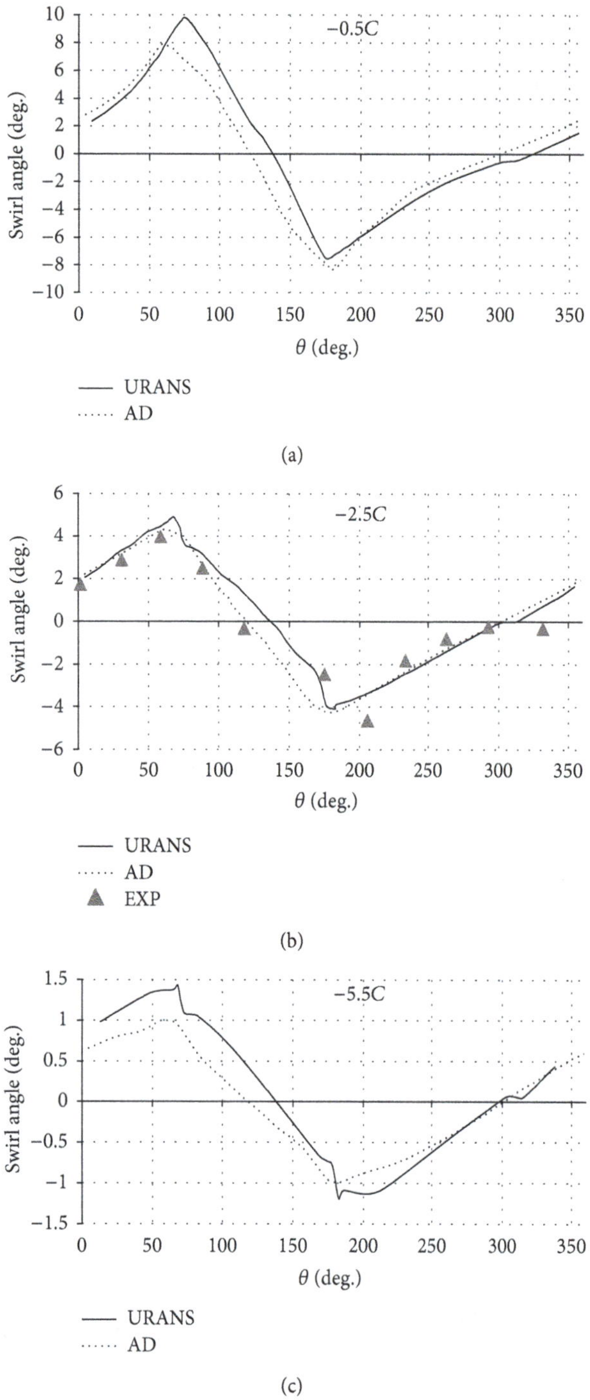

FIGURE 11: Comparison of the swirl angle for 3 reference planes.

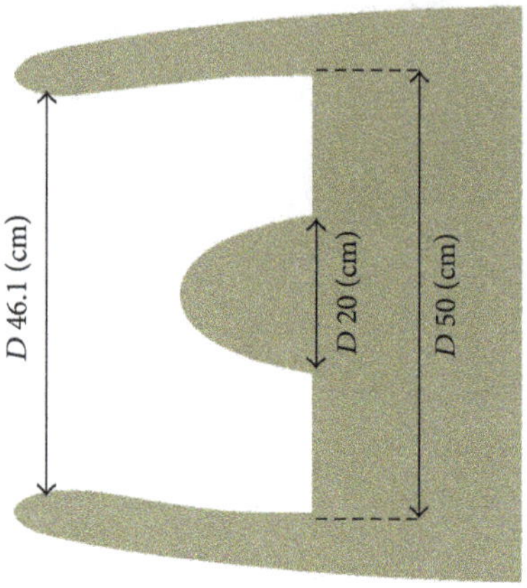

FIGURE 12: Intake geometry.

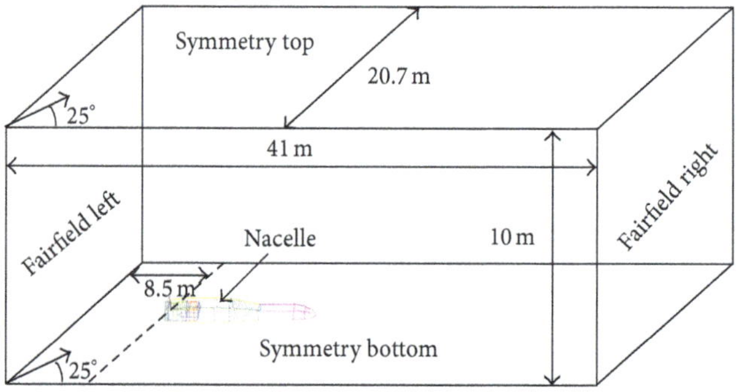

FIGURE 13: Intake field geometry.

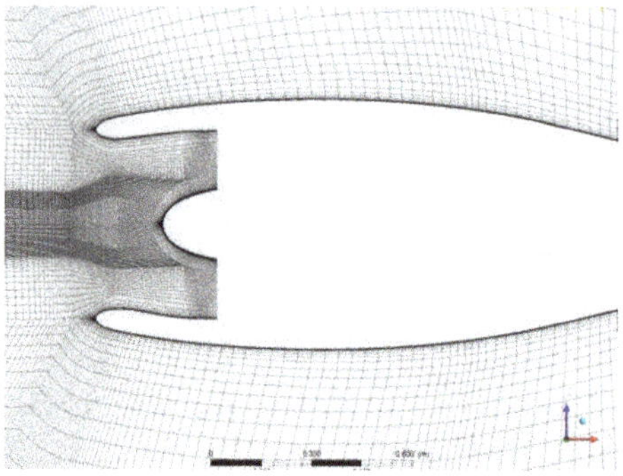

FIGURE 14: Nacelle meshing.

with the $K\omega$-sst turbulence model. In all cases, the general boundary conditions presented in Table 1 were applied. Note that in order to promote flow separation inside the inlet, the nacelle was placed at a 25° angle of incidence. The mass flow rate imposed in the engine is 33.6 kg/s at 100% of its maximum speed.

Four different approaches to capturing the interaction between the inlet and the fan are compared. The first is the classical mass flux sink approach, where a constant mass flux is imposed on a plane located downstream of the real fan face. The second and third models include the Rotor 67 in the solution: the second model uses the mixing plane approach and seeks a steady state solution of the coupled inlet and single rotor passage domains, and the third model uses URANS modeling of the full 22 Rotor 67 passages inside the nacelle. The fourth model is the AD-based approach currently proposed, where the boundary condition is placed directly on the fan face and the Rotor 67 is absent. In all four simulations, y^+ remains less than 2 on the nacelle's interior

TABLE 1: General boundary conditions.

Inlet fairfield	Total pressure: 101.325 (kPa) Total temperature: 288.15 (K) Flow direction: 25°
Exit fairfield	Total pressure: 97.7 (kPa)
Nacelle wall and jet surface	No slip

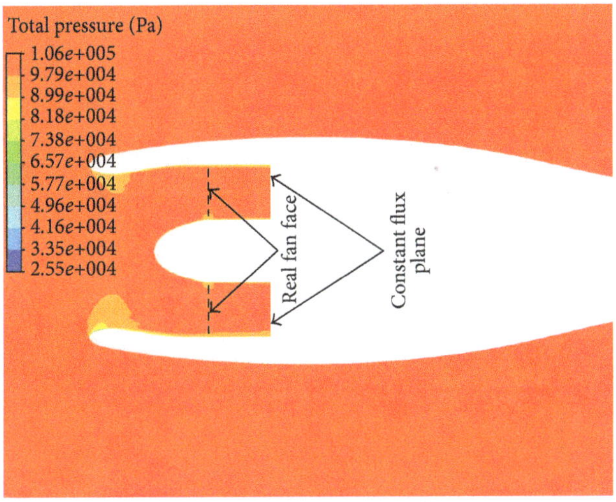

FIGURE 15: Contours of the total pressure for the constant mass flux model.

wall and less than 25 on the rotor blades. Based on the rotor's diameter, the Reynolds number is 2.5×10^6 and in term of spatial convergence, RMS $\leq 1 \times 10^5$ is guaranteed. The results for these four modeling approaches are presented in the following.

5.2.1. Constant Mass Flux. The first model is the classical mass flux boundary condition. As usual, it is positioned downstream of the real fan face to reduce its influence on the flow in the inlet. In the present run, the mass flux is imposed at 0.2 m behind the fan face, as illustrated in Figure 15. The total pressure contours obtained for this simulation are reproduced in Figure 15. One can observe a small region of lower total pressure on the inside of the lower lip of the nacelle, which is associated with a small separation and reattachment of the flow. With this model, the flow is reattached well ahead of the fan face, and so this simulation predicts only a small flow distortion at the fan face.

5.2.2. Steady State Mixing Plane Approach. To perform the steady state modeling, NASA Rotor 67 is inserted into the nacelle, and a "stage model" interface is placed between the rotor and the inlet. This interface performs a circumferential averaging of the flow properties through the interface. Steady state solutions are then obtained in each frame of reference. Downstream of the rotor, a static pressure is imposed and adjusted to obtain the required flow rate of 33.6 kg/s.

The calculation was performed on 8 2 GHz processors, and 4000 iterations were required to reach convergence. All

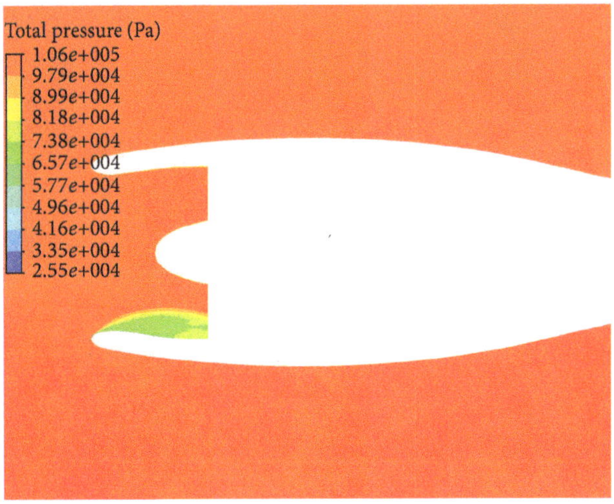

FIGURE 16: Contours of the total pressure obtained by a steady state calculation.

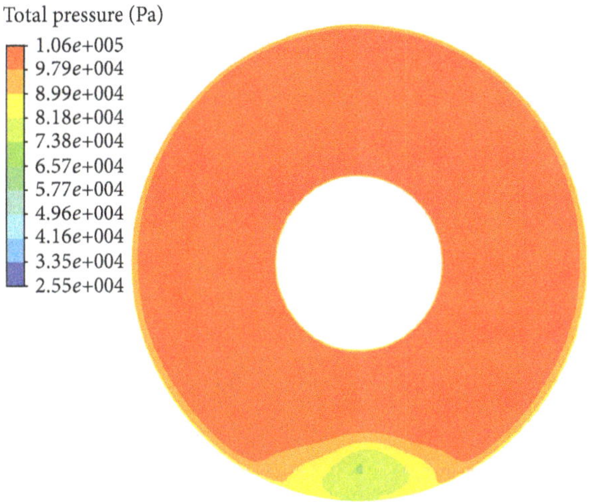

FIGURE 17: Contours of the total pressure on the fan face obtained by a steady state calculation.

the pressure contours obtained for this simulation are reproduced in Figures 16 and 17. One can observe an important region of separation on the inside of the lower lip of the nacelle, extending to the rotor fan face. As a result of this separation region, the distortion on the fan face is significant with this model. The fact that this result differs substantially from the constant mass flux performed above illustrates the importance of a proper boundary condition for such cases.

5.2.3. Transient Simulation with a Full Rotor. The third model uses URANS modeling with the full 22 passage Rotor 67 inserted into the nacelle. A "transient" interface is used between the rotor and the inlet. This interface transfers information at each time step between the two subdomains and makes it possible to obtain unsteady solutions. This model required much more in terms of CPU resources, as the simulation was performed for 10000 time steps, representing 8 complete revolutions of the fan. Note that periodic behavior

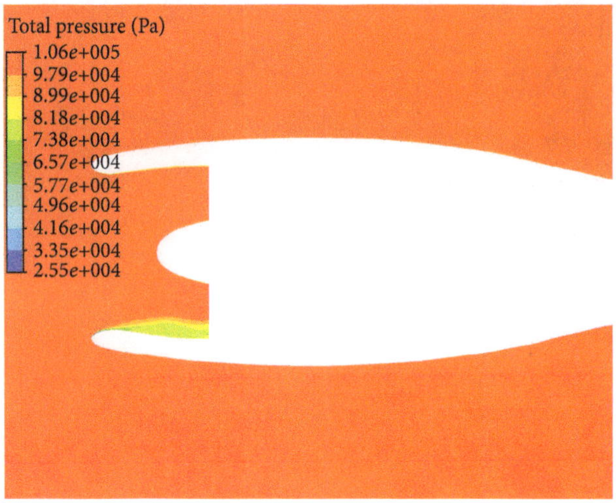

FIGURE 18: Contours of the total pressure obtained by a transient calculation time step: 600th.

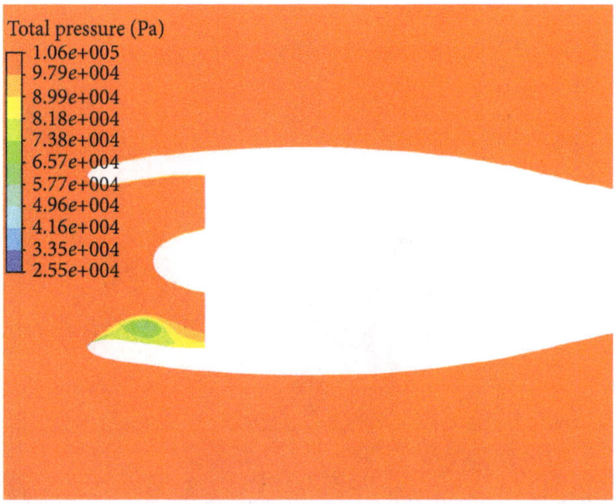

FIGURE 19: Contours of the total pressure obtained by a transient calculation time step: 2000th.

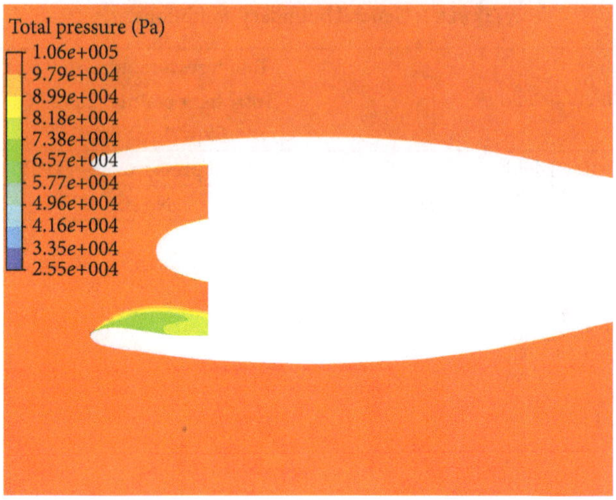

FIGURE 20: Contours of the total pressure obtained by the AD at the last iteration.

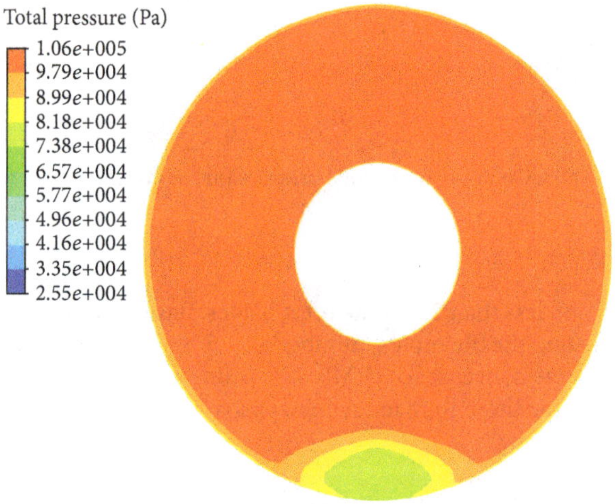

FIGURE 21: Contours of the total pressure at the fan face obtained by the AD at the last iteration.

was reached after 5 revolutions. The calculation was performed on 80 3 Ghz processors and took 2.2 days/revolution. The last 2000 steps were used for reporting the results.

The contours of the total pressure are presented in Figures 18 and 19 for two different time steps. It is obvious that the flow is unsteady, as the size of the flow separation varies in time. Averaged properties are presented in the following and compared to those of the other models.

5.2.4. Actuator Disc Model. The last model used is the AD-based approach proposed in this paper, which imposes our new boundary condition directly on the fan face. Figures 20 and 21 show the contours of the total pressure obtained with the AD-based boundary condition. One can see that the size of the separated region is important and compares well, qualitatively, to those of the steady state and unsteady models. More detailed quantitative comparisons are provided in the following.

5.3. Comparison of Models. Our final quantitative comparison of the four models tested involves plotting various flow properties on three different planes upstream of the fan face. The selected planes are the same as those used by Fidalgo et al. [1] and are illustrated in Figure 4. The flow properties compared are also the same as the Fidalgo properties, namely,

(i) axial speed multiplied by density,

(ii) static pressure,

(iii) angle between the absolute velocity vector and the axial direction.

Values are calculated at mid-span of the reference planes, presented in Figure 4 as black points.

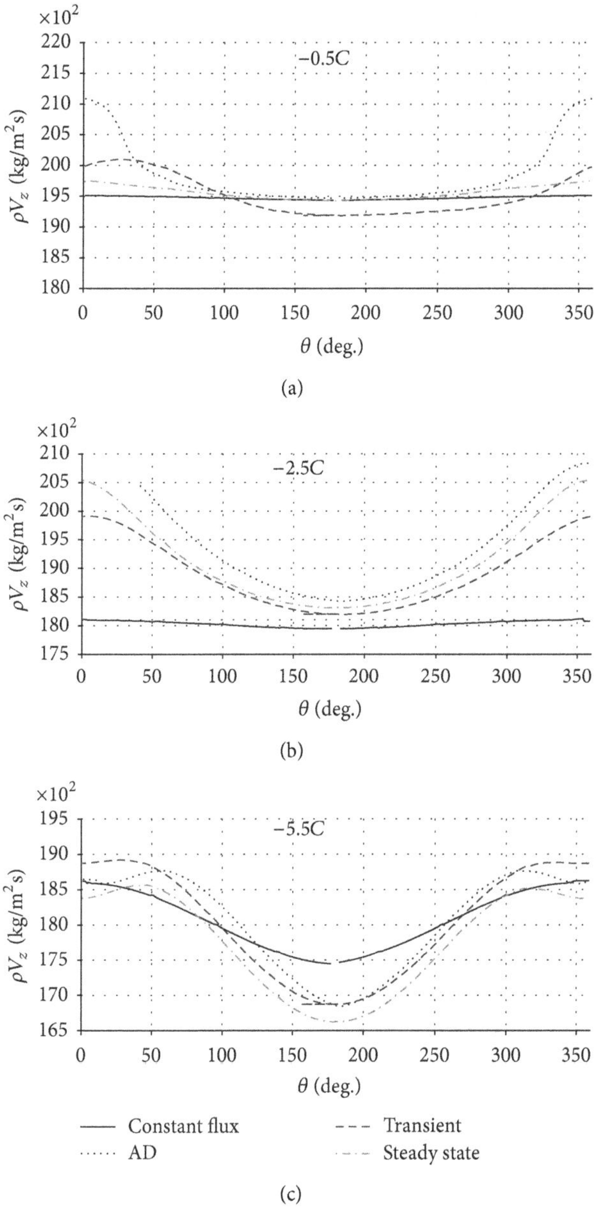

FIGURE 22: Comparison of unit mass flow rate of the first reference line.

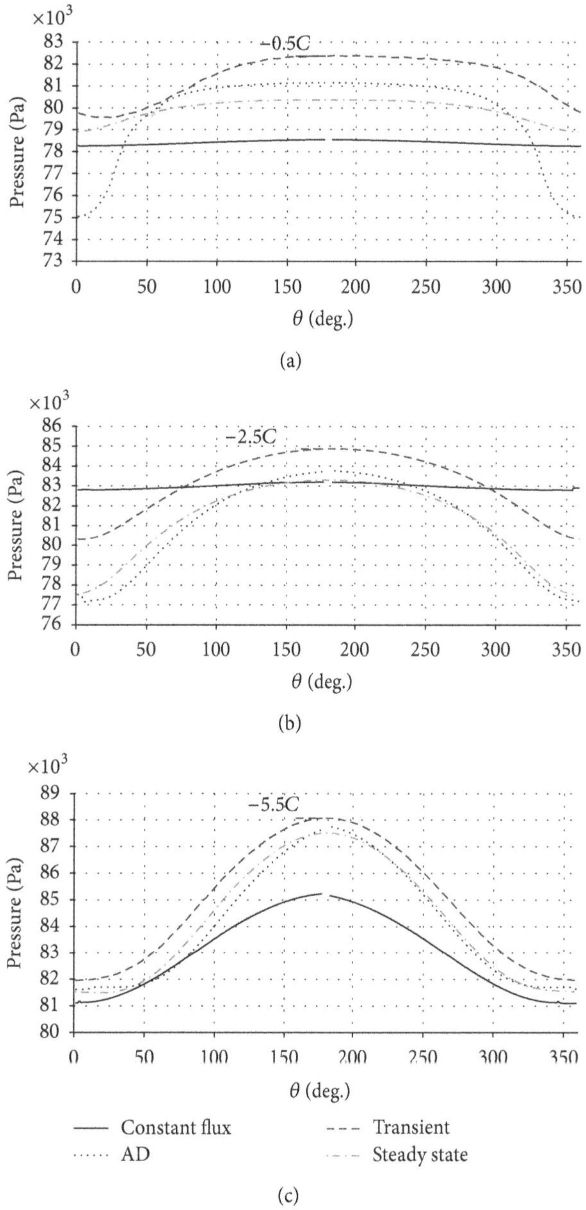

FIGURE 23: Static pressure comparison.

In these graphs, the transient case reported corresponds to a time average of the flow properties in the 8th revolution of the fan.

Figure 22 illustrates the evolution of unit mass flow rate field as the flow approaches the fan face. One can see that this property becomes more uniform as the flow approaches the fan face, indicating that the fan is attempting to eliminate the distortion. These figures clearly show that the uniform mass flux boundary condition produces the wrong results for the three planes reported. However, the other three models are similar, and the results obtained with the AD-based method are in good agreement with those of the transient and steady state models.

The static pressure profiles corresponding to these distributions are shown in Figure 23. Since the distortion is almost eliminated by imposing a constant flux, the pressure distribution corresponding to this case is almost constant when compared to those of the other models.

Figure 24 illustrates the angle between the velocity vector and the axial direction at the three planes. As expected, the magnitude of this angle is lower for constant flux, because the distortion is almost eliminated.

Overall, one can conclude that the results obtained with the AD-based boundary condition are, in general, close to those of the steady state "stage model" simulation and qualitatively represent the main trends observed in the transient case. Since the AD-based model is the cheapest in terms of CPU requirements, because the rotor is not included in

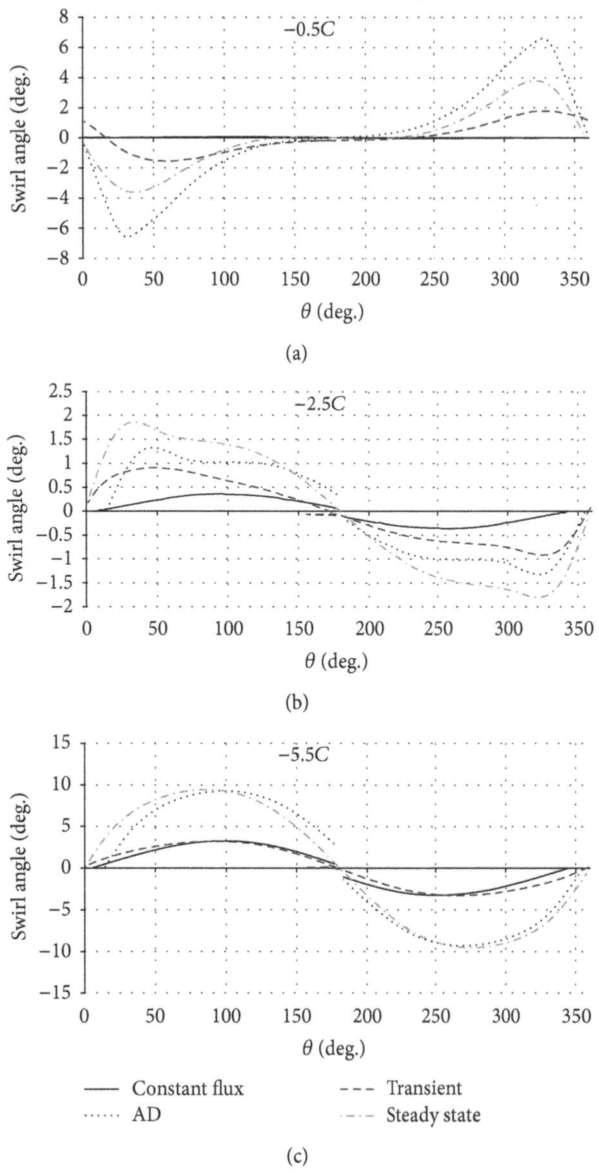

FIGURE 24: Swirl angle comparison.

the simulation but rather replaced by an AD, the AD-based approach seems to be an interesting compromise for inlet analysis and design.

6. Conclusion

In this work, a boundary condition at the engine fan face has been developed and tested to provide a better representation of the effect of the interaction between the fan and the inlet flows in cases with distortion at the fan face. First, an Actuator Disc model is used as a simplified model of the fan. Next, using the theory of parallel compressors, we divided the fan into several sectors, each of which contains an Actuator Disc. Then, we iteratively adjusted the static pressure in the various sectors of the fan face, in order to match the average outlet static pressure of a flow without distortion. The results obtained by our proposed AD-based approach were compared with the steady and unsteady RANS CFD results for two test cases involving NASA Rotor 67. The proposed AD-based approach gives results that are fairly similar to those of the RANS and URANS models and shows substantial improvement over those of the classical constant mass flux approach.

Nomenclature

a: Axial
AD: Actuator Disc
C: Blade Chord
c_p: Thermal capacity at constant pressure
h: Enthalpy
h_t: Total enthalpy
LE: Leading edge
$\dot{m}$: Mass flow rate
P: Static pressure
P_t: Total pressure
$\overline{P}$: Average static pressure
r: Radial
rel: Relative
RMS: Root mean square
s: Entropy
T: Static temperature
T.E.: Trailing edge
T_t: Total temperature
U: Rotating velocity
V: Absolute velocity
y^+: Dimensionless wall distance
α: Convergence factor
β: Relative angle
η: Efficiency
ω: Rotation
ρ: Density
θ: Angular position.

Acknowledgments

The authors thank Pratt & Whitney Canada, the J.-A. Bombardier Foundation, and the NSERC for their financial support of the IDEA Chair.

References

[1] V. Fidalgo, C. Hall, and Y. Colin, "A study of fan-distortion interaction within the nasa rotor 67 transonic stage," in *ASME Turboexpo Conference*, 2010.

[2] H. Pearson and A. McKenzie, "Wakes in axial compressors," *Journal of the Royal Aeronautical Society*, vol. 63, pp. 415–416, 1959.

[3] C. Reid, "The response of axial flow compressors to intake flow distortion," in *Proceedings of the International Gas Turbine and Aero-Engine Congress and Exhibition*, ASME, New York, NY, USA, 1969.

[4] R. S. Mazzawy, "Multiple segment parallel compressor model for circumferential flow distortion(in jet engines)," *Journal of*

Engineering for Power-Transactions of the ASME, vol. 99, no. 2, pp. 288–296, 1977.

[5] A. H. Stenning, "Inlet distortion effects in axial compressors," *Journal of Fluids Engineering, Transactions of the ASME*, vol. 102, no. 1, pp. 7–13, 1980.

[6] D. L. Whiteld and A. Jameson, "Three-dimensional Euler equation simulation of propeller-wing interaction in transonic flow," in *Proceedings of the 21st American Institute of Aeronautics and Astronautics, Aerospace Sciences Meeting*, Reno, Nev, USA, 1983.

[7] T. Q. Dang, "Calculations of propeller/airframe interference effects using the potential/multienergy flow method," *AIAA journal*, vol. 28, no. 5, pp. 771–777, 1990.

[8] E. Hsiao, M. Naimi, J. P. Lewis, K. Dalbey, Y. Gong, and C. Tan, "Actuator duct model of turbomachinery components for powered-nacelle Navier-Stokes calculations," *Journal of Propulsion and Power*, vol. 17, no. 4, pp. 919–927, 2001.

[9] A. Hale, M. Davis, and J. Sirbaugh, "A numerical simulation capability for analysis of aircraft inlet-engine compatibility," *Journal of Engineering for Gas Turbines and Power*, vol. 128, no. 3, pp. 473–481, 2006.

[10] J. Yao, S. E. Gorrell, and A. R. Wadia, "A time-accurate CFD analysis of inlet distortion induced swirl in multistage fans," in *Proceedings of the 43rd AIAA/ASME/SAE/ASEE Joint Propulsion Conference*, pp. 628–641, July 2007.

[11] R. I. Lewis, "Developments of actuator disc theory for compressible flow through turbo-machines," *International Journal of Mechanical Sciences*, vol. 37, no. 10, pp. 1051–1066, 1995.

[12] F. Mehdi Zadeh, *Amélioration de la Condition Frontière de Face de Soufflante Pour la Conception de L'admission D'Air Sous Distorsion*, École Polytechnique de Montréal, 2011.

[13] R. Lewis, *Turbomachinery Performance Analysis*, vol. 1, Butterworth-Heinemann, 1996.

12

Adaptive and Resilient Flight Control System for a Small Unmanned Aerial System

Gonzalo Garcia and Shahriar Keshmiri

Department of Aerospace Engineering, University of Kansas, 2120 Learned Hall 1530 W 15th Street, Lawrence, KS 66045, USA

Correspondence should be addressed to Gonzalo Garcia; garciagarreton@hotmail.com

Academic Editor: Nicolas Avdelidis

The main purpose of this paper is to develop an onboard adaptive and robust flight control system that improves control, stability, and survivability of a small unmanned aerial system in off-nominal or out-of-envelope conditions. The aerodynamics of aircraft associated with hazardous and adverse onboard conditions is inherently nonlinear and unsteady. The presented flight control system improves functionalities required to adapt the flight control in the presence of aircraft model uncertainties. The fault tolerant inner loop is enhanced by an adaptive real-time artificial neural network parameter identification to monitor important changes in the aircraft's dynamics due to nonlinear and unsteady aerodynamics. The real-time artificial neural network parameter identification is done using the sliding mode learning concept and a modified version of the self-adaptive Levenberg algorithm. Numerically estimated stability and control derivatives are obtained by delta-based methods. New nonlinear guidance logic, stable in Lyapunov sense, is developed to guide the aircraft. The designed flight control system has better performance compared to a commercial off-the-shelf autopilot system in guiding and controlling an unmanned air system during a trajectory following.

1. Introduction

The effectiveness of aircraft flight control systems in off-nominal conditions is heavily dependent upon the quality and accuracy of physics-based models and numerical simulations. The accuracy of these models depends on the mathematical formulation that describes the true physics, how uncertainty affects the outputs, and whether the model mismatch can be identified and estimated. In classical flight dynamics, the measured aerodynamic forces and moments are represented by differentiable functions and expanded in Taylor series. Although it is effective in benign portions of the flight envelope, this mathematical structure is incapable of modeling nonlinear and unsteady aerodynamics associated with hazardous and adverse onboard conditions. Adverse onboard conditions include vehicle impairment; system faults, failures, and errors and vehicle damage. External hazards and disturbances include wind shear and poor visibility; wake vortices; thunderstorms; snow and icing conditions. Aircraft robust control systems are designed to deal with uncertainty in dynamic models due to modeling errors, noise, and disturbances.

The motivation of this work is to develop a robust flight control system to improve control, stability, and survivability of a small unmanned aerial system (UAS) in hazardous and abnormal flight conditions. The presented flight control system consists of guidance, navigation, and control modules with adaptive control technologies and incorporated real-time learning-based system identification using artificial neural networks (ANNs) to update the physics-based model of UAS. An extended Kalman filter (EKF) is used to improve the measurements and to estimate inaccessible variables including the inertial wind components. The research is aimed at advancing flight control systems by the integration of functionalities required to update the physics-based model and adapt the control-law in presence of model uncertainties due to unsteady and nonlinear aerodynamics. The overall configuration comprises a set of stabilizing robust H-infinity controllers in the inner loop for adjacent trim points, to be scheduled when required. New nonlinear guidance logic, stable in Lyapunov sense, is developed to guide the aircraft. The new guidance law can achieve a stable and steady-state error-free guidance through the use of integral effect and by lifting constraints related to maximum distances and relative position with respect to the trajectory, expanding them to the entire state space.

Significant research has been carried out in areas of nonlinear guidance, navigation, adaptive control, and estimation. A broad range of research can be found on trajectory tracking, path following, and robust nonlinear controllers [1–9]. In [1], authors presented the design of guidance and control algorithms for autonomous vehicles through a simultaneous process that achieved a zero steady-state error about a trim point and, at the same time, guaranteed stability by avoiding possible cross-coupling between loops. In [2], a lateral track control-law for an autonomous aerial system that handles a larger range of wind disturbance was presented. A different approach is presented in [3] where the tracking control problem is formulated as a constrained nonlinear optimization problem. Unlike receding horizon methods, the minimization must be performed at the current instant. Reference [4] introduced nonlinear guidance logic for curved trajectory tracking reducing to a proportional-derivative control when applied to straight lines. In [5, 6], waypoint guidance was improved by including the horizontal wind estimation, allowing the aircraft to smoothly converge to a new course after switching to the next trajectory segment. A different approach for path following is presented in [7] where vector fields are used to represent desired ground track and to direct the aircraft to the desired track. More recently, in [8], a Lyapunov-based adaptive back stepping approach is used to design a flight-path controller for a nonlinear F-16 model. A nonlinear lateral guidance law, introduced in [4], is extended and modified by the authors to the longitudinal plane in [9] and tested in simulation.

Estimation of unmeasured states from noisy observations has become an essential part of flight control systems. The Kalman filter, either with an adaptive matrix gain or a fixed one, appears to be the most widely used observer (see [10–13]). Different versions of the Kalman filter ranging from linear to nonlinear, with static or time-varying Kalman gains, have been used. More refined filters such as Particle Filtering [14] and Sigma-Point [15] are designed to achieve better estimations. Another interesting approach is detailed in [16], where a nonlinear filter is designed using the state-dependent Riccati equation (SDRE) method. This technique is applicable whenever the UAS is describable in a time-varying linear format. Reference [17] estimates the external wind component in the horizontal plane to correct the velocity transformations between the respective coordinate systems. Reference [18] presents application of Kalman filters in estimation of wind components.

Different approaches using ANN for estimation of aerodynamic coefficients have been reported. In references [19–25], ANNs were propagated in a feedforward manner and their weights and biases were numerically searched by back propagating the estimation quadratic error. References [22, 23] extract the stability and control derivatives of an unstable aircraft under closed loop using a technique called Delta method. In [26–29], the fuzzy logic method was used as an alternative to ANN. The fuzzy logic modeling is used to capture nonlinear unsteadiness aerodynamic. In [26–28], fuzzy logic was used as an alternative to ANN to capture nonlinearities and unsteadiness in aerodynamics. In [29], using postflight fuzzy logic analysis, instability, high nonlinearity, and time-dependency were found in aircraft dynamics in crosswind.

The paper is structured as follows. Section 2 presents the nonlinear modeling of the aircraft and its linear equivalent at equilibrium points. Section 3 describes the inner loop and the H-infinity controller design, techniques for gain scheduling, the outer loop and its attitude guidance logics, and assessment of guidance logic inherent stability. Section 4 describes the design of the extended Kalman filter and its main properties, and Section 5 describes the ANN and its role in aerodynamic coefficient estimation. The paper is finalized with results from comparison of the designed flight control system (FCS) with flight test data obtained from a commercial off-the-shelf (COTS) autopilot.

2. Aircraft Modeling

Following [30], a 33% scale Yak-54, shown in Figure 1(a), is modeled using Newton-Euler equations for 6-DOF rigid-body motion, as detailed in Appendix A. Its dynamics, defined in the coordinate systems shown in Figure 1(b), is a function of the states $\mathbf{x}^T(t) = [V_T, \alpha, \beta, \phi, \theta, \psi, P, Q, R, \delta_t, \delta_e, \delta_a, \delta_r, p_N, p_E, p_H]$, that is, airflow angles α, β, airspeed V_T, attitude angles ϕ, θ, ψ, body angular rates P, Q, R, and inertial position p_N, p_E, p_H. Throttle, elevator, aileron, and rudder servos are modeled as a first-order system with time constants $\tau_t, \tau_e, \tau_a, \tau_r$ and states $\delta_t, \delta_e, \delta_a, \delta_r$, respectively. Wind components w_N, w_E, w_H are included as uncontrollable, but measurable, disturbances to the model and will be estimated by an extended Kalman filter.

Controllable inputs to the system are propulsive and aerodynamic forces and moments, and gravity is understood as an uncontrollable input. By assuming perfect alignment with X-axis in body frame $\overline{\mathbf{B}}$, propulsion or thrust $X^P(\mathbf{x}, t) = f_{X^P}(\delta_T, t)$ is modeled parametrically as a polynomial function on δ_T, as shown in Appendix B; consequently, no moments are induced. The aerodynamic forces $[\mathbf{F}_B(\mathbf{x}, t)]^A$ and moments $[\mathbf{M}_B(\mathbf{x}, t)]^A$, defined in body frame $\overline{\mathbf{B}}$, are not directly measurable, and they need to be estimated from accessible data. It is customary to express them in a normalized way as a function of coefficients C_X, C_Y, C_Z and C_l, C_m, C_n, as follows:

$$[\mathbf{F}_B(\mathbf{x}, t)]^A = \begin{bmatrix} X^A(\mathbf{x}, t) \\ Y^A(\mathbf{x}, t) \\ Z^A(\mathbf{x}, t) \end{bmatrix} = \begin{bmatrix} f_{X^A}(\mathbf{x}, t) \\ f_{Y^A}(\mathbf{x}, t) \\ f_{Z^A}(\mathbf{x}, t) \end{bmatrix} = \begin{bmatrix} \overline{q}(t) SC_X(\mathbf{x}, t) \\ \overline{q}(t) SC_Y(\mathbf{x}, t) \\ \overline{q}(t) SC_Z(\mathbf{x}, t) \end{bmatrix},$$

$$[\mathbf{M}_B(\mathbf{x}, t)]^A = \begin{bmatrix} L^A(\mathbf{x}, t) \\ M^A(\mathbf{x}, t) \\ N^A(\mathbf{x}, t) \end{bmatrix} = \begin{bmatrix} f_{L^A}(\mathbf{x}, t) \\ f_{M^A}(\mathbf{x}, t) \\ f_{N^A}(\mathbf{x}, t) \end{bmatrix} = \begin{bmatrix} \overline{q}(t) SbC_l(\mathbf{x}, t) \\ \overline{q}(t) S\overline{c}C_m(\mathbf{x}, t) \\ \overline{q}(t) SbC_n(\mathbf{x}, t) \end{bmatrix}, \quad (1)$$

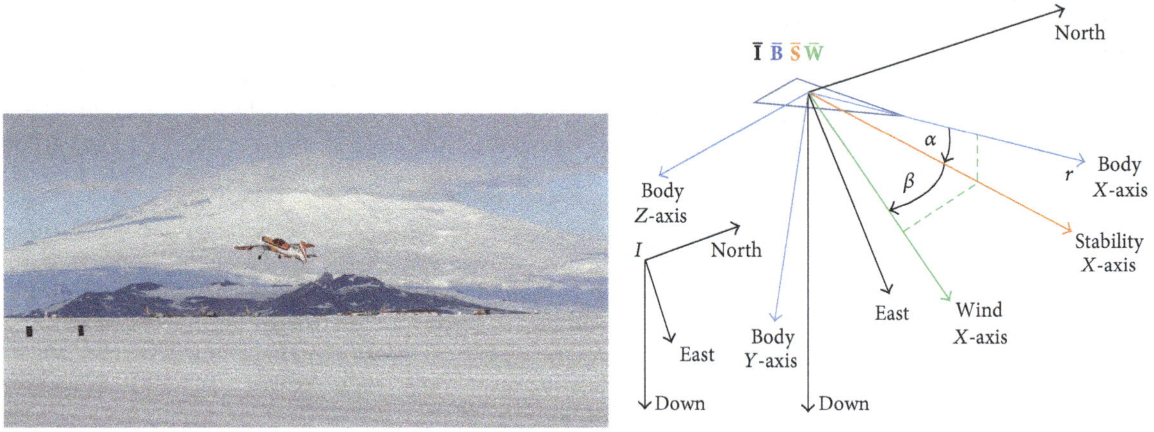

(a) (b)

FIGURE 1: 33% Scale Yak-54 UAS flight test at Pegasus Airfield, Antarctica (2010) (a). Coordinate Systems (b).

where b and $\bar{c}$ are the wing span and wing geometric mean chord, S is the wing reference area, and $\bar{q} = 0.5\rho V_T$ is the dynamic pressure, with air density ρ.

Although limited, a widely accepted parametric modeling is the component build-up approach, a time quasisteady and linear (in perturbed states $\Delta\mathbf{x}(t) = \mathbf{x}(t) - \mathbf{x}_{\text{trim}}$ around the trim condition $\mathbf{x}_{\text{trim}}$) model. Given its linearity and time-invariant structure, the model is valid in close proximity to the selected trim point. The present work chooses this method, based on [31], to assemble and compute these aerodynamic coefficients, as detailed in Appendix C. All constant coefficients, called stability and control derivatives, are obtained offline for selected trim conditions $\mathbf{x}_{\text{trim}}$. A useful tool, used to obtain stability and control derivatives in this work, is the engineering level software Advanced Aircraft Analysis (AAA) [32], developed by Design, Analysis, and Research Corporation (DARcorporation) that has a built-in aerodynamic database for different types of aircraft models. The AAA modeling is based on the chosen trim condition and the aircraft geometric parameters.

Designing a set of linear time-invariant controllers requires linear and time-invariant representations of the UAS in each equilibrium or trim point. For this purpose, a UAS nonlinear state space model derived from previous development and denoted $\dot{\mathbf{x}}(t) = \mathbf{f}(\mathbf{x}(t), \mathbf{u}(t)) \in R^n$ and $\mathbf{y}(t) = \mathbf{h}(\mathbf{x}(t), \mathbf{u}(t)) \in R^r$ with $\mathbf{u}(t) \in R^m$, with $\mathbf{u}^T(t) = [\delta_t^{\text{cmd}}, \delta_e^{\text{cmd}}, \delta_a^{\text{cmd}}, \delta_r^{\text{cmd}}]$ and $\mathbf{y}^T(t) = [V_T(t), \theta(t), \phi(t), \beta(t)]$, is linearized at each equilibrium $(\mathbf{x}_e^i, \mathbf{u}_e^i, \mathbf{y}_e^i)$, indexed by superscript i. Subscript e refers to the equilibrium point. The following truncated Taylor series expansion of the state space equation up to a first-order derivative, at each equilibrium point,

$$\dot{\mathbf{x}}(t) \approx \mathbf{f}(\mathbf{x}_e^i, \mathbf{u}_e^i) + \left.\frac{\partial \mathbf{f}}{\partial \mathbf{x}}\right|_{\substack{\mathbf{x}_e^i \\ \mathbf{u}_e^i}} (\mathbf{x}(t) - \mathbf{x}_e^i) + \left.\frac{\partial \mathbf{f}}{\partial \mathbf{u}}\right|_{\substack{\mathbf{x}_e^i \\ \mathbf{u}_e^i}} (\mathbf{u}(t) - \mathbf{u}_e^i),$$
$$\mathbf{y}(t) \approx \mathbf{h}(\mathbf{x}_e^i, \mathbf{u}_e^i) + \left.\frac{\partial \mathbf{h}}{\partial \mathbf{x}}\right|_{\substack{\mathbf{x}_e^i \\ \mathbf{u}_e^i}} (\mathbf{x}(t) - \mathbf{x}_e^i) + \left.\frac{\partial \mathbf{h}}{\partial \mathbf{u}}\right|_{\substack{\mathbf{x}_e^i \\ \mathbf{u}_e^i}} (\mathbf{u}(t) - \mathbf{u}_e^i), \tag{2}$$

gives the linear representation of the system at equilibrium. After cancellation of constant terms and defining the partial derivatives evaluated at each equilibrium point, the following linear time-invariant model, denominated $\mathbf{G}^i$, is obtained for control design purposes:

$$\Delta\dot{\mathbf{x}}(t) = \mathbf{A}_G^i \Delta\mathbf{x}(t) + \mathbf{B}_G^i \Delta\mathbf{u}(t),$$
$$\Delta\mathbf{y}(t) = \mathbf{C}_G^i \Delta\mathbf{x}(t) + \mathbf{D}_G^i \Delta\mathbf{u}(t). \tag{3}$$

From AAA, for a straight and level flight at a constant speed of 70 (knots) with $J_x = 1.0886\,(\text{slug} \cdot \text{ft}^2)$, $J_y = 2.1068(\text{slug} \cdot \text{ft}^2)$, $J_z = 3.0382\,(\text{slug} \cdot \text{ft}^2)$, $J_{xz} = 0.05\,(\text{slug} \cdot \text{ft}^2)$, $m = 28\,(\text{lb})$ and $g = 32.17\,(\text{ft/sec}^2)$, [33] calculates the stability and control derivatives of a 33% scale Yak54; shown in Table 1.

The work done in [33] employed the vortex lattice model (VLM) as a second method for estimation of some of the stability and control derivatives. For uncertainty analysis, results from the AAA and VLM estimations were compared with results from flight tests system identification. Table 2 shows stability and control derivatives, the level of uncertainty (maximum and minimum), and the resultant center values and half range of each uncertainty derived from comparison. In order to switch trim conditions using gain scheduling, the same set of information and analysis had to be generated at those different trim points. In this research, and for the sake of abstraction, only information for one trim point, at 70 (knots), is presented.

At the chosen trim condition, the linear model is stable in the roll, dutch roll, phugoid, and short period modes and is unstable in the spiral mode. Figure 2 shows the linearized system poles, zeros and modes.

3. Inner and Outer Loops Design

3.1. Inner Loop. The inner loop is constituted by a set of robust H-infinity controllers and a mechanism to perform gain scheduling among them based on current dynamic conditions. The linear time-invariant plant, given in (3), is

Table 1: AAA stability and control derivatives for 70 knots (from [33]).

Parameter	Value	Units
C_{D0}	0.0513	[−]
$C_{D\alpha}$	0.0863	[1/rad]
$C_{D\delta e}$	0	[1/rad]
$C_{D\dot{\alpha}}$	0	[1/rad]
C_{Dq}	0	[1/rad]
C_{Du}	0.0011	[1/rad]
C_{L0}	0	[−]
$C_{L\alpha}$	4.5465	[1/rad]
$C_{L\delta e}$	0.3792	[1/rad]
$C_{L\dot{\alpha}}$	1.8918	[1/rad]
C_{Lq}	5.5046	[1/rad]
C_{Lu}	0.0017	[1/rad]
C_{m0}	0.0020	[−]
$C_{m\alpha}$	−0.3937	[1/rad]
$C_{m\delta e}$	−0.8778	[1/rad]
$C_{m\dot{\alpha}}$	−4.3787	[1/rad]
C_{mq}	−8.0532	[1/rad]
C_{mu}	0.0002	[1/rad]
$C_{y\beta}$	−0.3602	[1/rad]
$C_{y\dot{\beta}}$	0	[1/rad]
C_{yp}	0.0085	[1/rad]
C_{yr}	0.2507	[1/rad]
$C_{y\delta a}$	0	[1/rad]
$C_{y\delta r}$	0.1929	[1/rad]
$C_{l\beta}$	−0.0266	[1/rad]
$C_{l\dot{\beta}}$	0	[1/rad]
C_{lp}	−0.3819	[1/rad]
C_{lr}	0.0514	[1/rad]
$C_{l\delta a}$	0.3490	[1/rad]
$C_{l\delta r}$	0.0154	[1/rad]
$C_{n\beta}$	0.1022	[1/rad]
$C_{n\dot{\beta}}$	0	[1/rad]
C_{np}	−0.0173	[1/rad]
C_{nr}	−0.1270	[1/rad]
$C_{n\delta a}$	−0.0088	[1/rad]
$C_{n\delta r}$	−0.0996	[1/rad]

Table 2: Stability and control derivatives uncertainty range knots (from [33]).

Parameter	Maximum	Minimum	Central value	Half range
C_{mq}	−4.3720	−16.1064	−10.2398	5.8666
$C_{m\delta e}$	−0.7572	−1.2289	−0.9930	0.2359
$C_{l\beta}$	−0.0220	−0.0314	−0.0267	0.0047
C_{nr}	−0.1156	−0.2890	−0.2023	0.0867
$C_{n\delta r}$	−0.0996	−0.1404	−0.1200	0.0204

augmented with design weighting matrices $\mathbf{W}_1$, $\mathbf{W}_2$, and $\mathbf{W}_3$, as shown in Figure 3(a), with inputs $\mathbf{u}_1 \in R^{m_1}$, $\mathbf{u}_2 \in R^{m_2}$ and outputs $\mathbf{y}_1 \in R^{p_1}$, $\mathbf{y}_2 \in R^{p_2}$, becoming the augmented plant $\mathbf{P}^i$.

This is done in order to harmonize otherwise conflicting performance objectives, as low frequency reference tracking and disturbance rejection, and high frequency noise cancellation, by appropriately shaping particular closed loop sensitivity functions through these frequency-dependent weighting matrices. Plant $\mathbf{P}^i$ is then expressed as

$$\begin{aligned}\dot{\mathbf{x}}_P &= \mathbf{A}^i\mathbf{x}_P + \mathbf{B}_1^i\mathbf{u}_1 + \mathbf{B}_2^i\mathbf{u}_2,\\ \mathbf{y}_1 &= \mathbf{C}_1^i\mathbf{x}_P + \mathbf{D}_{11}^i\mathbf{u}_1 + \mathbf{D}_{12}^i\mathbf{u}_2,\\ \mathbf{y}_2 &= \mathbf{C}_2^i\mathbf{x}_P + \mathbf{D}_{21}^i\mathbf{u}_1 + \mathbf{D}_{22}^i\mathbf{u}_2,\end{aligned} \tag{4}$$

where superscript i continues to be an index among the set of augmented plants corresponding to each linearized plant $\mathbf{G}^i$. One of the main requirements of the chosen gain scheduling procedure is to have a homogeneous augmented plant set in terms of same structure, states, inputs, and outputs. This is achieved by having a homogeneous linearized plant set, in terms of the similar nature of zeros and poles, same weighting matrices, and same controller calculation method.

The state vector $\mathbf{x}_P$ is constructed by adding the states of the linearized plant and the states of the weighting matrices. In particular, the inertial positions p_N, p_E, p_H are left out of the augmented state vector as they will be used in the guidance, and also yawing angle ψ is not included for being superfluous.

If the conditions detailed in Appendix D are met, then the suboptimal H_∞ control problem [34, 35] provides a controller $\mathbf{K}^i$ that internally stabilizes the closed-loop system such that

$$\left\|\mathbf{T}_{\mathbf{y}_1\mathbf{u}_1}(s)\right\|_\infty = \underbrace{\sup}_{jw}\bar{\sigma}\left[\mathbf{T}_{\mathbf{y}_1\mathbf{u}_1}(jw)\right] \le \gamma, \tag{5}$$

for a given $\gamma > 0$, where $\mathbf{T}_{\mathbf{y}_1\mathbf{u}_1}$ is the closed-loop transfer function from exogenous inputs $\mathbf{u}_1$ (including references, disturbances, and noises) to error responses $\mathbf{y}_1$ (including weighted error, control, and output signals). In the present work, $\mathbf{T}_{\mathbf{y}_1\mathbf{u}_1}$ is constructed as follows:

$$\left\|\mathbf{T}_{\mathbf{y}_1\mathbf{u}_1}(s)\right\|_\infty = \left\|\begin{matrix}\mathbf{W}_1(s)\mathbf{S}(s)\\ \mathbf{W}_2(s)\mathbf{R}(s)\\ \mathbf{W}_3(s)\mathbf{T}(s)\end{matrix}\right\|_\infty, \tag{6}$$

as a combination of desired weighted closed loop objectives, $\mathbf{S}(s)$, $\mathbf{R}(s)$, and $\mathbf{T}(s)$, that is, sensitivity, output sensitivity, and complementary sensitivity, respectively. Weighting allows their joint minimization. For an effective low frequency reference tracking and disturbance rejection, the weighting matrix $\mathbf{W}_1$ is built to have high singular values at the required range. For high frequency noise cancelation, $\mathbf{W}_3$ is designed to have high singular values in those higher frequencies. The weighting matrix $\mathbf{W}_2$ is designed as a constant matrix to uniformly penalize the control effort from getting unfeasible values and saturate actuators. Figure 3(b) shows singular values of the dynamic weightings chosen for the current trim condition.

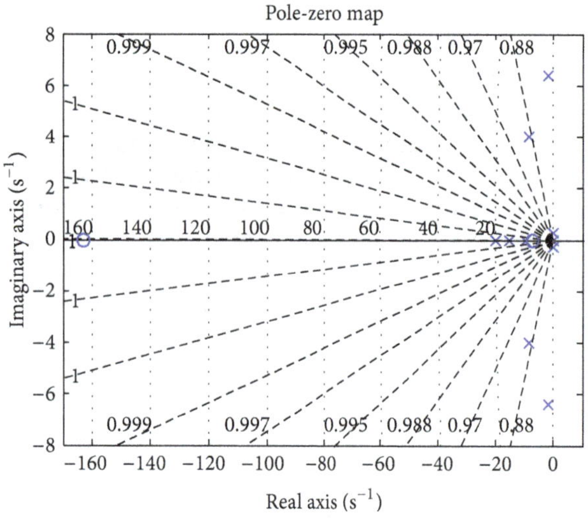

Mode	Roots	τ (s)	ζ	ω_n (rad/s)
Throttle servo	−10	0.1	—	—
Elevator servo	−20	0.05		
Phugoid	−0.09 ± 0.25i	—	0.34	0.26
Short period	−8.76 ± 3.99i	—	0.91	9.63

Mode	Roots	τ (s)	ζ	ω_n (rad/s)
Aileron servo	−20	0.05	—	—
Rudder servo	−20	0.05		
Roll	−15.50	0.06	—	—
Spiral	+0.005	180.5	—	—
Dutch-roll	−1.75 ± 6.38i	—	0.26	6.62

(b)

FIGURE 2: Linearized system poles: zeros (a) and modes (b).

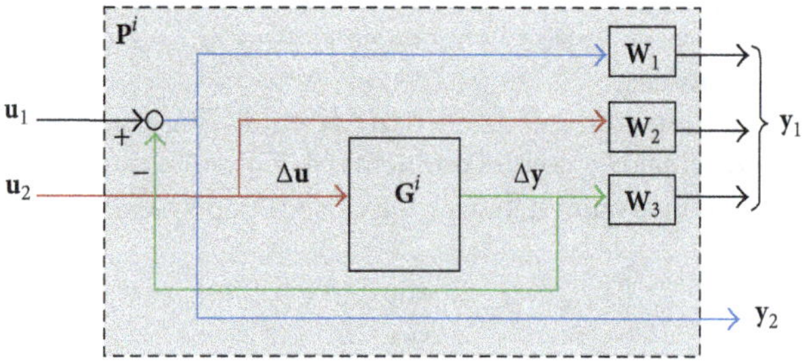

(a)

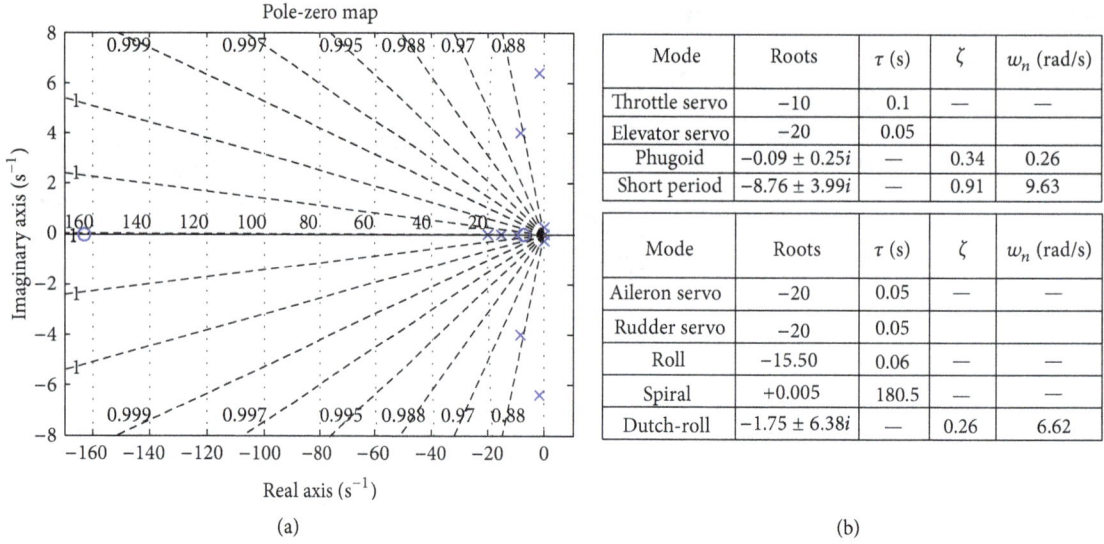

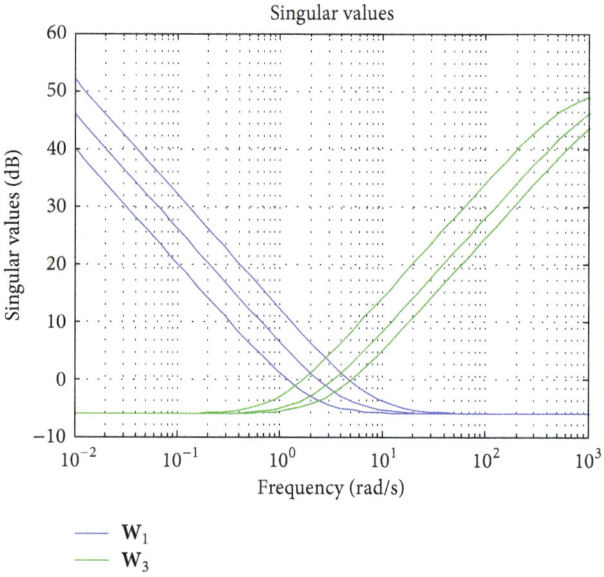

(b)

FIGURE 3: Augmented system (a). Singular values of weighting matrices $\mathbf{W}_1$ and $\mathbf{W}_3$ (b).

In order to characterize parameter uncertainty (see Table 2), a linear fractional transformation (LFT) is employed. Uncertainties are extracted and packed into $\boldsymbol{\Delta}^i$, as shown in Figure 4(a), leaving a deterministic plant $\widehat{\mathbf{P}}^i$, as opposed to the previous uncertain plant $\mathbf{P}^i$. This LFT interconnection can be defined as

$$\begin{bmatrix} \boldsymbol{v} \\ \mathbf{y}_1 \\ \mathbf{y}_2 \end{bmatrix} = \widehat{\mathbf{P}}^i \begin{bmatrix} \boldsymbol{\eta} \\ \mathbf{u}_1 \\ \mathbf{u}_2 \end{bmatrix} = \begin{bmatrix} \widehat{\mathbf{P}}^i_{11}(s) & \widehat{\mathbf{P}}^i_{12}(s) & \widehat{\mathbf{P}}^i_{13}(s) \\ \widehat{\mathbf{P}}^i_{21}(s) & \widehat{\mathbf{P}}^i_{22}(s) & \widehat{\mathbf{P}}^i_{23}(s) \\ \widehat{\mathbf{P}}^i_{31}(s) & \widehat{\mathbf{P}}^i_{32}(s) & \widehat{\mathbf{P}}^i_{33}(s) \end{bmatrix} \begin{bmatrix} \boldsymbol{\eta} \\ \mathbf{u}_1 \\ \mathbf{u}_2 \end{bmatrix}, \qquad (7)$$

$$\boldsymbol{\eta} = \Delta^i \boldsymbol{v}, \qquad \mathbf{u}_2 = \overline{\mathbf{K}}^i \mathbf{y}_2,$$

or as $\mathbf{y}_1 = \mathbf{F}_u(\mathbf{F}_l(\widehat{\mathbf{P}}^i, \overline{\mathbf{K}}^i), \boldsymbol{\Delta}^i)\mathbf{u}_1 = \mathbf{F}_l(\mathbf{F}_u(\widehat{\mathbf{P}}^i, \boldsymbol{\Delta}^i), \overline{\mathbf{K}}^i)\mathbf{u}_1$, where $\mathbf{F}_l$ corresponds to the lower LFT and $\mathbf{F}_u$ to the upper LFT. Assuming that $\|\Delta\|_\infty \leq 1$, then by the Small Gain Theorem, the interconnected system shown in Figure 4(a) is internally stable if and only if $\|\mathbf{F}_l(\widehat{\mathbf{P}}^i, \overline{\mathbf{K}}^i)\|_\infty < 1$, [34].

For each derivative given in Table 2, an uncertainty range is built considering the highest and lowest values. These extreme values are used to calculate the central value and half range value. With this information, a normalized uncertainty range is constructed and pulled out of the plant. Figure 4(b) shows the arrangement for $C_{l\beta}$ which is extensible to remaining uncertain derivatives. Rearrangement of $C_{l\beta}$ in terms of central value $C_{l\beta}{}^C$ and half range value $C_{l\beta}{}^H$ leaves its

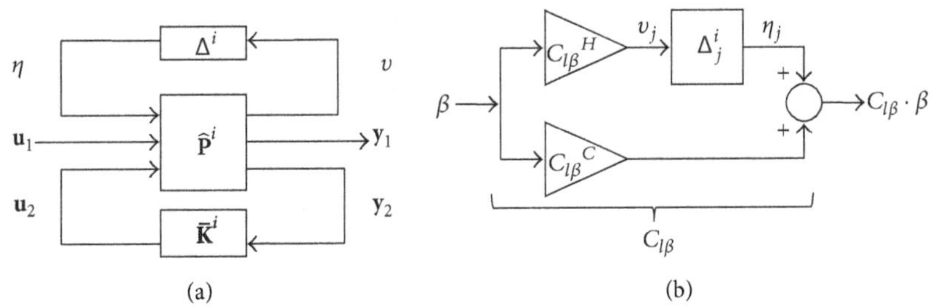

FIGURE 4: General LFT connection (a). Uncertainty normalization (b).

uncertainty block with $\|\Delta^i_j\|_\infty \le 1$. From pulling out of the system $\mathbf{P}^i$ each uncertainty block, an overall block $\boldsymbol{\Delta}^i$, from $\boldsymbol{v} = [v_1, \ldots, v_j, \ldots]^T$ to $\boldsymbol{\eta} = [\eta_1, \ldots, \eta_j, \ldots]^T$, is assembled as shown in Figure 4(a).

The 70 knots trim condition is labeled as $i = 1$, where, utilizing a set of weighting matrices as described before, a robust controller $\overline{\mathbf{K}}^1$, with $\|\mathbf{T}_{\mathbf{y}_1\mathbf{u}_1}(s)\|_\infty = \gamma^1 = 0.543$, is achieved. For this closed-loop configuration, after extracting the uncertainties blocks, the infinity norms $\|\mathbf{F}_l(\widehat{\mathbf{P}}^1, \overline{\mathbf{K}}^1)\|_\infty$, for each uncertain derivative, are shown in Table 3. The last row shows the worst case, the infinity norm for all uncertainties together; that is, $\|\mathbf{F}_l(\widehat{\mathbf{P}}^1, \overline{\mathbf{K}}^1)\|_\infty = 0.5617$. As by construction of $\|\Delta^1_j\|_\infty \le 1$, the Small Gain Theorem ensures that the controller robustness keeps stability facing the parameter uncertainty.

The flight control system's inner loop is equipped with a moderate fault tolerant capability. It is based on the ability to detect and identify dynamic changes due to faults that are characterized by changes in the aerodynamic forces and moment's causal functions. The implicitly assumed that static relation from aircraft states and state derivatives and controls to forces and moments is monitored through an online ANN. Changes in the network's parameters indicate unsteadiness in the aircraft dynamics, or an important trim condition transition. This information is utilized to adapt the current controller (and potentially the extended Kalman filter's model) for the new known dynamic condition, implementing a gain scheduling technique. The management of the ANN is addressed in Section 5.

Following the approach used in [36], the idea is to update the current controller $\overline{\mathbf{K}}^i$ based on an updated linear representation of the system $\mathbf{G}^i$, as described in (3). Its matrices $(\mathbf{A}^i_G, \mathbf{B}^i_G, \mathbf{C}^i_G, \mathbf{D}^i_G)$ are then used to compute a new robust controller through a new parameterization of $\mathbf{P}^i$, utilizing initially the same design parameters used in the computation of $\overline{\mathbf{K}}^i$ (i.e., same weighting matrices and same γ^i). Some simulation results are presented in Section 6. Before applying the new controller, assumptions in Appendix D should be assessed.

3.2. Outer Loop. The overall task of the inner loop, through the robust controller, is to drive the aircraft to follow external commands generated by the outer loop, by minimizing the error between the commanded and the actual state. This minimization is performed over the error in airspeed, pitch angle, bank angle, and side slip angle; that is,

$$e_{V_T}(t) = V_T{}^{\text{cmd}}(t) - V_T(t), \qquad e_\theta(t) = \theta^{\text{cmd}}(t) - \theta(t),$$
$$e_\phi(t) = \phi^{\text{cmd}}(t) - \phi(t), \qquad e_\beta(t) = \beta^{\text{cmd}}(t) - \beta(t). \tag{8}$$

The outer loop's task is to follow a trajectory path integrated by a set of waypoints $\{\vec{\mathbf{a}}, \vec{\mathbf{b}}, \vec{\mathbf{c}}, \ldots\}$, the straight segments subtended between them $\{\mathbf{ab}, \mathbf{bc}, \mathbf{cd}, \ldots\}$, and their assigned inertial speed $\{V_{\mathbf{ab}}, V_{\mathbf{bc}}, V_{\mathbf{cd}}, \ldots\}$. To this end, it has two main subsystems; one is to generate the attitude commands $\phi^{\text{cmd}}, \theta^{\text{cmd}}$ based on the current inertial position p_N, p_E, p_H and speed $v_N = \dot{p}_N, v_E = \dot{p}_E, v_H = \dot{p}_H$, and the relative location of the aircraft with respect to the path to follow; a second subsystem is to generate the airspeed command $V_T{}^{\text{cmd}}$ based on the path inertial speed corrected with the estimated wind. The sideslip command β^{cmd} does not require any specific logic, and it is set to zero for the entire flight.

The logic in the generation of attitude commands $\phi^{\text{cmd}}, \theta^{\text{cmd}}$, to keep the aircraft on path, is based on a nonlinear guidance logic developed in [4] for lateral guidance, which for small errors and straight lines approximates a proportional-derivative controller acting on the cross-track error. It generates the ϕ^{cmd}. To calculate the θ^{cmd}, a novel extension to the longitudinal plane from [9] is developed and employed. This extension also adds new features to the original guidance idea, ensuring its stability for any location and any attitude of the aircraft. Stability in the sense of Lyapunov is proven hereinafter for the extended logic. Figure 5 shows the quantities involved for each plane separately.

The guidance logic calculates the angles η_{Lat} and η_{Lon} between the lateral and longitudinal components of the inertial velocity vector $\vec{\mathbf{V}}_{\text{Lat}}$ and $\vec{\mathbf{V}}_{\text{Lon}}$ and the corresponding components vectors $\vec{\mathbf{L}}_{\text{Lat}}$ and $\vec{\mathbf{L}}_{\text{Lon}}$ from the position vector subtended from the current UAS position $\vec{\mathbf{p}}$ and an imaginary point $\vec{\mathbf{r}}$. Point $\vec{\mathbf{r}}$ is located in the current segment between $\vec{\mathbf{a}}$ to $\vec{\mathbf{b}}$; see Figure 5.

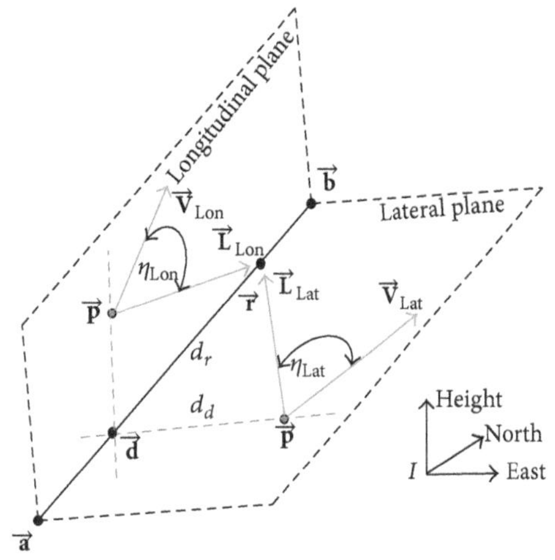

FIGURE 5: Attitude guidance logic geometry.

This point $\vec{\mathbf{r}}$ is also located at a fixed distance $|\vec{\mathbf{r}} - \vec{\mathbf{d}}| = d_r$, a numerical value which is closely dependent on the dynamic characteristics of the aircraft. The intersection between the perpendicular line from the UAS position to the current segment **ab** is called $\vec{\mathbf{d}}$. Point $\vec{\mathbf{r}}$ moves as $\vec{\mathbf{d}}$ moves in the direction of point $\vec{\mathbf{b}}$. From the geometry seen in Figure 5, it is evident that a small or zero angle η_{Lat} or η_{Lon} is equivalent to a small or zero lateral or longitudinal cross-track error (distance between points $\vec{\mathbf{p}}$ and $\vec{\mathbf{d}}$) and the UAS flying in the segment direction, or that the UAS is off-track, but flying toward the trajectory. Once point $\vec{\mathbf{r}}$ reaches a predetermined distance to the end of the segment, the segment switches to the next waypoint, now between $\vec{\mathbf{b}}$ to $\vec{\mathbf{c}}$, and the error angles are recalculated, based on the new position vectors.

These error angles η_{Lat} and η_{Lon} (being both negative as shown in Figure 5) are translated into attitude commands through the following nonlinear relations (see [9]):

$$\phi_{\text{cmd}} = \tan^{-1}\left\{k_{a_{\text{Lat}}} \cdot \eta_{\text{Lat}} \cdot \left(k_{p_{\text{Lat}}} + \frac{k_{i_{\text{Lat}}}}{s}\right)\right\},$$
$$\theta_{\text{cmd}} = \tan^{-1}\left\{k_{a_{\text{Lon}}} \cdot \eta_{\text{Lon}} \cdot \left(k_{p_{\text{Lon}}} + \frac{k_{i_{\text{Lon}}}}{s}\right)\right\}, \tag{9}$$

where gains k_p and k_i are convenient gains resembling PI controllers. The integral action has been added to force the steady state angle error to zero. The arctangent function acts as a smooth limiter, avoiding unrealistic growth of the commands. The adaptive gains $k_{a_{\text{Lat}}} = 2|\vec{\mathbf{V}}_{\text{Lat}}|^2/|\vec{\mathbf{L}}_{\text{Lat}}|$ and $k_{a_{\text{Lon}}} = 2|\vec{\mathbf{V}}_{\text{Lon}}|^2/|\vec{\mathbf{L}}_{\text{Lon}}|$ are used to adequately neutralize changes in the aircraft's speed due to external disturbances, such as a wind gusts; see [4]. For the particular geometry of the nonlinear logic, errors will be zero only when both vectors are aligned and the UAS is flying with zero cross-distance or when it is pointing toward the trajectory.

As the inner loop controller also works with the airspeed state error e_{V_T}, the predefined segment inertial speed is converted from inertial **I** to wind $\overline{\mathbf{W}}$ coordinate systems and adapted for the current wind estimation. So, when the aircraft, driven by the controller, is reaching the commanded airspeed and is on the track, it will also be flying closer to the desired inertial speed.

3.3. Stability in the Sense of Lyapunov. In order to test the stability of the guidance law (similar to [37]), a state space representation is developed in terms of the variables $\eta_{\text{Lat}}(t)$ and $d_d(t)$, the shortest distance from the UAS to the trajectory. Then, its stability in the sense of Lyapunov is assessed. From Figure 6(a), relations are defined among guidance variables, where the time dependence has been included for clarity. Rate of change of distance $d_d(t)$ is described as follows:

$$\dot{d}_d(t) = \left|\vec{\mathbf{V}}_{\text{Lat}}(t)\right| \sin\left(\eta_{\text{Lat}}(t) - \eta^c_{\text{Lat}}(t)\right), \tag{10}$$

where positive distance is as shown and $\eta^c_{\text{Lat}}(t)$ is defined as $\eta^c_{\text{Lat}}(t) = \tan^{-1}(d_d(t)/d_r)$. As a sign convention, angles measured counterclockwise are taken positive.

For the dynamics of $\eta_{\text{Lat}}(t)$ the centripetal acceleration $a_{\text{cent}}(t)$, shown in Figure 6(a), is defined as

$$\frac{d}{dt}\left(\eta_{\text{Lat}}(t) - \eta^c_{\text{Lat}}(t)\right) = -\frac{a_{\text{cent}}(t)}{\left|\vec{\mathbf{V}}_{\text{Lat}}(t)\right|}. \tag{11}$$

By assuming the inner controller accomplishes a close tracking of ϕ_{cmd}, and assuming $a_{\text{cent}} = g\tan\phi$, which is valid for coordinated turns, the following dynamics holds:

$$\dot{\eta}_{\text{Lat}}(t) = \frac{-\left|\vec{\mathbf{V}}_{\text{Lat}}(t)\right|\eta_{\text{Lat}}(t)}{\sqrt{d_d^2(t) + d_r^2}} + \frac{\left|\vec{\mathbf{V}}_{\text{Lat}}(t)\right| d_r}{\left(d_d^2(t) + d_r^2\right)} \sin\left(\eta_{\text{Lat}}(t) - \tan^{-1}\frac{d_d(t)}{d_r}\right), \tag{12}$$

where the speed is assumed to be constant; that is, $|\vec{\mathbf{V}}_{\text{Lat}}(t)| = 120$ [ft/sec] $\approx$ 70 knots, $k_{i_{\text{Lat}}} = 0$, $k_{p_{\text{Lat}}} = 1/2g$, and $d_r = 300$ [ft]. The dynamics from (10) and (12) are used to form an autonomous nonlinear state space system and to evaluate its stability. Using a numerical and graphical solution, these equations are propagated in time for different initial conditions. Figure 6(b) shows phase portrait of the autonomous system, where all trajectories converge to the origin. It is clear that the lateral guidance logic is asymptotically stable in the sense of Lyapunov for any initial distance d_d and $-\pi > \eta_{\text{Lat}} > \pi$.

4. Extended Kalman Filter

To improve the quality of measurements and to provide estimations of unmeasured quantities, an EKF is designed based on [38] (method briefly reviewed in Appendix E) for the following dynamic model $\dot{\hat{\mathbf{x}}}(t) = \hat{\mathbf{f}}(\hat{\mathbf{x}}(t), \mathbf{u}(t))$ composed of the aircraft modeling defined in Section 2. Biases and wind components are defined as new states

$$\widehat{\mathbf{f}}:\begin{cases}\dot{\widehat{p}}_N = \widehat{U}\cos\widehat{\phi}\cos\widehat{\psi} + \widehat{V}\left(-\cos\widehat{\phi}\sin\widehat{\psi} + \sin\widehat{\phi}\sin\widehat{\theta}\cos\widehat{\psi}\right) + \widehat{W}\left(\sin\widehat{\phi}\sin\widehat{\psi} + \cos\widehat{\phi}\sin\widehat{\theta}\cos\widehat{\psi}\right) - \widehat{\omega}_N,\\ \dot{\widehat{p}}_E = \widehat{U}\cos\widehat{\theta}\sin\widehat{\psi} + \widehat{V}\left(\cos\widehat{\phi}\cos\widehat{\psi} + \sin\widehat{\phi}\sin\widehat{\theta}\sin\widehat{\psi}\right) + \widehat{W}\left(-\sin\widehat{\phi}\cos\widehat{\psi} + \cos\widehat{\phi}\sin\widehat{\theta}\sin\widehat{\psi}\right) - \widehat{\omega}_E,\\ \dot{\widehat{p}}_H = \widehat{U}\sin\widehat{\theta} - \widehat{V}\sin\widehat{\phi}\cos\widehat{\theta} - \widehat{W}\cos\widehat{\phi}\cos\widehat{\theta} - \widehat{\omega}_H,\\ \dot{\widehat{V}}_T = \dfrac{\left(\widehat{U}\dot{U} + \widehat{V}\dot{\widehat{V}} + \widehat{W}\dot{\widehat{W}}\right)}{\widehat{V}_T},\\ \dot{\widehat{\alpha}} = \dfrac{\left(\widehat{U}\dot{\widehat{W}} - \widehat{W}\dot{U}\right)}{\left(\widehat{U}\widehat{U} + \widehat{W}\widehat{W}\right)},\\ \dot{\widehat{\beta}} = \dfrac{\left(\widehat{V}_T\dot{\widehat{V}} - \widehat{V}\dot{\widehat{V}}_T\right)}{\left(\widehat{V}_T^2\cos\widehat{\beta}\right)},\\ \dot{\widehat{\phi}} = \widehat{P} + \tan\widehat{\theta}\left(\widehat{Q}\sin\widehat{\phi} + \widehat{R}\cos\widehat{\phi}\right),\\ \dot{\widehat{\theta}} = \widehat{Q}\cos\widehat{\phi} - \widehat{R}\sin\widehat{\phi},\\ \dot{\widehat{\psi}} = \dfrac{\left(\widehat{R}\cos\widehat{\phi} + \widehat{Q}\sin\widehat{\phi}\right)}{\cos\widehat{\theta}},\\ \dot{\widehat{P}} = \left(c_2\widehat{P} + c_1\widehat{R}\right)\widehat{Q} + \bar{q}Sb\left(c_3\widehat{C}_l + c_4\widehat{C}_n\right),\\ \dot{\widehat{Q}} = c_5\widehat{P}\widehat{R} + c_6\left(\widehat{R}^2 - \widehat{P}^2\right) + \bar{q}S\bar{C}c_7\widehat{C}_m,\\ \dot{\widehat{R}} = c_8\widehat{P}\widehat{Q} - c_2\widehat{R}\widehat{Q} + \bar{q}Sbc_4\widehat{C}_l + \bar{q}Sbc_9\widehat{C}_n,\\ \dot{\widehat{\delta}}_T = -10\widehat{\delta}_T + 10\delta_{T\mathrm{cmd}},\\ \dot{\widehat{\delta}}_E = -14.30\widehat{\delta}_E + 14.30\delta_{E\mathrm{cmd}},\\ \dot{\widehat{\delta}}_A = -14.30\widehat{\delta}_A + 14.30\delta_{A\mathrm{cmd}},\\ \dot{\widehat{\delta}}_R = -5\widehat{\delta}_R + 5\delta_{R\mathrm{cmd}},\\ \dot{\widehat{\omega}}_N = \dot{\widehat{\omega}}_E = \dot{\widehat{\omega}}_H = \dot{\widehat{V}}_{Tb} = \dot{\widehat{\alpha}}_b = \dot{\widehat{\beta}}_b = \dot{\widehat{P}}_b = \dot{\widehat{Q}}_b = \dot{\widehat{R}}_b = \dot{\widehat{a}}_{Xb} = \dot{\widehat{a}}_{Yb} = \dot{\widehat{a}}_{Zb} = \dot{\widehat{H}}_{Xb} = \dot{\widehat{H}}_{Yb} = \dot{\widehat{H}}_{Zb} = 0,\end{cases} \tag{13}$$

where body velocities $\{\widehat{U}, \widehat{V}, \widehat{W}\}$ and their time rate of changes are defined as in Section 2 and Appendix A. The EKF model also utilizes the following output equation $\widehat{\mathbf{y}} = \widehat{\mathbf{h}}(\mathbf{x}(t), \widehat{\mathbf{u}}(t))$, designed based on available sensors including inertial position (GPS), Euler angles (IMU), body angular rates (IMU), airflow angles vane and airspeed probe, 3D magnetic flux, body angular rate (3D gyros), translational accelerations (3D accelerometers), and elevator, aileron, and rudder deflection sensors

$$\widehat{\mathbf{h}}:\begin{cases}\widehat{y}_1 = \widehat{p}_N, & \widehat{y}_2 = \widehat{p}_E, & \widehat{y}_3 = \widehat{p}_H, & \widehat{y}_4 = \widehat{\phi},\\ \widehat{y}_5 = \widehat{\theta}, & \widehat{y}_6 = \widehat{\psi}, & \widehat{y}_7 = \widehat{P}, & \widehat{y}_8 = \widehat{Q},\\ \widehat{y}_9 = \widehat{R}, & \widehat{y}_{10} = \widehat{V}_T + \widehat{V}_{T_b}, & \widehat{y}_{11} = \widehat{\alpha} + \widehat{\alpha}_b, & \widehat{y}_{12} = \widehat{\beta} + \widehat{\beta}_b,\\ \widehat{y}_{13} = \widehat{H}_X + \widehat{H}_{Xb}, & \widehat{y}_{14} = \widehat{H}_Y + \widehat{H}_{Yb}, & \widehat{y}_{15} = \widehat{H}_Z + \widehat{H}_{Zb}, & \widehat{y}_{16} = \widehat{P} + \widehat{P}_b,\\ \widehat{y}_{17} = \widehat{Q} + \widehat{Q}_b, & \widehat{y}_{18} = \widehat{R} + \widehat{R}_b, & \widehat{y}_{19} = \dfrac{\left(\widehat{X}^P + \widehat{X}^A\right)}{m} + \widehat{a}_{Xb}, & \widehat{y}_{20} = \dfrac{\widehat{Y}^A}{m} + \widehat{a}_{Yb},\\ \widehat{y}_{21} = \dfrac{\widehat{Z}^A}{m} + \widehat{a}_{Zb}, & \widehat{y}_{22} = \widehat{\delta}_E, & \widehat{y}_{23} = \widehat{\delta}_A, & \widehat{y}_{24} = \widehat{\delta}_R.\end{cases} \tag{14}$$

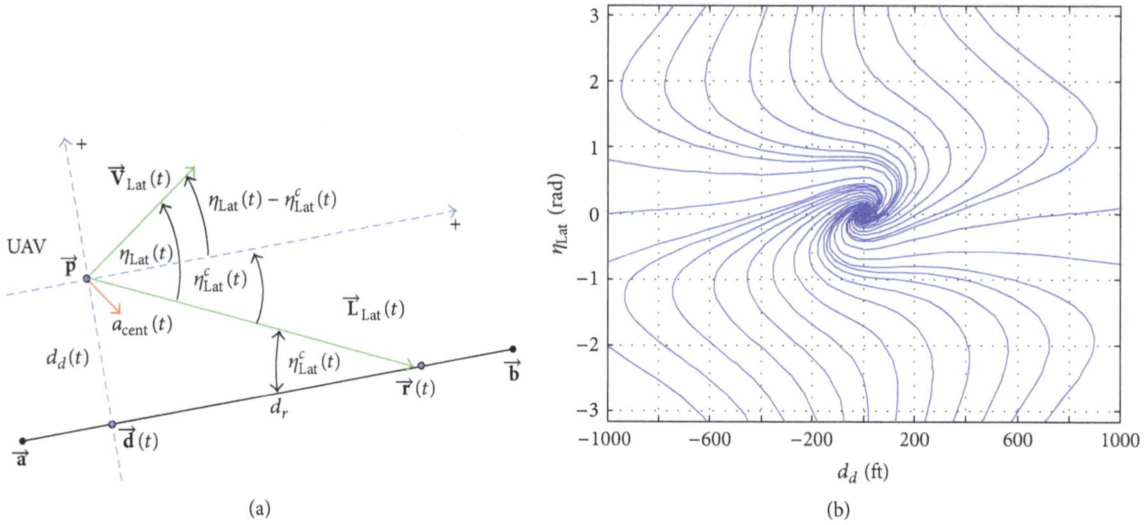

Figure 6: Lateral guidance geometry (a). Phase portrait of lateral guidance (b).

State vector $\widehat{\mathbf{x}}[k] \in R^{\widehat{n}}$, input vector $\mathbf{u}[k] \in R^{\widehat{m}}$, equivalent to the system input, and measurement vector $\widehat{\mathbf{y}}[k] \in R^{\widehat{r}}$ are defined as

$$
\begin{aligned}
\widehat{\mathbf{x}}[k] = \big[&\widehat{p}_N[k], \widehat{p}_E[k], \widehat{p}_H[k], \widehat{V}_T[k], \widehat{\alpha}[k], \widehat{\beta}[k], \\
&\widehat{\phi}[k], \widehat{\theta}[k], \widehat{\psi}[k], \widehat{P}[k], \widehat{Q}[k], \widehat{R}[k], \widehat{\delta}_T[k], \widehat{\delta}_E[k], \\
&\widehat{\delta}_A[k], \widehat{\delta}_R[k], \widehat{\omega}_N[k], \widehat{\omega}_E[k], \widehat{\omega}_H[k], \widehat{V}_{T_b}[k], \\
&\widehat{\alpha}_b[k], \widehat{\beta}_b[k], \widehat{P}_b[k], \widehat{Q}_b[k], \widehat{R}_b[k], \widehat{a}_{Xb}[k], \\
&\widehat{a}_{Yb}[k], \widehat{a}_{Zb}[k], \widehat{H}_{Xb}[k], \widehat{H}_{Yb}[k], \widehat{H}_{Zb}[k]\big]^T, \\
\mathbf{u}[k] = \big[&\delta_{T_{\text{cmd}}}[k], \delta_{E_{\text{cmd}}}[k], \delta_{A_{\text{cmd}}}[k], \delta_{R_{\text{cmd}}}[k]\big]^T, \\
\widehat{\mathbf{y}}[k] = \big[&\widehat{p}_N[k], \widehat{p}_E[k], \widehat{p}_H[k], \widehat{\phi}[k], \widehat{\theta}[k], \widehat{\psi}[k], \widehat{P}[k], \\
&\widehat{Q}[k], \widehat{R}[k], \widehat{V}_T[k] + \widehat{V}_{T_b}[k], \widehat{\alpha}[k] + \widehat{\alpha}_b[k], \\
&\widehat{\beta}[k] + \widehat{\beta}_b[k], H_{X0} + \widehat{H}_{Xb}[k], H_{Y0} + \widehat{H}_{Yb}[k], \\
&H_{Z0} + \widehat{H}_{Zb}[k], \widehat{P}[k] + \widehat{P}_b[k], \widehat{Q}[k] + \widehat{Q}_b[k], \\
&\widehat{R}[k] + \widehat{R}_b[k], \overline{a}_X[k] + \widehat{a}_{Xb}[k], \overline{a}_Y[k] + \widehat{a}_{Yb}[k], \\
&\overline{a}_Z[k] + \widehat{a}_{Zb}[k], \widehat{\delta}_E[k], \widehat{\delta}_A[k], \widehat{\delta}_R[k]\big]^T,
\end{aligned}
\tag{15}
$$

where the subscript b refers to bias, a refers to acceleration, and $\widehat{H}$ indicates magnetic flux. Subscripts $\{X, Y, Z\}$ are defined in body frame $\overline{\mathbf{B}}$ and the subscript zero denotes the inertial reference magnetic flux for the flying area, assumed fixed. Variable $\overline{a}$ is defined as a function of estimated states, as follows:

$$
\begin{bmatrix} \overline{a}x \\ \overline{a}y \\ \overline{a}z \end{bmatrix} = \begin{bmatrix} \widehat{R}\widehat{V} - \widehat{Q}\widehat{W} + \dfrac{\left(\widehat{T} + \widehat{\overline{q}}S\widehat{C}_X\right)}{m} \\ -\widehat{R}\widehat{U} + \widehat{P}\widehat{W} + \dfrac{\widehat{\overline{q}}S\widehat{C}_Y}{m} \\ \widehat{Q}\widehat{U} - \widehat{P}\widehat{V} + \dfrac{\widehat{\overline{q}}S\widehat{C}_Z}{m} \end{bmatrix}, \tag{16}
$$

where $\widehat{\overline{q}}$ is the estimated dynamic pressure and $\{\widehat{C}_X, \widehat{C}_Y, \widehat{C}_Z\}$ are the estimated body aerodynamic force coefficients. The calculated accelerations $\{\overline{a}_X, \overline{a}_Y, \overline{a}_Z\}$ plus the estimated biases $\{\widehat{a}_{Xb}, \widehat{a}_{Yb}, \widehat{a}_{Zb}\}$, as shown in (15), are used as the estimated accelerations. They do not include gravity effects to keep consistency to accelerometer measurements. As in Section 2 and Appendix A, the same relations apply to the estimated stability and body aerodynamic coefficients and to the estimated thrust $\widehat{T}$. The body magnetic flux $\{H_X, H_Y, H_Z\}$ is related to the inertial magnetic flux $\{H_{N0}, H_{E0}, H_{D0}\}$ as follows:

$$
\begin{bmatrix} H_X \\ H_Y \\ H_Z \end{bmatrix} = C_I^B(\phi, \theta, \psi) \begin{bmatrix} H_{N0} \\ H_{E0} \\ H_{D0} \end{bmatrix} + \begin{bmatrix} \widehat{H}_{Xb} \\ \widehat{H}_{Yb} \\ \widehat{H}_{Zb} \end{bmatrix}, \tag{17}
$$

where C_I^B corresponds to the direction cosine matrix for rotation from inertial reference $\mathbf{I}$ to body reference $\overline{\mathbf{B}}$. The inertial magnetic flux $\{H_{N0}, H_{E0}, H_{D0}\}$ is a known fixed data obtained for the area to be flown.

5. Artificial Neural Network

A feedforward ANN with a single hidden layer, containing a finite number of neuron cells with sigmoid-type activation functions and a linear activation functions at the output cells, intrinsically materializes the universal approximation theorem. It approximates any multivariate continuous function on a compact domain subset of R^m to any degree of accuracy. Based on this theorem, proven originally in [39], for any given continuous function $f_j(\mathbf{X}) \in R^1$, with $\mathbf{X} = [x_1, x_2, \ldots, x_n]^T \in R^n$, both normalized within the range $[-1, 1]$ and for $\delta > 0$, there exist an integer N and a set of parameters (weights and biases) $b_i, c_j \in R$, and $\mathbf{w}_i \in R^n$ and $a_{ji} \in R$ with $i = 1 \cdots N$ and j a fixed value from the set $\{1 \cdots p\}$ indicating a particular function within a set of related functions, as is the case of aircraft accelerations, such that

$$F_j(\mathbf{X}) = \sum_{i=1}^{N} a_{ji}\varphi\left(\mathbf{w}_i^T\mathbf{X} + b_i\right) + c_j \tag{18}$$

is as an approximation of the function $f_j(\mathbf{X})$; that is, $|F_j(\mathbf{X}) - f_j(\mathbf{X})| < \delta$, where $\varphi(\cdot)$ is a bounded monotonically increasing activation function. Figure 7 shows the general structure of the ANN and its interconnections.

To develop an iterative adaptive online algorithm, one learning epoch at each sample time is performed. This is done using a sliding window of data. This moving window has a constant length in time, but as new data is measured, the oldest data is discarded and the new is added as the most recent (concept similar to shift registering). Using the concept of sliding mode learning, the ANN is kept permanently in the training stage adapting its parameters to reflect changes in the dynamics of the aircraft.

For vertical acceleration estimation $\hat{a}_Z = \bar{a}_Z + \hat{a}_{Zb}$ (reducing the ANN to one single output), the sliding windows of input and target data, $\widehat{\mathbf{X}}[k] \in R^{N_A \times n}$ and $\widehat{\mathbf{Y}}[k] \in R^{N_A}$, are defined as follows, where N_A sets the window's length:

$$\widehat{\mathbf{X}}^T[k] = \begin{bmatrix} \widehat{V}_T[k-N_A+1] & \widehat{V}_T[k-N_A+2] & \cdots & V_T[k-1] & V_T[k] \\ \hat{\alpha}[k-N_A+1] & \hat{\alpha}[k-N_A+2] & \cdots & \hat{\alpha}[k-1] & \alpha[k] \\ \hat{\beta}[k-N_A+1] & \hat{\beta}[k-N_A+2] & \cdots & \hat{\beta}[k-1] & \beta[k] \\ \widehat{Q}[k-N_A+1] & \widehat{Q}[k-N_A+2] & \cdots & \widehat{Q}[k-1] & Q[k] \\ \dot{\hat{\alpha}}[k-N_A+1] & \dot{\hat{\alpha}}[k-N_A+2] & \cdots & \dot{\hat{\alpha}}[k-1] & \dot{\alpha}[k] \\ \hat{\delta}_T[k-N_A+1] & \hat{\delta}_T[k-N_A+2] & \cdots & \hat{\delta}_T[k-1] & \delta_T[k] \\ \hat{\delta}_E[k-N_A+1] & \hat{\delta}_E[k-N_A+2] & \cdots & \hat{\delta}_E[k-1] & \delta_E[k] \end{bmatrix},$$
$$\widehat{\mathbf{Y}}^T[k] = \left[\hat{a}_Z[k-N_A+1] \;\; \hat{a}_Z[k-N_A+2] \;\; \cdots \;\; \hat{a}_Z[k-1] \;\; \hat{a}_Z[k]\right]. \tag{19}$$

The training is done by minimizing the accumulated quadratic error between $\hat{a}_Z = \bar{a}_Z + \hat{a}_{Zb}$ and the estimated ANN vertical accelerations $\underline{a}_Z$, that is, minimizing the cost function $W(\boldsymbol{\eta}) = \boldsymbol{\varepsilon}^T(\boldsymbol{\eta}) \cdot \boldsymbol{\varepsilon}(\boldsymbol{\eta}) \in \mathbf{R}$, where

$$\boldsymbol{\varepsilon}(\boldsymbol{\eta}) = \begin{bmatrix} \varepsilon[k-N_A+1,\boldsymbol{\eta}] \\ \vdots \\ \varepsilon[k,\boldsymbol{\eta}] \end{bmatrix} \times \begin{bmatrix} \hat{a}_z[k-N_A+1] - \underline{a}_z[k-N_A+1,\boldsymbol{\eta}] \\ \vdots \\ \hat{a}_z[k] - \underline{a}_z[k,\boldsymbol{\eta}] \end{bmatrix}, \tag{20}$$

where the line subscript is used to identify the ANN estimations, and where $\boldsymbol{\eta} \in R^m$ is the parameter vector

$$\begin{aligned} \boldsymbol{\eta}^T &= [\eta_1, \ldots, \eta_m] \\ &= [a_{11}, \ldots, a_{p1}, \ldots, a_{1N}, \ldots, a_{pN}, b_1, \ldots, b_N, \\ &\quad w_{11}, \ldots, w_{1n}, \ldots, w_{N1}, \ldots, w_{Nn}, c_1, \ldots, c_p] \end{aligned} \tag{21}$$

with $m = p \times n + N \times n + N + p$. Using a Newton-type search method, the following iterative update is defined as

$$\boldsymbol{\eta}[k+1] = \boldsymbol{\eta}[k] - \left[\nabla^2 W(\boldsymbol{\eta}[k])\right]^{-1} \cdot \nabla^T W(\boldsymbol{\eta}[k]), \tag{22}$$

where $\nabla W(\boldsymbol{\eta}[k])$ and $\nabla^2 W(\boldsymbol{\eta}[k])$ are the gradient and the Hessian of the cost function W with respect to the parameter vector, defined as follows:

$$\nabla W(\boldsymbol{\eta}) = \left[\frac{\partial W}{\partial \eta_1} \cdots \frac{\partial W}{\partial \eta_m}\right],$$
$$\nabla^2 W(\boldsymbol{\eta}) = \begin{bmatrix} \dfrac{\partial^2 W}{\partial \eta_1^2} & \cdots & \dfrac{\partial^2 W}{\partial \eta_1 \partial \eta_m} \\ \vdots & \ddots & \vdots \\ \dfrac{\partial^2 W}{\partial \eta_m \partial \eta_1} & \cdots & \dfrac{\partial^2 W}{\partial \eta_m^2} \end{bmatrix}. \tag{23}$$

These quantities are approximated numerically at each iteration where the error vector $\boldsymbol{\varepsilon}(\boldsymbol{\eta})$ can be approximated by a first-order Taylor series

$$\boldsymbol{\varepsilon}(\boldsymbol{\eta} + \Delta\boldsymbol{\eta}) \approx \boldsymbol{\varepsilon}(\boldsymbol{\eta}) + J(\boldsymbol{\eta}) \cdot \Delta\boldsymbol{\eta}, \tag{24}$$

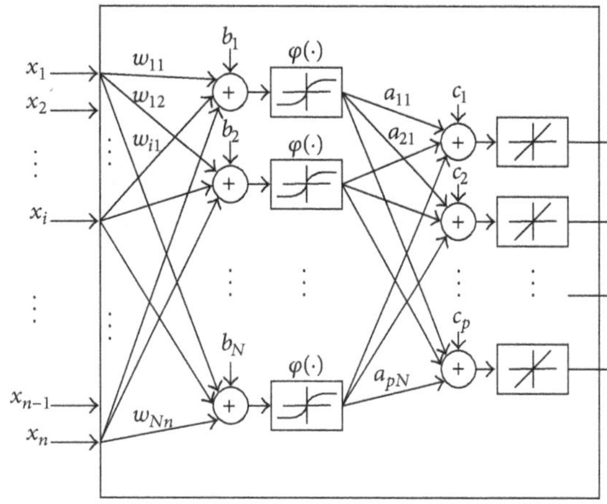

Figure 7: Structure of ANN.

where the rows of $\mathbf{J}(\boldsymbol{\eta})$, the Jacobian of $\boldsymbol{\varepsilon}(\boldsymbol{\eta})$, include the gradients of each element of $\boldsymbol{\varepsilon}(\boldsymbol{\eta})$ with respect to the parameter vector $\boldsymbol{\eta}$ as follows:

$$\mathbf{J}(\boldsymbol{\eta}) = \begin{bmatrix} \nabla\varepsilon[k-N_A+1,\boldsymbol{\eta}] \\ \vdots \\ \nabla\varepsilon[k,\boldsymbol{\eta}] \end{bmatrix} = \begin{bmatrix} \dfrac{\partial\varepsilon[k-N_A+1,\boldsymbol{\eta}]}{\partial\eta_1} & \cdots & \dfrac{\partial\varepsilon[k-N_A+1,\boldsymbol{\eta}]}{\partial\eta_m} \\ \vdots & \ddots & \vdots \\ \dfrac{\partial\varepsilon[k,\boldsymbol{\eta}]}{\partial\eta_1} & \cdots & \dfrac{\partial\varepsilon[k,\boldsymbol{\eta}]}{\partial\eta_m} \end{bmatrix}. \tag{25}$$

Then, the cost function $W(\boldsymbol{\eta})$, evaluated at $\boldsymbol{\eta} + \Delta\boldsymbol{\eta}$, is expressed as

$$W(\boldsymbol{\eta}+\Delta\boldsymbol{\eta}) \approx \overbrace{\boldsymbol{\varepsilon}^T(\boldsymbol{\eta})\cdot\boldsymbol{\varepsilon}(\boldsymbol{\eta})}^{W(\boldsymbol{\eta})} + \overbrace{2\boldsymbol{\varepsilon}^T(\boldsymbol{\eta})\cdot J(\boldsymbol{\eta})}^{\nabla W(\boldsymbol{\eta})}\cdot\Delta\boldsymbol{\eta} + \frac{1}{2}\Delta\boldsymbol{\eta}^T\cdot\underbrace{2J^T(\boldsymbol{\eta})\cdot J(\boldsymbol{\eta})}_{\nabla^2 W(\boldsymbol{\eta})}\cdot\Delta\boldsymbol{\eta}. \tag{26}$$

From (26), the cost function's gradient $\nabla W(\boldsymbol{\eta}[k])$ and Hessian $\nabla^2 W(\boldsymbol{\eta}[k])$ are then obtained as functions of the error vector $\boldsymbol{\varepsilon}(\boldsymbol{\eta})$ and its Jacobian $\mathbf{J}(\boldsymbol{\eta})$. Then, (22) is rewritten as

$$\boldsymbol{\eta}[k+1] = \boldsymbol{\eta}[k] - \left[2\mathbf{J}^T(\boldsymbol{\eta}[k])\cdot\mathbf{J}(\boldsymbol{\eta}[k])\right]^{-1} \cdot\left[2\boldsymbol{\varepsilon}^T(\boldsymbol{\eta}[k])\cdot\mathbf{J}(\boldsymbol{\eta}[k])\right]^T. \tag{27}$$

As a way to improve the numerical search, the Levenberg approach is chosen (see [40]), where a factor called Levenberg adaptive parameter λ is introduced. This method permits a transition from a steepest descent to a quadratic method when closer to the minimum. Based on this approach, (27) is modified as

$$\boldsymbol{\eta}[k+1] = \boldsymbol{\eta}[k] - \left[2\mathbf{J}^T(\boldsymbol{\eta}[k])\cdot\mathbf{J}(\boldsymbol{\eta}[k]) + \lambda I\right]^{-1} \cdot\left[2\boldsymbol{\varepsilon}^T(\boldsymbol{\eta}[k])\cdot\mathbf{J}(\boldsymbol{\eta}[k])\right]^T. \tag{28}$$

Table 3: Closed-loop infinity norm associated to each uncertainty.

Parameter	$\|\mathbf{F}_l(\widehat{\mathbf{P}}^1, \overline{\mathbf{K}}^1)\|_\infty$
C_{mq}	0.3846
$C_{m\delta e}$	0.2451
$C_{l\beta}$	0.0100
C_{nr}	0.3816
$C_{n\delta r}$	0.1674
All	0.5617

Under Levenberg's logic, this adaptive parameter is adjusted during each iteration based on the cost function evolution. In this paper, the logic is modified to reset the Levenberg's adaptive parameter, that is, $\lambda[k]$, when reaching predefined extreme values to avoid unbounded growing specially when the cost function has reached a minimum and consecutive iterations remain closer to it (a common occurrence in one-shot offline minimizations). This modification is required to keep the ANN under permanent training. The modification is as follows:

$$\begin{aligned} &\text{Start with } \lambda[k] = \lambda_0 \\ &\text{IF } W([k+1]) < W([k]) \longrightarrow \lambda[k+1] = \frac{\lambda[k]}{\lambda_{\text{dec}}} \\ &\text{IF } \lambda[k+1] < \lambda_{\min} \longrightarrow \lambda[k+1] = \lambda_0 \\ &\text{IF } W([k+1]) \geq W([k]) \longrightarrow \lambda[k+1] = \lambda[k]\cdot\lambda_{\text{inc}} \\ &\text{IF } \lambda[k+1] > \lambda_{\max} \longrightarrow \lambda[k+1] = \lambda_0, \end{aligned} \tag{29}$$

where $0 < \lambda_{\min} < \lambda_{\max}$ are the minimum and maximum allowed values, and $\lambda_{\text{dec}} > 0$ and $\lambda_{\text{inc}} > 0$ are the decrement and increment correction factors. All these factors are empirically obtained after extensive simulations.

6. Simulation and Testing

This section is devoted to test previous designs in simulation, by employing stored flight test data, and also by comparing the designed FCS performance in contrast with a COTS autopilot under similar conditions, for the same aircraft. The manufacture of COTS system cannot be revealed due to contractual obligations.

6.1. ANN Simulation Results. The ANN estimation capability is evaluated for a window length of $N_A = 400$ (samples), which corresponds to 10 seconds of past flight for a sampling

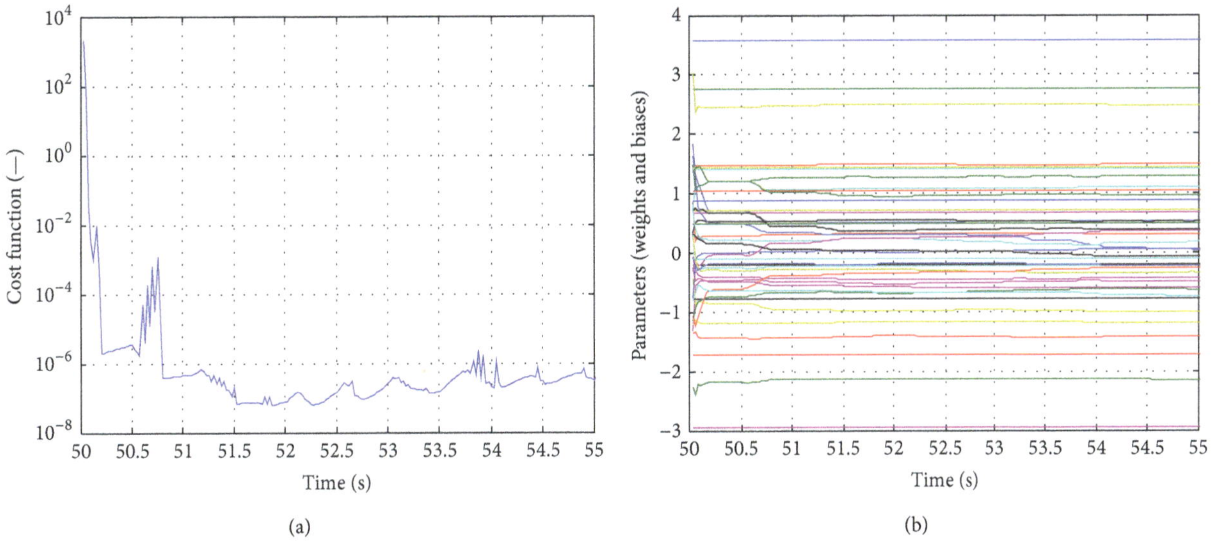

FIGURE 8: Cost function minimization, no noise (a). Parameter evolution, no noise (b).

interval $dt = 1/40$ seconds. For the logic defined in (29), $\lambda_0 = 0.001$, $\lambda_{max} = 1$, $\lambda_{min} = 0.00001$, $\lambda_{dec} = 3$, and $\lambda_{inc} = 2$.

Two simulations are run for a 33% Yak-54 UAS, with and without measurement noise. Figures 8 and 9 show the convergence of the parameters, the cost function minimization, the Levenberg adaptive parameter, and the measured and estimated vertical acceleration, considering noise. Figures 10 and 11 show the same information for the case without noise.

By using the Delta method, which is a numerical technique to approximate partial derivatives, stability and control derivatives can be extracted.

The standard application of this technique, in previous cited articles, is generally offline postprocessing analysis. The idea presented here is to allocate a specific logic and to compute these approximate partial derivatives online. Two identical structures of the ANN are kept in parallel. They are fed incoming measurements, similar to the ANN under training. Also, these two structures are updated with the current weight and biases from the ANN. One of these structure is fed with a perturbed input (i.e., measured angle of attack plus a small added perturbation), while the second structure is kept fed by unperturbed inputs. By taking the quotient between the differences of the two structures outputs, with respect to the magnitude of the added perturbation, an approximation of the partial derivative is obtained in real time.

For this case, the partial derivative of the vertical acceleration with respect to some input of the ANN is obtained. These quantities can then be used to calculate more specific stability and control derivatives by converting from body to stability coordinates. This allows the capability of detecting dynamic changes due to internal or external changing conditions, or due to failures, by monitoring the time tendency of these values.

By comparing Figures 8 and 9, with Figures 10 and 11, the effect of measurement noise is clear. Cost function values are considerably higher. Also, the parameter vector does not settle permanently and it adjusts continuously due to the inherent variability condition of the Levenberg parameter.

As shown in Figure 11(b), a fairly good prediction is achieved. Further precaution needs to be observed to avoid deep changes, as seen between seconds 54 and 55 when a steep readjustment occurred. Using a desktop, each iteration took less than 1.2 seconds; this means that a dedicated microprocessor must be used for real-time estimations.

6.2. Adaptive Artificial Neural Network Testing. To verify the performance of ANN's adaptive capability in the modeling of unsteady aerodynamic forces, actual 33% scale Yak-54 UAS flight test data from the 2009 flight test at Pegasus Airfield in Antarctica is used. The data was collected with a sampling frequency of 10 (Hz). For comparison and validation purposes, the same ANN for Z-axis acceleration is trained and tested with and without adaptiveness. This ANN is constructed from twenty cells ($M = 20$) in the hidden layer and fed with same inputs as defined earlier for online ANN. Figure 12 shows the flight test data and the estimations with and without adaptiveness.

The flight had a well-defined pattern. From a prolonged circular autonomous period of flight, the aircraft is switched to remote control for landing around second 1145. During autonomous flight, the nonadaptive ANN is trained. It is then tested in both the remaining autonomous flight segment and during the remote controlled period. The adaptive ANN is tested in the same interval as the nonadaptive ANN.

Figure 12 shows a close estimation of both types of ANN before mode switching. After changing control sources, a significant dynamic behavior experienced by the aircraft left nonadaptive ANN outdated.

6.3. Gain Scheduling Simulation Results. To assess the gain scheduling techniques, hypothetical failures are induced through changes in the numerical value of stability and

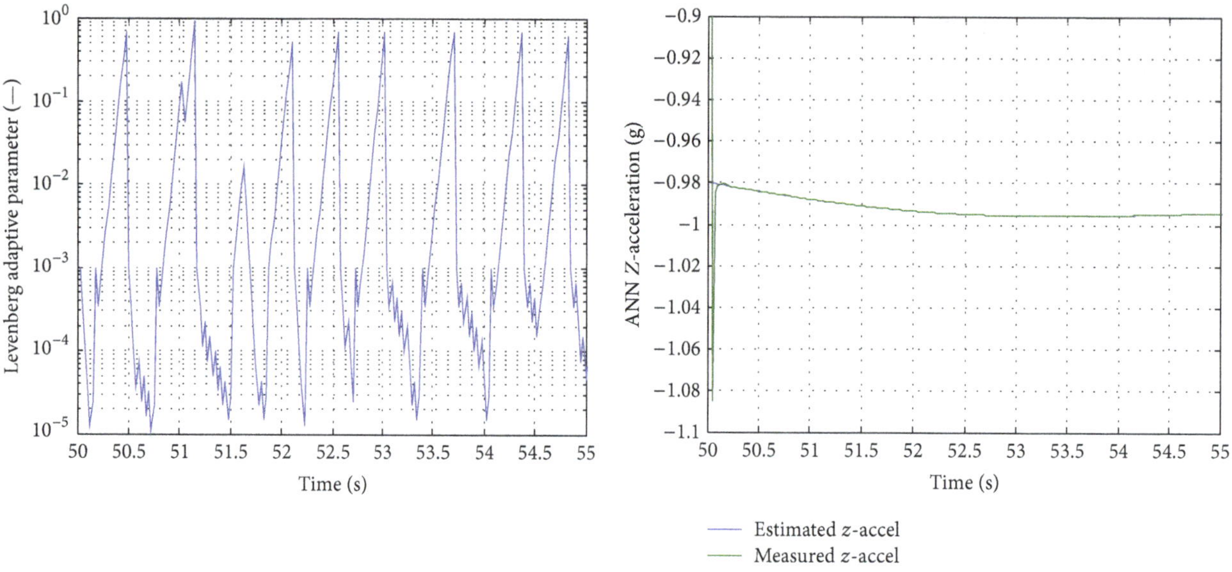

(a) (b)

FIGURE 9: Levenberg adaptive parameter, no noise (a). Vertical acceleration, no noise (b).

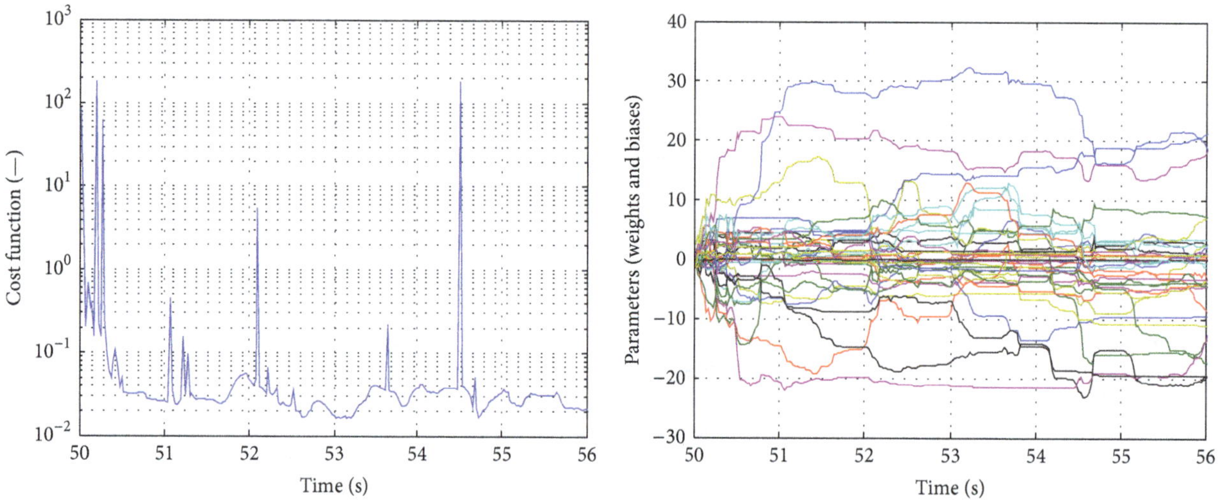

(a) (b)

FIGURE 10: Cost function minimization, with noise (a). Parameter evolution, with noise (b).

control derivatives. The gain scheduling adapts controllers to guarantee the aircraft stability and performance in the event of an adverse condition (e.g., servo failure or damage). Extreme changes could be a result of system failure or a result of damage to the aircraft. An adaptive flight control system must have the potential to adjust for failure or damage to maintain control in flight and meet the required handling qualities that allow for safe recovery of the airplane. As a condition for successful gain scheduling, the FCS is expected to achieve a similar level of performance and required uncertainty management before and after detection of dynamic changes.

To replicate a major change in the lateral-directional dynamics of UAS due to a failure or unsteady aerodynamics, at 120 seconds of a horizontal level flight, the numerical value of C_{n_β} (Table 1) is reduced by a factor of 4. This reflects a sudden change in the UAS dynamics, affecting the stability of UAS directionally. After inducing this change, the current controller loses its stabilizing capability and oscillation can be traced in the aircraft's lateral-directional states. As it was demonstrated in Section 6.2, the adaptive ANN allows identification of sudden changes in the aircraft dynamics in real time.

At 138 seconds and after detection of an unexpected change in the aircraft's stability and control derivative (i.e., change in C_{n_β}), the adaptive controller replaced the old controller with a new one and successfully mitigated sudden

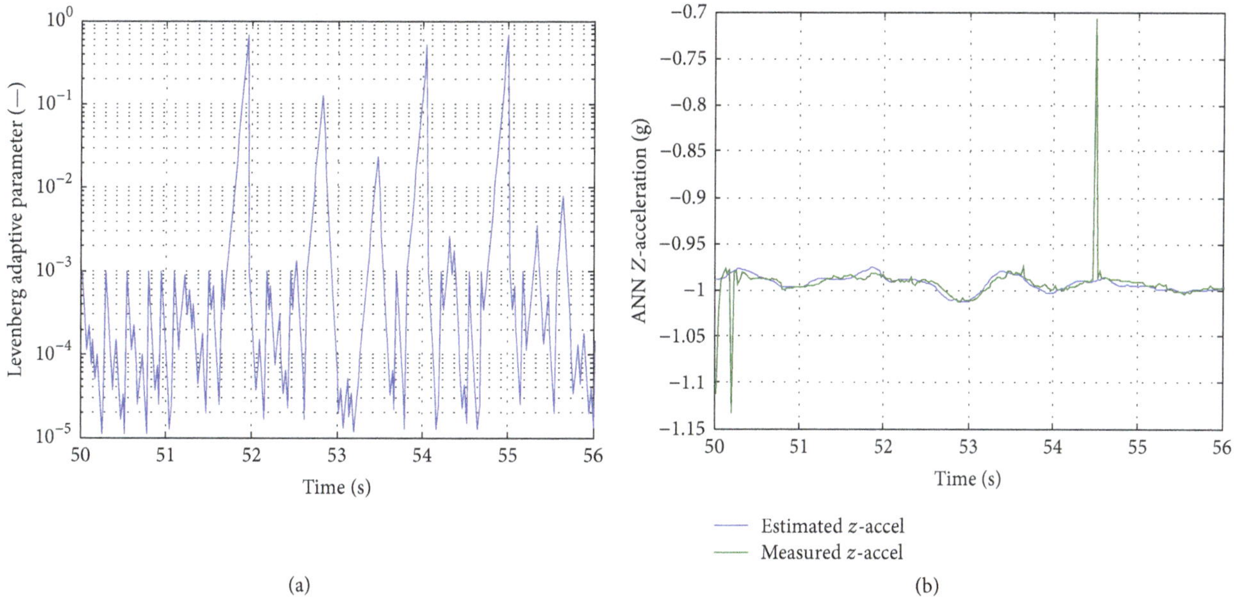

FIGURE 11: Levenberg adaptive parameter, with noise (a). Vertical acceleration, with noise (b).

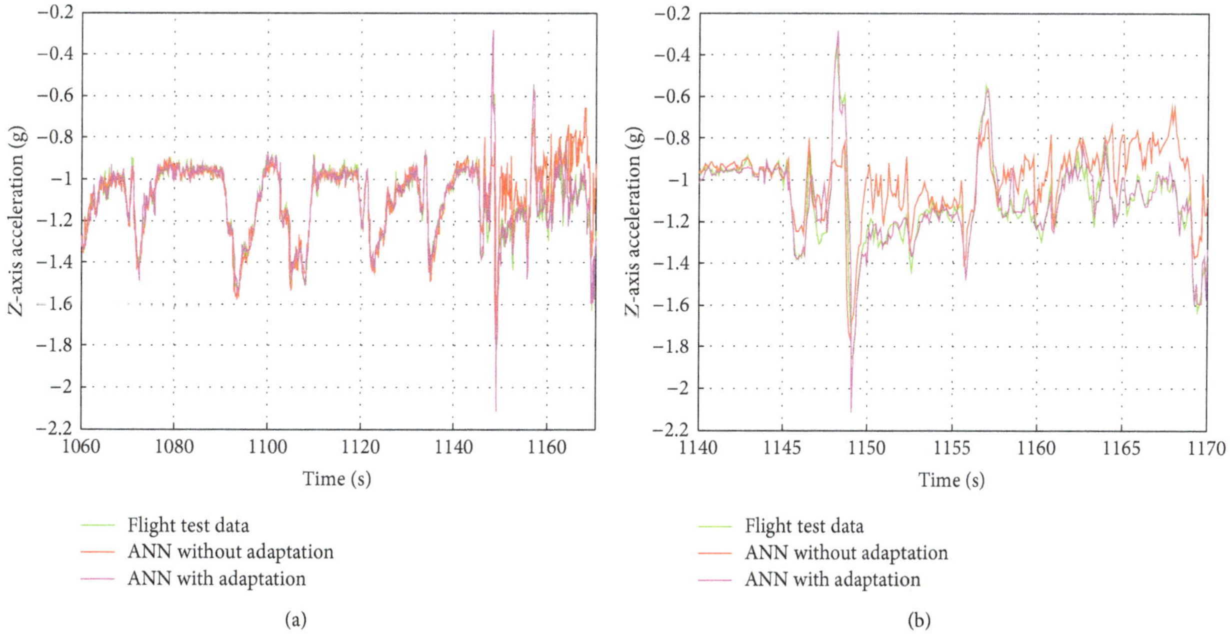

FIGURE 12: Z-axis measured and predicted acceleration: larger flight portion (a); zoom after switching control (b).

change in the aircraft's dynamics. The growing amplitude of oscillation is damped out and the roll and yaw rates and sideslip angle are effectively controlled.

The resultant period of time taken to implement the gain scheduling (18 seconds) is affected by the fact that the failure does not show up immediately and the vehicle remains mostly unaffected. The induced failure is an alteration of the rate of change in yaw rate with respect to changes in sideslip angle, that is, C_{n_β}, so for an original constant and close to zero sideslip (horizontal level flight), no immediate effect is observed. After sideslip angle starts to change, the failure has a multiplicative effect and grows faster. This is evident after second 130.

It is important to point out that although the closed loop remains (linearly) stable, the uncertainty robustness is lost, as the infinity norm grows over the unity $\|\mathbf{F}_l(\widehat{\mathbf{P}}^1, \overline{\mathbf{K}}^1)\|_\infty = 4.6355$. In other words, a well-designed flight control system is robust to a fairly large range of uncertainty or changes in aircraft dynamics, however, if the changes become more extreme the performance degrades and the aircraft may

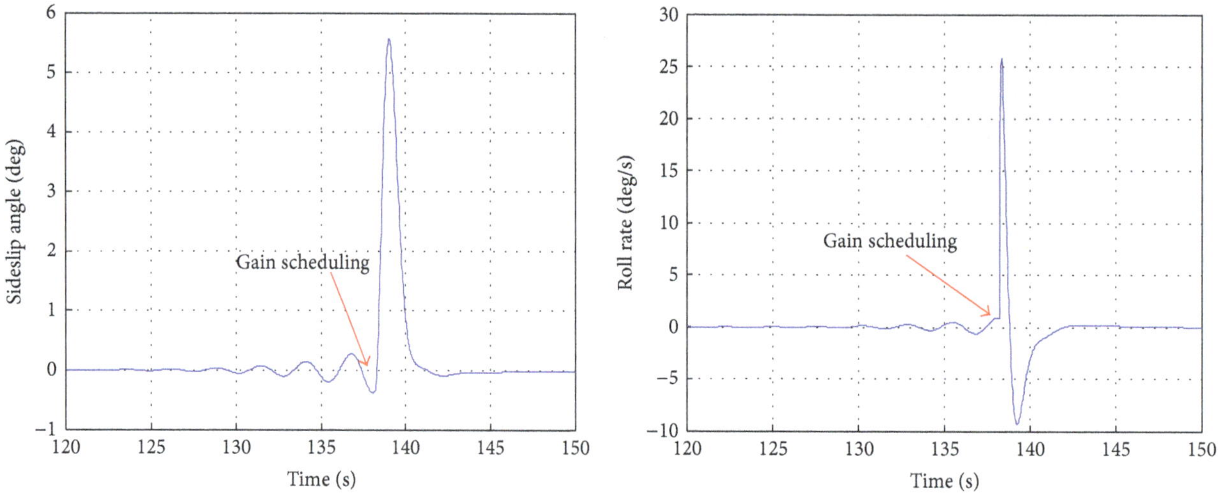

(a) (b)

FIGURE 13: Sideslip angle (a). Roll rate (b).

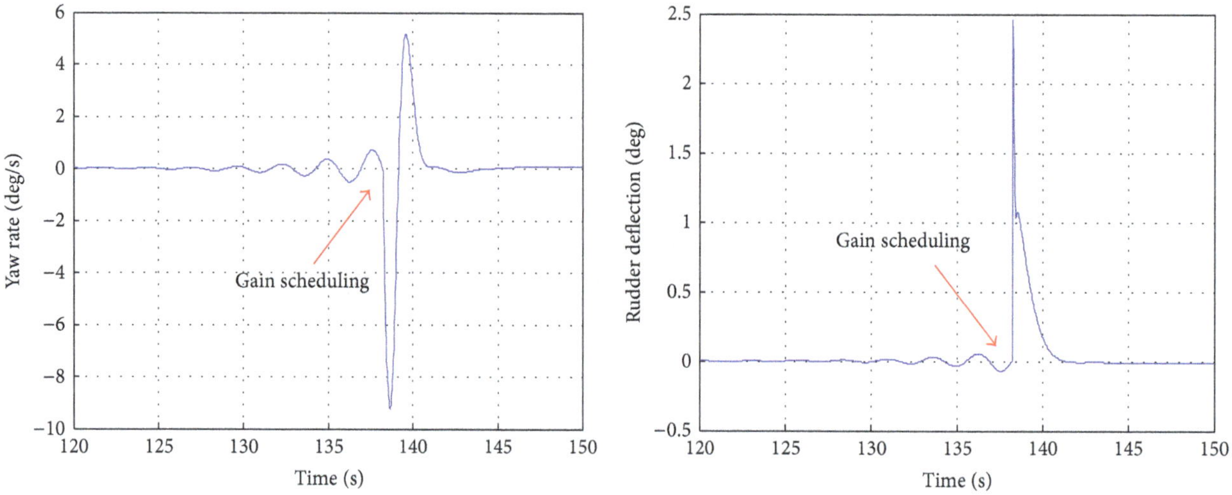

(a) (b)

FIGURE 14: Yaw rate (a). Rudder deflection (b).

go unstable. An adaptive system has the ability to readjust the controller to achieve desired performance or regain robustness about the new point.

In transition between controllers and to avoid significant sensitivity, the gain scheduling must be coupled with the ANN output. Figures 13 and 14 show the failure and consequent gain scheduling effects, on primarily involved states. All simulations were performed with the full FCS.

To broaden the applicability of FCS technologies and to extend assessment process, a failure which degrades aircraft's longitudinal stability is modeled. It is assumed that at 200 seconds of a horizontal level flight, a failure in the aircraft caused a major change in the magnitude of C_{m_α} (Table 1). The numerical value of C_{m_α} is reduced by a factor of 10. As it is shown in Figures 15, 16, and 17 and after the failure took place, the longitudinal robust stability is severely affected; that is, $\|\mathbf{F}_l(\widehat{\mathbf{P}}^1, \overline{\mathbf{K}}^1)\|_\infty = 0.8895$.

Detection of the failure, calculation of a new controller, and gain scheduling are assumed to happen after 5 seconds (200–205 (sec)). At 205 seconds, the new controller is online modifying control surface commands. Severe impact is seen in all major longitudinal variables, as shown in Figures 15, 16, and 17.

The abrupt change in the angle of attack stability coefficient (C_{m_α}) increases the pitch angle and subsequently reduces the airspeed. To mitigate these effects, the elevator is commanded to quickly recover the equilibrium about pitch axes and restoring the pitch angle. To restore the airspeed, the throttle is also commanded. As demonstrated with the simulated failure, the system without adaptation cannot achieve the desired tracking or performance.

6.4. Functionality Tests and Comparison with Flight Test Data. The purpose of this subsection is to assess different

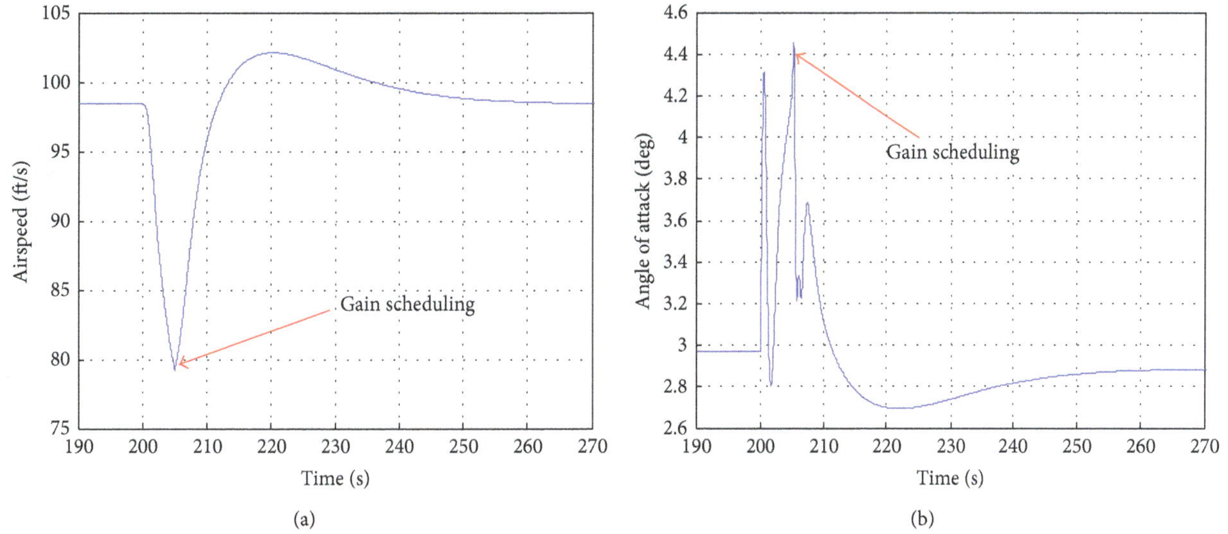

FIGURE 15: Airspeed (a). Angle of attack (b).

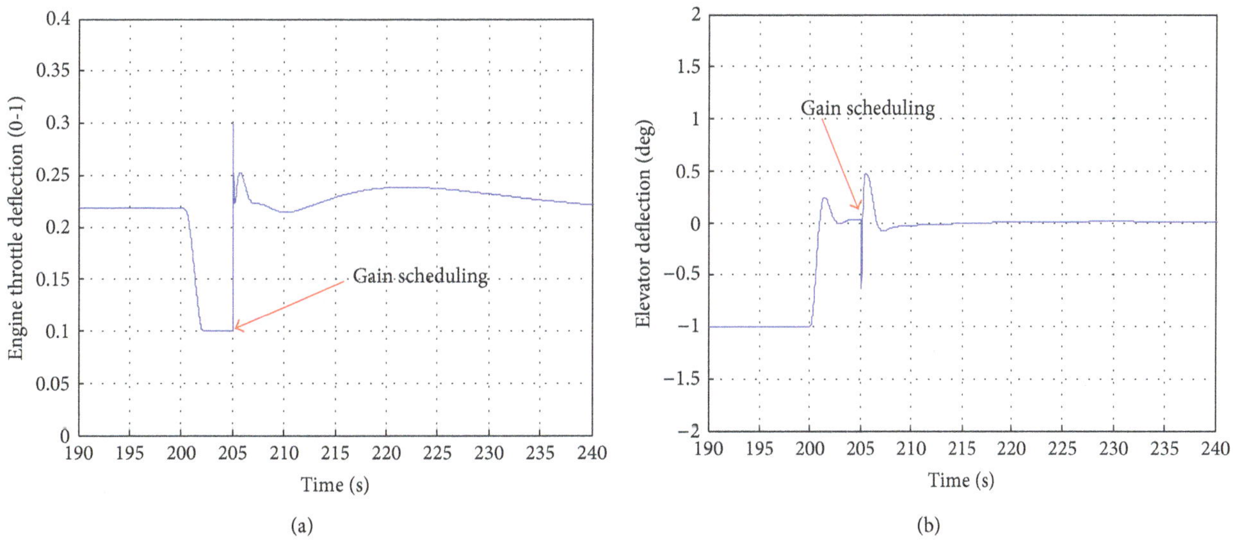

FIGURE 16: Engine throttle (a). Elevator deflection (b).

functionalities of the designed FCS and to compare its performance in closed loop with autonomous flight test data obtained using a COTS autopilot system. It is known that the COTS system uses an H-infinity controller. For comparison purposes, similar flight conditions are induced in the simulation, and same waypoints/trajectory are used for path following. Superficially, performances of the designed FCS and the COTS autopilot are compared in the following categories:

(i) extended Kalman filter performance (flight measurements versus EKF estimation);

(ii) inner loop: H-infinity robust controller performance (e.g., the maximum commanded bank angle);

(iii) outer loop: accuracy of guidance logic and path following in tight turns in presence of crosswind (e.g., deviation from the desired trajectory);

(iv) airspeed and altitude control;

(v) saturation of controls.

6.4.1. Extended Kalman Filter Testing. A slow update rate GPS position sampled at 2 (Hz) is employed to test the smooth interpolation capability of the EKF. Figure 18 shows the measured and estimated geographical position and altitude.

6.4.2. Flight Control System Testing. As part of the comparison, inner and outer loops and extended Kalman filter, as defined before, are included in testing. In this subsection, no gain scheduling and no ANN estimations are performed. These two features are tested independently and presented in previous two subsections.

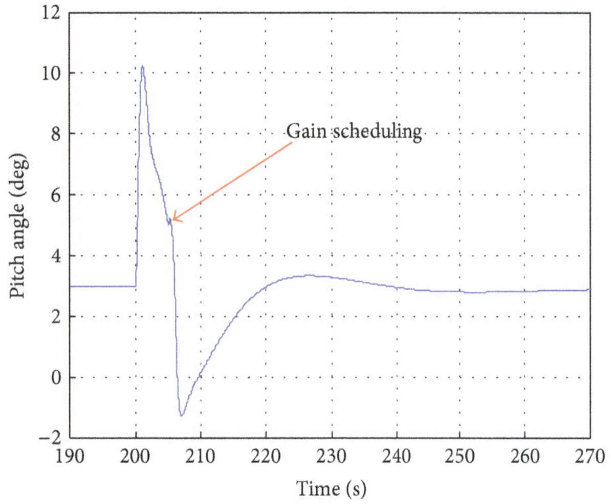

FIGURE 17: Pitch angle.

The FCS controls the aircraft to follow a trajectory autonomously flown by a 33% Yak-54 UAS equipped with a COTS autopilot. This system contains an H-infinity robust controller in the inner loop, an outer loop logic for waypoint following, and an extended Kalman filter, running at 10 (Hz). Its sensors include a GPS, an IMU, a 3D accelerometer, a 3D gyro, a 3D magnetometer, a barometer, and a pitot tube.

For the similarity of the flight test and simulation comparisons, same initial conditions are extracted and used from the flight test data. To replicate external disturbance, COTS estimated wind information was also injected.

Figure 19 shows 3D views, and Figures 20 and 21 show the control signals for both the designed FCS and COTS autopilot. The top view of 3D simulation in Figure 19 demonstrates superior performance of new nonlinear guidance logic in guiding UAS on curved and tight trajectories in presence of a crosswind. The wind was blowing from the west, and the maximum deviation from the desired trajectory happened when the UAS equipped with a COTS autopilot was in a crosswind (magenta line Figure 19).

By design, the COTS autopilot does not command rudder. This imposed extra demand on the aileron causing higher banking angles when initiating turns. Despite dissimilarity of COTS system and the designed FCS, some basic but important comparisons can be made. As it is shown in Figures 20 and 21, the COTS autopilot and the designed FCS are not subject of control saturation issues.

Figure 22(a) shows the comparison between the COTS autopilot and the designed FCS in airspeed command following. For the sake of unbiased comparison, the COTS autopilot wind estimation is used in the simulation 14.1 (ft/sec) from 291°. The airspeed command has a step increment around 380 seconds.

The ground speed is shown in Figure 22(b) in a wider portion of the flight. The error in airspeed control of the COTS autopilot is seen as the result of bias. This presumption is confirmed by adding a bias of about 10–12 (ft/sec) to the measured airspeed and disabling the FCS's Kalman filter for an even comparison. After inducing the bias, ground speeds achieve higher similarity.

Figure 23 shows the attitude angles. It is important to notice that the extended Kalman filter takes a few seconds to converge from the given initial conditions affecting the first part of the estimation, but showing a good command following after possible transients.

7. Conclusions

An adaptive control comprising a robust inner controller, stable outer nonlinear guidance logic, and neural network fault detection is designed to provide increased resiliency to mitigate effects of on-board adverse conditions and unsteady aerodynamics. The aerodynamic models of UAS associated with hazardous weather or on-board adverse conditions are inherently nonlinear and unsteady. Under these conditions, although the closed-loop H-infinity is still linearly stable, the controller does not have uncertainty robustness and the infinity norm can grow over unity. A modified artificial neural network is successfully implemented for real-time nonlinear estimations of the aerodynamic forces. The Levenberg's adaptive parameter resets when reaching predefined extreme values. This modification allows avoiding unbounded growing when the cost function has reached a minimum. The sliding mode learning concept is used to keep the artificial neural network in learning phase continuously. A dynamic learning process of artificial neural networks increases the robustness against on-board adverse conditions and external disturbances. New nonlinear guidance logic which is extended from the lateral plane to the longitudinal plane has demonstrated superior performance in guiding unmanned air systems on curved trajectories in presence of external disturbances such as crosswind. The new logic adds new features ensuring its stability for any location and any attitude of the aircraft. The flight control system shows a satisfactory level of adaptability and resilience which is achieved by the adaptive nonlinear modeling, noise rejection, and biases attenuation. The robustness is assured by the proper gain scheduling between predefined trim points and for identified dynamic changes. It was found that in transition between gains and to avoid significant sensitivity, the gain scheduling must be coupled with the ANN output.

Appendices

A. Aircraft Dynamic Modeling

Based on a Newton-Euler modeling approach, the aircraft dynamic model comprises a 6-DOF rigid-body set of equations for a fixed-wing aircraft in flat earth. Referred to the aircraft's center of gravity CG, as shown in Figure 1(b), it is defined in four interrelated coordinate systems, that is, the inertial coordinate system $\overline{\mathbf{I}}$ aligned with the inertial local flat earth coordinate system $\mathbf{I}$, the body coordinate system $\overline{\mathbf{B}}$ fixed to the aircraft, the wind coordinate system $\overline{\mathbf{W}}$ oriented

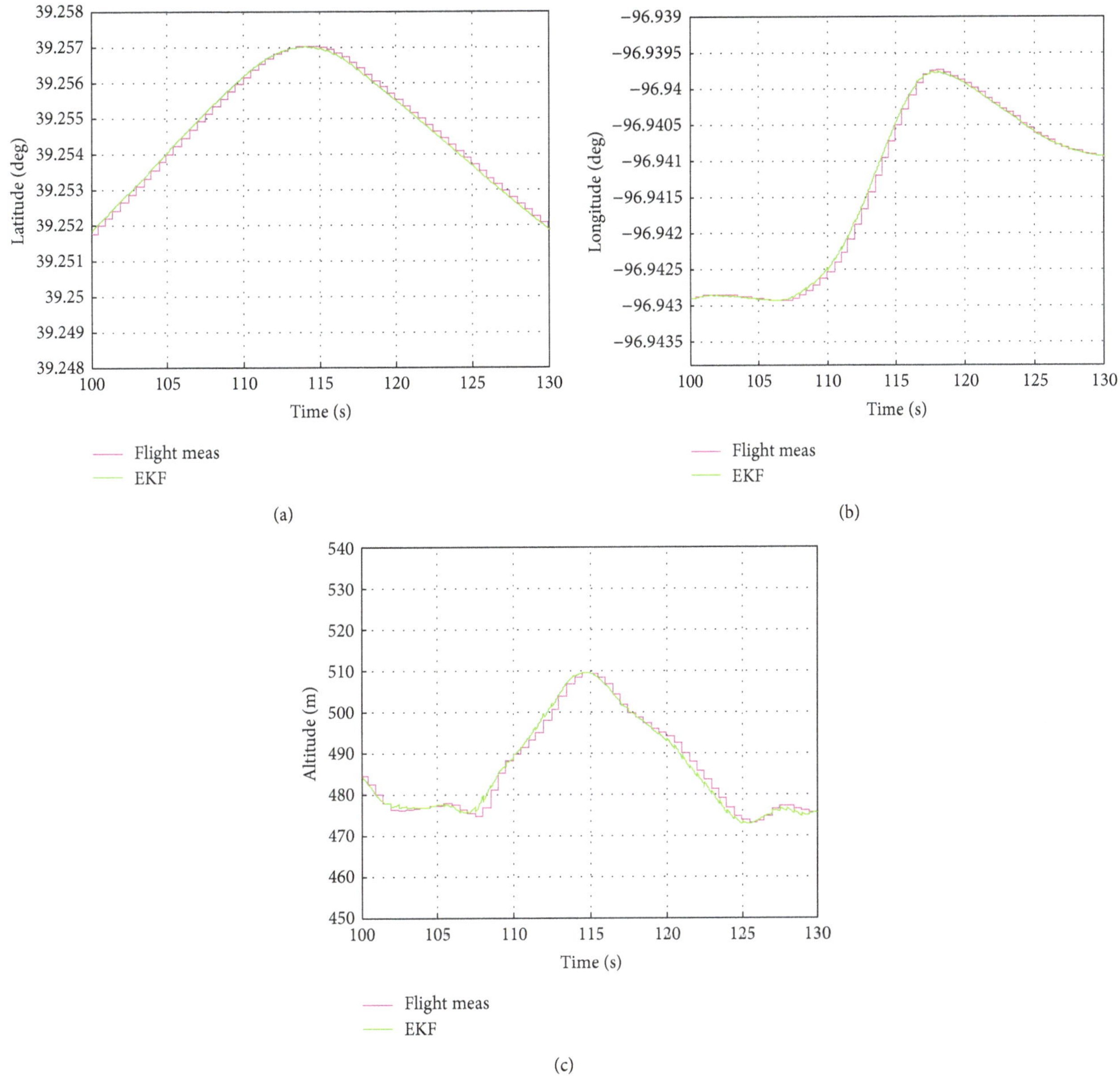

FIGURE 18: Latitude (a). Longitude (b). Altitude (c).

toward the relative wind, and the stability coordinate system $\overline{\mathbf{S}}$ oriented as shown. Consider

$$\dot{V}_T = \frac{\left(U\dot{U} + V\dot{V} + W\dot{W}\right)}{V_T},$$

$$\dot{\alpha} = \frac{\left(U\dot{W} - W\dot{U}\right)}{(UU + WW)},$$

$$\dot{\beta} = \frac{\left(V_T\dot{V} - V\dot{V}_T\right)}{(V_T^2 \cos\beta)},$$

$$\dot{\phi} = P + \tan\theta\left(Q\sin\phi + R\cos\phi\right),$$

$$\dot{\theta} = Q\cos\phi - R\sin\phi,$$

$$\dot{\psi} = \frac{(Q\sin\phi + R\cos\phi)}{\cos\theta},$$

$$\dot{P} = \left(J_{xz}\left[J_x - J_y + J_z\right]PQ - \left[J_z\left(J_z - J_y\right) + J_{xz}^2\right]QR + J_zL_A + J_{xz}N_A\right) \times \left(J_xJ_z - J_{xz}^2\right)^{-1},$$

$$\dot{Q} = \frac{\left[J_z - J_x\right]PR - J_{xz}\left[P^2 - R^2\right] + M_A}{J_xJ_z - J_{xz}^2},$$

$$\dot{R} = \left(\left[\left(J_x - J_y\right)J_x + J_{xz}^2\right]PQ - J_{xz}\left[J_x - J_y + J_z\right]QR + J_{xz}L_A + J_xN_A\right) \times \left(J_xJ_z - J_{xz}^2\right)^{-1},$$

$$\dot{\delta}_t = \frac{\delta_t}{\tau_t} + \frac{\delta_t^{\text{cmd}}}{\tau_t},$$

$$\dot{\delta}_e = \frac{\delta_e}{\tau_e} + \frac{\delta_e^{\text{cmd}}}{\tau_e},$$

$$\dot{\delta}_a = \frac{\delta_a}{\tau_a} + \frac{\delta_a^{\text{cmd}}}{\tau_a},$$

$$\dot{\delta}_r = \frac{\delta_r}{\tau_r} + \frac{\delta_r^{\text{cmd}}}{\tau_r},$$

$$\begin{aligned}\dot{p}_N &= U\cos\theta\cos\psi + V\left(-\cos\phi\sin\psi + \sin\phi\sin\theta\cos\psi\right)\\ &\quad + W\left(\sin\phi\sin\psi + \cos\phi\sin\theta\cos\psi\right) - w_N,\\ \dot{p}_E &= U\cos\theta\sin\psi + V\left(\cos\phi\cos\psi + \sin\phi\sin\theta\sin\psi\right)\\ &\quad + W\left(-\sin\phi\cos\psi + \cos\phi\sin\theta\sin\psi\right) - W_E,\\ \dot{p}_H &= U\sin\theta - V\sin\phi\cos\theta - W\cos\phi\cos\theta - w_H,\end{aligned} \tag{A.1}$$

where moments of inertia J_x, J_y, J_z, product of inertia J_{xz}, mass m, and gravity acceleration g are assumed constant. Body velocities U, V, W in body frame $\overline{\mathbf{B}}$ are define by $U = V_T\cos\alpha\cos\beta$, $V = V_T\sin\beta$, and $W = V_T\sin\alpha\cos\beta$, and their rate of change as

$$\begin{aligned}\dot{U} &= RV - QW - g\sin\theta + \frac{(X_T + X)}{m},\\ \dot{V} &= -RU + PW + g\sin\phi\cos\theta + \frac{Y}{m},\\ \dot{W} &= QU - PV + g\cos\phi\cos\theta + \frac{Z}{m},\end{aligned} \tag{A.2}$$

where aerodynamic forces X^A, Y^A, Z^A and moments L^A, M^A, N^A, together with the engine thrust X_T, act in the system as controllable inputs.

B. Polynomial Time-Invariant Engine Thrust Modeling

Highly dependent on the type and configuration of the engine, thrust X^P is modeled parametrically as a polynomial function on δ_T as

$$X^P(\mathbf{x}, t) = f_{X^P}(\delta_T, t) = T_X^2 \cdot {\delta_T}^2(t) + T_X^1 \cdot \delta_T(t) + T_X^0, \tag{B.1}$$

where T_X^2, T_X^1, T_X^0 are constant parameters to be estimated from flight test data for the trim point $\mathbf{x}_{\text{trim}}$.

C. Linear and Time-Invariant Aerodynamic Coefficient Modeling

The following relations, defined in the stability axis $\overline{\mathbf{S}}$, are used to model the aerodynamic coefficients $C_D, C_L, C_m, C_y, C_l, C_n$ as

$$\begin{aligned}C_D &= C_{D0} + C_{D\alpha}\alpha + C_{D\delta_e}\delta e + C_{D\dot{\alpha}}\frac{\dot{\alpha}\bar{c}}{2U_{\text{trim}}}\\ &\quad + C_{Dq}\frac{q\bar{c}}{2U_{\text{trim}}} + C_{Du}\frac{u}{U_{\text{trim}}},\\ C_L &= C_{L0} + C_{L\alpha}\alpha + C_{L\delta_e}\delta e + C_{L\dot{\alpha}}\frac{\dot{\alpha}\bar{c}}{2U_{\text{trim}}}\\ &\quad + C_{Lq}\frac{q\bar{c}}{2U_{\text{trim}}} + C_{Lu}\frac{u}{U_{\text{trim}}},\\ C_m &= C_{m0} + C_{m\alpha}\alpha + C_{m\delta_e}\delta e + C_{m\dot{\alpha}}\frac{\dot{\alpha}\bar{c}}{2U_{\text{trim}}}\\ &\quad + C_{mq}\frac{q\bar{c}}{2U_{\text{trim}}} + C_{mu}\frac{u}{U_{\text{trim}}},\\ C_y &= C_{y\beta}\beta + C_{y\dot{\beta}}\frac{\dot{\beta}b}{2U_{\text{trim}}} + C_{yp}\frac{pb}{2U_{\text{trim}}} + C_{yr}\frac{rb}{2U_{\text{trim}}}\\ &\quad + C_{y\delta_a}\delta a + C_{y\delta_r}\delta r,\\ C_l &= C_{l\beta}\beta + C_{l\dot{\beta}}\frac{\dot{\beta}b}{2U_{\text{trim}}} + C_{lp}\frac{pb}{2U_{\text{trim}}} + C_{lr}\frac{rb}{2U_{\text{trim}}}\\ &\quad + C_{l\delta_a}\delta a + C_{l\delta_r}\delta r,\\ C_n &= C_{n\beta}\beta + C_{n\dot{\beta}}\frac{\dot{\beta}b}{2U_{\text{trim}}} + C_{np}\frac{pb}{2U_{\text{trim}}} + C_{nr}\frac{rb}{2U_{\text{trim}}}\\ &\quad + C_{n\delta_a}\delta a + C_{n\delta_r}\delta r,\end{aligned} \tag{C.1}$$

where p, q, r are the deviations of the body angular velocities from their constant trim values $P_{\text{trim}}, Q_{\text{trim}}, R_{\text{trim}}$, and u is the deviation from the constant trim speed U_{trim}. Aerodynamic coefficients in the stability coordinate system $\overline{\mathbf{S}}$ are transformed into body axes $\overline{\mathbf{B}}$ coefficients by a rotation in the amount of angle of attack trim value α_{trim}.

D. H-Infinity Robust Controller Design Conditions

Conditions to obtain a stabilizing H-infinity controller are

(a) $(\mathbf{A}^i, \mathbf{B}_2^i)$ stabilizable and $(\mathbf{C}_2^i, \mathbf{A}^i)$ detectable;

(b) $\text{rank}(\mathbf{D}_{12}^i) = m_2$ and $\text{rank}(\mathbf{D}_{21}^i) = p_2$;

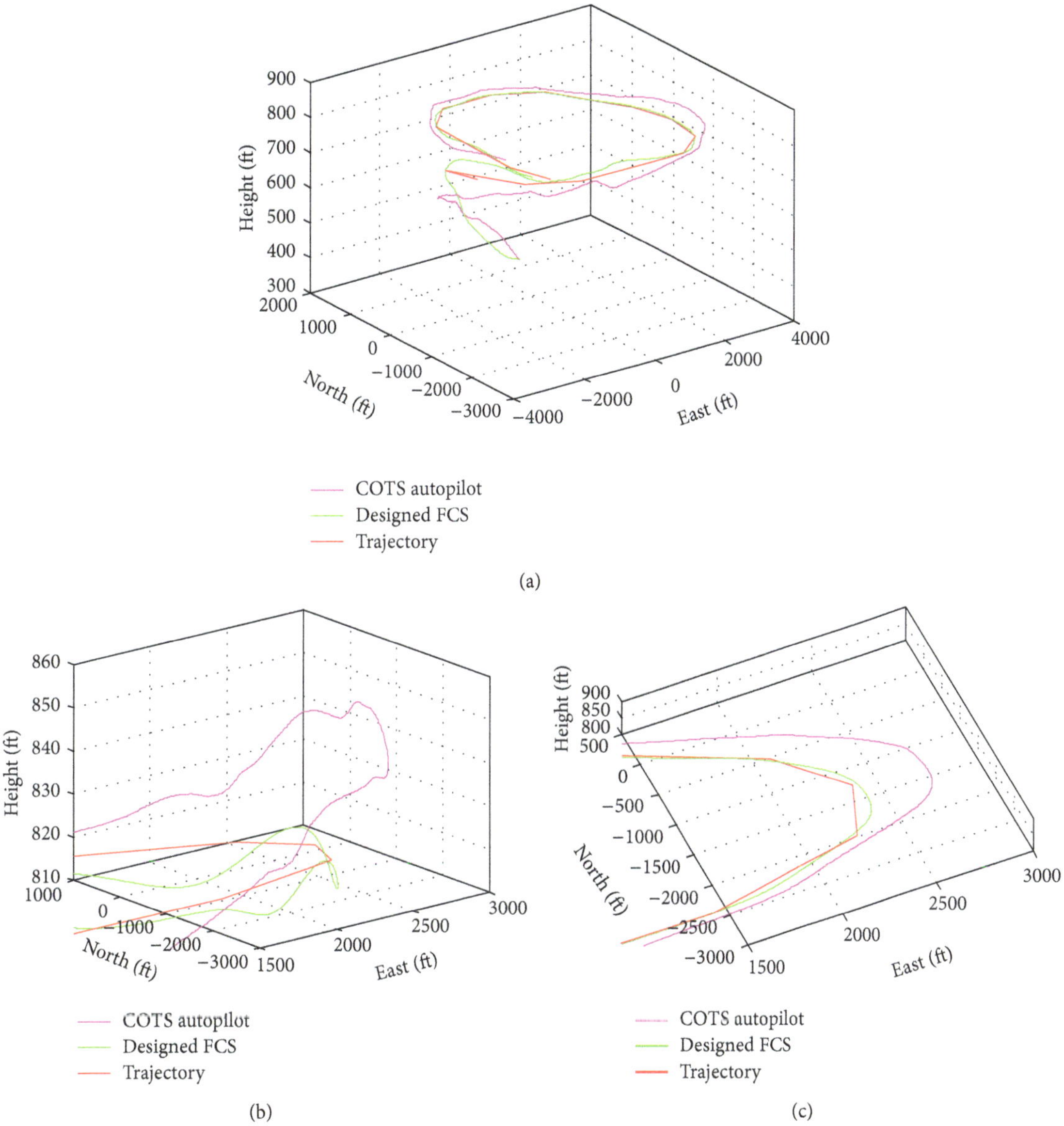

FIGURE 19: 3D full view path following (a). 3D zoom view path following (b) and (c).

(c) $\mathbf{D}^i_{12} = [0 \;\; I_{m_2}]^T$ and $\mathbf{D}^i_{21} = [0 \;\; I_{p_2}]$, and $\mathbf{D}^i_{11}$ is expressible as

$$\mathbf{D}^i_{11} = \begin{bmatrix} \underbrace{\mathbf{D}^i_{1111}}_{(p_1-m_2)x(m_1-p_2)} & \underbrace{\mathbf{D}^i_{1112}}_{(p_1-m_2)x(p_2)} \\ \underbrace{\mathbf{D}^i_{1121}}_{(m_2)x(m_1-p_2)} & \underbrace{\mathbf{D}^i_{1122}}_{(m_2)x(p_2)} \end{bmatrix}; \qquad \text{(D.1)}$$

(d) $\mathbf{D}^i_{22} = 0$;

(e) $\operatorname{rank}\begin{bmatrix} \mathbf{A}^i - jwI & \mathbf{B}^i_2 \\ \mathbf{C}^i_1 & \mathbf{D}^i_{12} \end{bmatrix} = n + m_2$;

(f) $\operatorname{rank}\begin{bmatrix} \mathbf{A}^i - jwI & \mathbf{B}^i_1 \\ \mathbf{C}^i_2 & \mathbf{D}^i_{21} \end{bmatrix} = n + p_2$,

where assumption c, if not achieved by the chosen weighting matrices, can be obtained via scaling of $\mathbf{u}_2$ and $\mathbf{y}_2$ together with a unitary transformation of $\mathbf{u}_1$ and $\mathbf{y}_1$. After previous assumptions are met, there is $\gamma^i > 0$, for which the following relations holds:

$$\max\left(\overline{\sigma}\left[\mathbf{D}^i_{1111} \;\; \mathbf{D}^i_{1112}\right], \overline{\sigma}\left[\mathbf{D}^{i\;T}_{1111} \;\; \mathbf{D}^{i\;T}_{1112}\right]\right) < \gamma^i,$$
$$\mathbf{X}^i \geq 0, \;\; \mathbf{Y}^i \geq 0, \qquad \text{(D.2)}$$
$$\rho\left(\mathbf{X}^i \cdot \mathbf{Y}^i\right) < \left(\gamma^i\right)^2.$$

This is equivalent to the existence of an internal stabilizing controller $\overline{\mathbf{K}}^i$

$$\overline{\mathbf{K}}^i = \begin{bmatrix} \overline{\mathbf{A}}^i & \overline{\mathbf{B}}^i_1 & \overline{\mathbf{B}}^i_2 \\ \overline{\mathbf{C}}^i_1 & \overline{\mathbf{D}}^i_{11} & \overline{\mathbf{D}}^i_{12} \\ \overline{\mathbf{C}}^i_2 & \overline{\mathbf{D}}^i_{21} & 0 \end{bmatrix} \qquad \text{(D.3)}$$

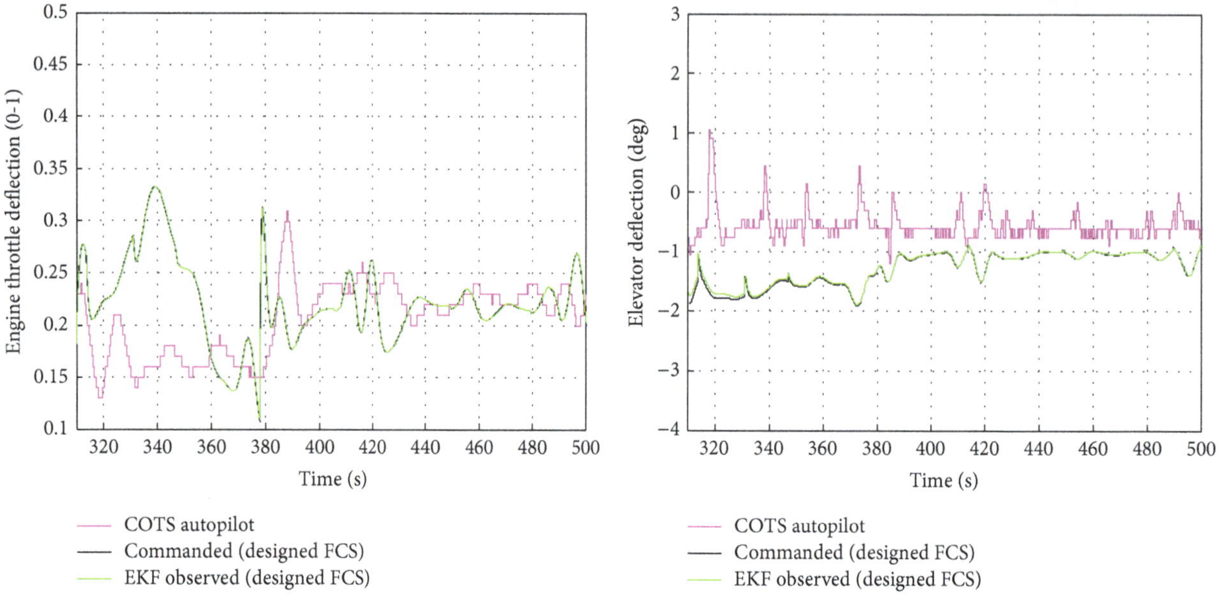

(a) (b)

FIGURE 20: Engine throttle deflection (a). Elevator deflection (b).

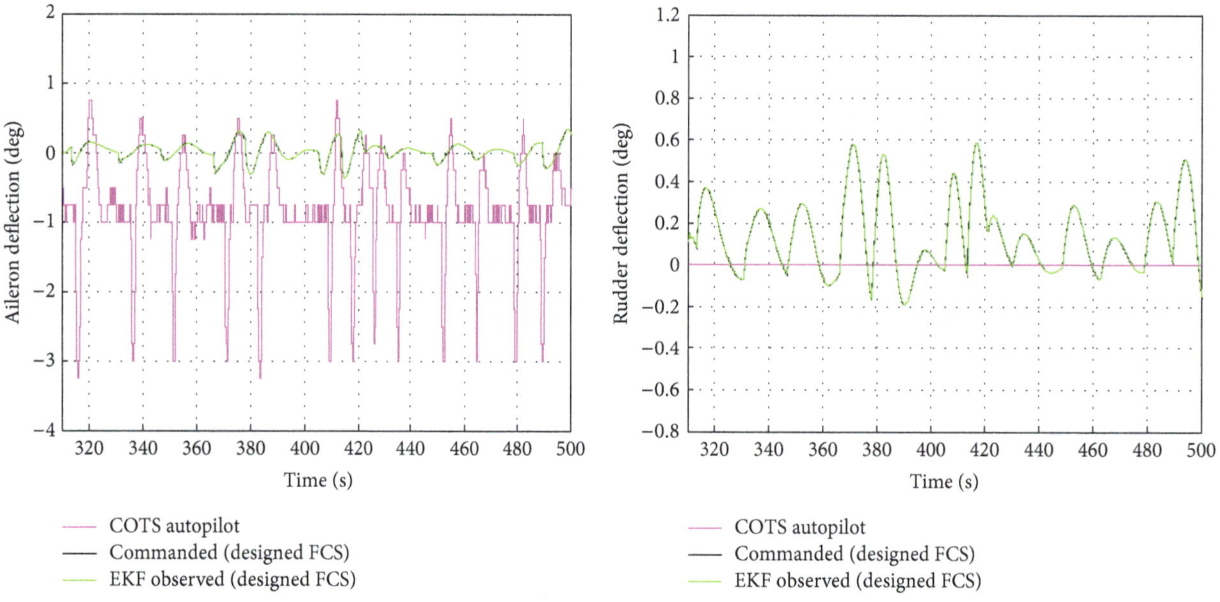

(a) (b)

FIGURE 21: Aileron deflection (a). Rudder deflection (b).

whose components are given by

$$\overline{\mathbf{A}}^i = \mathbf{A}^i + \mathbf{H}^i\mathbf{C}^i + \overline{\mathbf{B}}_2^i\left(\overline{\mathbf{D}}_{12}^i\right)^{-1}\overline{\mathbf{C}}_1^i,$$

$$\overline{\mathbf{B}}_1^i = -\mathbf{H}_2^i + \overline{\mathbf{B}}_2^i\left(\overline{\mathbf{D}}_{12}^i\right)^{-1}\overline{\mathbf{D}}_{11}^i,$$

$$\overline{\mathbf{B}}_2^i = \left(\mathbf{B}_2^i + \mathbf{H}_{12}^i\right)\overline{\mathbf{D}}_{12}^i,$$

$$\overline{\mathbf{C}}_1^i = \mathbf{F}_2^i\mathbf{Z}^i + \overline{\mathbf{D}}_{11}^i\left(\overline{\mathbf{D}}_{21}^i\right)^{-1}\overline{\mathbf{C}}_2^i,$$

$$\overline{\mathbf{C}}_2^i = -\overline{\mathbf{D}}_{21}^i\left(\mathbf{C}_2^i + \mathbf{F}_{12}^i\right)\mathbf{Z}^i,$$

$$\overline{\mathbf{D}}_{11}^i = -\mathbf{D}_{1121}^i{\mathbf{D}_{1111}^i}^T\left(\left(\gamma^i\right)^2\mathbf{I} - \mathbf{D}_{1111}^i{\mathbf{D}_{1111}^i}^T\right)^{-1}\mathbf{D}_{1112}^i - \mathbf{D}_{1122}^i, \tag{D.4}$$

and $\overline{\mathbf{D}}_{12}^i$ and $\overline{\mathbf{D}}_{21}^i$ are any matrices satisfying the relation $\overline{\mathbf{D}}_{12}^i{\overline{\mathbf{D}}_{12}^i}^T = \mathbf{I} - \overline{\mathbf{D}}_{1121}^i((\gamma^i)^2 I - {\overline{\mathbf{D}}_{1111}^i}^T\overline{\mathbf{D}}_{1111}^i)^{-1}{\overline{\mathbf{D}}_{1121}^i}^T$ and the

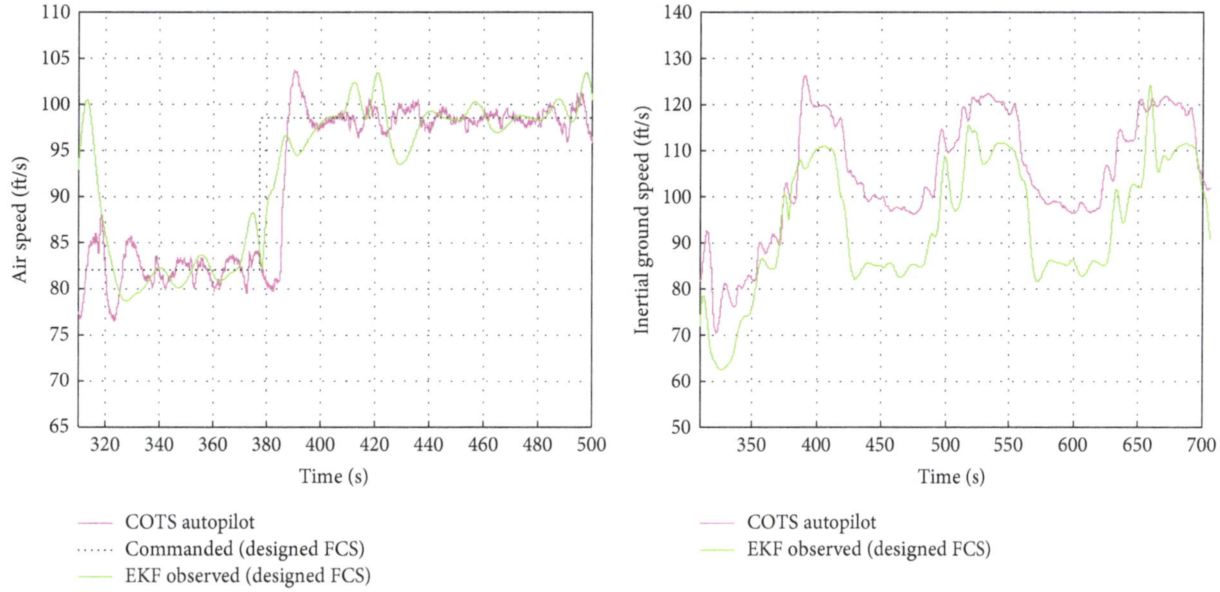

(a) (b)

Figure 22: Airspeed (a). Ground speed (b).

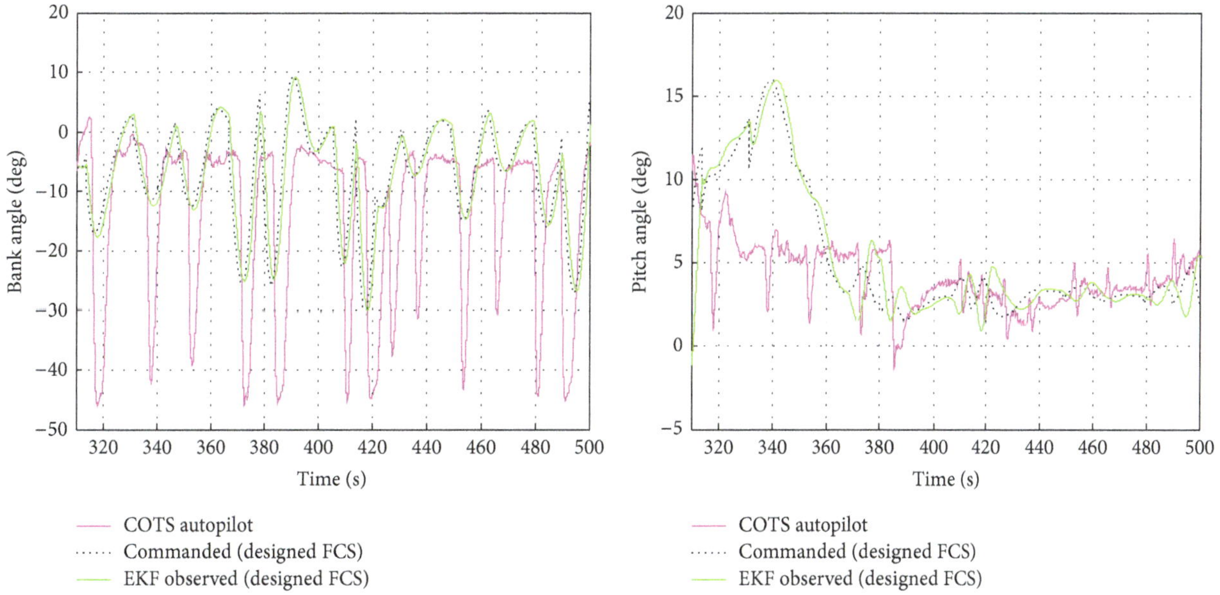

(a) (b)

Figure 23: Bank angle (a). Pitch angle (b).

relation $\overline{\mathbf{D}}^i_{21}\overline{\mathbf{D}}^{i\,T}_{21} = \mathbf{I} - \overline{\mathbf{D}}^i_{1112}((\gamma^i)^2 I - \overline{\mathbf{D}}^{i\,T}_{1111}\overline{\mathbf{D}}^i_{1111})^{-1}\overline{\mathbf{D}}^{i\,T}_{1112}$. The feedback $\mathbf{F}^i$ and state feedback $\mathbf{H}^i$ matrices are

$$\mathbf{F}^i = \begin{bmatrix} \mathbf{F}^i_{11}\}_{(m_1-p_2)x(n)} \\ \mathbf{F}^i_{12}\}_{(p_2)x(n)} \\ \mathbf{F}^i_{2}\}_{(m_2)x(n)} \end{bmatrix} = -\left(\mathbf{R}^i_a\right)^{-1}\left[\mathbf{D}^{i\,T}_{1a}\mathbf{C}^i_{1i} + \mathbf{B}^{i^T}\mathbf{X}^i\right],$$

$$\mathbf{H}^i = \Big[\underbrace{\mathbf{H}^i_{11}}_{(n)x(p_1-m_2)} \quad \underbrace{\mathbf{H}^i_{12}}_{(n)x(m_2)} \quad \underbrace{\mathbf{H}^i_{2}}_{(n)x(p_2)}\Big] = -\left[\mathbf{B}^i\mathbf{D}^{i\,T}_{1b} + \mathbf{Y}^i\mathbf{C}^{i^T}\right]\left(\mathbf{R}^i_b\right)^{-1}, \tag{D.5}$$

where

$$\mathbf{B}^i = \left[\mathbf{B}^i_1 \quad \mathbf{B}^i_2\right], \qquad \mathbf{C}^i = \begin{bmatrix}\mathbf{C}^i_1 \\ \mathbf{C}^i_2\end{bmatrix},$$

$$\mathbf{Z}^i = \left(\mathbf{I} - \left(\gamma^i\right)^{-2}\mathbf{Y}^i \cdot \mathbf{X}^i\right)^{-1},$$

$$\mathbf{D}_{1a}^i = \begin{bmatrix}\mathbf{D}_{11}^i & \mathbf{D}_{12}^i\end{bmatrix}, \qquad \mathbf{D}_{1b}^i = \begin{bmatrix}\mathbf{D}_{11}^i \\ \mathbf{D}_{21}^i\end{bmatrix},$$

$$\mathbf{R}_a^i = \mathbf{D}_{1a}^{i\,T}\mathbf{D}_{1a}^i - \begin{bmatrix}(\gamma^i)^2\mathbf{I}_{m_1} & 0 \\ 0 & 0\end{bmatrix}, \tag{D.6}$$

$$\mathbf{R}_b^i = \mathbf{D}_{1b}^i\mathbf{D}_{1b}^{i\,T} - \begin{bmatrix}(\gamma^i)^2\mathbf{I}_{p_1} & 0 \\ 0 & 0\end{bmatrix}.$$

Matrices $\mathbf{X}^i$ and $\mathbf{Y}^i$ are the solutions of the following Riccati equations:

$$\begin{aligned}
&\left(\mathbf{A}^i - \mathbf{B}^i(\mathbf{R}_a^i)^{-1}\mathbf{D}_{1a}^{i\,T}\mathbf{C}_1^i\right)^T\mathbf{X}^i + \mathbf{X}^i\left(\mathbf{A}^i - \mathbf{B}^i(\mathbf{R}_a^i)^{-1}\mathbf{D}_{1a}^{i\,T}\mathbf{C}_1^i\right) \\
&\quad - \mathbf{X}^i\left(\mathbf{B}^i(\mathbf{R}_a^i)^{-1}\mathbf{B}^{iT}\right)\mathbf{X}^i + \mathbf{C}_1^{i\,T}\left(\mathbf{I} - \mathbf{D}_{1a}^i(\mathbf{R}_a^i)^{-1}\mathbf{D}_{1a}^{i\,T}\right)\mathbf{C}_1^i \\
&= 0, \\
&\left(\mathbf{A}^i - \mathbf{B}_1^i\mathbf{D}_{1b}^{i\,T}(\mathbf{R}_b^i)^{-1}\mathbf{C}^i\right)\mathbf{Y}^i + \mathbf{Y}^i\left(\mathbf{A}^i - \mathbf{B}_1^i\mathbf{D}_{1b}^{i\,T}(\mathbf{R}_b^i)^{-1}\mathbf{C}^i\right)^T \\
&\quad - \mathbf{Y}^i\left(\mathbf{C}^{iT}(\mathbf{R}_b^i)^{-1}\mathbf{C}^i\right)\mathbf{Y}^i + \mathbf{B}_1^i\left(\mathbf{I} - \mathbf{D}_{1b}^{i\,T}(\mathbf{R}_b^i)^{-1}\mathbf{D}_{1b}^i\right)\mathbf{B}_1^{i\,T} \\
&= 0.
\end{aligned} \tag{D.7}$$

E. Extended Kalman Filter Algorithm

The filter's functions $\widehat{\mathbf{f}}$ and $\widehat{\mathbf{h}}$ represent the observed system. They incorporate all available information of the aircraft characterized by functions $\mathbf{f}$ and $\mathbf{h}$ and other variables to be estimated as biases and wind; that is, $\widehat{\mathbf{f}} \supseteq \mathbf{f}$ and $\widehat{\mathbf{h}} \supseteq \mathbf{h}$. Model mismatch $\mathbf{v}(t)$ and sensor noise $\mathbf{w}(t)$ are included through the Kalman gain $\mathbf{K}(t)$ calculation as shown later. The continuous-time observer is defined as

$$\begin{aligned}
\dot{\widehat{\mathbf{x}}}(t) &= \widetilde{\mathbf{f}}(\widehat{\mathbf{x}}(t), \mathbf{u}(t)) + \widetilde{\mathbf{K}}(t)(\mathbf{y}(t) - \widehat{\mathbf{y}}(t)), \\
\widehat{\mathbf{y}}(t) &= \widetilde{\mathbf{h}}(\widehat{\mathbf{x}}(t)),
\end{aligned} \tag{E.1}$$

where $\widehat{\mathbf{x}}(t) \in R^{\widehat{n}}$ is the filter state vector and $\widehat{\mathbf{u}}(t) \in R^{\widehat{m}}$ is the filter input vector.

Aircraft dynamics and observation equations $\dot{\mathbf{x}}(t) = \mathbf{f}(\mathbf{x}(t), \mathbf{u}(t)) + \mathbf{v}(t)$ and $\mathbf{y}(t) = \mathbf{h}(\mathbf{x}(t), \mathbf{u}(t)) + \mathbf{w}(t)$ are assumed to be additively corrupted by uncorrelated Gaussian noise signals $\mathbf{v}$ and $\mathbf{w}$. Applying the Euler discretization method, the $\widehat{n}$ order discrete version is obtained $\widehat{\mathbf{x}}[k+1] = \widehat{\mathbf{x}}[k] + \Delta T[\widehat{\mathbf{f}}(\widehat{\mathbf{x}}[k], \widehat{\mathbf{u}}[k])$ and $\widehat{\mathbf{y}}[k] = \widehat{\mathbf{h}}(\widehat{\mathbf{x}}[k], \widehat{\mathbf{u}}[k])$, valid at each sampling time interval ΔT, where index $[k]$ corresponds to time (t) sampled at $t = k\Delta T$. The discrete extended Kalman filter algorithm is defined in two phases:

$$\begin{aligned}
&\text{prediction}\begin{cases}
\widehat{\mathbf{x}}[k/k-1] = \widehat{\mathbf{f}}(\widehat{\mathbf{x}}[k-1/k-1], \widehat{\mathbf{u}}[k-1]), \\
\mathbf{P}[k/k-1] = \mathbf{F}[k-1]\mathbf{P}[k-1/k-1]\mathbf{F}^T[k-1] + \mathbf{Q}[k-1],
\end{cases} \\
&\text{correction}\begin{cases}
\mathbf{K}[k] = \mathbf{P}[k/k-1]\mathbf{H}^T[k]\left(\mathbf{H}[k]\mathbf{P}[k/k-1]\mathbf{H}^T[k] + \mathbf{R}[k]\right)^{-1}, \\
\widehat{\mathbf{x}}[k/k] = \widehat{\mathbf{x}}[k/k-1] + \mathbf{K}[k]\left(\mathbf{y}[k] - \widehat{\mathbf{h}}(\widehat{\mathbf{x}}[k/k-1], \widehat{\mathbf{u}}[k])\right), \\
\mathbf{P}[k/k] = [I - \mathbf{K}[k]\mathbf{H}[k]]\mathbf{P}[k/k-1].
\end{cases}
\end{aligned} \tag{E.2}$$

No new information is considered in the prediction stage, and the one-step ahead prediction state vector $\widehat{\mathbf{x}}[k/k-1]$ is obtained by propagating the equation $\widehat{\mathbf{f}}$ from previous state and input vectors at time $[k-1]$. The correction phase delivers $\widehat{\mathbf{x}}[k/k]$ by adjusting $\widehat{\mathbf{x}}[k/k-1]$ based on the difference between measurements and prediction, that is, $\mathbf{y}[k] - \widehat{\mathbf{h}}(\widehat{\mathbf{x}}[k/k-1], \widehat{\mathbf{u}}[k])$, through the Kalman gain $\mathbf{K}[k]$. In (E.2), $\mathbf{P}$ acts as the state error $(\mathbf{x} - \widehat{\mathbf{x}})$ covariance, and matrices $\mathbf{F}$ and $\mathbf{H}$ constitute the Jacobian of the propagation equation $\widehat{\mathbf{f}}$ and observation equation $\widehat{\mathbf{h}}$, respectively; that is, $\widetilde{\mathbf{F}}[k-1] = \partial\widetilde{\mathbf{f}}/\partial\mathbf{x}_{\widehat{\mathbf{x}}[k-1/k-1],\mathbf{u}[k-1]}$ and $\widetilde{\mathbf{H}}[k] = \partial\widehat{\mathbf{h}}/\partial\mathbf{x}_{\widehat{\mathbf{x}}[k/k-1]}$. $\widetilde{\mathbf{Q}}$ and $\widetilde{\mathbf{R}}$ are the covariance matrices of $\mathbf{v}$ and $\mathbf{w}$, respectively.

Acknowledgment

This research was supported by the NSF Center for Remote Sensing of Ice Sheet (CReSIS) Grant NSF-0066685, at the University of Kansas.

References

[1] I. Kaminer, A. Pascoal, E. Hallberg, and C. Silvestre, "Trajectory tracking for autonomous vehicles: an integrated approach to guidance and control," *Journal of Guidance, Control, and Dynamics*, vol. 21, no. 1, pp. 29–38, 1998.

[2] M. Niculescu, "Lateral track control law for Aerosonde UAV," in *Proceedings of the 39th AIAA Aerospace Sciences Meeting and Exhibit*, Reno, Nev, USA, 2001.

[3] D. Boyle and G. Chamitoff, "Robust nonlinear lasso control: a new approach for autonomous trajectory tracking," in *Proceedings of the AIAA Guidance, Navigation, and Control Conference and Exhibit*, Austin, Tex, USA, 2003.

[4] S. Park, J. Deyst, and J. How, "A new nonlinear guidance logic for trajectory tracking," in *Proceedings of the AIAA Guidance, Navigation, and Control Conference and Exhibit*, Providence, RI, USA, 2004.

[5] J. Osborne and R. Rysdyk, "Waypoint guidance for small UAVs in wind," in *Proceedings of the AIAA Infotech, Aerospace Conference*, Arlington, Va, USA, 2005.

[6] R. Rysdyk, "Unmanned aerial vehicle path following for target observation in wind," *Journal of Guidance, Control, and Dynamics*, vol. 29, no. 5, pp. 1092–1100, 2006.

[7] D. R. Nelson, D. B. Barber, T. W. McLain, and R. W. Beard, "Vector field path following for miniature air vehicles," *IEEE Transactions on Robotics*, vol. 23, no. 3, pp. 519–529, 2007.

[8] L. Sonneveldt, E. R. Van Oort, Q. P. Chu, and J. A. Mulder, "Nonlinear adaptive trajectory control applied to an F-16 model," *Journal of Guidance, Control, and Dynamics*, vol. 32, no. 1, pp. 25–39, 2009.

[9] G. Garcia, S. Keshmiri, and R. Colgren, "Advanced H-infinity trainer autopilot," in *Proceedings of the AIAA Modeling and Simulation Technologies Conference*, Toronto, Canada, 2010.

[10] J. A. Rios and E. White, "Fusion filter algorithm enhancements for a MEMS GPS/IMU," in *Proceedings of the 14th International Technical Meeting of the Satellite Division of the Institute of Navigation (ION GPS '01)*, pp. 1382–1393, Salt Lake City, Utah, USA, September 2001.

[11] P. Zhang, J. Gu, E. E. Milios, and P. Huynh, "Navigation with IMU/GPS/digital compass with unscented Kalman filter," in *Proceedings of the IEEE International Conference on Mechatronics and Automation (ICMA '05)*, pp. 1497–1502, Niagara Falls, Canada, August 2005.

[12] Y. Li, J. Wang, C. Rizos, P. Mumford, and W. Ding, "Low-cost tightly coupled GPS/INS integration based on a nonlinear Kalman filtering design," in *Proceedings of the Institute of Navigation, National Technical Meeting (NTM '06)*, pp. 958–966, Monterey, Calif, USA, January 2006.

[13] X. Lin-lin, W. Jian-guo, Z. Li-hui, and G. Li-jun, "Nonlinear gaussian filter algorithm enhacements for los-cost integrated navigations systems," in *Proceedings of the IEEE International Conference on Mechatronics and Automation*, pp. 4566–4571, Changchun, China, August 2009.

[14] Y. Ren and X. Ke, "Particle filter data fusion enhancements for MEMS-IMU/GPS," *Intelligent Information Management*, vol. 2, pp. 417–421, 2010.

[15] R. Van der Merwe and E. Wan, "Sigma-point kalman filters for integrated navigation," in *Proceedings of the 60th Annual Meeting of the Institute of Navigation (ION '04)*, Dayton, Ohio, USA, 2004.

[16] A. Nemra and N. Aouf, "Robust INS/GPS sensor fusion for UAV localization using SDRE nonlinear filtering," *IEEE Sensors Journal*, vol. 10, no. 4, pp. 789–798, 2010.

[17] A. Eldredge, *Improved state estimation for miniature air vehicles [M.S. thesis]*, Department of Mechanical Engineering, Brigham Young University, Provo, Utah, USA, 2006.

[18] M. Pachter, N. Ceccarelli, and P. R. Chandler, "Estimating MAV's heading and the wind speed and direction using GPS, inertial, and air speed measurements," in *Proceedings of the AIAA Guidance, Navigation and Control Conference and Exhibit*, Honolulu, Hawaii, USA, August 2008.

[19] D. J. Linse and R. F. Stengel, "Identification of aerodynamic coefficients using computational neural networks," *Journal of Guidance, Control, and Dynamics*, vol. 16, no. 6, pp. 1018–1025, 1993.

[20] W. E. Faller and S. J. Schreck, "Real-time prediction of unsteady aerodynamics: application for aircraft control and maneuverability enhancement," *IEEE Transactions on Neural Networks*, vol. 6, no. 6, pp. 1461–1468, 1995.

[21] M. Larson, P. De Raedt, and M. Hedlund, "Aerodynamic identification using neural networks," Reptort LiTH-154-1937, Department of Electrical Engineering, Linköping University, Linköping, Sweden, 1997.

[22] S. C. Raisinghani, G. V. S. Reddy, and A. K. Ghosh, "Parameter estimation of an augmented aircraft using neural networks," in *Proceedings of the AIAA Atmospheric Flight Mechanics Conference and Exhibit*, Portland, Ore, USA, 1999.

[23] S. C. Raisinghani and A. K. Ghosh, "Parameter estimation of an unstable aircraft in turbulent atmosphere using neural networks," in *Proceedings of the AIAA Atmospheric Flight Mechanics Conference and Exhibit*, Monterey, Calif, USA, 2002.

[24] N. K. Peyada and A. K. Ghosh, "Aircraft parameter estimation using a new filtering technique based upon a neural network and Gauss-Newton method," *Aeronautical Journal*, vol. 113, no. 1142, pp. 243–252, 2009.

[25] S. Das, R. A. Kuttieri, M. Sinha, and R. Jategaonkar, "Neural partial differential method for extracting aerodynamic derivatives from flight data," *Journal of Guidance, Control, and Dynamics*, vol. 33, no. 2, pp. 376–384, 2010.

[26] Z. Wang, C. E. Lan, and J. M. Brandon, "Fuzzy logic modeling of nonlinear unsteady aerodynamics," in *Proceedings of the AIAA Atmospheric Flight Mechanics Conference and Exhibit*, Boston, Mass, USA, 1998.

[27] Z. Wang, C. E. Lan, and J. M. Brandon, "Fuzzy logic modeling of lateral-directional unsteady aerodynamics," in *Proceedings of the AIAA Atmospheric Flight Mechanics Conference and Exhibit*, Boston, Mass, USA, 1998.

[28] A. Hossain, A. Rahman, J. Hossen, A. K. M. P. Iqbal, and M. I. Zahirul, "Prediction of aerodynamic characteristics of an aircraft model with and without winglet using fuzzy logic technique," *Aerospace Science and Technology*, vol. 15, no. 8, pp. 595–605, 2011.

[29] C. E. Lan, S. Keshmiri, and R. Hale, "Fuzzy-logic modeling of a rolling unmanned vehicle in antarctica wind shear," *AIAA Journal of Guidance, Control and Dynamics*, vol. 35, no. 5, pp. 1538–1547, 2012.

[30] B. Stevens and F. Lewis, *Aircraft Control and Simulation*, John Wiley & Sons, Hoboken, NJ, USA, 2nd edition, 2003.

[31] J. Roskam, *Airplane Flight Dynamics and Automatic Flight Controls (Part I)*, DARcorporation, Lawrence, Kan, USA, 2003.

[32] DARcorporation, *Advanced Aircraft Analysis*, Software Package, Wichita, Kan, USA, 2009.

[33] E. Leong, *Development of A 6DOF Nonlinear Simulation Model Enhanced with Fine Tuning Procedures*, The University of Kansas, 2008.

[34] K. Glover and J. C. Doyle, "State-space formulae for all stabilizing controllers that satisfy an H∞-norm bound and relations to relations to risk sensitivity," *Systems and Control Letters*, vol. 11, no. 3, pp. 167–172, 1988.

[35] K. Zhou and J. Doyle, *Essentials of Robust Control*, Prentice Hall, Upper Saddle River, NJ, USA, 1998.

[36] G. Garcia, S. Keshmiri, and R. Colgren, "H-infinity gain scheduling design for the Meridian UAS for a broader range of operation and for fault tolerant applications," in *Proceedings of the 9th IEEE International Conference on Control and Automation (ICCA '11)*, pp. 1174–1180, December 2011.

[37] S. Park, J. Deyst, and J. P. How, "Performance and lyapunov stability of a nonlinear path-following guidance method," *Journal of Guidance, Control, and Dynamics*, vol. 30, no. 6, pp. 1718–1728, 2007.

[38] D. Simon, *Optimal State Estimation*, John Wiley & Sons, Hoboken, NJ, USA, 2006.

[39] G. Cybenko, "Approximation by superpositions of a sigmoidal function," *Mathematics of Control, Signals, and Systems*, vol. 2, no. 4, pp. 303–314, 1989.

[40] J. Nocedal and S. J. Wright, *Numerical Optimization*, Springer, New York, NY, USA, 1999.

13

High-Velocity Impact Behaviour of Prestressed Composite Plates under Bird Strike Loading

Sebastian Heimbs and Tim Bergmann

Department of Structures Engineering, Production and Aeromechanics, EADS Innovation Works, Willy-Messerschmitt-Straße, 81663 Munich, Germany

Correspondence should be addressed to Sebastian Heimbs, sebastian.heimbs@eads.net

Academic Editor: Mark Price

An experimental and numerical analysis of the response of laminated composite plates under high-velocity impact loads of soft body gelatine projectiles (artificial birds) is presented. The plates are exposed to tensile and compressive preloads before impact in order to cover realistic loading conditions of representative aeronautic structures under foreign object impact. The modelling methodology for the composite material, delamination interfaces, impact projectile, and preload using the commercial finite element code Abaqus are presented in detail. Finally, the influence of prestress and of different delamination modelling approaches on the impact response is discussed and a comparison to experimental test data is given. Tensile and compressive preloading was found to have an influence on the damage pattern. Although this general behaviour could be predicted well by the simulations, further numerical challenges for improved bird strike simulation accuracy are highlighted.

1. Introduction

The application of carbon fibre-reinforced plastic (CFRP) materials in aircraft structures is ever-expanding. Besides established utilisation in control surfaces or wing structures, the latest generation of commercial airliners also feature a fuselage made of carbon fibre composite material. Besides their well-known advantages in terms of weight-specific mechanical properties and fatigue tolerance, such structures are vulnerable against transversal impact loads, which can lead to undetectable internal damage and cracks that can reduce the strength and grow under load. Typical examples of such impact load cases with frequent occurrence are bird strike loads on composite wing leading edges or hard body impact loads from stones or runway debris being thrown against lower fuselage panels by the aircraft tires.

Most studies that have been performed so far to investigate the impact performance of composite laminates are based on unloaded structures. However, this simplification may be far off reality as the structure in flight may typically be under a state of prestress before impact. For example, the lower fuselage panels are typically exposed to compressive loads during takeoff, when stone impact is likely to occur. There is a strong interest in investigating the influence of preloads on the impact response of aeronautic structures.

Most published studies on this topic are based on low-velocity impact loads, which can experimentally be achieved on a drop tower test rig with a falling mass impacting the preloaded sample [1–3] or using an instrumented pendulum test rig [4–6]. During such tests, the investigation of tensile preloads is comparably easy to perform and has been conducted in [7–16]. Compressive preloads, in contrast, introduce the complexity of plate buckling, which was analysed by Heimbs et al. [17], Zhang et al. [18], and Morlo and Kunz [19]. Choi [20], Sun and Chen [21], Sun and Chattopadhyay [22] and Khalili et al. [23] present analytical studies of the low-velocity impact response of composite plates under compressive and tensile preload.

The high-velocity impact response of preloaded laminated composite structures was investigated in only very few studies. For this purpose, a gas cannon is typically used, which accelerates the projectile to velocities of approximately 50–300 m/s. In the study of Herszberg et al. [24] a 13 mm

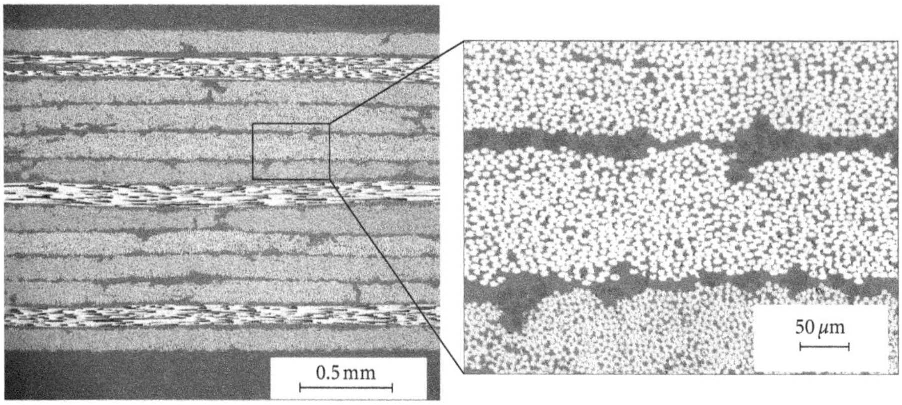

FIGURE 1: Micrograph of T800S/M21 composite laminate plate.

steel sphere was fired with velocities from 20 m/s to 70 m/s against a 2 mm carbon fibre/epoxy plate with tensile preload leading to a change in damage pattern for higher preloads. Schueler et al. [25] present an experimental and numerical study of the high-velocity impact response of 17-ply CFRP plates under tensile preload using a 21 g glass sphere projectile and an impact velocity of 64 m/s, obtaining less delamination compared to the unloaded case. Their numerical model in Abaqus was based on stacked continuum shell elements with cohesive element interfaces in between. General trends found during the experiments could also be drawn from the simulation results. Vaidya and Shafiq [26] performed ballistic impact tests on 4.5 mm thick woven E-glass/vinylester plates with compressive preload and velocities roughly in the range of 100–300 m/s and report a detrimental effect on the residual strength of the composite plate. Garcia-Castillo et al. [27, 28] performed ballistic impact tests on 3.19 mm thick woven E-glass/polyester plates with biaxial tensile preload using a 7.5 mm steel sphere and velocities from 140 m/s to 525 m/s. They report an increase of the ballistic limit for the preloaded plates, correlating to their analytical and numerical Abaqus model.

An investigation on the effect of prestress on the high-velocity impact behaviour of laminated composite structures under soft body impacts has not been performed yet. This effect can be of great importance for bird strike analyses and bird-proof design of composite aircraft structures, though. There is a big difference between the spread contact area of soft body impact loads resulting from bird or hail strike compared to hard body impact loads, which are resulting from bird strike compared much more localised. For simplification and generalisation of this study, the soft body impact on a flat composite plate was analysed. The intention was to assess the influence of preload on the damage behaviour both experimentally and numerically, using the commercial finite element (FE) code Abaqus V6.10.

2. Experiments

This study was based on an experimental test campaign in order to obtain a reliable data basis for model validations. The central part of this test campaign was high-velocity impact tests with soft body projectiles on simple flat composite plates under compressive or tensile preload.

The carbon fibre/epoxy laminated plates were made from 13 plies of Hexcel T800S/M21 prepreg with a stacking sequence of [+45/90/−45/+45/−45/0/90/0/−45/+45/−45/90/+45] and a nominal thickness of 1.625 mm. The micrograph in Figure 1 shows a cross-section of the laminate with the dark areas between the plies not being cracks or manufacturing imperfections but conglomerations of thermoplastic interleaf particles of the resin system for enhanced impact damage tolerance. The plates were cured in an autoclave at 180°C and 7 bar pressure for 120 minutes. Standard coupon tests of this material were performed as part of the test campaign in order to obtain the elastic, strength, and damage properties that are necessary for the numerical material modelling.

The final specimens for the impact tests had a size of 550 mm × 200 mm with bonded metallic end tabs of 125 mm length for the fixation in the preloading device, reducing the free specimen length to 300 mm × 200 mm (Figure 2). A total of six strain gauges were used for the plates with compressive preload in order to accurately determine the prestrain and buckling pattern. Three strain gauges were used for the plates with tensile preload or without preload.

The bird impact tests were performed at the DLR Stuttgart gas cannon test facility using a specially designed test rig that allows for uniaxial tensile or compressive preloading. The longitudinal ends with the metallic end tabs were clamped to the machine, and the lateral ends were simply supported from both sides. A small gelatine projectile of 32 g weight and a cylindrical geometry with one hemispherical end cap was used as a typical small artificial bird impactor (Figure 3(a)). Impact velocities of 93–171 m/s or kinetic energies of 135–443 J were tested. Three different preloading conditions were applied:

(i) tensile prestrain of 0.25%, which is a typical value for the limit load of this structure,

(ii) unloaded,

(iii) compressive preload of 11 kN, which is 2.15 times higher than the experimentally determined buckling load and represents the ultimate load for this structure.

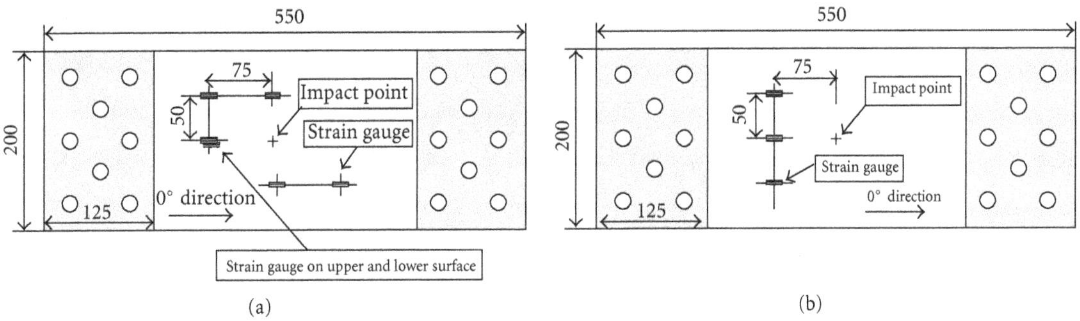

FIGURE 2: Impact specimen dimensions (in mm) and position of strain gauges for (a) compressive preload and (b) tensile preload.

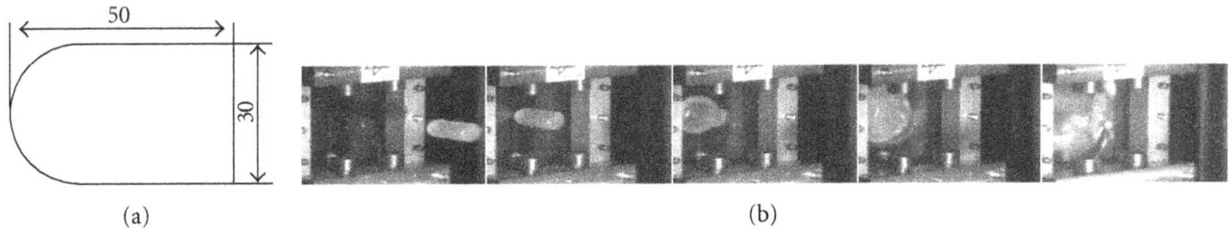

FIGURE 3: (a) Geometry of artificial bird impactor (in mm) and (b) high-speed video images of bird strike test (by DLR).

A differentiation for tensile and compressive preloading was made here as limit loads and ultimate loads are often given as strain values (microstrains) for the tensile load case and as a relation to the buckling load for the compressive load case.

The bird projectile was shot against the plate centre with the soft body impactor material flowing and spreading across the target surface (Figure 3(b)).

Posttest damage assessment in the first instance was performed with nondestructive damage inspection methods like ultrasonic C-scanning and microcomputer tomography. However, as the damage inside the plates even for the highest impact velocity was rather small and the exact type of damage was difficult to determine, micrographs were taken in order to get a detailed look into the damaged laminate.

For the lowest velocity of 90–100 m/s (approx. 140 J) there was almost no damage in none of the test plates, which shows that the deformation of the impacted plate remains in the elastic domain and the biggest part of the initial kinetic energy remains as residual kinetic energy of the spreading impactor material. For higher impact velocities the amount of internal damage increases, which is shown in Figure 4 for the specimens with tensile preload. The only damage mode occurring in this case is matrix cracking in the upper and lower plies due to tensile and compressive loads during plate bending and impactor contact forces, which was verified in the micrographs (Figure 5(a)). No delamination occurred in the plates with tensile preload for the given range of impact velocities.

It is worth noting that the type of damage changes for the three preload cases. In general, the amount of matrix cracking appeared to be higher for the tensile preloaded specimens. The specimen with compressive preload showed more localised internal matrix cracks and fibre blowout on the back surface (Figure 5(b)). This might be due to a softening effect of the compressive preload, which allows for higher bending deformations, in contrast to a stiffening effect of the tensile preload. This corresponds to results that were also reported in other studies [3, 25]. The plate bending deformation was unfortunately not measured during the high-velocity impact tests and can only be assessed in the numerical simulation model.

3. Model Development

In aerospace engineering there is a strong interest in predictive numerical methods for damage assessment in vulnerability analyses of composite structures under foreign object impact [29]. In this context, a numerical approach was developed and investigated in this study to predict the impact damage of preloaded composite plates under bird strike loading. This is a very complex task in terms of accurate composite material modelling with intra- and interlaminar damage, hydrodynamic soft body impactor modelling, and fluid-structure interaction as well as implicit-explicit coupling for efficient preload modelling.

3.1. Composite Material Modelling. Different damage mechanisms occur in composite plates under impact loading that absorb part of the initial kinetic energy of the impactor and reduce the residual strength. Although they strongly depend on factors like plate thickness, boundary conditions, impact velocity, impactor mass, and geometry, they can often be identified as matrix cracking, delaminations, and finally fibre rupture. It is desired that all these potential failure modes are covered by the numerical model enabling their occurrence in the simulation as well.

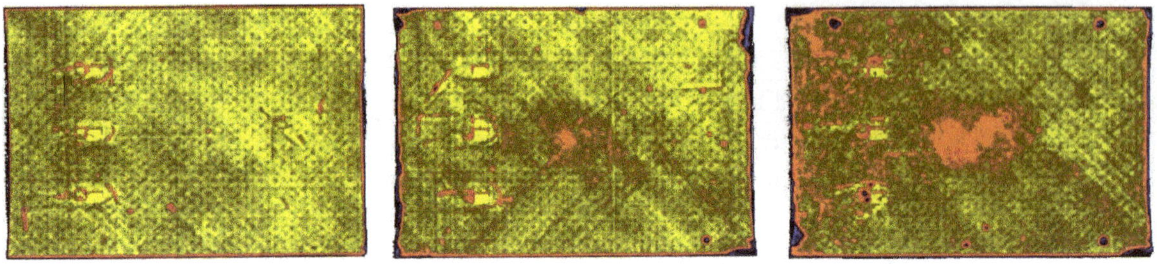

(a) 97 m/s, 142 J (b) 142 m/s, 311 J (c) 171 m/s, 443 J

FIGURE 4: Increasing damage with increasing impact velocity visualised by ultrasonic C-scans of specimens with tensile preload (no delaminations but only matrix cracking).

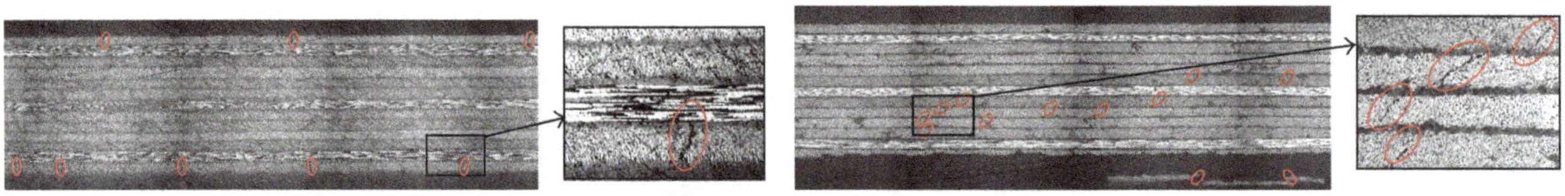

(a) (b)

FIGURE 5: Posttest micrographs of specimen centre (a) with tensile preload and impact velocity of 171 m/s (443 J) and (b) with compressive preload and impact velocity of 170 m/s (449 J), matrix cracks are highlighted.

State of the art of intralaminar stiffness and failure modelling of the individual ply is either the assumption of linear elastic stiffness behaviour in combination with failure criteria or the utilisation of continuum-damage-mechanics-(CDM-) based models with a continuous stiffness degradation under increasing load. In the current study, the standard composite material model in the commercial FE code Abaqus was used, which is based on an orthotropic linear elastic formulation and Hashin failure criteria for damage initiation:

(i) tensile failure in fibre direction ($\hat{\sigma}_{11} \geq 0$):

$$F_f^t = \left(\frac{\hat{\sigma}_{11}}{X_t}\right)^2 + \alpha\left(\frac{\hat{\tau}_{12}}{S_L}\right)^2, \quad \geq 1 \text{ failure}, < 1 \text{ elastic}, \quad (1)$$

(ii) compressive failure in fibre direction ($\hat{\sigma}_{11} \leq 0$):

$$F_f^c = \left(\frac{\hat{\sigma}_{11}}{X_c}\right)^2, \quad \geq 1 \text{ failure}, < 1 \text{ elastic}, \quad (2)$$

(iii) tensile failure in matrix direction ($\hat{\sigma}_{22} \geq 0$):

$$F_m^t = \left(\frac{\hat{\sigma}_{22}}{Y_t}\right)^2 + \left(\frac{\hat{\tau}_{12}}{S_L}\right)^2, \quad \geq 1 \text{ failure}, < 1 \text{ elastic}, \quad (3)$$

(iv) compressive failure in matrix direction ($\hat{\sigma}_{22} \leq 0$):

$$F_m^c = \left(\frac{\hat{\sigma}_{22}}{2S_T}\right)^2 + \left[\left(\frac{Y_c}{2S_T}\right)^2 - 1\right]\frac{\hat{\sigma}_{22}}{Y_c} + \left(\frac{\hat{\tau}_{12}}{S_L}\right)^2, \quad (4)$$
$$\geq 1 \text{ failure}, < 1 \text{ elastic},$$

with $\hat{\sigma}_{11}$, $\hat{\sigma}_{22}$, and $\hat{\tau}_{12}$ as the components of the effective stress tensor, X_t and X_c as the tensile and compressive strength values in fibre direction, Y_t and Y_c as the tensile and compressive strength values in matrix direction, S_L and S_T as the shear strength values in longitudinal and transversal direction, and α as the shear stress interaction coefficient.

Damage evolution until complete erosion of the ply is controlled by fracture energies in fibre and matrix direction for compression and tension with a linear stiffness degradation. All material parameters used for this model were taken from the coupon test results mentioned before or from data sheets of the manufacturer.

8-node continuum shell elements of type SC8R with reduced integration and enhanced hourglass formulation were used for the composite material modelling. An element size of 2.5 mm was chosen as the a result of a convergence study in terms of results accuracy and computational efficiency.

3.2. Delamination Modelling. The separation of adjacent plies due to normal or shear loads, referred to as delamination, absorbs impact energy and decreases the laminate stiffness and therefore needs to be covered by the model as well. Because delaminations cannot be represented inside the continuum shell elements, the laminate was divided into a certain number of sublaminates with cohesive interfaces in between, which can fail during the simulation according to a specified failure law. This approach is referred to as "stacked shell" approach and consists of continuum shell elements with cohesive elements in between. The utilisation of a cohesive contact formulation as an alternative was also investigated but could finally not be adopted due to lack of stability. The cohesive elements of type COH3D8R were modelled with zero initial thickness so that the total laminate thickness is not influenced by the additional element layers. The number of these cohesive interfaces inside the 13-ply

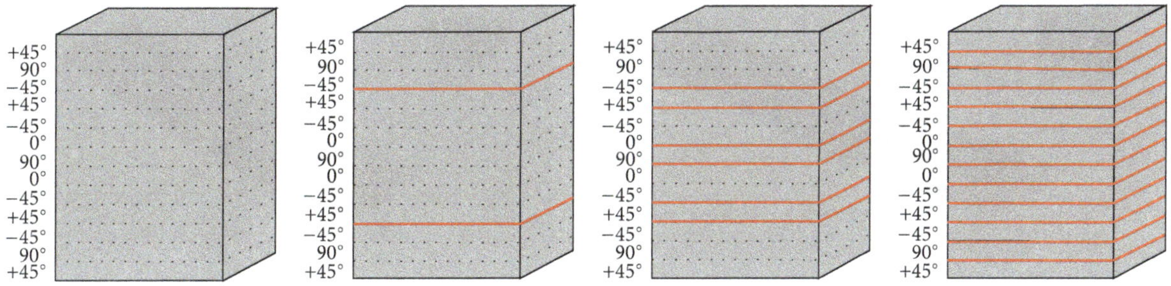

(a) No cohesive interfaces (b) 2 cohesive interfaces (c) 6 cohesive interfaces (d) 12 cohesive interfaces

Figure 6: Illustration of investigated delamination modelling approaches.

laminate was a parameter of investigation and the following options were analysed:

(i) no cohesive interfaces (no delamination, Figure 6(a));

(ii) 2 cohesive interfaces as an efficient, simplified "macrorepresentation" of delaminations that was also used in [17] (Figure 6(b));

(iii) 6 cohesive interfaces, between all plies with a difference of orientation of 90° (Figure 6(c)) as they are prone to delaminations due to large differences in stiffness;

(iv) 12 cohesive interfaces, between all plies, most realistic and expensive modelling approach (Figure 6(d)).

On the one hand, a higher number of delamination interfaces are desired as the model becomes more realistic, but, on the other hand, the computational cost increases significantly due to the higher number of model degrees of freedom due to more cohesive and shell elements and the reduced explicit time step due to shorter element lengths. A simplification is typically necessary today to obtain industry-relevant calculation times. The accurate representation of the correct bending stiffness of all four approaches was assessed in a preliminary cantilever beam study and proved consistency within 5% of the analytical reference solution.

The failure law of the cohesive elements is based on the classical cohesive zone model (CZM) [30] with a bilinear traction-separation approach, characterised by the critical energy release rates for mode I (G_{IC}) and mode II (G_{IIC}) as the area under the bilinear curves. These critical energy release rates for T800S/M21 material were taken from Ilyas et al. [31] as $G_{IC} = 765\,J/m^2$ and $G_{IIC} = 1250\,J/m^2$. Besides these measurable parameters, further penalty parameters need to be defined in the model, like the maximum interface stresses t_i^0 and the stiffness values K_{xx} for the elastic interface behaviour. They were determined using the approach proposed by Diehl [32] and validated by performing double cantilever beam (DCB) test simulations for mode I and end-notched flexure (ENF) test simulations for mode II and comparing the results with available coupon tests (Figure 7) [33, 34]. Following a mesh convergence study, a final mesh size of 1.0 mm for the cohesive elements was selected, which were attached to the continuum shell elements of the sublaminates by tie constraints.

3.3. Soft Body Impactor Modelling. The projectile used in the current study was made of gelatine, which is a standard material in aerospace engineering for artificial birds used in precertification bird strike tests in order to improve convenience, cost, and reproducibility compared to tests with real birds, which can exhibit large scatter due to their irregular shape. When a bird hits the surface of an aeronautical structure at the velocity of interest, it behaves like a fluid and flows along the surface with a relatively large contact area. Such a pressure loading was found to be well represented by using water-based gelatine projectiles and simplified geometries like cylinders with hemispherical ends [35].

Since real birds and artificial gelatine birds are mostly composed of water, a water-like hydrodynamic response can be considered as a valid approximation for a constitutive model for bird strike analyses. An equation of state (EOS) is typically used for the hydrodynamic modelling, which describes the pressure-volume relationship with parameters of water at room temperature. The high deformations of the spreading material are a major challenge for computational simulations of soft body impacts. Two different soft body impactor modelling methods were investigated and compared in the framework of this study: the Lagrangian and the Eulerian approach.

The Lagrangian modelling method is the standard approach for most structural finite element analyses with the nodes of the Lagrangian mesh being associated with the material and therefore following the material under motion and deformation. The major problem of the Lagrangian bird impactor models is the severe mesh deformation. Large distortions of the elements may lead to inaccurate results, severe hourglassing, reduced time steps, and even error termination, which has to be prevented with adequate element erosion criteria [36]. The Lagrangian bird impactor model in this study was meshed with 3 mm C3D8R solid elements leading to 1600 elements in total. A tabular EOS with parameters given in [37] was used with element erosion being controlled by a damage initiation criterion at 400% element strain.

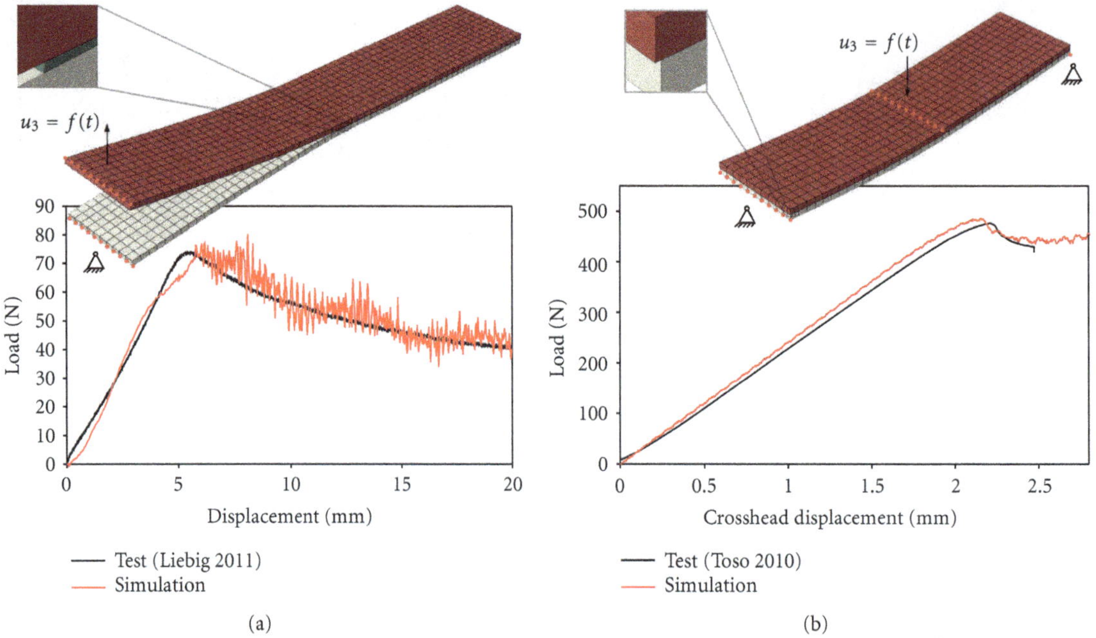

FIGURE 7: Validation of delamination model in simulations of (a) DCB and (b) ENF tests of T800S/M21 specimens.

A promising alternative to the Lagrangian impactor is the Eulerian modelling technique, where the mesh remains fixed in space and the material flows through the mesh. Because the mesh does not move, mesh deformations do not occur and the explicit time step is not influenced. Stability problems due to excessive element deformation do not occur as well. Since in a bird strike simulation typically only the impactor is modelled as a fluid-like body with the Eulerian elements and the target as a solid structure with the Lagrangian elements, a coupled Eulerian-Lagrangian (CEL) approach is used for this fluid-structure interaction problem. Because the mesh in the classical Eulerian technique is fixed in space, the computational domain should cover not only the region where the material currently exists but also additional void space to represent the region where material may exist at a later time of interest. Thus, the computational domain for structural analyses with the classical Eulerian technique is relatively large, leading to high computational cost due to the high number of elements and the cost-intensive calculation of element volume fractions and interactions. Typically, the element size of the Eulerian mesh has to be defined very small in order to achieve accurate results. In this study, a mesh size of 3 mm was selected for the impact zone and a coarser mesh of 15 mm in a larger distance. The Eulerian elements of the type EC3D8R were used combined with a Mie-Grüneisen EOS (u_s-u_p approach) adopted from [38]. In contrast to the Lagrangian bird model, no element erosion criterion is necessary here.

In order to validate the bird impactor models, a common approach is to use experimental bird strike test data from impacts either on instrumented rigid surfaces to compare the pressure-time history with the numerical results (Figure 8) or on flexible metallic plates to compare the residual plastic deformation after impact (Figure 9). Aircraft manufacturers typically use internal test data, but a large set of publicly accessible experimental bird impact test data can also be found in the literature [39–41]. For both load cases, both the Lagrangian and Eulerian bird models led to promising results. However, the initial contact force peak in the impact simulation on a rigid surface seems to be overpredicted for both models (Figure 8). The experimentally measured plastic deformation of the 6.35 mm thick flexible aluminium plate of 42.86 mm [40] was predicted by the Lagrangian model with a value of 47.58 mm and by the Eulerian model with a value of 45.22 mm, which is an overprediction of 11% and 5.5% (Figure 9). The accuracy of the Eulerian model was slightly better in both validation studies. However, the Eulerian impact simulations took 9-10 times longer to complete. All in all, these results gave good confidence to use the bird impactor models for the following bird strike simulations on composite plates.

3.4. Preload Modelling. In the numerical simulation, there are different possibilities to how to model the preloading before impact. In most studies in the literature the preloading was also performed within the explicit calculation step [9, 17]. If oscillations can be avoided, this approach is working well, but it is relatively expensive. Typically half of the computational cost is ascribed to the preloading, half to the impact simulation. A much more elegant approach, which is straight-forward in Abaqus and was applied in this study, is implicit-explicit coupling. The preloading is performed during an implicit calculation step in Abaqus/Standard, which takes only a few minutes, and then the model and stress state are transferred to a calculation with Abaqus/Explicit for the impact loading.

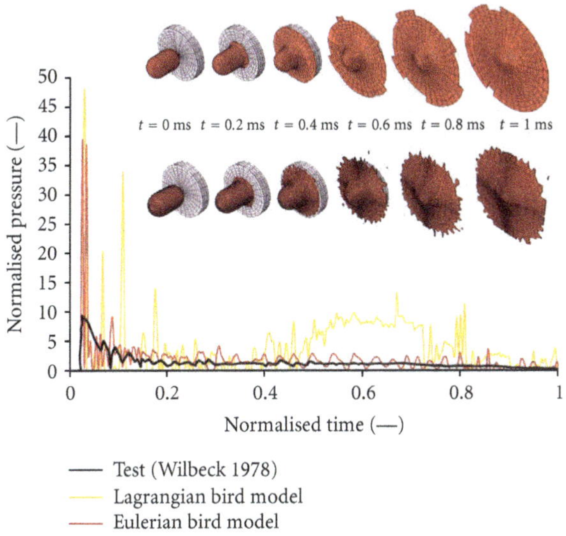

FIGURE 8: Validation of bird impactor models, impact on rigid plate: pressure-time-plots.

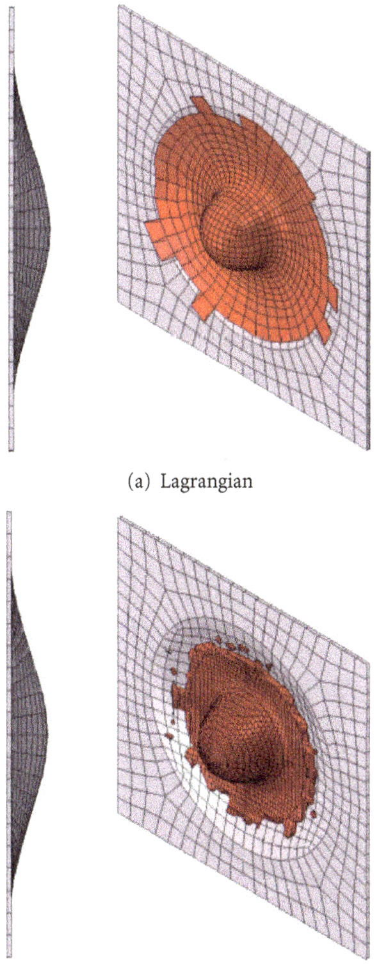

(a) Lagrangian

(b) Eulerian

FIGURE 9: Validation of bird impactor models, impact on flexible metallic plate: plate deformation.

4. Simulation Results

After successful separate validation of all model components, that is, composite and delamination model as well as bird impactor model, everything was combined in bird strike simulation models on preloaded composite plates. A large number of bird strike simulations was performed in order to represent the whole test matrix and to analyse the influence of different parameters, that is,

(i) preload: compression, unloaded, and tension;

(ii) impact velocity: 100 m/s, 140 m/s, and 170 m/s;

(iii) impactor model: Lagrangian and Eulerian;

(iv) delamination interfaces: none, 2, 6, and 12.

Figure 10 gives an impression of the numerical bird spreading behaviour of the Lagrangian and Eulerian impactor models, which appears to be similar to the high-speed test video. The calculation time for the Eulerian model was about 2.5 times higher than that of the Lagrangian model. Specific output values have been used for the damage analysis in the simulation model. Interlaminar damage was visualised using the variable SDEG, which illustrates the stiffness degradation of the cohesive elements. Of course, complete delamination appears by eroded cohesive elements, but the variable SDEG highlights cohesive elements that are already damaged by not completely failed yet. Intralaminar damage can best be visualised by plotting the variable DAMAGEMT, which corresponds to matrix tensile failure. This failure mode is typically the first and in this study also the only failure mode occurring in the model, similar to the matrix cracking observed in the posttest micrographs. An example of postimpact damage status is shown in Figure 11 for a sample with tensile preload and an impact velocity of 100 m/s, modelled with two delamination interfaces.

It can be noted that the extent of damage is slightly higher than in the experimental test sample, especially for the Eulerian impactor. The physical sample showed no damage and a pure elastic response, but in the simulation model there are already matrix cracks and even delaminations visible. This overprediction of damage was observed in all simulations.

As a first measure, the influence of different numbers of cohesive interfaces was investigated. Indeed, the increase from two to six or even twelve delamination interfaces improved the results in a sense that the extent of interlaminar damage was smaller in the individual interface and the global response was much more realistic. This improvement of results quality, however, had to be paid by a significant increase in computational cost and frequent model instabilities due to very complex contact calculations.

The damage overprediction was finally attributed to the high initial peak contact force of the bird impact loading, which was already pointed out in the bird validation study and which was also reported in several other papers [42–45]. Therefore, the model validation by just comparing the plastic deformation of a flexible metallic plate might not be sufficient as this peak load has no crucial effect here. However, during the impact simulation on a laminated

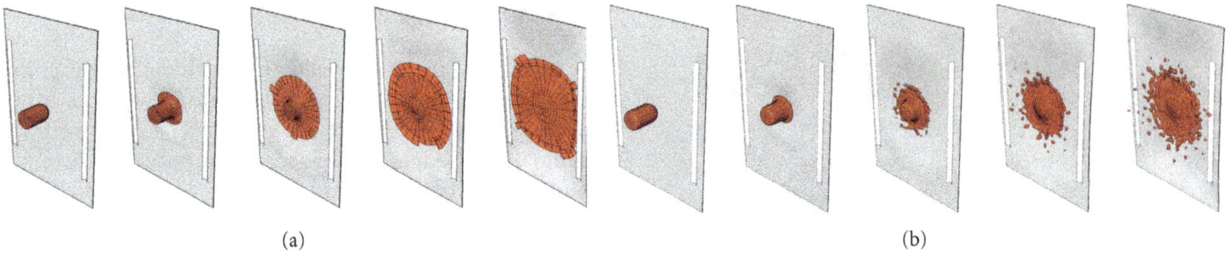

FIGURE 10: Bird strike simulation on composite plate with the (a) Lagrangian and (b) Eulerian impactor models.

composite plate, significant intralaminar and delamination damage is already initiated in the moment of contact, which grows during the remaining pressure loading phase, leading to an overprediction of damage. One measure to obtain more realistic results can be the improvement of the EOS of the bird impactor model. Ivancevic and Smojver [44], for example, modified the EOS as a user-defined material law to account for 10% porosity and obtained much lower peak pressure forces compared to a standard Mie-Grüneisen EOS formulation without porosity used before [46], although still overpredicting the test data from reference [39]. Furthermore, the contact algorithm, typically a penalty contact formulation, and its contact stiffness adjustment appear to be one of the key factors on the way to realistic bird strike simulations on composite plates.

Finally, the influence of prestress on the impact response was assessed in the simulation. The tensile preloading led to less bending deformation of the plate compared to the unloaded case (Figure 12). Consequently, the interlaminar damage is slightly smaller. The intralaminar damage was found to be a little higher, which corresponds to the experimental observations.

In case of compressive preload, plate buckling becomes an issue due to the small thickness of the composite plates. Different buckling modes occurred depending on the level of compressive prestrain, which can lead to an initial deflection of the plate centre towards or away from the impactor before impact. In this study, the deflection was always selected to be away from the impactor. The assessment of the impact simulation results showed that the global deflection of the preloaded plate is higher than that of the unloaded plate (Figure 12), which is explained both by the initial buckling deformation and by the compressive preloading. This higher bending deformation leads to slightly higher interlaminar damage, which was visible for all impact velocities. The compressive preload was found to initiate more delamination and more localized failure upon impact. The experimental results in terms of ultrasonic C-scans and micrographs support these pretest simulation results, highlighting the benefit of such numerical tools.

5. Conclusions

An experimental and numerical study of the high-velocity impact of a soft body projectile on preloaded composite plates was performed. This complex simulation task in terms of nonlinear composite damage modelling with inter- and intralaminar failure modes, implicit-explicit coupling for efficient preload modelling, and fluid-structure interaction calculation with hydrodynamic bird impactor models showed that current commercial software tools are efficient and robust for the prediction of foreign object impact damage, but the process of accurate model validation is complex and further improvements to increase simulation accuracy are necessary.

The major conclusions of this study are summarised as follows.

(i) The water-like soft body projectile flows along the surface leading to a distributed, nonlocal contact area, resulting in a mostly elastic response of the relatively thin plate and only few matrix cracks for the complete range of velocities (up to 171 m/s) and impact energies (up to 443 J).

(ii) Tensile preload leads to less bending deformation of the target plate and consequently less interlaminar but more intralaminar matrix cracking damage, which primarily appeared in the two outer layers.

(iii) Compressive preload allows for higher plate bending deformation and therefore more interlaminar damage. The intralaminar damage in terms of matrix cracking appeared to be more localised in the plate centre and distributed through several internal layers.

(iv) Both the Lagrangian and Eulerian impactor models lead to promising results, as long as an appropriate strain-based element erosion criterion was used for the Lagrangian model. The Eulerian model was significantly more expensive and led to slightly more damage in the composite plates.

(v) In correlation to numerous other research papers, both the Lagrangian and Eulerian impactor models overpredicted the initial contact pressure, leading to increased inter- and intralaminar damage in the composite target structure when used with a classical Mie-Grüneisen or tabular equation of state for water at room temperature. Improvements to obtain lower peak pressures beyond the increase of porosity are mandatory.

(vi) The contact algorithm is a major part of the bird strike simulation, and it is desirable that the software allows the user to influence the penalty stiffness calculation to avoid overstiff contact reaction forces.

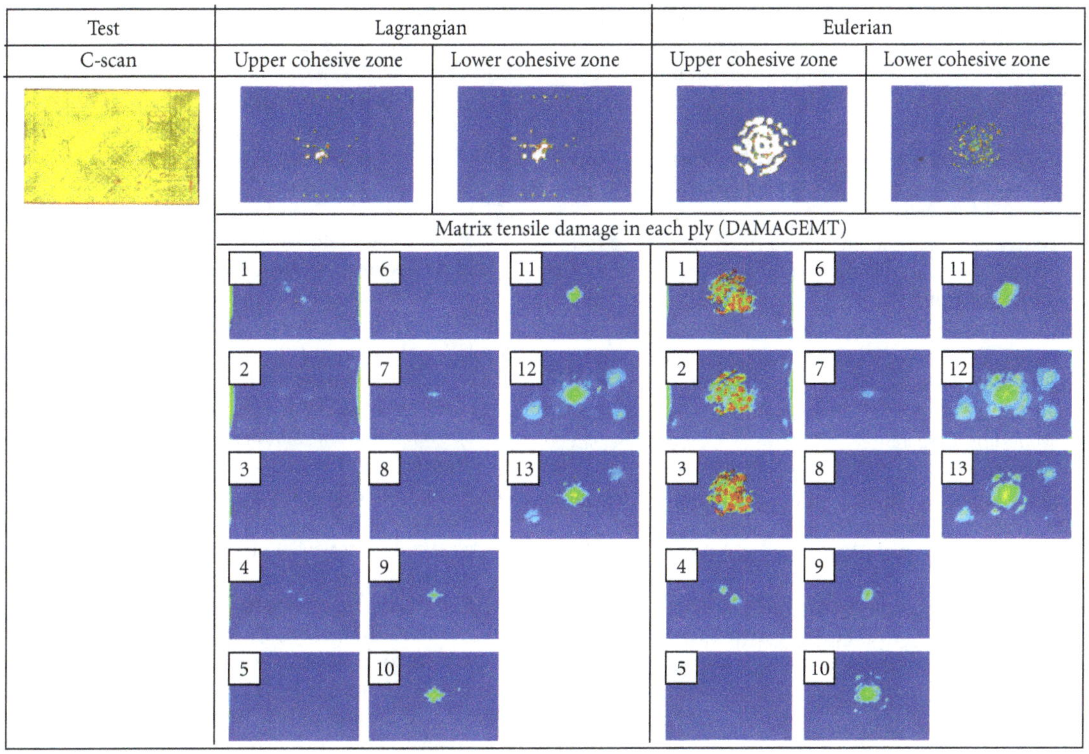

Figure 11: Interlaminar and intralaminar damage after bird impact with 100 m/s.

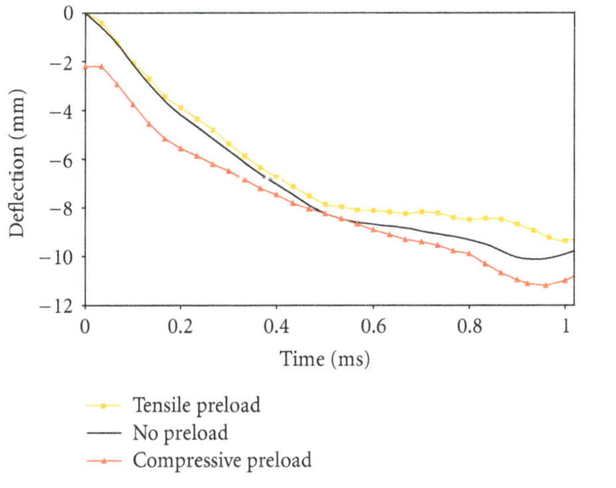

Figure 12: Predicted deflection of plate centre versus time under bird impact with 100 m/s.

(vii) The more the delamination interfaces are used, the more realistic the results appeared. The extent of delaminations with only two interfaces was much overpredicted. However, a model with delamination interface between each ply was very expensive and showed a lack of stability.

Acknowledgments

This study has been funded by the European Commission FP7/2007–2013 under grant agreement no. 213371 within the European project MAAXIMUS (http://www.maaximus.eu/). The authors would like to thank Nathalie Toso, Dominik Schueler, and Albert Reiter from DLR Stuttgart for performing the bird strike tests and Sindy Engel for manufacturing the composite plates that were used in this study.

References

[1] S. T. Chiu, Y. Y. Liou, Y. C. Chang, and C. L. Ong, "Low velocity impact behavior of prestressed composite laminates," *Materials Chemistry and Physics*, vol. 47, no. 2, pp. 268–272, 1997.

[2] M. D. Robb, W. S. Arnold, and I. H. Marshall, "The damage tolerance of GRP laminates under biaxial prestress," *Composite Structures*, vol. 32, no. 1–4, pp. 141–149, 1995.

[3] A. Kursun and M. Senel, "Investigation of the effect of low-velocity impact on composite plates with preloading," *Experimental Techniques*. In press.

[4] J. H. Tweed, R. J. Lee, R. J. Dyson, N. L. Hancox, and J. C. McCarthy, "Impact performance of stressed composites," in *Proceedings of the 7th European Conference on Composite Materials (ECCM '96)*, pp. 111–116, London, UK, May 1996.

[5] B. R. Butcher, "The impact resistance of unidirectional CFRP under tensile stress," *Fibre Science and Technology*, vol. 12, no. 4, pp. 295–326, 1979.

[6] B. R. Butcher and P. J. Fernback, "Impact resistance of unidirectional CFRP under tensile stress: further experimental variables," *Fibre Science and Technology*, vol. 14, no. 1, pp. 41–58, 1981.

[7] K. M. Mikkor, R. S. Thomson, I. Herszberg, T. Weller, and A. P. Mouritz, "Finite element modelling of impact on preloaded composite panels," *Composite Structures*, vol. 75, no. 1–4, pp. 501–513, 2006.
[8] B. Whittingham, I. H. Marshall, T. Mitrevski, and R. Jones, "The response of composite structures with pre-stress subject to low velocity impact damage," *Composite Structures*, vol. 66, no. 1–4, pp. 685–698, 2004.
[9] A. K. Pickett, M. R. C. Fouinneteau, and P. Middendorf, "Test and modelling of impact on pre-loaded composite panels," *Applied Composite Materials*, vol. 16, no. 4, pp. 225–244, 2009.
[10] A. D. Kelkar, J. Sankar, K. Rajeev, R. J. Aschenbrenner, and G. Schoeppner, "Analysis of tensile preloaded composites subjected to low-velocity impact loads," in *Proceedings of the 39th AIAA/ASME/ASCE/AHS/ASC Structures, Structural Dynamics, and Materials Conference*, pp. 1978–1987, Long Beach, Calif, USA, April 1998.
[11] A. D. Kelkar, J. Sankar, and C. Grace, "Behavior of tensile preloaded composites subjected to low-velocity impact loads," *American Society of Mechanical Engineers*, vol. 369, pp. 39–46, 1997.
[12] T. Mitrevski, I. H. Marshall, R. S. Thomson, and R. Jones, "Low-velocity impacts on preloaded GFRP specimens with various impactor shapes," *Composite Structures*, vol. 76, no. 3, pp. 209–217, 2006.
[13] A. Nettles, V. Daniel, and C. Branscomb, "The effects of tensile preloads on the impact response of carbon/epoxy laminates," in *Proceedings of the 40th International SAMPE Symposium*, pp. 1019–1025, Anaheim, Calif, USA, May 1995.
[14] D. C. Phillips, N. Park, and R. J. Lee, "The impact behaviour of high performance, ceramic matrix fibre composites," *Composites Science and Technology*, vol. 37, no. 1–3, pp. 249–265, 1990.
[15] N. Park, "The impact response of composites under stress," in *Proceedings of the 4th International Conference, Fibre Reinforced Composites (FRC '90)*, pp. 137–143, Institution of Mechanical Engineers, March 1990.
[16] B. V. Sankar and C. T. Sun, "Low-velocity impact damage in graphite-epoxy laminates subjected to tensile initial stresses," *AIAA journal*, vol. 24, no. 3, pp. 470–471, 1986.
[17] S. Heimbs, S. Heller, P. Middendorf, F. Hähnel, and J. Weiße, "Low velocity impact on CFRP plates with compressive preload: test and modelling," *International Journal of Impact Engineering*, vol. 36, no. 10-11, pp. 1182–1193, 2009.
[18] X. Zhang, G. A. O. Davies, and D. Hitchings, "Impact damage with compressive preload and post-impact compression of carbon composite plates," *International Journal of Impact Engineering*, vol. 22, no. 5, pp. 485–509, 1999.
[19] H. Morlo and J. Kunz, "Impact behaviour of loaded composites," in *Proceedings of the 4th European Conference on Composite Materials*, Stuttgart, Germany, September 1990.
[20] I. H. Choi, "Low-velocity impact analysis of composite laminates under initial in-plane load," *Composite Structures*, vol. 86, no. 1–3, pp. 251–257, 2008.
[21] C. T. Sun and J. K. Chen, "On the impact of initially stressed composite laminates," *Journal of Composite Materials*, vol. 19, no. 6, pp. 490–504, 1985.
[22] C. T. Sun and S. Chattopadhyay, "Dynamic response of anisotropic laminated plates under initial stress to impact of a mass," *Journal of Applied Mechanics, Transactions ASME*, vol. 42, no. 3, pp. 693–698, 1975.
[23] S. M. R. Khalili, R. K. Mittal, and N. M. Panah, "Analysis of fiber reinforced composite plates subjected to transverse impact in the presence of initial stresses," *Composite Structures*, vol. 77, no. 2, pp. 263–268, 2007.
[24] I. Herszberg, T. Weller, K. H. Leong, and M. K. Bannister, "The residual tensile strength of stitched and unstitched carbon/epoxy laminates impacted under tensile load," in *Proceedings of the 1st Australasian Congress on Applied Mechanics*, pp. 309–314, Melbourne, Australia, February 1996.
[25] D. Schueler, N. Toso-Pentecote, and H. Voggenreiter, "Modelling of high velocity impact on preloaded composite panels," in *Proceedings of the 3rd ECCOMAS Thematic Conference on the Mechanical Response of Composites*, pp. 587–594, Composites, Hannover, Germany, September 2011.
[26] U. K. Vaidya and B. Shafiq, "Dynamic response of navy relevant laminated and sandwich composites subjected to complex impact loads," in *Workshop on Dynamic Failure of Composite and Sandwich Structures*, pp. 39–44, Toulouse, France, June 2011.
[27] S. K. Garcia-Castillo, S. Sanchez-Saez, J. Lopez-Puente, E. Barbero, and C. Navarro, "Impact behaviour of preloaded glass/polyester woven plates," *Composites Science and Technology*, vol. 69, no. 6, pp. 711–717, 2009.
[28] S. K. Garcia-Castillo, S. Sanchez-Saez, E. Barbero, and C. Navarro, "Response of pre-loaded laminate composite plates subject to high velocity impact," *Journal de Physique IV*, vol. 134, no. 1, pp. 1257–1263, 2006.
[29] J. M. Guimard and S. Heimbs, "Towards the industrial assessment of bird strike simulations on composite laminate structures," in *Proceedings of the 3rd ECCOMAS Thematic Conference on the Mechanical Response of Composites*, Composites, Hannover, Germany, September 2011.
[30] R. Borg, L. Nilsson, and K. Simonsson, "Simulation of low velocity impact on fiber laminates using a cohesive zone based delamination model," *Composites Science and Technology*, vol. 64, no. 2, pp. 279–288, 2004.
[31] M. Ilyas, F. Lachaud, C. Espinosa, and M. Salaün, "Dynamic delamination of aeronautic structural composites by using cohesive finite elements," in *Proceedings of the 17th International Conference on Composite Materials (ICCM '09)*, Edinburgh, UK, July 2009.
[32] T. Diehl, "On using a penalty-based cohesive-zone finite element approach, part I: elastic solution benchmarks," *International Journal of Adhesion and Adhesives*, vol. 28, no. 4-5, pp. 237–255, 2008.
[33] W. Liebig, "TUHH: MAAXIMUS achievements and way forward, status at M35," in *Proceedings of the MAAXIMUS Technical Meeting*, Technical University Hamburg-Harburg, Ottobrunn, Germany, February 15.
[34] N. Toso, D. Kohlgrüber, D. Schueler, M. Waimer, and A. Johnson, "MAAXIMUS WP 3.5—vulnerability, DLR test activities, DLR stuttgart," in *Proceedings of the MAAXIMUS Technical Meeting*, Hamburg, Germany, November 2010.
[35] J. S. Wilbeck and J. L. Rand, "Development of a substitute bird model," *Journal of Engineering for Power*, vol. 103, no. 4, pp. 725–730, 1981.
[36] S. Heimbs, "Computational methods for bird strike simulations: a review," *Computers & Structures*, vol. 89, no. 23-24, pp. 2093–2112, 2011.
[37] Dassault Systems Simulia Corp, A Strategy for Bird Strike Simulations Using Abaqus/Explicit. Simulia Online Support Dokument, Answer-ID 4493 (Best Practices for Birdstrike Analysis), 2011.
[38] S. Heimbs, "Bird strike simulations on composite aircraft structures," in *Proceedings of the SIMULIA Customer Conference (SCC '11)*, pp. 73–86, Barcelona, Spain, May 2011.

[39] J. S. Wilbeck, "Impact behaviour of low strength projectiles," Report AFML-TR-77-134, US Air Force, 1978.
[40] C. J. Welsh and V. Centonze, "Aircraft transparency testing—artificial birds," Report AEDC-TR-86-2, US Air Force, 1986.
[41] M. A. Lavoie, A. Gakwaya, M. N. Ensan, D. G. Zimcik, and D. Nandlall, "Bird's substitute tests results and evaluation of available numerical methods," *International Journal of Impact Engineering*, vol. 36, no. 10-11, pp. 1276–1287, 2009.
[42] S. Kari, J. Gabrys, and D. Lincks, "Birdstrike analysis of radome and wing leading edge using LS-DYNA," in *Proceedings of the 5th International LS-DYNA Users Conference*, Southfield, Mich, USA, September 1998.
[43] M. A. Lavoie, A. Gakwaya, M. N. Ensan, and D.G. Zimcik, "Validation of available approaches for numerical bird strike modelling tools," *International Review of Mechanical Engineering*, vol. 1, no. 4, pp. 380–389, 2007.
[44] D. Ivancevic and I. Smojver, "Hybrid approach in bird strike damage prediction on aeronautical composite structures," *Composite Structures*, vol. 94, no. 1, pp. 15–23, 2011.
[45] R. Hedayati and S. Ziaei-Rad, "Effect of impact orientation on bird strike analysis," *International Journal of Vehicle Structures & Systems*, vol. 3, no. 3, pp. 184–191, 2011.
[46] I. Smojver and D. Ivancevic, "Numerical simulation of bird strike damage prediction in airplane flap structure," *Composite Structures*, vol. 92, no. 9, pp. 2016–2026, 2010.

Permissions

The contributors of this book come from diverse backgrounds, making this book a truly international effort. This book will bring forth new frontiers with its revolutionizing research information and detailed analysis of the nascent developments around the world.

We would like to thank all the contributing authors for lending their expertise to make the book truly unique. They have played a crucial role in the development of this book. Without their invaluable contributions this book wouldn't have been possible. They have made vital efforts to compile up to date information on the varied aspects of this subject to make this book a valuable addition to the collection of many professionals and students.

This book was conceptualized with the vision of imparting up-to-date information and advanced data in this field. To ensure the same, a matchless editorial board was set up. Every individual on the board went through rigorous rounds of assessment to prove their worth. After which they invested a large part of their time researching and compiling the most relevant data for our readers. Conferences and sessions were held from time to time between the editorial board and the contributing authors to present the data in the most comprehensible form. The editorial team has worked tirelessly to provide valuable and valid information to help people across the globe.

Every chapter published in this book has been scrutinized by our experts. Their significance has been extensively debated. The topics covered herein carry significant findings which will fuel the growth of the discipline. They may even be implemented as practical applications or may be referred to as a beginning point for another development. Chapters in this book were first published by Hindawi Publishing Corporation; hereby published with permission under the Creative Commons Attribution License or equivalent.

The editorial board has been involved in producing this book since its inception. They have spent rigorous hours researching and exploring the diverse topics which have resulted in the successful publishing of this book. They have passed on their knowledge of decades through this book. To expedite this challenging task, the publisher supported the team at every step. A small team of assistant editors was also appointed to further simplify the editing procedure and attain best results for the readers.

Our editorial team has been hand-picked from every corner of the world. Their multi-ethnicity adds dynamic inputs to the discussions which result in innovative outcomes. These outcomes are then further discussed with the researchers and contributors who give their valuable feedback and opinion regarding the same. The feedback is then collaborated with the researches and they are edited in a comprehensive manner to aid the understanding of the subject.

Apart from the editorial board, the designing team has also invested a significant amount of their time in understanding the subject and creating the most relevant covers. They scrutinized every image to scout for the most suitable representation of the subject and create an appropriate cover for the book.

The publishing team has been involved in this book since its early stages. They were actively engaged in every process, be it collecting the data, connecting with the contributors or procuring relevant information. The team has been an ardent support to the editorial, designing and production team. Their endless efforts to recruit the best for this project, has resulted in the accomplishment of this book. They are a veteran in the field of academics and their pool of knowledge is as vast as their experience in printing. Their expertise and guidance has proved useful at every step. Their uncompromising quality standards have made this book an exceptional effort. Their encouragement from time to time has been an inspiration for everyone.

The publisher and the editorial board hope that this book will prove to be a valuable piece of knowledge for researchers, students, practitioners and scholars across the globe.

List of Contributors

Bruce Chehroudi
Advanced Technology Consultants, Laguna Niguel, CA 92677, USA

Radoslav Bozinoski
Thermal/Fluid Science & Engineering Department, Sandia National Laboratories, Livermore, CA 945501, USA

Roger L. Davis
Mechanical and Aero Engineering Department, University of California at Davis, Davis, CA 95616, USA

Francesco Simeoni and Lorenzo Casalino
Dipartimento di Ingegneria Meccanica e Aerospaziale, Politecnico di Torino, Corso Duca degli Abruzzi 24, 10129 Torino, Italy

Alessandro Zavoli and Guido Colasurdo
Dipartimento di Ingegneria Meccanica e Aerospaziale, Sapienza Universit`a di Roma, Via Eudossiana 18, 00184 Roma, Italy

J. X. Hu, Z. X. Xia, W. H. Zhang, Z. B. Fang, D. Q. Wang and L. Y. Huang
College of Aerospace and Materials Engineering, National University of Defense Technology, Changsha 410073, China

Uzair Ahmed Dar, Weihong Zhang and Yingjie Xu
Laboratory of Engineering Simulation and Aerospace Computing,Northwestern PolytechnicalUniversity, Xi'an, Shaanxi 710072, China

Uzair Ahmed Dar, Weihong Zhang and Yingjie Xu
Laboratory of Engineering Simulation and Aerospace Computing,Northwestern PolytechnicalUniversity, Xi'an, Shaanxi 710072, China

Giorgio Guglieri
Dipartimento di Ingegneria Meccanica e Aerospaziale, Politecnico di Torino, Corso Duca degli Abruzzi 24, 10129 Torino, Italy

David W. Hoeppner and Carlos A. Arriscorreta
Department of Mechanical Engineering, University of Utah, 50 South Central Campus Drive, Room 2010, Salt Lake City, UT 84112, USA

Leiyong Jiang and Andrew Corber
Aerospace Research, Gas Turbine Laboratory, The National Research Council of Canada, 1200 Montreal Road, M-10, Ottawa, ON, Canada K1A 0R6

Francisco Miranda
Center for Research and Development in Mathematics and Applications, University of Aveiro, Portugal
School of Technology and Management, Polytechnic Institute of Viana do Castelo, Avenida do Atl^antico, 4900-348 Viana do Castelo, Portugal

Foad Mehdi Zadeh, Jean-Yves Trépanier and Eddy Petro
Department of Mechanichal Engineering, ´Ecole Polytechnique de Montr´eal, 2500, Chemin de Polytechnique, Montr´eal, QC, Canada H3T 1J4

Gonzalo Garcia and Shahriar Keshmiri
Department of Aerospace Engineering, University of Kansas, 2120 Learned Hall 1530W15th Street, Lawrence, KS 66045, USA

Sebastian Heimbs and Tim Bergmann
Department of Structures Engineering, Production and Aeromechanics, EADS Innovation Works, Willy-Messerschmitt-Straße, 81663 Munich, Germany

www.ingramcontent.com/pod-product-compliance
Lightning Source LLC
Chambersburg PA
CBHW050445200326
41458CB00014B/5071
* 9 7 8 1 6 3 2 4 0 3 8 0 3 *